Novel Therapeutic Approaches for Cancer Treatment

Novel Therapeutic Approaches for Cancer Treatment

Guest Editor

Marzia Bruna Gariboldi

MDPI

Basel • Beijing • Wuhan • Barcelona • Belgrade • Novi Sad • Cluj • Manchester

Guest Editor
Marzia Bruna Gariboldi
Department of Biotechnology
and Life Sciences
University of Insubria
Varese
Italy

Editorial Office
MDPI AG
Grosspeteranlage 5
4052 Basel, Switzerland

This is a reprint of the Special Issue, published open access by the journal *Cancers* (ISSN 2072-6694), freely accessible at: https://www.mdpi.com/journal/cancers/special_issues/B4864480C0.

For citation purposes, cite each article independently as indicated on the article page online and as indicated below:

Lastname, A.A.; Lastname, B.B. Article Title. *Journal Name* **Year**, *Volume Number*, Page Range.

ISBN 978-3-7258-7068-4 (Hbk)
ISBN 978-3-7258-7069-1 (PDF)
https://doi.org/10.3390/books978-3-7258-7069-1

Contents

About the Editor

Marzia Bruna Gariboldi

Marzia Bruna Gariboldi is an Associate Professor in the Department of Biotechnology and Life Sciences, University of Insubria, Italy. Her research interests focus on the molecular mechanisms of cancer cell response to anticancer therapies, including multidrug resistance, signal transduction pathways regulating cell survival and apoptosis, tumor hypoxia, and the development of innovative therapeutic approaches such as photodynamic therapy and novel platinum-based and natural anticancer compounds.

Preface

Cancer remains a major global health challenge, driven by genetic and epigenetic alterations as well as environmental and lifestyle factors. Although conventional treatments such as surgery, chemotherapy, and radiotherapy remain central to cancer care, their limitations have spurred the development of innovative and more selective therapeutic strategies.

This Special Issue, 'Novel Therapeutic Approaches for Cancer Treatment', brings together original research articles and reviews highlighting recent advances in targeted therapies, nanotechnology-based approaches, light-activated treatments, novel bioactive compounds, and advanced experimental models, including 3D systems and organoids. Together, these contributions reflect the multidisciplinary and translational nature of current cancer research.

As Guest Editor, I sincerely thank all authors and reviewers for their valuable contributions and commitment, which made this Special Issue possible. I hope this collection will inspire further research and collaboration toward more effective and personalized cancer therapies.

Marzia Bruna Gariboldi
Guest Editor

cancers

MDPI

icle

nticancer Effect of Nature-Inspired Indolizine-Based entathiepines in 2D and 3D Cellular Model

berto Tallarita [1,†], Federica Randisi [2,†], Lukas Manuel Jacobsen [1], Emanuela Marras [2], Mattia Riva [2], ulia Modoni [2], Johannes Fimmen [1], Siva Sankar Murthy Bandaru [1], Carola Schulzke [1,*] d Marzia Bruna Gariboldi [2,*]

[1] Bioinorganic Chemistry, Institute of Biochemistry, University of Greifswald, Felix-Hausdorff-Str. 4, 17489 Greifswald, Germany; roberto.tallarita@uni-greifswald.de (R.T.); lukasmanuel.jacobsen@stud.uni-greifswald.de (L.M.J.); johannes.fimmen@stud.uni-greifswald.de (J.F.); siva.bandaru@uni-greifswald.de (S.S.M.B.)

[2] Department of Biotechnology and Life Sciences (DBSV), University of Insubria, 21100 Varese, Italy; frandisi1@studenti.uninsubria.it (F.R.); emanuela.marras@uninsubria.it (E.M.); mriva8@studenti.uninsubria.it (M.R.); gmodoni@studenti.uninsubria.it (G.M.)

* Correspondence: carola.schulzke@uni-greifswald.de (C.S.); marzia.gariboldi@uninsubria.it (M.B.G.); Tel.: +49-3034-420-4321 (C.S.); +39-0331339418 (M.B.G.)

† These authors contributed equally to this work.

Simple Summary: Drug resistance and limited selectivity remain major challenges in cancer therapy, underscoring the urgent need for the development of novel, more effective chemotherapeutic agents and treatment strategies. In this context, 1,2,3,4,5-pentathiepines (PTEs) have recently emerged as promising anticancer compounds. Multiple mechanisms have been proposed to contribute to the antineoplastic activity of PTEs, including the induction of intracellular reactive oxygen species (ROS), inhibition of glutathione peroxidase, disruption of cell cycle progression, and activation of cell death pathways. Despite these promising effects, PTEs often display limited target specificity and are associated with off-target activities, highlighting the requirement for more selective, less toxic, and more efficacious analogs. Moreover, further studies are required to gain a deeper understanding of their anticancer mechanisms of action.

Abstract: Background: 1,2,3,4,5-pentathiepines (PTEs) are compounds originally identified in marine ascidians and are currently under investigation for their promising pharmacological properties, particularly as potential antineoplastic agents. Objectives: In this study, we investigated the antineoplastic properties of a series of ten indolizine-based PTEs, comprising eight previously reported compounds and two newly synthesized derivatives. Methods: These compounds were evaluated against a panel of human cancer cell lines of diverse tissue origins, as well as, for the first time, on non-cancerous CR9 fibroblasts to assess their cytotoxic selectivity. In addition, their effects were tested on 3D spheroid models, providing preliminary insights into their potential in vivo efficacy. Initial screening focused on cell viability, followed by a more detailed characterization of the most active compounds in terms of their ability to induce apoptosis, necrosis, cell cycle arrest, and reactive oxygen species (ROS) generation. The anti-migratory activity of PTEs and a newly adapted assay to confirm sulfur species release in the cells were also performed for the first time. Results and Conclusions: Our findings reveal that four PTEs bearing hydrophilic, hydrogen-bonding functional groups, particularly the two inspired by natural analogs, exhibited the most potent anticancer activity.

 https://doi.org/10.3390/cancers17142393

Keywords: pentathiepines; polysulfides; molybdenum; S-heterocycle; indolizines; 2D and 3D cell models; cell death; migration; sulfide production

1. Introduction

Despite substantial advances in identifying novel risk factors, early diagnostic biomarkers, and alternative therapeutic strategies, malignant neoplasms remain the second leading cause of mortality globally, representing a critical global challenge with profound implications for public health, societal well-being, and economic stability in the 21st century [1,2]. Current cancer therapies, including targeted ones, face various challenges, with drug resistance and low selectivity as the primary obstacles to effective treatment [1,3]. Therefore, novel, advanced, and more potent chemotherapeutic drugs and strategies are necessary.

Recent studies have highlighted the remarkable biological activities of penta- thiepines (PTEs), a class of heterocyclic compounds characterized by a seven-membered ring structure comprising five sulfur and two sp^2 carbon atoms [4,5].

Although the first synthetic representatives of PTEs were reported as early as 1971, their biological evaluation did not commence until the early 1990s, following the isolation of the natural PTE varacin from marine *Ascidiacea* (ascidians or sea squirts) [6]. Varacin and a few other analogs isolated later (Figure 1) were reported to possess potent cytotoxic activity, sparking renewed interest in this compound class [6,7].

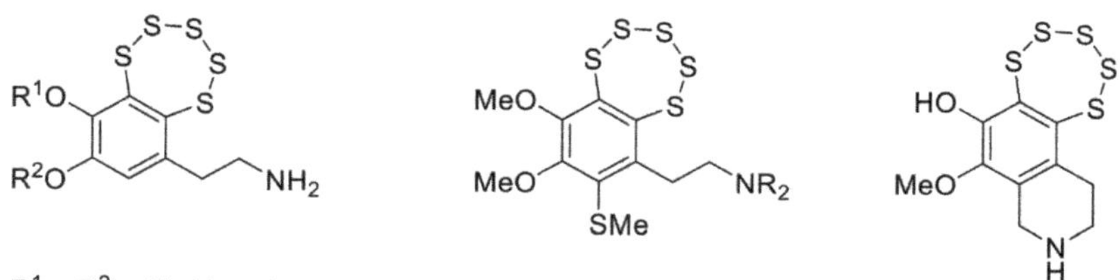

$R^1 = R^2$ = Me Varacin
R^1 = H, R^2 = Me Lissoclinotoxin A
R^1 = Me R^2 = H Isolissoclinotoxin A
$R^1 = R^2$ = H 3,4-desmethylvaracin

R = H 5-(methylyhio)varacin
R = Me *N,N*-dimethyl-5-(methylthio)varacin

Lissoclinotoxin B

Figure 1. Representative natural products containing the pentathiepine motif, including varacin and related analogs isolated from marine ascidians. These compounds served as structural inspiration for the design of the synthetic indolizine-based PTEs investigated in this study.

Since then, intensified research efforts have revealed a broad spectrum of intriguing biological activities associated with this unique family of compounds, including antifungal, antiviral, antibacterial, and DNA-cleaving properties, as well as cytotoxicity in cancer cell lines [4,8,9]. Furthermore, PTEs have been identified as specific inhibitors of several key enzymes involved in the pathogenesis of Alzheimer's disease, as well as in DNA repair mechanisms related to several cancers and neurodegenerative disorders [4,10–12].

Recently, synthesized indolizine-based PTEs, with the molybdenum-mediated double cyclisation reaction [13] have exhibited potent and selective inhibition of glutathione peroxidase 1 (GPx1), a key enzyme involved in the regulation of cellular redox homeostasis [14,15]. The role of GPx1 in oncogenesis, tumor progression, and therapeutic response remains a subject of ongoing debate. Several studies have reported a correlation between elevated GPx1 expression and poor clinical outcomes, whereas others have suggested a potential protective function for the enzyme [9,16–18]. Nonetheless, upregulation of GPx1 has been implicated in mechanisms of chemotherapeutic resistance [19–21]. Notably, recent evidence indicates that GPx1 knockout cells display markedly increased sensitivity to conventional chemotherapeutic agents compared to their parental counterparts [22]. These findings support the hypothesis that pharmacological inhibition of GPx1, achievable through the use of PTEs, may represent a promising strategy to enhance the efficacy of chemotherapy, particularly in drug-resistant cancer cells.

Additional mechanisms have been proposed to underlie the anticancer activity of PTEs, among which is the rise in intracellular reactive oxygen species (ROS). This increase has been partially attributed to the generation of hydrogen peroxide (H_2O_2) due to the presumed interaction between the polysulfur ring system of PTEs and intracellular thiols [9,15]. Furthermore, the accumulation of ROS may be exacerbated by the reduced enzymatic capacity of GPx1, whose activity is selectively inhibited by PTEs. The elevation of ROS ultimately contributes to enhanced oxidative stress within cancer cells, thereby promoting the induction of apoptotic cell death [9,14]. In addition, cell cycle alterations and DNA damage have also been involved in their anticancer effect [9,23]. Despite these promising effects, PTEs often exhibit limited target specificity and are associated with off-target activities, underscoring the necessity for the development of more selective, non-toxic, and efficacious analogs.

While previous studies have explored derivatives bearing sterically demanding side chains, designed to enhance pharmacological properties through structural complexity [9], this work shifted the focus toward a more streamlined approach. For the first time, we investigated the potential anticancer effect of the simplest core scaffold **1**, along with selectively substituted analogs at positions C-6 ($\mathbf{2^{Me}}$, $\mathbf{2^{Ph}}$, $\mathbf{2^{Gly}}$, and $\mathbf{2^{PG}}$) and C-9 ($\mathbf{3^{Br}}$, $\mathbf{3^{NO}}_2$, $\mathbf{3^{Me}}$, $\mathbf{3^{CHO}}$, and $\mathbf{3^{CN}}$) (Figure 2). The C-6 derivatives, plus $\mathbf{3^{Br}}$, $\mathbf{3^{NO}}_2$, and $\mathbf{3^{Me}}$ are previously characterized compounds, originating from earlier synthetic optimization efforts [24], whereas the $\mathbf{3^{CHO}}$ and $\mathbf{3^{CN}}$ analogs represent novel entities, synthesized and characterized for the first time in the present study.

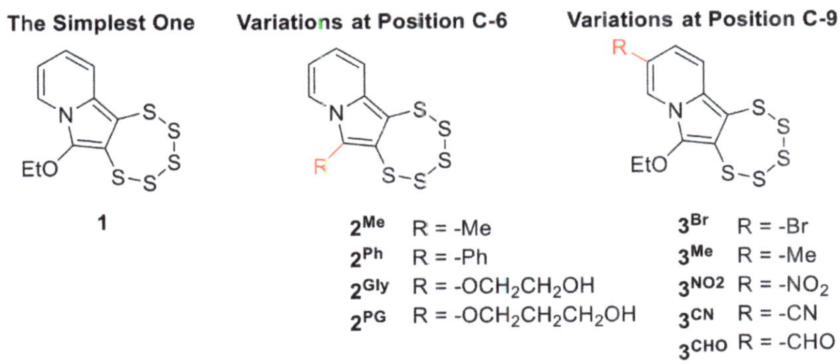

Figure 2. Structures of indolizine-based PTEs selected for biological evaluation. Compounds are grouped based on substitution patterns at positions C-6 (**2** series) and C-9 (**3** series), with compound **1** representing the parent scaffold with the typical –OEt at C-6.

PTEs are known to be intrinsically reactive in the presence of nucleophilic species. All known natural analogs possess an ethylamino side chain ($-CH_2CH_2NH_2$) as represented by the naturally occurring motif **4** (Figure 3), the terminal nucleophilic group of which is believed to play a key role in their biological function [4]. Its proximity to the polysulfide core is thought to promote sulfur release through intramolecular interaction, offering a plausible explanation for their observed pharmacological effects. $\mathbf{2^{Gly}}$ and $\mathbf{2^{PG}}$ comprise therefore compounds of particular interest in this series.

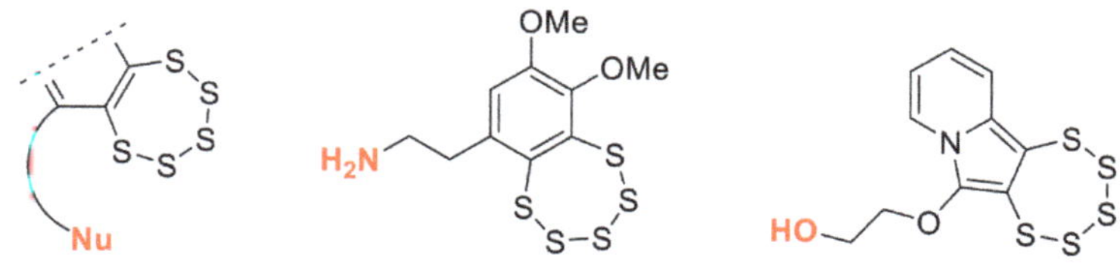

Figure 3. Illustrative comparison of the core structural motif **4** common in bioactive PTEs, the natural reference compound varacin, regarded as the prototypical member of the PTE family, and the synthetic analog $\mathbf{2^{Gly}}$, bearing a newly introduced hydroxyl-terminated polar substituent.

We also investigated the most potent PTEs for their ability to induce apoptotic and/or necrotic cell death, alter cell cycle progression, and promote ROS generation across a panel of four cancer cell lines. Moreover, building on prior studies [25], we established a method to evaluate the ability of PTEs to stimulate sulfide production in viable cells. This effect may contribute to the pharmacological activity of PTEs, potentially linking sulfide signaling to their anticancer properties. It is worth highlighting that to the best of our knowledge, this study represents the first attempt to explore this topic.

Further novel aspects of the present study involve investigating pentathiepines' effects on cancer cell migration and their anticancer efficacy in a three-dimensional (3D) tumor model that more accurately recapitulates the in vivo tumor microenvironment.

2. Materials and Methods

2.1. Reagents and Chemicals

All standard chemicals and cell culture reagents used, unless otherwise indicated, were obtained from Sigma-Aldrich Chemie GmbH (Taufkirchen Germany), Euroclone (Pero, Milan, Italy), and VWR International GmbH (Darmstadt, Germany).

2.2. Pentathiepine Synthesis and Characterization

2.2.1. General Experimental Procedures

All PTEs evaluated for biological activity in this study were previously reported [24,26], except for the newly synthesized compounds $\mathbf{3^{CN}}$ and $\mathbf{3^{CHO}}$. These two derivatives were prepared following the optimized protocol recently described (Figure 4) [24].

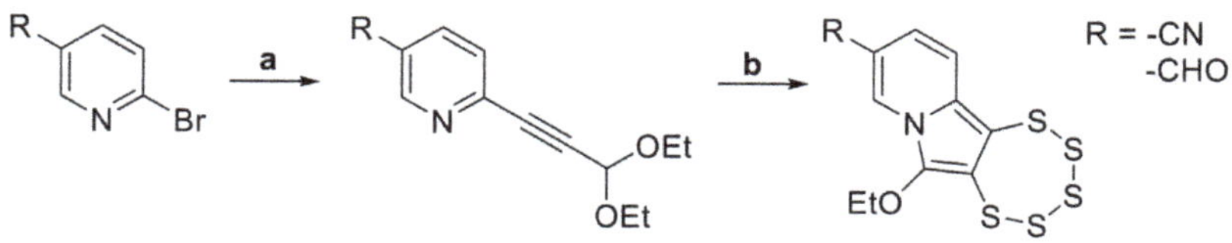

Figure 4. Synthetic route to indolizine-based PTEs $\mathbf{3^R}$, starting from $\mathbf{5^R}$ via Sonogashira cross-coupling to afford intermediates $\mathbf{6^R}$, followed by cyclization to form the final polysulfide-containing scaffold (a. degassed DMF, $Pd(OAc)_2$ (5%), CuI (10%), NEt_3 (10 equiv.), 3,3-diethoxy-1-propyne (1.2 equiv.), 50 °C, under N_2, 48 h. b. DMF, $(NEt_4)_2[MoO(S_4)_2]$ (0.5 equiv.), S_8 (1 equiv.), 50 °C, air, 48 h).

Reagents, solvents, and starting materials were used as purchased without further purification. ^1H-, ^{13}C-NMR spectra were recorded on a Bruker Avance II 300 spectrometer (300, 75.5, and 282.4 MHz, respectively) using CD_2Cl_2 and $CDCl_3$, which were dried over activated zeolites, as solvents. The chemical shifts (δ) are given in parts per million (ppm). The ^1H-NMR spectra were referenced to the peaks of residual protons of the deuterated solvent; ^{13}C-NMR spectra were referenced to the deuterated solvent itself. Multiplicities are abbreviated as follows: s, singlet; d, doublet; t, triplet; q, quartet; m, multiplet; *J*, coupling constant (Hertz). The mass spectra were recorded with an Advion Expression CMS (Europe branch: Advion, Ltd. Harlow Enterprise Hub Kao Hockham Building Edinburgh Way, Harlow CM20 2NQ, UK). The ^{1}H and ^{13}C NMR spectra of **3**^{CN}, **3**^{CHO}, **6**^{CN}, **6**^{CHO}, DNS-N_3, mass spectra of **3**^{CN}, **3**^{CHO}, **6**^{CN}, **6**^{CHO}, DNS-N_3 and an IR spectrum of DNS-N_3 are provided as Supplementary Information (Figures S1–S16). High-resolution mass spectrometric data were acquired using a Bruker Elute UHPLC system coupled to a Bruker Compact QTOF-MS equipped with an APCI ionization source (Bruker Daltonics, Bremen, Germany) for **3**^{CN} and **3**^{CHO}.

2.2.2. Sonogashira Cross-Coupling Reactions

To an oven-dried Schlenk tube bearing a side-arm stopcock and holding a magnetic stirring bar, DMF (5.00 mL) and NEt_3 (8.75 mL, 63.00 mmol, 10.0 equiv.) were added. The solution was degassed by bubbling nitrogen through it for 10 min. The system was kept under a nitrogen atmosphere throughout the synthesis. Then CuI (120 mg, 0.630 mmol, 0.10 equiv.) was added first, followed by 3,3-diethoxyprop-1-yne (1.10 mL, 7.56 mmol, 1.20 equiv.), upon which the solution turned yellow. PPh_3 (330 mg, 1.26 mmol, 0.20 equiv.) and the appropriate 5-substituted 2-halopyridine (6.30 mmol, 1.00 equiv.) were then introduced. Finally, $Pd(OAc)_2$ (71 mg, 0.315 mmol, 0.05 equiv.) was added, resulting in a rapid color change to reddish brown. The reaction mixture was stirred at 50 °C for 48 h. Progress was monitored by TLC (Hex/EtOAc, 70:30 *v*/*v*; UV 254 nm) and APCI mass spectrometry. After completion, the reaction solvent was removed under reduced pressure using a cooling trap. The crude residue was extracted with DCM and washed three times with saturated NH_4Cl solution. The resulting organic layer was dried and purified by column chromatography.

5-cyano-2-(3,3-diethoxyprop-1-yn-1-yl)pyridine (**6**^{CN}): purified via column chromatography (Hex/EtOAc = 75/25), yield: 81%. ^{1}H NMR (300 MHz, $CDCl_3$) δ 8.85 (dd, *J* = 2.0, 1.0 Hz, 1H), 7.95 (dd, *J* = 8.2, 2.1 Hz, 1H), 7.59 (dd, *J* = 8.1, 1.0 Hz, 1H), 5.52 (s, 1H), 4.00–3.64 (m, 4H), 1.28 (t, *J* = 7.1 Hz, 6H). ^{13}C{^{1}H} NMR (75 MHz, $CDCl_3$) δ 152.52, 145.57, 139.37, 127.12, 116.16, 109.17, 91.47, 88.72, 82.88, 61.44, 15.05. MS (APCI) *m*/*z*: $[M + H]^+$ calculated for $C_{13}H_{15}N_2O_2$ 231.11; found 232.3.

5-formyl-2-(3,3-diethoxyprop-1-yn-1-yl)pyridine (**6**^{CHO}): purified via column chromatography (Hex/EtOAc = 70/30), yield: 77%. ^{1}H NMR (300 MHz, $CDCl_3$) δ 10.12 (s, 1H), 9.10–8.89 (m, 1H), 8.15 (dd, *J* = 8.1, 2.1 Hz, 1H), 7.64 (d, *J* = 8.1 Hz, 1H), 5.53 (s, 1H), 4.07–3.49 (m, 4H), 1.28 (t, *J* = 7.1 Hz, 6H). ^{13}C{^{1}H} NMR (75 MHz, $CDCl_3$) δ 189.76, 152.02, 146.99, 135.83, 130.18, 127.58, 91.45, 87.92, 83.48, 76.57, 61.30, 14.98.MS (APCI) *m*/*z*: $[M + H]^+$ 235.4. MS (APCI) *m*/*z*: $[M + H]^+$ calculated for $C_{13}H_{16}NO_3$ 234.11; found 235.4.

2.2.3. Pentathiepine-Pyrrole Double Ring Closing Reactions

To a 100mL round bottom flask, holding a magnetic stirring bar, the appropriate alkyne (1.00 mmol, 1.0 equiv.), bistetraethylammonium oxo-bistetrasulfido molybdate (0.50 mmol, 0.5 equiv.), and elemental sulfur (1.00 mmol, 1.0 equiv.) were dissolved in ambient atmosphere in commercial DMF (5 mL). The resulting heterogeneous mixture was stirred at 50 °C for 48 h. After completion, the reaction was cooled to room temperature, and the solvent was removed under reduced pressure using a cooling trap. The crude residue was purified by gravity column chromatography, with Hex/EtOAc as eluent mixture. Residual sulfur removal (Rf ≈ 0.95) was monitored by TLC (UV 254 nm).

6-ethoxy-9-formyl-[1,2,3,4,5]pentathiepino [6,7-*a*]indolizine (**3**^{CHO}): purified via gradient column chromatography (starting mix Hex/EtOAc = 95/5, ending mix Hex/EtOAc = 90/10), yield 26%, yellow crystalline powder, R_f = 0.3 (Hex/EtOAc = 85/15, *v*/*v*). ^{1}H NMR (300 MHz, CD2Cl2) δ 9.75 (q, *J* = 1.0 Hz, 1H), 8.20 (dt, *J* = 1.6, 0.9 Hz, 1H), 7.45 (dd, *J* = 9.4, 1.0 Hz, 1H), 7.17 (dd, *J* = 9.4, 1.5, 1H), 4.67–4.37 (m, 2H), 1.60–1.34 (m, 3H). ^{13}C{^{1}H} NMR (75 MHz, CD_2Cl_2): δ 188.11, 141.88, 130.60, 130.44, 124.01, 118.68, 116.73, 114.19, 112.40, 72.64, 15.37. HRMS (APCI/Q-TOF) *m*/*z*: $[M + H]^+$ calculated for $C_{11}H_{10}NO_2S_5$ 347.9310; found 347.9314.

The compound's chemical structure was confirmed by SCXRD. A drawing of the molecular structure is shown in the SI together with crystal and refinement data (Table S1, also see Supporting Materials Reference [1]).

6-ethoxy-9-cyano-[1,2,3,4,5]pentathiepino [6,7-*a*]indolizine) ($\mathbf{3^{CN}}$): purified via column chromatography (Hex/EtOAc = 95/5), yield: 29%, yellow crystalline powder, R_f = 0.3 (Hex/EtOAc = 90/10, *v*/*v*). ^{1}H NMR (300 MHz, CD_2Cl_2): δ 8.14 (t, *J* = 1.3 Hz, 1H), 7.46 (dd, *J* = 9.4, 1.1 Hz, 1H), 6.79 (dd, *J* = 9.4, 1.5 Hz, 1H), 4.49 (qd, *J* = 7.1, 1.8 Hz, 2H), 1.41 (t, *J* = 7.0 Hz, 4H). ^{13}C{^{1}H} NMR (75 MHz, CD_2Cl_2): δ 140.82, 128.42, 128.18, 119.59, 119.26, 116.73, 114.48, 112.61, 98.42, 72.76, 15.33. HRMS (APCI/Q-TOF) *m*/*z*: $[M + H]^+$ calculated for $C_{11}H_9N_2OS_5$ 344.9313; found 344.9311.

The compound's chemical structure was confirmed by SCXRD. A drawing of the molecular structure is shown in the SI together with crystal and refinement data (Table S2).

2.2.4. Synthesis and Solid-State Characterization of Dansyl Azide

The synthesis of dansyl azide (DNS-N_3) followed a previously described protocol [25]. Unlike the original report, which described DNS-N_3 as an oil, we successfully isolated it as a crystalline solid, likely due to the higher purity of the final product. The compound was fully characterized by single-crystal X-ray diffraction (SCXRD), providing additional structural information and unambiguous confirmation.

5-(dimethylamino)naphthalene-1-sulfonyl azide (DNS-N_3): a suspension of 5-(dimethylamino)naphthalene-1-sulfonyl chloride (250 mg, 0.93 mmol, 1.0 equivalent) in 20 mL of ethanol was added gradually to a stirred solution of NaN_3 (181 mg, 2.79 mmol, 3.0 equivalents) in 10 mL of a 1:1 mixture of water and ethanol at room temperature. The reaction mixture was then kept under stirring for 12 h. The organic solvent was removed under vacuum, and the remaining aqueous phase was extracted with CH_2Cl_2. The combined organic layers were washed with brine and dried over Na_2SO_4. After solvent removal, the crude product was obtained and purified by flash column chromatography in gradient (starting with Hex 100%, ending with Hex/EtOAc = 80/20), yield 43%, yellow crystals. ^{1}H NMR (300 MHz, CD_2Cl_2) δ 8.69 (dt, *J* = 8.6, 1.1 Hz, 1H), 8.35 (dd, *J* = 7.4, 1.3 Hz, 1H), 8.20 (dt, *J* = 8.7, 0.9 Hz, 1H), 7.63 (m, 2H), 7.26 (dd, *J* = 7.6, 1.0 Hz, 1H), 2.90 (s, 6H). ^{13}C{^{1}H} NMR (75 MHz, CD_2Cl_2) δ 152.59, 133.84, 133.09, 130.51, 130.33, 129.89, 129.54, 123.29, 118.75, 116.07, 45.48. MS (APCI) *m*/*z*: $[M + H]^+$ calculated for $C_{12}H_{13}N_4O_2S$ 277.08; found 277.32. IR (KBr disk, cm^{-1}) [27,28]: ν = 3292.55w, 2989.72w, 2952.10w (asymmetric -CH_3 stretching), 2871.09w, 2837.34w, 2793.94w (symmetric -CH_3 stretching), 2144.88s (N_3 stretching), 1980.93w, 1955.85w, 1924.03w, 1866.16w, 1782.26w, 1762.00w, 1688.71w, 1609.62m, 1585.51m, 1566.23m, 1500.64m, 1477.50m, 1451.46m, 1409.99m, 1393.59m, 1369.48m, 1352.12m (C-N stretching), 1307.76m, 1233.50m, 1192.99m, 1166.95s (-SO_2- stretching), 1147.67m, 1093.66m, 1062.80m, 1044.47m, 978.89m, 940.31m, 893.06m, 839.05m, 796.61s (C-H naphthyl out-of-plane bending), 773.47m, 741.64w, 678.95w, 630.73w, 596.01w, 562.26w, 514.04w. The compound's chemical structure was confirmed by SCXRD. A drawing of the molecular structure is shown in the SI together with crystal and refinement data (Table S3).

2.2.5. Single-Crystal X-Ray Diffraction

Single-crystal X-ray diffraction (SCXRD) data of $\mathbf{3^{CN}}$, $\mathbf{3^{CHO}}$ and DNS-N_3 were recorded at 100 K on an XtaLAB Synergy diffractometer from Rigaku, with mirror monochromated

Cu-*K*α-radiation (λ = 1.54184 Å). The instrument is equipped with a hybrid pixel array detector (HyPix). The samples were mounted on LithoLoops made by Molecular Dimensions fixed on pins made by Hampton Research. The samples were mounted on glass fibers. Absorption corrections were carried out using CrysAlisPro 1.171.42.61a (Rigaku OD, 2022, Tokyo, Japan). All structures were solved by dual methods (SHELXT-2018/3) and refined by full-matrix least-squares techniques employing the SHELXL-2019/2 executable and the WingX GUI [29–31]. All non-hydrogen atoms were refined with anisotropic displacement parameters. All hydrogen atoms were refined isotropically at calculated positions using a riding model, with their U_{iso} values constrained to 1.2 times U_{eq} of their pivot atoms for aromatic or methylene hydrogen atoms and to 1.5 times U_{eq} for the methyl atoms (which were allowed to rotate but not tip). General crystallographic, crystal, and refinement data are provided as Supplementary Information for $\mathbf{3^{CN}}$, $\mathbf{3^{CHO}}$, and DNS-N_3 (Tables S1–S3). Crystallographic data were deposited with the Cambridge Crystallographic Data Centre, CCDC, 12 Union Road, Cambridge CB21EZ, UK. These data can be obtained free of charge on quoting the depository numbers 2463816 ($\mathbf{3^{CN}}$), 2463818 ($\mathbf{3^{CHO}}$), and 2463817 (DNS-N_3) per email (deposit@ccdc.cam.ac.uk), or their web interface (at http://www.ccdc.cam.ac.uk).

2.3. *In Vitro Biological Studies*

2.3.1. Cell Lines and Culture Conditions

The human cancer cell lines MCF7 (HTB-22, breast adenocarcinoma), MDA-MB-231 (HTB-26, breast adenocarcinoma), HCT116 (CCL-247, colorectal carcinoma), and A549 (CRM-CCL-185, lung carcinoma) were obtained from the American Type Culture Collection (ATCC, Manassas, VA, USA). The fibroblast cell line CR9, kindly provided by Dr. Guven, was maintained in ISCOVE medium [32]. HCT116 cells were cultured in Dulbecco's Modified Eagle Medium (DMEM), whereas A549, MCF7, and MDA-MB-231 cells were maintained in Roswell Park Memorial Institute 1640 (RPMI-1640) medium. All media were supplemented with 10% fetal bovine serum (FBS), 1% L-glutamine, and 0.5% antibiotic mixture (penicillin, streptomycin, and neomycin). Cell cultures were maintained at 37 °C in a humidified atmosphere with 5% CO_2, following standard culture protocols. Additionally, for HCT116 cells, the culture medium was further supplemented with 1% sodium pyruvate and 1% non-essential amino acids.

To generate spheroids, HCT116 cells were seeded into 96-well ultra-low attachment plates (Nunclon™ Sphera, Thermo Fisher Scientific, Milan, Italy, 5×10^3 cells per well). The plates were incubated at 37 °C in a humidified atmosphere containing 5% CO_2, and spheroids were utilized for experiments on day 4 post-seeding.

2.3.2. Cell Viability Assay

To evaluate the impact of the tested PTEs on cell viability, a colorimetric MTT assay was performed as previously described [31]. Briefly, cells were seeded in 96-well microplates at initial densities of 5×10^3 cells per well for the MCF7, MDA-MB231, and A549 cell lines, and 4×10^3 cells per well for HCT116 cells. After an initial 24 h incubation period to allow for cell attachment and growth, the cultures were treated with different concentrations of PTEs (0.25–20 μM). Following 48 h of treatment, 50 μL of MTT solution (2 mg/mL in PBS) was added to each well, and the plates were incubated at 37 °C for 3 h to allow for MTT reduction by metabolically active cells. The resulting formazan crystals were solubilized, and cell viability was measured by recording the absorbance at 570 nm using an iMark Microplate Reader (BIORAD, Segrate, Milan, Italy). The cytotoxic effectiveness of the compounds was expressed as IC_{50} values, which were calculated from dose–response curves fitted with non-linear regression analysis using Calcusyn version 1.1.4. software (Biosoft, Cambridge, UK).

The cytotoxic effects of PTEs on HCT116 spheroids were evaluated by monitoring spheroid growth using a dye exclusion assay. The spheroids were treated with concentrations of the various PTEs equivalent to their IC_{50} values, as previously determined from MTT assays performed on HCT116 cells cultured under 2D conditions. At 24 h intervals, three spheroids per treatment group were independently harvested, enzymatically dissociated using trypsin-EDTA, and viable cells were quantified via Trypan Blue exclusion using a Bürker hemocytometer. Control spheroids were treated only with culture medium and incubated under the same conditions as the treated groups.

2.3.3. Flow Cytometric Analysis

The potential of the tested PTE derivatives to induce apoptotic and/or necrotic cell death, as well as to cause alterations in cell cycle distribution, was evaluated following a 48 h exposure to their respective IC_{50} concentrations, using flow cytometric analysis as previously described. Briefly, the indicated cell lines were seeded in 12-well plates (3×10^5 cells/well) and treated 24 h post-seeding. At the end of the treatment period, following harvesting, cells were washed with PBS and fixed in ice-cold 70% ethanol at −20 °C for a minimum duration of 45 min, to assess apoptotic cell death and perform cell cycle analysis. After fixation, cells were washed again with PBS and stained with a propidium iodide (PI)/RNase solution (50 µg/mL PI and 30 U/mL RNase) in PBS for 15 min at room temperature.

Flow cytometric analysis was performed using a FACScalibur flow cytometer (Becton Dickinson, Milan, Italy), equipped with a 15 mW, 488 nm air-cooled argon ion laser. Data were acquired and analyzed with CellQuestPRO version 5.1 software. PI fluorescence was detected using a 575 nm band-pass filter. The percentage of apoptotic cells was determined by identifying sub-G1 peaks in monoparametric histograms obtained in logarithmic mode. For cell cycle distribution, PI fluorescence was also analyzed with monoparametric histograms, but data were collected in linear mode.

Necrotic cell death was assessed by omitting the ethanol fixation step from the staining protocol, thereby enabling propidium iodide (PI) to penetrate membrane-compromised cells, indicative of necrosis.

Intracellular ROS levels were quantified using 2′,7′-dichlorodihydrofluorescein diacetate (DCFH-DA) as a fluorescent probe. After 48 h of treatment, cells were collected, rinsed with PBS, and incubated in a PBS solution containing 10 µM DCFH-DA. Following incubation, fluorescence corresponding to oxidized fluorescein was measured using a 530 nm band-pass filter. ROS production was expressed in arbitrary units based on the median fluorescence intensity (MFI).

PTEs-induced sulfide production was evaluated using dansyl azide (DNS-N_3) as a probe. DNS-N_3 is non-fluorescent itself, while in the presence of hydrosulfide, the azide transforms into the dansyl amide and fluorescence is switched on [25]. After 48 h of treatment, cells were harvested, washed with PBS, and incubated in a PBS solution containing 200 µM DNS-N_3. Following 30 min incubation at 37 °C, fluorescence increase was measured through a 575 nm band-pass filter. The levels of hydrosulfide production were assessed and reported in arbitrary units, calculated from the median fluorescence intensity (MFI) measurements. For all flow cytometric analyses, untreated cells (i.e., not exposed to PTE compounds) were used as negative controls.

2.3.4. Cleavage Assay

DNA single- and double-strand break formation was assessed using a plasmid cleavage assay. In this method, single-strand breaks convert supercoiled (SC) plasmid DNA into its relaxed open circular (OC) form, while double-strand breaks produce linear (L) DNA fragments. Notably, the presence of a thiol compound, in this case reduced glutathione (GSH), is essential for the induction of DNA cleavage. Briefly, pGEM7 plasmid (0.2 µg) was incubated with the studied pentathiepines (25 µM) in the presence or absence of glutathione (GSH, 2 mM) in a sodium phosphate buffer at pH 5.5 for 20 h at 37 °C. Afterwards, the samples were separated on a 1.5% agarose gel at 90 V (6 V/cm) for 1.5 h, stained with Ethidium Bromide, and the image captured.

2.3.5. Scratch Wound Healing Assay

To investigate the potential effects of the studied pentathiepines on cell migration in the MDA-MB231 and A549 cell lines, we performed the scratch wound healing assay, as previously reported [33]. Cells were seeded in 12-well plates at densities of 2×10^5 cells per well for MDA-MB231 and 6×10^4 cells per well for A549; they were allowed to grow for 24 h until reaching near confluence. A uniform scratch was then generated in the cell monolayer using a sterile pipette tip, after which the cells were treated with subtoxic concentrations of pentathiepine derivatives, corresponding to their respective IC_{20} values. Images of the scratch area were collected immediately after scratch formation (t_0) and after 24 h of incubation, using a camera connected to an Olympus IX81 microscope. The percentage of open wound area was quantified at both time points using TScratch version 1.1.2. software.

2.3.6. Western Blot Analysis

Metalloproteinases MMP2 and MMP9 protein levels were assessed by Western blotting in MDA-MB-231 and A549 cells following treatment with PTE derivatives at concentrations corresponding to their respective IC_{20} values. After treatment, cells were lysed in a buffer containing 120 mM NaCl, 25 mM NaF, 5 mM EDTA, 6 mM EGTA, 25 mM sodium pyrophosphate, 20 mM Tris-HCl (pH 7.4), supplemented with 2 mM PMSF, 1 mM Na_3VO_4, 1 mM phenylarsine oxide, 1% NP-40, and 10% protease inhibitor cocktail. Following the addition of SDS (final concentration 0.1%), samples were incubated on ice for 10 min. Lysates were clarified by centrifugation at 12,800 rpm for 20 min at 4 °C.

Protein concentrations were determined using the BCA protein assay (Pierce, Italy). Equal amounts of protein (50 µg per lane) were separated by SDS-PAGE on 8% polyacrylamide gels under denaturing conditions and transferred to Hybond-P PVDF membranes (Amersham Biosciences, Milan, Italy). Membranes were blocked and probed using mouse monoclonal antibodies specific to MMP2 and MMP9 (Santa Cruz Biotechnology, Inc., Segrate, Milan, Italy). To verify equal protein loading, membranes were re-probed with a mouse monoclonal anti-actin antibody (Santa Cruz Biotechnology, Inc.). Immunoreactive bands were visualized using the Bio-Rad ChemiDoc™ Imaging System, with HRP-conjugated anti-mouse secondary antibodies and Westar Supernova chemiluminescent substrate (Cyanagen). Densitometric quantification of protein bands was performed using ImageJ version 1.53e software.

2.3.7. Statistical Analysis

Intergroup comparisons were conducted using one- or two-way analysis of variance (ANOVA), followed by Bonferroni post hoc correction for multiple comparisons. A significance level of 0.05 was applied. All statistical analyses were performed using GraphPad Prism version 8.4.

3. Results

3.1. Effects on Cell Viability

The effects of **1**, **2**^{Me}, **2**^{Gly}, **2**^{Ph}, **2**^{PG}, **3**^{Me}, **3**^{Br}, **3**^{NO2}, **3**^{CHO}, and **3**^{CN} on the viability of MCF7, MDA-MB-231, A549, and HCT116 cells were assessed using the MTT assay after 48 h of treatment with increasing concentrations of the compounds. IC_{50} values, derived from the respective dose–response curves, are reported in Table 1. These values are also graphically represented as histograms in Figure S17 for easier interpretation.

Table 1. IC_{50} values obtained in MCF7, MDA-MB231, A549, HCT116 and CR9 cell lines following 48h treatment with **1**, **2**^{Me}, **2**^{Gly}, **2**^{Ph}, **2**^{PG}, **3**^{Me}, **3**^{Br}, **3**^{NO2}, **3**^{CHO}, and **3**^{CN}, and MTT assay. IC50 values obtained following treatment with doxo and cisPt were also reported (mean ± S.D. of 3/5 independent experiments; *** $p < 0.001$ vs. **1**, **2**^{Me}, **2**^{Gly}, **2**^{PG}, **3**^{NO2}, **3**^{CHO}, and **3**^{CN} in the same cell line; # $p < 0.01$ vs. same PTE in all the other cell lines; @ $p < 0.01$ vs. same PTE in MDA-MB231, A549, and HCT116 cells).

IC_{50} (μM)	MCF7	MDA-MB231	A549	HCT116	CR9
1	2.68 ± 0.31	0.63 ± 0.04	1.537 ± 0.08	0.83 ± 0.04	7.32 ± 0.15 #
2^{Me}	4.77 ± 0.14	1.45 ± 0.04	2.54 ± 0.27	2.27 ± 0.34	7.91 ± 0.32 #
2^{Gly}	1.13 ± 0.02	0.30 ± 0.03	0.79 ± 0.06	0.603 ± 0.04	1.61 ± 0.13 @
2^{Ph}	14.93 ± 1.74 ***	4.41 ± 0.22 ***	12.22 ± 0.66 ***	4.79 ± 0.45 ***	26.88 ± 0.78 #
2^{PG}	0.90 ± 0.07	0.35 ± 0.06	0.72 ± 0.08	0.50 ± 0.01	2.04 ± 0.12 #
3^{Me}	13.66 ± 1.51 ***	2.80 ± 0.12 ***	4.50 ± 0.69 ***	3.28 ± 0.22 ***	75.92 ± 0.16 #
3^{Br}	8.35 ± 0.35 ***	3.62 ± 0.21 ***	6.42 ± 0.31 ***	5.83 ± 0.07 ***	10.9 ± 0.89 @
3^{NO2}	3.21 ± 0.13	0.63 ± 0.75	3.22 ± 0.39	1.59 ± 0.15	2.87 ± 0.14
3^{CHO}	1.70 ± 0.03	0.39 ± 0.04	1.03 ± 0.04	0.74 ± 0.06	1.9 ± 0.12 @
doxo	4.5 ± 0.51	17.8 ± 1.96	1.5 ± 0.21	3.5 ± 0.60	-

The results revealed variability in response to the tested PTEs, depending on the cell line. Nonetheless, a trend of increased statistical potency of the compounds was observed across the cell lines. Specifically, **2**^{Gly}, **2**^{PG}, and **3**^{CHO} exhibited the highest cytotoxic potency, displaying submicromolar IC_{50} values in the majority of the tested cell lines. Notably, the first two show structural features reminiscent of the natural product varacin, particularly in their extended polar substituents with a nucleophilic group located adjacent to the pentathiepino ring.

Among the remaining compounds, including those bearing -H, $-NO_2$, and -CN groups at the C9 position (namely, **1**, **3**^{NO2}, and **3**^{CN}), as well as **2**^{Me}, also demonstrated noteworthy efficacy. **2**^{Ph}, **3**^{Me}, and **3**^{Br} exhibited the lowest cytotoxic activity across all tested cell lines. This reduced efficacy can likely be attributed to the inherently low water solubility of pentathiepines, which is further compromised by the presence of hydrophobic substituents.

Interestingly, all the PTEs evaluated exhibited their highest cytotoxic potency against the TNBC MDA-MB231 cell line, as evidenced by the consistently lower IC_{50} values obtained.

The effects of PTEs on cell viability were further compared to those induced by doxorubicin (doxo), a standard chemotherapeutic agent. The IC_{50} values for doxo, determined under the same experimental conditions, are reported in Table 1. Overall, the most active PTEs exhibited comparable or superior potency to doxorubicin across all tested cell lines.

To evaluate potential off-target effects, the cytotoxic activity of the PTEs was also assessed in CR9 human fibroblast cells. As reported in Table 1, the IC_{50} values indicate significantly reduced cytotoxicity in this cell line, compared to the others, thus underscoring a degree of selectivity toward malignant phenotypes.

Overall, the data highlight **1**, **2**^{Me}, **2**^{Gly}, **2**^{PG}, **3**^{NO2}, **3**^{CHO}, and **3**^{CN} as the most promising candidates for further investigation, due to their enhanced cytotoxic activity. The remaining compounds were excluded from subsequent analyses because of their limited efficacy.

The effects of PTEs were further evaluated in three-dimensional (3D) spheroid cultures derived from HCT116 cells, following 48 h treatments with compound concentrations corresponding to the IC_{50} values previously determined in two-dimensional (2D) monolayer cultures of the same cell line. Analysis of spheroid morphology revealed that control spheroids exhibited a progressive increase in diameter over the incubation period, consistent with typical proliferative behavior in the absence of PTE. In contrast, spheroids exposed to PTEs demonstrated a substantial reduction in size by 48 h post-treatment, indicative of pronounced antiproliferative effects. These morphological observations were confirmed by quantitative analyses of viable cell numbers obtained from spheroid disaggregation at 24 and 48 h post-treatment. As illustrated in Figure 5, all PTE derivatives induced a statistically significant reduction in cell viability within the spheroids, further supporting their cytotoxic efficacy in a 3D tumor model.

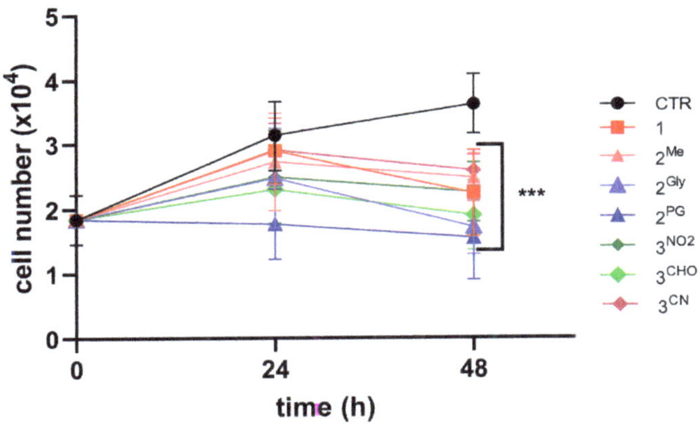

Figure 5. Growth curves of HCT116 spheroids following treatment with PTEs for 48h at concentrations corresponding to the IC_{50} values obtained in monolayer-cultured cells. Viable cell counts were performed immediately before (0) and every 24h, up to 2 days, using a Burker chamber (*** $p < 0.001$ vs. CTR).

3.2. Cell Death Induction

To gain further insight into the mode of action of the most promising compounds, additional studies were carried out to assess their ability to induce apoptotic and/or necrotic cell death in previously characterized cancer cell lines. To this aim, flow cytometric analysis, following 48 h of treatment with equitoxic concentrations corresponding to each compound's respective IC_{50} value, was performed.

Overall, pentathiepine compounds demonstrated a greater propensity to induce apoptosis rather than necrosis. Precisely, treatment with PTEs resulted in a statistically significant increase in apoptotic cell death compared to untreated controls across all cell lines tested, although the extent of the response varied. Notably, the apoptotic response was cell-type dependent, with MDA-MB-231 cells exhibiting the highest sensitivity to pentathiepine-induced apoptosis. Furthermore, while some variation was observed, the apoptotic responses were generally consistent with the cytotoxicity profiles obtained from the MTT assay, with compounds 1, 2^{Gly}, 2^{PG}, 3^{NO2}, and 3^{CHO} emerging as the most potent inducers of apoptosis (Figure 6).

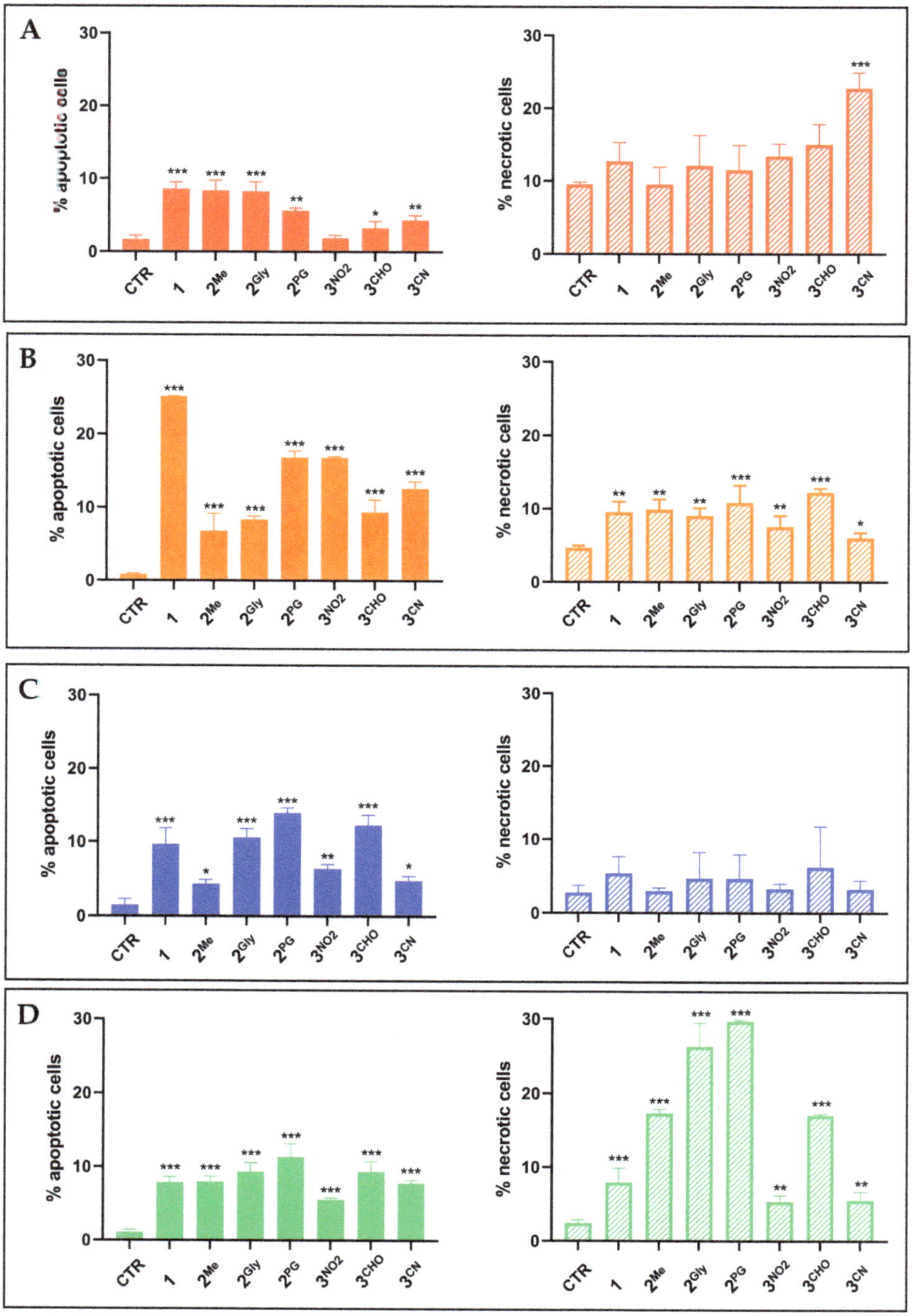

Figure 6. Percentage of apoptotic and necrotic MCF7 (**A**), MDA-MB231 (**B**), A549 (**C**) and HCT116 (**D**) cells following 48 h treatment with **1**, $\mathbf{2^{Me}}$, $\mathbf{2^{Gly}}$, $\mathbf{2^{PG}}$, $\mathbf{3^{NO2}}$, $\mathbf{3^{CHO}}$, and $\mathbf{3^{CN}}$ at concentrations corresponding to their IC_{50} values (mean ± S.D. of 3/4 independent experiments; * $p < 0.05$, ** $p < 0.01$, and *** $p < 0.001$ vs. CTR).

Significant increases in necrotic cell death were also observed in MCF7, MDA-MB231, and HCT116 cells following treatment with PTEs. However, compounds **2**^{Me}, **2**^{Gly}, **2**^{PG}, and **3**^{CHO}, and **3**^{CN} induced higher levels of necrosis relative to apoptosis in HCT116 and MCF-7 cells, respectively.

3.3. Cell Cycle Alterations

The impact of PTEs on DNA content distribution across cell cycle phases was evaluated by flow cytometry after 48 h of treatment at equitoxic concentrations. Consistent with previous observations, the effects of PTEs exhibited cell type-specific variability (Figure 7). Nevertheless, in all instances where the treatment induced alterations in cell cycle progression, a consistent decrease in the G0/G1 phase population was mainly observed, accompanied by a corresponding G2/M phase accumulation. As shown in Figure 7, compound **1** induced this characteristic cell cycle shift across all examined cell lines. In contrast, compounds **2**^{Me}, **2**^{Gly}, and **2**^{PG} triggered similar effects in both MDA-MB231 and HCT116 cells, while **3**^{CHO} exerted this effect exclusively in MDA-MB231 cells.

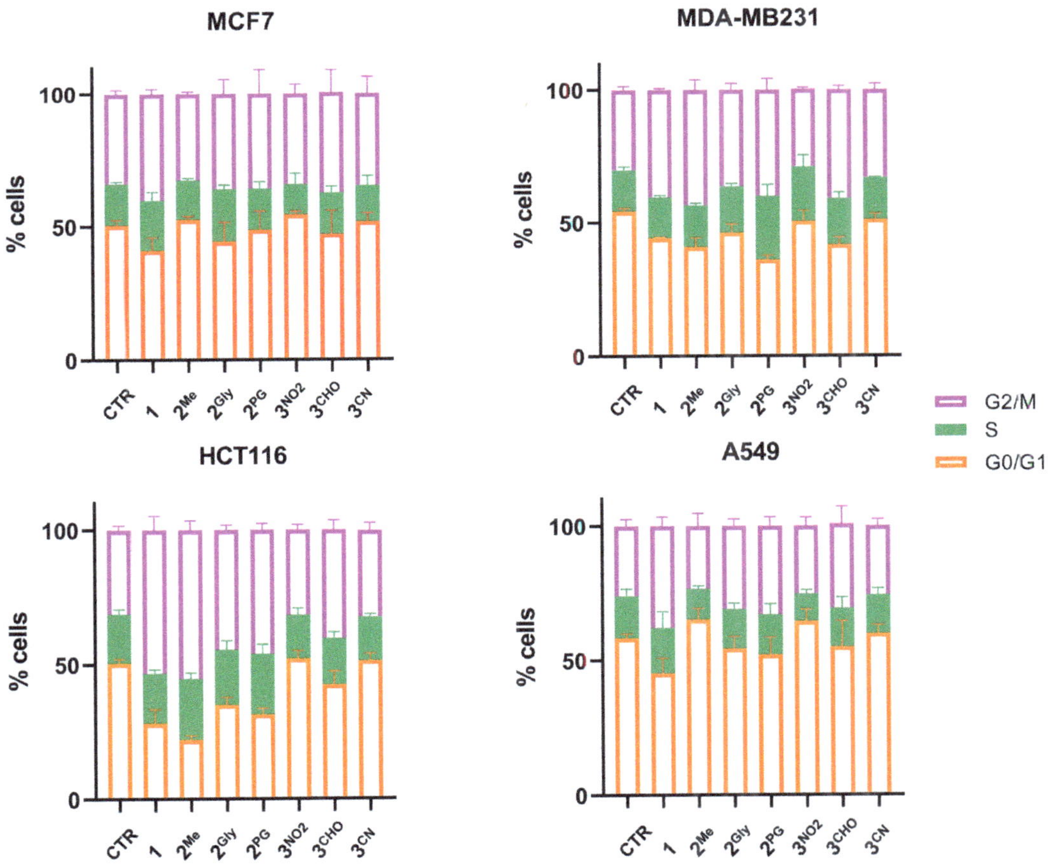

Figure 7. Cell cycle distribution of MCF7, MDA-MB231, A549 and HCT116 cells following 48 h treatment with **1**, **2**^{Me}, **2**^{Gly}, **2**^{PG}, **3**^{NO2}, **3**^{CHO}, and **3**^{CN} at concentrations corresponding to their IC_{50} values (mean ± S.D. of 3/4 independent experiments).

3.4. ROS Production

To evaluate the oxidative stress induced by PTEs, intracellular ROS levels were assessed following a 48 h treatment of the four cell lines included in this study with each compound at its respective IC_{50} concentration, using 2′,7′-dichlorodihydrofluorescein diacetate (DCFH-DA) as a fluorescent probe. The results demonstrate a general ability of the pentathiepine compounds to increase intracellular ROS levels, although the magnitude of this effect varied across the different cell lines (Figure 8). In MCF7, MDA-MB231, and HCT116 cells, only a modest increase in ROS production was observed. Nonetheless, this increase was statistically significant for compounds **1**, $\mathbf{2^{Me}}$, $\mathbf{2^{PG}}$, and $\mathbf{3^{NO2}}$, and $\mathbf{3^{CHO}}$ in both MCF7 and MDA-MB231 cells; for $\mathbf{2^{Gly}}$, and $\mathbf{3^{CN}}$ in MDA-MB231 cells; and for **1**, in HCT116 cells.

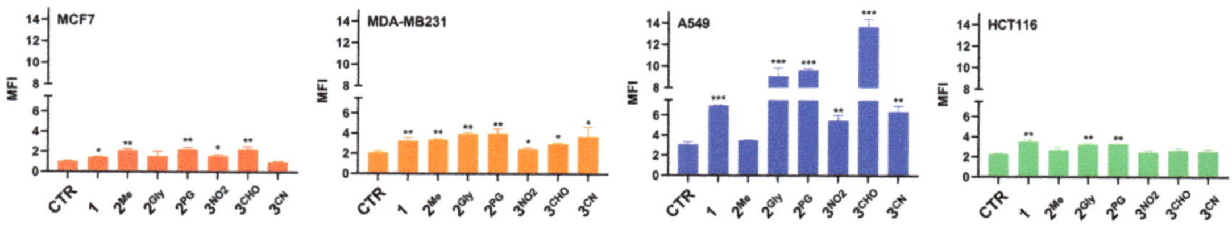

Figure 8. Intracellular ROS levels following 48 h treatment with **1**, $\mathbf{2^{Me}}$, $\mathbf{2^{Gly}}$, $\mathbf{2^{PG}}$, $\mathbf{3^{NO2}}$, $\mathbf{3^{CHO}}$, and $\mathbf{3^{CN}}$ at concentrations corresponding to their respective IC_{50} values (mean ± S.D. of 3 independent experiments; * $p < 0.05$, ** $p < 0.01$, and *** $p < 0.001$ vs. CTR).

In contrast, all compounds, except $\mathbf{2^{Me}}$, induced significant and marked increases in ROS levels in A549 cells.

Interestingly, in agreement with their effects on cell viability and/or their ability to induce cell death, **1**, $\mathbf{2^{Gly}}$, $\mathbf{2^{PG}}$, and $\mathbf{3^{CHO}}$ were the better ROS producers.

3.5. Hydrosulfide Production

The selective reactivity of DNS-N_3 towards hydrosulfide ions (HS^-) was previously validated, resulting in the release of the corresponding fluorescent sulfuric amide derivative bearing an -$S(O)_2NH_2$ group(DNS-NH_2; Figure 9) [25]. Originally developed to quantify HS^- ions in blood samples, we successfully adapted and extended this method for the quantitative detection of hydrosulfide ions in cultured cells following exposure to PTE.

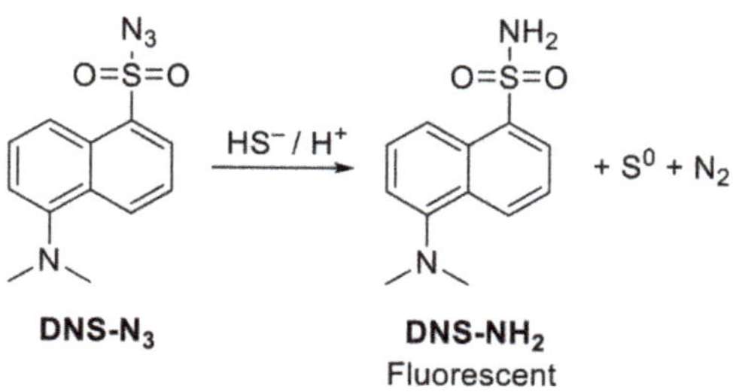

Figure 9. Reaction of DNS-N_3 with HS^- leading to the formation of the fluorescent amide derivative DNS-NH_2. This transformation exemplifies the reduction in organic azides (R-N_3) in aqueous environments, where HS^- is oxidized to S^0 by two electrons, which may further convert into its stable allotrope, cyclo-S_8, under high local concentrations. Concurrently, one nitrogen of the azide moiety is partially reduced (formally N^- to N^{3-}) to become part of the sulfuric amide -$S(O)_2NH_2$ moiety. In contrast, the remaining nitrogen atoms form molecular nitrogen (N_2) without changing their oxidation states. Hydrogen atoms, which are present in the hydrosulfide (the prevalent species at and around neutral pH) and the -NH_2 moiety of the amide, are most likely derived from medium water and/or acidic amino acids of proteins in this study.

To confirm that DNS-N_3 does not react with PTEs, an equimolar amount of compound **1** and DNS-N_3 was dissolved in methanol and stirred at room temperature in the dark for 48 h. No reaction was observed, confirming the probe's stability and specificity.

DNS-N_3 was employed as a fluorescent pro-probe to assess hydrosulfide production induced by PTEs in MCF7, MDA-MB231, A549 and HCT116 cell lines following 48 h of treatment with equitoxic concentrations, corresponding to the respective IC_{50} values. As shown in Figure 10, an increase in DNS-NH_2 fluorescence, measured as median fluorescence intensity and indicative of hydrosulfide presence, was observed across all cell lines. PTE exposure resulted in a general increase in intracellular sulfide levels, although variations were observed among cell types. Among the tested compounds, **1**, $\mathbf{2^{Gly}}$, $\mathbf{2^{PG}}$, $\mathbf{3^{CHO}}$, and $\mathbf{3^{CN}}$ were the most potent inducers of sulfide production. Only $\mathbf{3^{CN}}$ and $\mathbf{2^{Me}}$ did not induce sulfide increase in A549 and HCT116 cells, respectively.

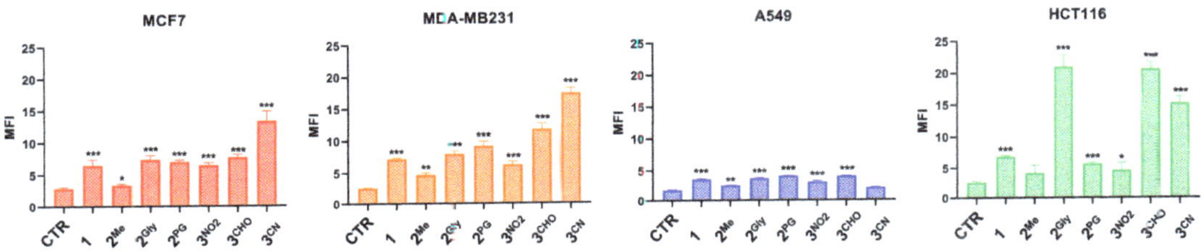

Figure 10. Intracellular hydrosulfide production following 48 h treatment with **1**, $\mathbf{2^{Me}}$, $\mathbf{2^{Gly}}$, $\mathbf{2^{PG}}$, $\mathbf{3^{NO2}}$, $\mathbf{3^{CHO}}$, and $\mathbf{3^{CN}}$ at concentrations corresponding to their respective IC_{50} values (mean ± S.D. of 3 independent experiments; * $p < 0.05$, ** $p < 0.01$, and *** $p < 0.001$ vs. CTR).

3.6. DNA Cleavage

DNA single- and double-strand break formation was assessed using a plasmid cleavage assay. In this method, single-strand breaks convert supercoiled (SC) plasmid DNA into its relaxed open circular (OC) form, while double-strand breaks produce linear (L) DNA fragments. Notably, the presence of a thiol compound, in this case reduced glutathione (GSH), is essential for the induction of DNA cleavage.

Treatment of the plasmid DNA with the different PTEs in the presence of GSH resulted in an increased generation of OC DNA and a concomitant reduction in SC DNA compared to the PTEs without GSH (Figure 11). These findings confirm that PTEs are capable of inducing DNA strand cleavage under reducing conditions provided by GSH.

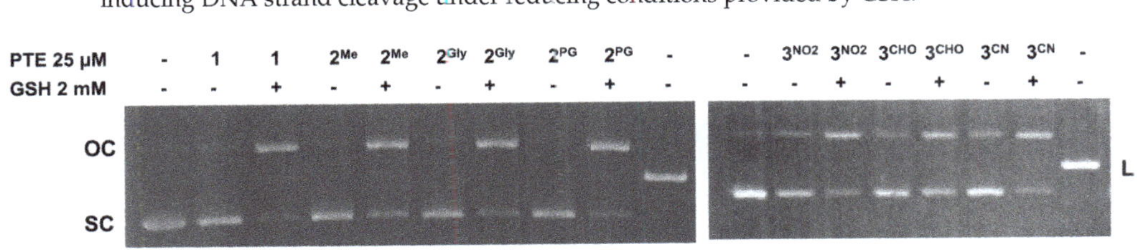

Figure 11. Plasmid DNA cleavage assay after 20 h incubation at 37 °C with compounds **1**, $\mathbf{2^{Me}}$, $\mathbf{2^{Gly}}$, $\mathbf{2^{PG}}$, $\mathbf{3^{NO2}}$, $\mathbf{3^{CHO}}$, and $\mathbf{3^{CN}}$ at a concentration of 25 µM in the absence and the presence of GSH (2 mM). The linear DNA control band (L) corresponds to plasmid DNA digested with the restriction enzyme BamHI (uncropped images of the gels are shown in Figure S18).

3.7. Cell Migration and Matrix Metalloproteinase Inhibition

Cell migration plays a key role in the metastatic spread of cancer and represents an important target in the development of anticancer therapies. To investigate whether the selected pentathiepine derivatives exert anti-migratory effects, a scratch wound healing assay was performed on MDA-MB231 and A549 cells, two well-established models with high intrinsic motility. In contrast, MCF7 and HCT116 cells, which lack significant migratory capacity under standard conditions, were excluded from this analysis. To specifically evaluate migration-related effects independent of cytotoxicity, compound concentrations were carefully selected to fall within a subtoxic range, exerting minimal influence on cell viability.

Images of the scratches taken at 0 and 24 h post-treatment (Figure S19) showed that all the tested compounds reduced cell migration in both MDA-MB231 and A549 cells. To better quantify these observations, the percentage of open scratch areas was measured using T-Scratch software 1.0. The results confirmed the ability of the pentathiepine compounds to significantly reduce cellular migration in the two cell lines tested, as indicated by the statistically significant increase in the percentage of open scratches (Figure 12).

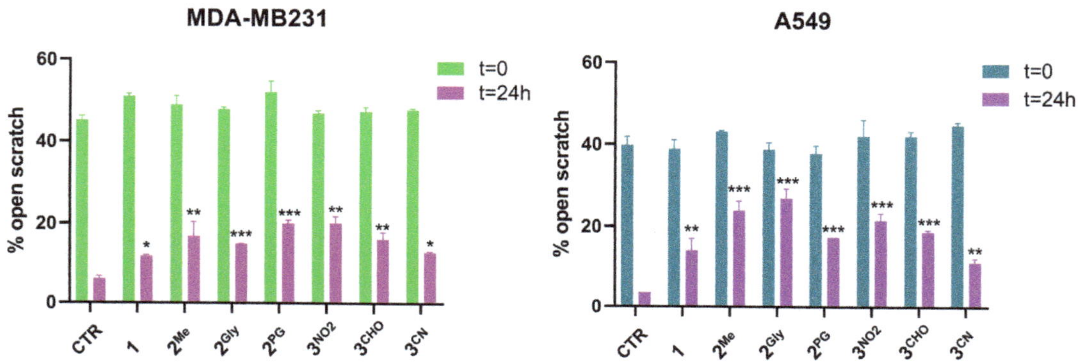

Figure 12. Percentages of open scratches in MDA-MB231 and A549 cells following treatment with **1**, $\mathbf{2^{Me}}$, $\mathbf{2^{Gly}}$, $\mathbf{2^{PG}}$, $\mathbf{3^{NO2}}$, $\mathbf{3^{CHO}}$, and $\mathbf{3^{CN}}$ (mean ± S.D. of 2 independent experiments, with two replicates for each treatment; * $p < 0.05$; ** $p < 0.001$; *** $p < 0.001$ vs. CTR 24 h).

Western blot analyses were subsequently performed to assess the effects of PTEs on the expression levels of matrix metalloproteinases 2 and 9 (MMP2 and MMP9), which are key mediators of cancer progression through their roles in promoting tumor invasion, metastasis, and angiogenesis. Given their involvement in these processes, inhibition of MMP2 and MMP9 may also contribute to the observed anti-migratory effects of PTE treatment. Figure 13 presents representative Western blot images and the corresponding densitometric analyses of protein expression following 48 h treatment with the same concentrations of the tested PTEs employed in the scratch wound healing assay. Densitometric quantification of MMP2 and MMP9 levels was normalized to actin. Overall, all compounds induced a reduction in MMP2 and/or MMP9 expression.

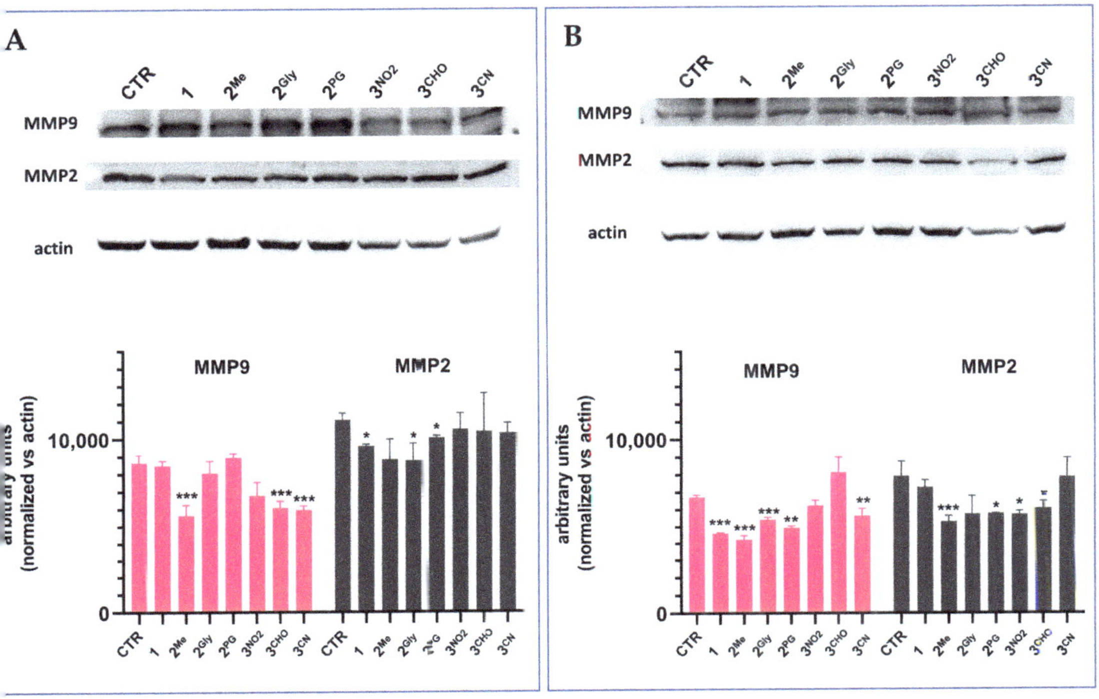

Figure 13. MMP9 and MMP2 protein in MDA-MB231 (**A**) and A459 (**B**) cell lines following treatment with **1**, $\mathbf{2}^{\mathbf{Me}}$, $\mathbf{2}^{\mathbf{Gly}}$, $\mathbf{2}^{\mathbf{PG}}$, $\mathbf{3}^{\mathbf{NO2}}$, $\mathbf{3}^{\mathbf{CHC}}$, and $\mathbf{3}^{\mathbf{CN}}$. Representative Western blot images out of two independent experiments with similar results (the uncropped images of the blots are shown in Figure S20) and densitometric analysis of both experiments are reported. Densitometric analysis was performed using the ImageJ software (mean ± S.D. of 2 independent experiments; *** $p < 0.001$; ** $p < 0.01$; * $p < 0.05$ vs. CTR).

4. Discussion

Pentathiepines have attracted growing attention owing to their promising anticancer potential. Nevertheless, many of these compounds exhibit limited target specificity and are associated with off-target effects, underscoring the necessity for the development of more selective, non-toxic, and efficacious pentathiepine-based therapeutics. Moreover, investigations into the molecular mechanisms underlying their anticancer activity remain limited, highlighting a critical gap in the current understanding of their mode of action.

In the present study, we investigated for the first time the anticancer properties of the simplest core scaffold **1** (Figure 2), along with selectively substituted analogs at positions C-6 and C-9, which were characterized previously [24,26], and two novel C-9 substituted entities, synthesized and characterized for the first time for this study (namely, $\mathbf{3}^{\mathbf{CHO}}$ and $\mathbf{3}^{\mathbf{CN}}$). Their synthesis was driven not only by the observation that electron-withdrawing groups enhance reaction yields at the C-9 position but also by the potential of these polar functionalities to engage in hydrogen bonding. Such interactions may strengthen binding to biological targets, improve aqueous solubility and bioavailability, and potentially reduce the likelihood of drug resistance [34,35]. The C-6 position is particularly interesting, as it is adjacent to the pentathiepino ring and reflects the natural substitution pattern observed in marine-derived analogs. Notably, the functionalization of this position has only recently become synthetically accessible in indolizine-based PTEs, thereby opening new avenues for structural diversification and potential enhancement of biological activity [26]. Its pharmacological relevance has been underscored by previous studies demonstrating that substitution at this site plays a pivotal role in modulating the compounds' biological activity [36,37].

Our results demonstrate that most of the tested PTEs exhibit significant cytotoxic activity against the cancer cell lines included in this study. Interestingly, these compounds showed a degree of selectivity, exerting greater cytotoxicity toward malignant cells compared to non-tumorigenic fibroblast controls. Certain chemical features may partially explain the enhanced potency observed for compounds $\mathbf{2}^{\mathbf{Gly}}$, $\mathbf{2}^{\mathbf{PG}}$, and $\mathbf{3}^{\mathbf{CHO}}$. Notably, these three molecules possess hydrophilic functional groups capable of forming hydrogen bonds, which may increase aqueous solubility and promote more effective interactions with biological targets. For example, the aldehyde moiety in $\mathbf{3}^{\mathbf{CHO}}$ may undergo hydration in aqueous environments, thereby increasing its polarity and potentially improving bioavailability. Interestingly, the position of the methyl group appears to influence cytotoxicity: methyl substitution at the C6 position does not significantly impair biological activity, whereas substitution at the C9 position is associated with a marked reduction in efficacy.

Monolayer (2D) cell cultures provide several practical advantages, including ease of handling, cost-effectiveness, and compatibility with high-throughput functional assays. However, growing evidence highlights their limitations in faithfully replicating the in vivo tumor microenvironment, resulting in the limited translational success of several drug candidates that exhibited promising results in 2D systems but failed to meet efficacy benchmarks in clinical trials [38,39]. Several in vitro three-dimensional (3D) cell culture models, including multilayers, spheroids, and organoids, have provided valuable platforms for assessing the efficacy of anticancer agents [40]. Our results indicate that the PTEs under investigation maintained their efficacy in three-dimensional spheroid models derived from the HCT116 cell line.

Emerging evidence suggests that the anticancer properties of PTEs are associated with their capacity to trigger apoptotic and/or necrotic pathways of cell death as well as to induce cell cycle alterations [9,14,23]. Although a strong apoptotic or necrotic response was not prominently observed under our experimental conditions, the data suggest a general tendency of the studied PTEs to promote apoptosis over necrosis. It is worth noting that in previous studies conducted on analogous PTEs synthesized via molybdenum-mediated methods, apoptosis was assessed under substantially more aggressive conditions, namely IC_{90} concentrations and extended exposure times (48 h) [9,14]. Even under such drastic treatments, the apoptotic response reported was limited, suggesting that the induction of classical apoptosis may not be the primary mechanism of action for pentathiepines. This observation supports the need to reconsider or broaden the framework through which cell death mechanisms are interpreted in the context of these compounds. Interestingly, among the tested cell lines, MDA-MB231, which represents a particularly aggressive and treatment-resistant cancer phenotype, namely triple-negative breast cancer (TNBC), exhibited the highest sensitivity to PTE-induced apoptosis. Moreover, the overall pattern of apoptotic response closely mirrored the cytotoxicity profiles previously obtained through the MTT assay, confirming the most promising compounds to be **1**, $\mathbf{2}^{\mathbf{Me}}$, $\mathbf{2}^{\mathbf{Gly}}$, $\mathbf{2}^{\mathbf{PG}}$, $\mathbf{3}^{\mathbf{NO2}}$, $\mathbf{3}^{\mathbf{CHO}}$, and $\mathbf{3}^{\mathbf{CN}}$.

The observed ability of PTEs to induce G2/M phase arrest in cell cycle progression is in agreement with prior studies, further supporting the notion that these compounds interfere with cell cycle regulation through conserved mechanisms [9]. Cell cycle progression is tightly regulated by genome integrity, with DNA damage triggering arrest at specific checkpoints to allow for repair or, in cases of severe damage, to initiate programmed cell death pathways [41,42]. Accordingly, the perturbations in cell cycle observed following PTE exposure may be attributed to their capacity to induce DNA damage, likely mediated by oxidative stress. Supporting this hypothesis, the present study confirmed literature data showing that the tested PTEs promote both ROS generation and DNA damage.

Recent studies have demonstrated a direct connection between sulfide species, such as S^{2-} and S^{1-}, and the key reactive site of PTEs, being the neutral sulfur S^0 atoms of the polysulfide moiety [43,44]. Polysulfidation, referring to the set of transformations that give rise to neutral sulfur (S^0) embedded within structures such as R–S–$S^0{}_n$–S–R, originates from species holding S^{2-} and S^{1-}-based functional groups and has been implicated in the regulation of oxidative stress, redox signaling, and mitochondrial bioenergetics [45,46]. Literature reports on PTEs have described specific interactions with key functional motifs, including the selenol group (–SeH) of glutathione peroxidase (GPx), zinc finger domains essential for protein folding and stability, and the deprotonated cysteine residue within the catalytic site of striatal-enriched phosphatase (STEP) [12,14,47]. While these studies provide detailed and well-supported insights into the potential pharmacological targets of PTEs, they are often limited in scope and do not fully capture the broader spectrum of their biological activity. Therefore, to bridge these observations by proposing a broader, unified interpretation with the literature regarding sulfur mobilization in cells and its functional biochemical consequences, we established a method to evaluate the ability of PTEs to stimulate sulfide production in viable cells. Specifically, the process of polysulfidation might be postulated, whereby thiols, persulfides, or enzyme active site residues are targeted and saturated by sulfur fragments released by PTEs. Thus, our results suggest that the detection of sulfide ion S^{2-} release following PTE treatment could serve as a valuable indicator of its activity within cells. Nevertheless, considering the variability among different cell lines, a greater number of replicates is required to perform robust linear regression analysis. The obtained saturated groups, namely polysulfides R-S_n-R or hydropolysulfides R-S_n-H (where $n \geq 3$ and R represents a carbon-based residue [48]), may disrupt cellular metabolism through two potential mechanisms: (i) a collapse of essential cellular functions due to excessive polysulfidation of critical biomolecules, or (ii) a compensatory cellular response aimed at degrading the –S_n– chains, leading to the overproduction and intracellular accumulation of sulfide ions (S^{2-}), which may reach cytotoxic concentrations [49]. Therefore, the increased sulfide production observed in all cell lines may partially explain the cytotoxic effect observed.

To the best of our knowledge, the anti-migration effects of PTEs were evaluated for the first time in the present work. Cellular migration is a critical hallmark of cancer, contributing significantly to tumor invasion and metastasis, which are the primary causes of cancer-related mortality [50]. This process involves dynamic cytoskeletal rearrangements, the modulation of cell–matrix and cell–cell adhesion, and the degradation of the extracellular matrix, enabling malignant cells to invade surrounding tissues and disseminate to distant sites via the bloodstream or lymphatic system [51,52]. As such, targeting cell migration has emerged as a promising approach to hinder metastasis and improve therapeutic outcomes. In this context, recent studies have shown that the migratory and invasive capabilities of metastatic cancer cells can be significantly impaired by therapeutic strategies targeting MMP activity, thereby highlighting MMPs as promising targets for anti-metastatic interventions [53]. Matrix metalloproteinases (MMPs) are zinc-dependent endopeptidases that function like molecular scissors, degrading extracellular matrix components to facilitate cancer cell movement [54]. Several observations, including the correlation between MMP activity and advanced tumor stage and reduced survival, or their altered activity in a wide range of malignancies, underscore the prognostic significance of MMPs and their potential utility as biomarkers for cancer progression and therapeutic response monitoring [55,56]. Finally, our findings indicate that the PTEs under investigation exhibit notable anti-migratory activity, which appears to be associated, at least in part, with their capacity to suppress the expression of MMP2 and MMP9. By inhibiting MMP2 and MMP9, the PTEs may effectively "dull the blades," thereby limiting the cells' ability to invade surrounding tissues and metastasize.

This study has enhanced the understanding of the biological mechanisms underlying the antitumor effects of PTEs, particularly by demonstrating for the first time their ability to induce sulfide accumulation, inhibit cell migration, and retain activity in a three-dimensional cellular model (i.e., spheroids) that more closely mimics the in vivo tumor environment. Further investigations are warranted to clarify their mechanisms of action in greater detail. Specifically, in light of the observed downregulation of MMPs, assessing the effects of PTEs on cell adhesion could help determine whether these compounds also possess anti-angiogenic properties, thereby supporting their potential in developing vascular-targeted (immuno)therapies. Moreover, the ability of PTEs to induce intracellular stress, such as DNA damage, oxidative stress, and the generation of sulfide species, may enhance the expression or presentation of tumor-associated antigens, potentially increasing tumor cell immunogenicity. This hypothesis deserves further investigation.

5. Conclusions

In this study, we investigated the anticancer properties of selectively substituted analogs at the C-6 and C-9 positions of scaffold **1**, encompassing both previously characterized and newly synthesized pentathiepine derivatives (PTEs). These compounds were evaluated in two-dimensional (2D) monolayer cultures representing various tumor types, including a particularly aggressive and therapeutically underserved model. Furthermore, to our knowledge, PTEs were tested for the first time in three-dimensional (3D) cancer cell spheroids. The results underscore the potential of PTEs as promising candidates for anticancer therapy. Mechanistically, several novel insights into their antineoplastic activity were obtained. Notably, this study provides the first evidence of sulfide species production induced by PTEs, alongside the inhibition of cancer cell migration and the suppression of matrix metalloproteinase (MMP) activity. Although a low-to-moderate cytotoxicity was observed in the non-tumorigenic fibroblast cell line CR9, indicating possible off-target effects, all PTEs exhibited significantly higher cytotoxic potency against cancer cells, suggesting a preferential antiproliferative effect. Notably, the most active PTEs exhibited cytotoxic effects that were comparable to, or in some cases exceeded, those of doxorubicin across all four cancer cell lines investigated. This finding suggests that these PTEs may possess therapeutic potential similar to, or greater than, that of the reference chemotherapeutic agent under the tested conditions, supporting the further development of PTEs as anticancer agents. In particular, encapsulation strategies or targeted delivery systems may enhance tumor specificity and reduce systemic toxicity, thereby improving their therapeutic potential.

Supplementary Materials: The following supporting information can be downloaded at https://www.mdpi.com/article/10.3390/cancers17142393/s1. Figures S1–S10. ^{1}H and ^{13}CNMR spectra of **3**^{CN}, **3**^{CHO}, **6**^{CN}, **6**^{CHO}, DNS-N_3; Figure S11. IR spectrum of DNS-N_3; Figures S12–S16. APCI Mass spectra of **3**^{CN}, **3**^{CHO}, **6**^{CN}, **6**^{CHO}; Tables S1–S3. Crystal data and structure refinement for **3**^{CN}, **3**^{CHO}, DNS-N_3; Figure S17. IC_{50} values obtained in MCF7, MDA-MB231, A549, and HCT116 cell lines following 48h treatment with the studied compounds and MTT assay (mean ± S.D. 3/4 independent experiments; *** $p < 0.001$ vs. other compounds); Figure S18. Uncropped gel images showing the complete results of the plasmid cleavage assay, corresponding to the data presented in the main figure; Figure S19. Effects of subtoxic concentrations of PTEs, on the migratory capacity of MDA-MB231 and A549 cells. Figure S20. Uncropped Western blot images, corresponding to the data presented in the main figure. One reference is cited in the Supplementary Materials [57].

Author Contributions: Conceptualization, R.T., C.S. and M.B.G.; methodology, F.R., M.R., G.M. and E.M.; formal analysis, C.S., E.M., F.R. and M.B.G.; investigation, R.T., L.M.J., J.F., F.R., M.R., G.M. and E.M.; resources, C.S.; data curation, R.T., S.S.M.B., F.R., M.R., G.M., E.M. and M.B.G.; writing—original draft preparation, R.T., F.R., M.R., G.M., E.M. and M.B.G.; writing—review and editing, C.S. and M.B.G.; visualization, R.T., C.S. and M.B.G.; supervision, C.S. and M.B.G.; project administration, C.S. and M.B.G.; funding acquisition, L.M.J. All authors have read and agreed to the published version of the manuscript.

Funding: This research was funded by a Ph.D. scholarship from the county of Mecklenburg-Vorpommern (Landesgraduiertenstipendium granted to Lukas M. Jacobsen).

Institutional Review Board Statement: Not applicable.

Informed Consent Statement: Not applicable.

Data Availability Statement: Data is contained within the article or Supplementary Material.

Conflicts of Interest: The authors declare no conflicts of interest.

ferences

Liu, B.; Zhou, H.; Tan, L.; Siu, K.T.H.; Guan, X.-Y. Exploring Treatment Options in Cancer: Tumor Treatment Strategies. *Signal Transduct. Target. Ther.* **2024**, *9*, 175. [CrossRef] [PubMed]

Pal, M.; Muinao, T.; Boruah, H.P.D.; Mahindroo, N. Current Advances in Prognostic and Diagnostic Biomarkers for Solid Cancers: Detection Techniques and Future Challenges. *Biomed. Pharmacother.* **2022**, *146*, 112488. [CrossRef] [PubMed]

Khan, S.U.; Fatima, K.; Aisha, S.; Malik, F. Unveiling the Mechanisms and Challenges of Cancer Drug Resistance. *Cell Commun. Signal.* **2024**, *22*, 109. [CrossRef] [PubMed]

Jacobsen, L.M.; Tallarita, R.; Bandaru, S.S.M.; Schulzke, C. Synthesis of Pharmacologically Significant Pentathiepins: A Journey from Harsh to Mild Conditions. In *Non-Conventional Synthesis: Bioactive Heterocycles*; De Gruyter Brill: Berlin, Germany, 2023.

Pozzetti, L.; Asquith, C.R.M. Pentathiepins Are an Understudied Molecular Prism of Biological Activities. *Arch. Pharm.* **2024**, *357*, e2400646. [CrossRef] [PubMed]

Davidson, B.S.; Molinski, T.F.; Barrows, L.R.; Ireland, C.M. Varacin: A Novel Benzopentathiepin from *Lissoclinum vareau* That Is Cytotoxic Toward a Human Colon Tumor. *J. Am. Chem. Soc.* **1991**, *113*, 4709–4710. [CrossRef]

Zhou, J.; Li, W.L.; Wang, Z.X.; Chen, N.Y.; Tang, Y.; Hu, X.X.; Deng, J.H.; Lu, Y.; Lu, G. Dong varacin-1, a Novel Analog of Varacin C, Induces P53-Independent Apoptosis in Cancer Cells through ROS-Mediated Reduction of XIAP. *Acta Pharmacol. Sin.* **2019**, *40*, 222–230. [CrossRef] [PubMed]

Konstantinova, L.S.; Oleg, A. Rakitin Pentathiepins. *Chem. Rev.* **2004**, *104*, 2617–2630. [CrossRef] [PubMed]

Wolff, L.; Bandaru, S.S.M.; Eger, E.; Lam, H.N.; Napierkowski, M.; Baecker, D.; Schulzke, C.; Bednarski, P.J. Comprehensive Evaluation of Biological Effects of Pentathiepins on Various Human Cancer Cell Lines and Insights into Their Mode of Action. *Int. J. Mol. Sci.* **2021**, *22*, 7631. [CrossRef] [PubMed]

Baguley, T.D.; Nairn, A.C.; Lombroso, P.J.; Ellman, J.A. Synthesis of Benzopentathiepin Analogs and Their Evaluation as Inhibitors of the Phosphatase STEP. *Bioorg. Med. Chem. Lett.* **2015**, *25*, 1044–1046. [CrossRef] [PubMed]

Zhang, Y.; Kurup, P.; Xu, J.; Carty, N.; Fernandez, S.M.; Nygaard, H.B.; Pittenger, C.; Greengard, P.; Strittmatter, S.M.; Nairn, A.C.; et al. Genetic Reduction of Striatal-Enriched Tyrosine Phosphatase (STEP) Reverses Cognitive Andcellular Deficits in an Alzheimer's Disease Mouse Model. *Proc. Natl. Acad. Sci. USA* **2010**, *107*, 19014–19019. [CrossRef] [PubMed]

Xu, J.; Chatterjee, M.; Baguley, T.D.; Brouillette, J.; Kurup, P.; Ghosh, D.; Kanyo, J.; Zhang, Y.; Seyb, K.; Ononenyi, C.; et al. Inhibitor of the Tyrosine Phosphatase Step Reverses Cognitive Deficits in a Mouse Model of Alzheimer's Disease. *PLoS Biol.* **2014**, *12*, e1001923. [CrossRef] [PubMed]

Zubair, M.; Ghosh, A.C.; Schulzke, C. The Unexpected and Facile Molybdenum Mediated Formation of Tri- and Tetracyclic Pentathiepins from Pyrazine-Alkynes and Sulfur. *Chem. Commun.* **2013**, *49*, 4343–4345. [CrossRef] [PubMed]

Behnisch-Cornwell, S.; Bandaru, S.S.M.; Napierkowski, M.; Wolff, L.; Zubair, M.; Urbainsky, C.; Lillig, C.; Schulzke, C.; Bednarski, P.J. Pentathiepins: A Novel Class of Glutathione Peroxidase 1 Inhibitors That Induce Oxidative Stress, Loss of Mitochondrial Membrane Potential and Apoptosis in Human Cancer Cells. *ChemMedChem* **2020**, *15*, 1515–1528. [CrossRef] [PubMed]

Napierkowski, M.; Schöne, T.; Bandaru, S.S.M.; Judernatz, J.; Schulig, L.; Schmidt, L.; Schulzke, C.; Bednarski, P.J. Structure-Activity-Relationships of the Stability of Six Pentathiepins Towards Glutathione: Possible Correlations with Biological Activities. *ChemMedChem* **2025**, *20*, e202400727. [CrossRef] [PubMed]

Lee, J.R.; Roh, J.L.; Lee, S.M.; Park, Y.; Cho, K.J.; Choi, S.H.; Nam, S.Y.; Kim, S.Y. Overexpression of Glutathione Peroxidase 1 Predicts Poor Prognosis in Oral Squamous Cell Carcinoma. *J. Cancer Res. Clin. Oncol.* **2017**, *143*, 2257–2265. [CrossRef] [PubMed]

17. Wieczorek, E.; Jablonowski, Z.; Tomasik, B.; Gromadzinska, J.; Jablonska, E.; Konecki, T.; Fendler, W.; Sosnowski, M.; Wasow W.; Reszka, E. Different Gene Expression and Activity Pattern of Antioxidant Enzymes in Bladder Cancer. *Anticancer. Res.* **2** *37*, 841–848. [CrossRef] [PubMed]
18. Cullen, J.J.; Mitros, F.A.; Oberley, L.W. Expression of Antioxidant Enzymes in Diseases of the Human Pancreas: Another L between Chronic Pancreatitis and Pancreatic Cancer. *Pancreas* **2003**, *26*, 23–27. [CrossRef] [PubMed]
19. Chen, B.; Shen, Z.; Wu, D.; Xie, X.; Xu, X.; Lv, L.; Dai, H.; Chen, J.; Gan, X. Glutathione Peroxidase 1 Promotes NSCLC Resista to Cisplatin via ROS-Induced Activation of PI3K/AKT Pathway. *BioMed Res. Int.* **2019**, *2019*, 7640547. [CrossRef] [PubMed]
20. Handy, D.E.; Loscalzo, J. The Role of Glutathione Peroxidase-1 in Health and Disease. *Free Radic. Biol. Med.* **2022**, *188*, 146– [CrossRef] [PubMed]
21. Wei, R.; Qiu, H.; Xu, J.; Mo, J.; Liu, Y.; Gui, Y.; Huang, G.; Zhang, S.; Yao, H.; Huang, X.; et al. Expression and Prognostic Poten of GPX1 in Human Cancers Based on Data Mining. *Ann. Transl. Med.* **2020**, *8*, 124. [CrossRef] [PubMed]
22. Schulz, R.; Emmrich, T.; Lemmerhirt, H.; Leffler, U.; Sydow, K.; Hirt, C.; Kiefer, T.; Link, A.; Bednarski, P.J. Identification c Glutathione Peroxidase Inhibitor That Reverses Resistance to Anticancer Drugs in Human B-Cell Lymphoma Cell Lines. *Bio Med. Chem. Lett.* **2012**, *22*, 6712–6715. [CrossRef] [PubMed]
23. Napierkowski, M.; Janke, U.; Rong, A.; Delcea, M.; Bandaru, S.S.M.; Schulzke, C.; Bednarski, P.J. Liposomal Formulation of Mo Pentathiepin Improves Solubility and Stability toward Glutathione While Preserving Anticancer Activity. *Arch. Pharm.* **2023**, e2300087. [CrossRef] [PubMed]
24. Tallarita, R.; Jacobsen, L.M.; Elvers, B.J.; Richter, S.; Bandaru, S.S.M.; Correia, J.V.; Schulzke, C. Synthesis of Seven Indolizi Derived Pentathiepines: Strong Electronic Structure Response to Nitro Substitution in Position C-9. *Molecules* **2024**, *29*, [CrossRef] [PubMed]
25. Peng, H.; Cheng, Y.; Dai, C.; King, A.L.; Predmore, B.L.; Lefer, D.J.; Wang, B. A Fluorescent Probe for Fast and Quantitat Detection of Hydrogen Sulfide in Blood. *Angew. Chem.* **2011**, *123*, 9846–9849. [CrossRef]
26. Tallarita, R.; Jacobsen, L.M.; Bandaru, S.S.M.; Elvers, B.J.; Schulzke, C. The Role of –OEt Substituents in Molybdenum-Assis Pentathiepine Formation—Access to Diversely Functionalized Azines. *Molecules* **2024**, *29*, 3806. [CrossRef] [PubMed]
27. Long, D.A. Infrared and Raman Characteristic Group Frequencies. Tables and Charts George Socrates John Wiley and Sons, L Chichester, Third Edition, 2001. Price £135. *J. Raman Spectrosc.* **2004**, *35*, 905. [CrossRef]
28. Cantor, S.E. Nonmigrating Antioxidants via Sulfonyl Azide Intermediates. In *Stabilization and Degradation of Polymers*; Americ Chemical Society: Washington, DC, USA, 1978; Volume 169, pp. 253–260. [CrossRef]
29. Farrugia, L.J. WinGX and ORTEP for Windows: An Update. *J. Appl. Crystallogr.* **2012**, *45*, 849–854. [CrossRef]
30. Sheldrick, G.M. SHELXT—Integrated Space-Group and Crystal-Structure Determination. *Acta Crystallogr. Sect. A Fou Crystallogr.* **2015**, *71*, 3–8. [CrossRef] [PubMed]
31. Sheldrick, G.M. Crystal Structure Refinement with SHELXL. *Acta Crystallogr. Sect. C Struct. Chem.* **2015**, *71*, 3–8. [CrossR [PubMed]
32. Alexeyenko, A.; Alkasalias, T.; Pavlova, T.; Szekely, L.; Kashuba, V.; Rundqvist, H.; Wiklund, P.; Egevad, L.; Csermely, Korcsmaros, T.; et al. Confrontation of Fibroblasts with Cancer Cells In Vitro: Gene Network Analysis of Transcriptome Chan and Differential Capacity to Inhibit Tumor Growth. *J. Exp. Clin. Cancer Res.* **2015**, *34*, 62. [CrossRef] [PubMed]
33. Stamm, A.; Reimers, K.; Strauß, S.; Vogt, P.; Scheper, T.; Pepelanova, I. In Vitro Wound Healing Assays—State of the Art. *BioNanoMaterials*; De Gruyter Brill: Berlin, Germany, 2016; Volume 17.
34. Fleming, F.F.; Yao, L.; Ravikumar, P.C.; Funk, L.; Shook, B.C. Nitrile-Containing Pharmaceuticals: Efficacious Roles of the Nitr Pharmacophore. *J. Med. Chem.* **2010**, *53*, 7902–7917. [CrossRef] [PubMed]
35. Gampe, C.; Verma, V.A. Curse or Cure? A Perspective on the Developability of Aldehydes as Active Pharmaceutical Ingredie *J. Med. Chem.* **2020**, *63*, 14357–14381. [CrossRef] [PubMed]
36. Brzostowska, E.M.; Greer, A. The Role of Amine in the Mechanism of Pentathiepin (Polysulfur) Antitumor Agents. *J. Am. Ch Soc.* **2003**, *125*, 396–404. [CrossRef] [PubMed]
37. Sato, R.; Ohyama, T.; Ogawa, S. Efficient Synthesis and Biological Properties of New Benzopentathiepins. *Heterocycles* **1995**, 893. [CrossRef]
38. Kapałczyńska, M.; Kolenda, T.; Przybyła, W.; Zajączkowska, M.; Teresiak, A.; Filas, V.; Ibbs, M.; Bliźniak, R.; Łuczewski, Lamperska, K. 2D and 3D Cell Cultures—A Comparison of Different Types of Cancer Cell Cultures. *Arch. Med. Sci.* **2018**, 910–919. [CrossRef] [PubMed]
39. Jensen, C.; Teng, Y. Is It Time to Start Transitioning From 2D to 3D Cell Culture? *Front. Mol. Biosci.* **2020**, *7*, 33. [CrossR [PubMed]
40. Nunes, A.S.; Barros, A.S.; Costa, E.C.; Moreira, A.F.; Correia, I.J. 3D Tumor Spheroids as in Vitro Models to Mimic in Vivo Hum Solid Tumors Resistance to Therapeutic Drugs. *Biotechnol. Bioeng.* **2019**, *116*, 206–226. [CrossRef] [PubMed]
41. Yam, C.Q.X.; Lim, H.H.; Surana, U. DNA Damage Checkpoint Execution and the Rules of Its Disengagement. *Front. Cell Dev. B* **2022**, *10*, 1020643. [CrossRef] [PubMed]

2. Barnum, K.J.; O'Connell, M.J. Cell Cycle Regulation by Checkpoints. In *Methods in Molecular Biology*; Humana Press: Totowa, NJ, USA, 2014; pp. 29–40. [CrossRef]
3. Mueller, E.G. Trafficking in Persulfides: Delivering Sulfur in Biosynthetic Pathways. *Nat. Chem. Biol.* **2006**, *2*, 185–194. [CrossRef] [PubMed]
4. Fujii, S.; Sawa, T.; Motohashi, H.; Akaike, T. Persulfide Synthases That Are Functionally Coupled with Translation Mediate Sulfur Respiration in Mammalian Cells. *Br. J. Pharmacol.* **2019**, *176*, 607–615. [CrossRef] [PubMed]
5. Ida, T.; Sawa, T.; Ihara, H.; Tsuchiya, Y.; Watanabe, Y.; Kumagai, Y.; Suematsu, M.; Motohashi, H.; Fujii, S.; Matsunaga, T.; et al. Reactive Cysteine Persulfides and S-Polythiolation Regulate Oxidative Stress and Redox Signaling. *Proc. Natl. Acad. Sci. USA* **2014**, *111*, 7606–7611. [CrossRef] [PubMed]
6. Akaike, T.; Ida, T.; Wei, F.Y.; Nishida, M.; Kumagai, Y; Alam, M.M.; Ihara, H.; Sawa, T.; Matsunaga, T.; Kasamatsu, S.; et al. Cysteinyl-TRNA Synthetase Governs Cysteine Polysulfidation and Mitochondrial Bioenergetics. *Nat. Commun.* **2017**, *8*, 1177. [CrossRef] [PubMed]
7. Asquith, C.R.M.; Laitinen, T.; Konstantinova, L.S.; Tizzard, G.; Poso, A.; Rakitin, O.A.; Hofmann-Lehmann, R.; Hilton, S.T. Investigation of the Pentathiepin Functionality as an Inhibitor of Feline Immunodeficiency Virus (FIV) via a Potential Zinc Ejection Mechanism, as a Model for HIV Infection. *ChemMedChem* **2019**, *14*, 454–461. [CrossRef] [PubMed]
8. Moos, G.P.; Smith, P.A.S.; Tavernier, D. Glossary of Class Names of Organic Compounds and Reactivity Intermediates Based on Structure. *Pure Appl. Chem.* **1995**, *67*, 1307–1375. [CrossRef]
9. Lindahl, S.; Xian, M. Recent Development of Polysulfides: Chemistry and Biological Applications. *Curr. Opin. Chem. Biol.* **2023**, *75*, 102325. [CrossRef] [PubMed]
0. Shi, X.; Wang, X.; Yao, W.; Shi, D.; Shao, X.; Lu, Z.; Chai, Y.; Song, J.; Tang, W.; Wang, X. Mechanism Insights and Therapeutic Intervention of Tumor Metastasis: Latest Developments and Perspectives. *Signal Transduct. Target. Ther.* **2024**, *9*, 192. [CrossRef] [PubMed]
1. Friedl, P.; Wolf, K. Tumour-Cell Invasion and Migration Diversity and Escape Mechanisms. *Nat. Rev. Cancer* **2003**, *3*, 362–374. [CrossRef] [PubMed]
2. Wu, J.S.; Jiang, J.; Chen, B.J.; Wang, K.; Tang, Y.L.; Liang, X. Hua plasticity of Cancer Cell Invasion: Patterns and Mechanisms. *Transl. Oncol.* **2021**, *14*, 100899. [CrossRef] [PubMed]
3. Ashja Ardalan, A.; Khalili-Tanha, G.; Shoari, A. Shaping the Landscape of Lung Cancer: The Role and Therapeutic Potential of Matrix Metalloproteinases. *Int. J. Transl. Med.* **2024**, *4*, 661–679. [CrossRef]
4. Lamouille, S.; Xu, J.; Derynck, R. Molecular Mechanisms of Epithelial-Mesenchymal Transition. *Nat. Rev. Mol. Cell Biol.* **2014**, *15*, 178–196. [CrossRef] [PubMed]
5. Quintero-Fabián, S.; Arreola, R.; Becerril-Villanueva, E.; Torres-Romero, J.C.; Arana-Argáez, V.; Lara-Riegos, J.; Ramírez-Camacho, M.A.; Alvarez-Sánchez, M.E. Role of Matrix Metalloproteinases in Angiogenesis and Cancer. *Front. Oncol.* **2019**, *9*, 1370. [CrossRef] [PubMed]
6. Izdebska, M.; Zielińska, W.; Krajewski, A.; Hałas-Wiśniewska, M.; Mikołajczyk, K.; Gagat, M.; Grzanka, A. Downregulation of Mmp-9 Enhances the Anti-Migratory Effect of Cyclophosphamide in Mda-Mb-231 and Mcf-7 Breast Cancer Cell Lines. *Int. J. Mol. Sci.* **2021**, *22*, 12783. [CrossRef] [PubMed]
7. Bruno, I.J.; Cole, J.C.; Kessler, M.; Luo, J.; Motherwell, W.S.; Purkis, L.H.; Smith, B.R.; Taylor, R.; Cooper, R.I.; Harris, S.E.; et al. Retrieval of crystallographically-derived molecular geometry information. *J. Chem. Inf. Comput. Sci.* **2004**, *44*, 2133–2144. [CrossRef] [PubMed]

cancers

MDPI

Review

The Evolution of Anticancer 3D In Vitro Models: The Potential Role of Machine Learning and AI in the Next Generation of Animal-Free Experiments

Carolina Momoli †, Beatrice Costa †, Lorenzo Lenti, Matilde Tubertini, Marco Daniele Parenti, Elisa Martella *, Greta Varchi * and Claudia Ferroni

Institute for the Organic Synthesis and Photoreactivity—Italian National Research Council, 40129 Bologna, Italy; carolina.momoli@isof.cnr.it (C.M.); beatrice.costa@isof.cnr.it (B.C.); lorenzo.lenti@isof.cnr.it (L.L.); matilde.tubertini@isof.cnr.it (M.T.); marcodaniele.parenti@isof.cnr.it (M.D.P.); claudia.ferroni@isof.cnr.it (C.F.)
* Correspondence: elisa.martella@isof.cnr.it (E.M.); greta.varchi@isof.cnr.it (G.V.)
† These authors contributed equally to this work.

Simple Summary: This review explores the evolution of advanced 3D in vitro models, which better replicate the tumor microenvironment than traditional 2D cultures. It highlights significant advancements in these models and their role in cancer research. The integration of machine learning (ML) and artificial intelligence (AI) is discussed as a way to enhance predictive accuracy, analyze complex data, and optimize experimental conditions. These technologies can reduce reliance on animal testing and improve therapeutic outcome predictions. Together, ML, AI, and 3D models have the potential to revolutionize anticancer drug development.

Abstract: The development of anticancer therapies has increasingly relied on advanced 3D in vitro models, which more accurately mimic the tumor microenvironment compared to traditional 2D cultures. This review describes the evolution of these 3D models, highlighting significant advancements and their impact on cancer research. We discuss the integration of machine learning (ML) and artificial intelligence (AI) in enhancing the predictive power and efficiency of these models, potentially reducing the dependence on animal testing. ML and AI offer innovative approaches for analyzing complex data, optimizing experimental conditions, and predicting therapeutic outcomes with higher accuracy. By leveraging these technologies, the next generation of 3D in vitro models could revolutionize anticancer drug development, offering effective alternatives to animal experiments.

Keywords: three-dimensional cancer models; spheroids; organoids; tumor-on-a-chip; machine learning; artificial intelligence

1. Introduction

Despite significant advancements in early detection and targeted treatments, cancer remains a leading global health burden and cause of death [1]. The current challenge in oncology is shifting from traditional "one-size-fits-all" treatments to personalized "one dose–one patient" approaches [2]. Tumors have been shown to be complex ecosystems, consisting not only of cancer cells but also of a variety of host cells, soluble factors, and altered extracellular matrix components, collectively known as the tumor microenvironment (TME). The dynamic crosstalk between cancer cells and the TME significantly influences cancer initiation, progression, metastasis progression, and drug resistance, making it a key therapeutic target and a crucial asset in terms of treatment development [3]. Notably, clinical diagnoses primarily rely on tumor biopsies, which often fail to capture the full extent of intratumoral heterogeneity and may overlook newly emerging, highly aggressive tumor clones. Furthermore, patients with the same cancer subtype can exhibit distinct tumor phenotypes that evolve dynamically during disease progression and treatment, leading to highly variable therapeutic responses, including both natural and acquired resistance to therapy [4].

Given their cost effectiveness, availability, high-throughput, easy replication and results' interpretation, two-dimensional (2D) cell cultures, or monolayers, are commonly used in cellular biology to study diseases and for drug screening. However, a large body of evidence indicates that 2D in vitro cell cultures have various limitations (Figure 1A). First, isolating and maintaining cancer cell lines from patient biopsies can be challenging and often inefficient. Second, once cultured, these cells adhere to and grow on flat synthetic surfaces, losing their original morphology and polarization, which may disrupt critical cellular signaling pathways or alter their responses to external stimuli. Third, cells in 2D cultures often undergo extensive clonal selection, leading to the establishment of derived cell lines that no longer reflect the genetic heterogeneity of the original tumors. Additionally, in vitro cancer cell lines are rarely paired with a patient-matched 2D normal tissue counterpart. Most critically, they fail to replicate the complex network of dynamic interactions present in the three-dimensional TME of living patient tumors, which can significantly influence the effectiveness of cancer therapies [5].

The approach to research in the 21st century is experiencing a significant shift, particularly in how animal models are used as a distinctive method for analysis and investigation. It has been several decades since the three Rs principles (replacement, reduction and refinement) were introduced by Russell and Burch in 1959 [6].

First, researchers should make every effort to substitute animal models with alternative non-animal methods whenever possible (replacement). In the second step, researchers should aim to minimize (reduction) the number of subjects used in an experimental protocol. This can be achieved, for instance, by employing suitable statistical techniques that help determine the minimum number of animals required for a specific experimental design, ensuring that the results are statistically significant with the chosen statistical test [7]. Finally, refinement comes into play when full replacement methods are not feasible, and all strategies in theory and practice have been exhausted to minimize the number of animals used in each experiment. Russell and Burch defined refinement as "any decrease in the incidence or severity of inhumane procedures applied to those animals which still are to be used" [6]. This entirely new approach to designing experiments has prompted the scientific community to explore new and alternative avenues for advancing science in the future [8].

The use of animal models in cancer research aids in comprehending the genetic foundations of cancer, the functions of genes and mutations in cancer formation and contributes to the advancement and evaluation of anticancer drugs [9]. As cancer research advances, an increasing number of animals are being utilized in the creation of animal models. Currently, most of the animal models utilized in cancer research are small animals like mice, rats, zebrafish, fruit flies, and other similar species. Although animal models have been extensively used in cancer-drug development, the low success rates in translating preclinical findings from animals to effective clinical treatments have raised growing concerns about their reliability as predictors of human responses [10,11].

Despite the continued necessity of animal models for progressing to clinical phases, their inherent limitations underscore the need for alternative validation approaches, such as physiologically relevant human in vitro models to study cancer biology and advance therapeutic development [8].

In this view, there is a growing focus on 3D in vitro systems, better mimicking the structure and function of tumors in vivo (Figure 1B). These 3D cultures are gaining prominence in cancer research, serving three key purposes: (i) studying the pathophysiology of cancer progression and resistance [12]; (ii) conducting in vitro screening of anticancer therapies [13]; and (iii) replicating the unique characteristics of an individual patient's tumors in vitro, enabling personalized screening for the most effective treatments [14].

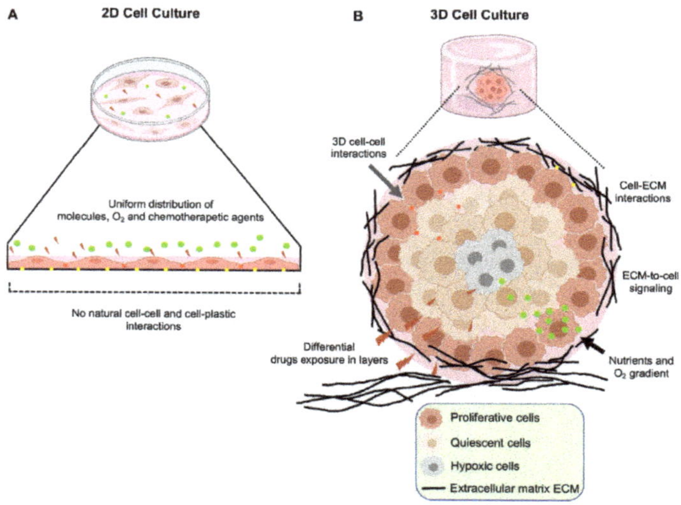

Figure 1. Illustration depicting the key distinctions between 2D and 3D cell cultures. (**A**) Conventional 2D cell cultures, where cells grow in a flattened monolayer on the surface of plastic plates. These cultures are characterized by reduced cell–cell interactions and unlimited exposure to nutrients, oxygen, and drugs, which are significant limitations. (**B**) 3D cell culture systems, which enhance cell–cell and cell–extracellular matrix interactions. These systems have restricted access to nutrients and oxygen, along with heterogeneous drug interactions, allowing for a more accurate mimicry of the tumor microenvironment observed in vivo. Reproduced with permission from [15].

The 3D models available for validating anticancer therapies are numerous and vary in complexity and application. Tumor spheroids are simple 3D cell aggregates, useful for studying cell–cell and cell–matrix interactions, but limited by their static nature and inability to replicate immune response, fibroblast presence, or vascularization. Organoids are more complex, mimicking tumor architecture but still have similar limitations. To overcome these, tumor-on-a-chip (ToC) models have been developed, incorporating microfluidic technology to better simulate the TME. ToC models offer advantages like precise control of conditions, compatibility with analytical techniques, and faster experimental timelines.

The introduction of these models into industrial high-throughput drug screening, where thousands of compounds are tested at the same time, launch the challenge to establish an accurate, automated system for their analysis providing key information on their potential optimization [16].

In this perspective, the incorporation of artificial intelligence (AI) has the potential to provide huge improvements in terms of predictive accuracy, data integration, model optimization, and the ability to analyze complex datasets, ultimately enhancing the design, personalization, and effectiveness of 3D models for anticancer therapies. AI refers to systems programmed to perform tasks that typically require human abilities, such as learning and problem-solving. Machine learning (ML), a subset of AI, involves developing algorithms that analyze data to identify patterns of behavior [17,18]. ML models are particularly effective when the analyzed system is highly complex and presents numerous variables with hidden relationships between them, including the complex and multifaceted disorder characterized by thousands of genetic and epigenetic variations typical of the tumor [19].

Building on these preliminary considerations, this review aims to provide an overview of the latest applications of 3D models in the cancer therapy discovery pathway, with a particular focus on their integration with artificial intelligence. We will highlight how these models have the potential to revolutionize the preclinical testing landscape by offering more accurate, human-relevant data compared to traditional 2D cultures or animal models. The synergistic combination of 3D models and AI could drive a transformative shift toward more efficient, targeted, and patient-specific cancer therapies, bridging gaps in drug efficacy prediction and personalized treatment design.

Although this is not a systematic review, the literature search was conducted using multiple databases, including PubMed, Scopus and Web of Science to ensure a comprehensive coverage of interdisciplinary advancements. A non-exhaustive list of keywords and phrases employed for the search is as follows: "3D cancer models", "3D bioprinted tumor models", "organoid-based drug screening", "tumor-on-a-chip systems", "patient-derived organoids", "tumor microenvironment (TME) modeling", "artificial intelligence in cancer research", "machine learning for drug discovery", "deep learning in oncology", "AI-driven biomarker identification", "predictive oncology algorithms", "neural networks for tumor profiling", "computational drug repurposing".

These terms were used to identify pivotal studies, emerging trends, and unresolved challenges in the field, with emphasis on publications from 2012 to 2024.

2. Three-Dimensional Models

2.1. Development and Relevant TME Features

Since the initial discovery of the ability to replicate an in vivo-like environment, the 3D approach has emerged as a groundbreaking innovation in the field of alternative methods. A 3D cell culture is defined as "a cell culture that can mimic a living organ's organization and microarchitecture" [20].

As illustrated in Figure 2, a significant milestone was achieved in 1951 with the isolation of HeLa cells from human uterine cervical cancer tissue. The use of HeLa cells enhanced the adoption of in vitro techniques globally, as they offered consistent and repeatable outcomes due to their immortalized and uniform nature. Spheroids were introduced in the early 1970s by Sutherland and collaborators [21], and since then various models and techniques for their generation have been developed.

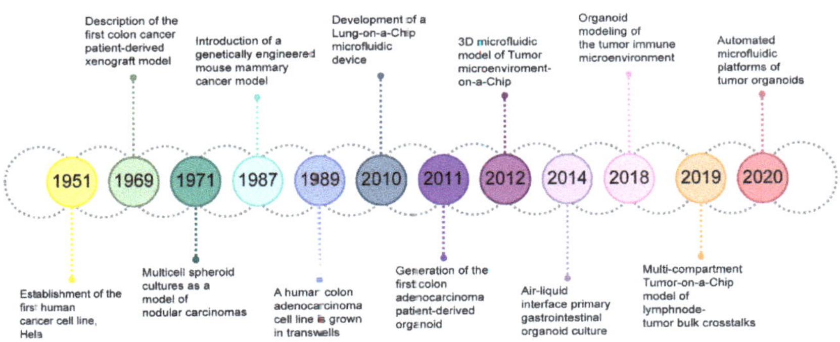

Figure 2. Timeline of milestones in the development of 3D cancer models. Reproduced with permission of [5].

In 2010, Huh and colleagues developed the first lung-on-a-chip microfluidic device that reconstituted the critical functional alveolar–capillary interface of the human lung. This bioinspired microdevice mimics complex, integrated organ-level responses to bacteria and inflammatory cytokines introduced into the alveolar space [22].

In recent years, 3D cell models have attracted increasing attention in cancer research due to their ability to closely mimic several hallmarks of in vivo tumors, including the TME. The TME is distinguished by several key characteristics, including low extracellular pH, elevated levels of reactive oxygen species (ROS), hypoxia, and a predominantly immunosuppressive niche compared to normal tissue [23]. As shown in Figure 3, the neoplastic cells are heterogeneous assemblies of infiltrating or resident host non-neoplastic cells—such as T and B lymphocytes, natural killer cells, dendritic cells, monocytes, endothelial cells, pericytes, cancer-associated fibroblasts (CAFs), mesenchymal stromal cells, and adipocytes—alongside niche-specific soluble factors (e.g., cytokines, growth factors, metabolites, enzymes, miRNAs) and a modified extracellular matrix (ECM).

Mounting evidence underscores the critical role of the dynamic interplay between cancer cells and components of the TME in shaping disease progression, including cancer initiation, metastasis, and drug resistance, making the TME a promising therapeutic target. To achieve more durable, side-effect-limited or, ideally, side-effect-free personalized anti-cancer strategies, it will be essential to accurately model both cancer heterogeneity and the complex interactions within the TME [24].

Hypoxia, a condition of low oxygen and nutrients common in larger solid tumors, significantly impacts cancer progression by upregulating the glucose metabolism, activating angiogenic growth factors and receptors, recruiting pro-tumor immune cells, and tumor-associated macrophages [23]. This hypoxic condition cannot be mimicked in 2D systems but can be recreated in various 3D culture platforms, such as spheroids, organoids and ToC [25].

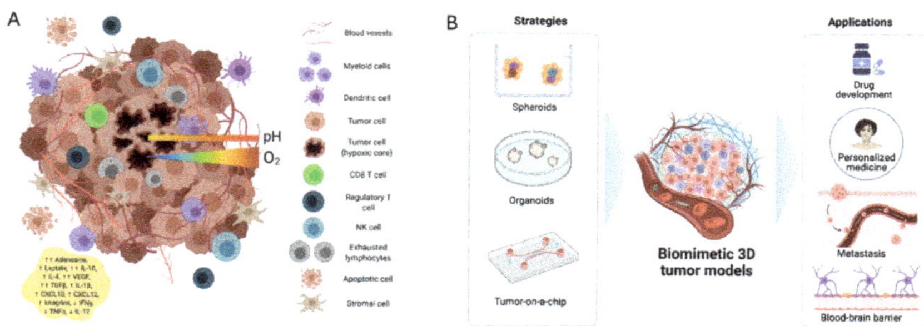

Figure 3. (**A**) The tumor microenvironment (TME). Reproduced with permission of [26]. (**B**) Schematic overview illustrating in vitro strategies for replicating the human TME in three dimensions, including spheroids, organoids, tumor-on-a-chip models and some of their applications. Reproduced with permission of [27].

Numerous initiatives have taken place in recent years to create 3D models that are physiologically relevant and can completely mimic the functioning of tissues and organs (Figure 3B). Overall, 3D cultures present a robust platform for exploring critical tumor processes that mirror in vivo conditions. They offer an alternative to using laboratory animals; indeed, the toxicity levels observed in certain 3D cell cultures closely match those found in animal studies.

The following sections aim to provide a comprehensive and up-to-date overview of spheroids, organoids, and ToC models in the development of cancer treatments, covering the methodologies for creating these 3D models, their respective strengths and limitations, and their contributions to advancing our understanding of tumor biology and improving therapeutic strategies.

2.2. Spheroids

Since their initial introduction in the 70s, various spheroid models and techniques for their generation have been developed. Spheroids, typically grown as free-floating clusters of spherical cell units, are regarded as having a relatively simple structure for modeling tumors [28]. Spheroids mimic key tumor characteristics such as diverse cell types, signaling pathways, growth rates, ECM production as well as its interactions with cells [29], gene expression profiles resembling in vivo settings, and a multi-layered tumor structure [30]. Techniques like hanging drop, agitation-based technique and the liquid overlay techniques can be utilized to create cell spheroids (Figure 4).

The liquid overlay technique (Figure 4a) consists of cells seeded on non-adhesive surfaces to avoid attachment. This method has many benefits: it is simple and generally reproducible as equal numbers of cells can be seeded into each well to produce consistent spheroids. As a result, spheroid sizes can be easily adjusted—larger spheroids can be obtained by seeding a greater number of cells [31]. The primary drawbacks of these methods include the variability in the size of the resulting spheroids, inconsistent cell distribution and composition, along with demanding labor needs.

Mayer and his colleagues developed multicellular gastric cancer spheroids utilizing this technique and they compared them with the corresponding xenografts in immunodeficient mice. They showed that 12 out of 17 gastric cancer cell lines could replicate the characteristics of the original carcinoma when spheroids were cultured utilizing this method [32]. In conclusion, this technique offers a simple, scalable, and cost-effective approach for 3D spheroid culture, suitable for drug screening and other applications. While adaptations to analysis methods are required, this model provides a reliable tool for studying cancer biology and treatment responses.

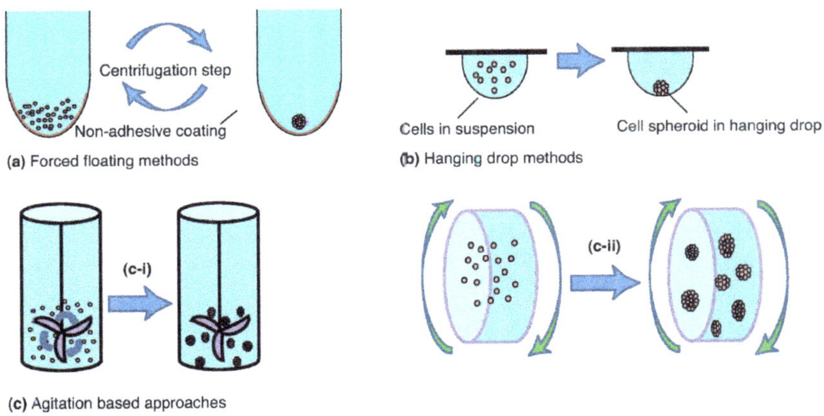

Figure 4. Most used methods to generate multicellular 3D spheroids. (**a**) Forced-floating of cells, (**b**) hanging drop method and (**c**) agitation-based approaches: (**c-i**) spinner flask bioreactors and (**c-ii**) rotational culture systems. Adapted with permission of [33].

The hanging drop technique (Figure 4b) is based on the use of a small aliquot (typically 20 μL) of a single cell suspension which is then seeded in the lid of a Petri dish; after seeding, the lid is then turned over and the aliquots of cell suspension turn into hanging drops that are kept in place due to surface tension. A saline solution is added to the bottom of the Petri dish to prevent evaporation of the drops [33]. This method is relatively simple and has been reported to have a reproducibility of almost 100%. Kelm et al. demonstrated that 3D spheroids generated using this technique with HepG2 liver cancer cells and MCF-7 breast cancer cells were patho-physiologically relevant. The resulting 3D structures were highly organized, produced their own ECM, and were therefore described as "tissue-like" [34]. Since this method relies on the natural tendency of cells to adhere to one another, rather than depending on matrices or scaffolds, it eliminates concerns about the potential effects that these external elements may have on the formed 3D structures. A potential drawback of the hanging drop method is the limited volume of the liquid drop containing the cells. Typically, this method can accommodate a maximum volume of 50 μL (including the drug test medium), as the surface tension that holds the liquid on the culture surface is not sufficient to support larger volumes [33]. Another limitation of this approach is related to the culture medium changing; indeed, it is challenging and time-consuming, leading to a restricted production of spheroids [35].

Another method for creating spheroids is the agitation-based technique, as shown in Figure 4c. Here, cells are placed in a container that is continuously stirred to prevent adherence to surfaces, promoting cell–cell interactions instead [33]. As a result, this technique is only suitable for cell lines that can tolerate high shear stress. The primary advantage of this method is its ability to generate uniform-sized 3D tumor models (spheroids) in high yields. The fluid movement within the bioreactor creates a controlled environment that facilitates the transport of nutrients and waste from the spheroid surface, thereby mimicking the transport mechanisms found in vivo [36]. Agitation-based methods for generating 3D spheroids can generally be grouped into two categories: spinner flask bioreactors (Figure 4c-i) and rotational culture systems (Figure 4c-ii). The first category includes a container to hold the cell suspension and a stirring mechanism to ensure the suspension is constantly mixed [37]. Rotating cell culture bioreactors operate similarly to spinner flask bioreactors. However, instead of agitating the cell suspension with a stirring bar or rod, the entire culture container itself is rotated [33]. Brancato et al. recently validated the consistency of a 3D model with in vivo conditions using biodegradable gelatin-based microcarriers in a spinner flask bioreactor [38]. In the proposed system, pancreatic cancer cells were co-cultured with CAFs, leading to the formation of human pancreatic ductal adenocarcinoma microtissues. The cancer cells were found to be the primary proliferating cell population, aligning with observations from other 3D co-culture models. Analysis of ECM gene and protein expression revealed that fibroblasts co-cultured with pancreatic cancer cells transformed into myofibroblasts, expressing desmoplastic markers. This platform highlighted the critical role of stromal–cancer cell interactions in influencing both cell types.

An important characteristic of this 3D model is that tumor spheroids that reach a size over 400 μm can produce oxygen and nutrient gradients similar to in vivo conditions, including a necrotic core, a quiescent area, and a thick outer layer that is actively proliferating [39]. These conditions make them a suitable model for investigating the impact of hypoxia on cancer progression, blood vessel formation, and drug effectiveness. Moreover, larger spheroids mimic the passive perfusion of drug better in a structured stroma.

Tumorspheres, a specific type of spheroid, serve as a floating sphere model to evaluate cancer stem cell (CSC)-related characteristics in vitro. A key feature of these tumor-derived spheroids is their enrichment in CSCs or cells exhibiting stem cell-like properties. Recent models of CSCs propose that cancer cells are hierarchically organized, with CSCs representing a distinct subset that is crucial for cancer propagation due to their capabilities for self-renewal and differentiation, allowing them to regenerate complete cancer structures [40]. In the 1990s, Dick and colleagues discovered CSCs in hematopoietic cancers [41]. Subsequent research indicated that cells with CSC-like properties are also found in solid tumors, including gliomas, colon and breast cancers. This discovery prompted researchers to develop techniques for cultivating CSCs from solid tumors in vitro. Since normal neural cells had already been successfully grown in sphere cultures, it was not surprising that the first successful spheroid cultures of cancer cells involved glioma CSCs [42]. Following these initial studies, tumorspheres were also developed for cancers of the breast, colon, ovary, and prostate [43]. For instance, tumorspheres generated from breast cancer cell lines were enriched in early progenitor/stem cells and had the capacity to differentiate into all three mammary epithelial lineages. The established sphere cells are considered cancer-initiating cells because they express stem-cell markers and have the ability to form xenograft tumors in immunocompromised mice [40].

Another type of tumorspheres is colorectal cancer spheroids (colonspheres); they were initially derived from CD133-positive colon cancer cells cultured in serum-free medium. The resulting sphere cells retained the ability to replicate the histopathological features of the original tumor when implanted in immunocompromised mice [44]. These colonspheres were used to study CSC-related traits, including chemoresistance, metastatic potential, and tumorigenicity at the single-cell level. Additionally, colonspheres were employed to assess the chemosensitivity of novel compounds targeting the Wnt pathway [45].

Although not exhaustive, these examples of tumor spheroids used to study various aspects of cancers highlight the potential of these models to enhance our understanding of the key mechanisms involved in tumor progression and dissemination by providing an environment that better mimics in vivo conditions compared to traditional 2D cultures.

2.3. *Organoids*

While tumor spheroids from cell lines are commonly used to study cancer growth, invasion, and drug screening, they do not accurately mimic the intricate biological and clinical characteristics of primary tumor tissues. This hinders their ability to accurately predict how individual patients will respond to therapy [27]. To address this restriction, organoids have become valuable resources in disease modeling, drug testing, and personalized medicine advancement [46]. The term "organoid" refers to clusters of cells that grow in a defined three-dimensional in vitro environment, where they self-organize and differentiate into functional cell types. These mini-clusters mimic both the structure and function of a natural organ/tumor in vivo, and are sometimes referred to as "mini-organs" [30]. Organoids can be derived from various types of stem cells, including embryonic stem cells, induced pluripotent stem cells (iPSCs), or neonatal/adult stem cells, through a process that mirrors how an organ develops its unique structure [47]. The key point is that tumor organoids retain the essential structural and functional characteristics of real tumors, potentially offering a highly predictive in vitro model for guiding clinical decisions [27]. Their applications range from high-throughput drug screening to sophisticated disease modeling, and some have even progressed to clinical translation (Figure 5). For instance, organoids have been utilized to recreate complex TME by co-culturing with immune cells, significantly enhancing our understanding of cancer biology [48].

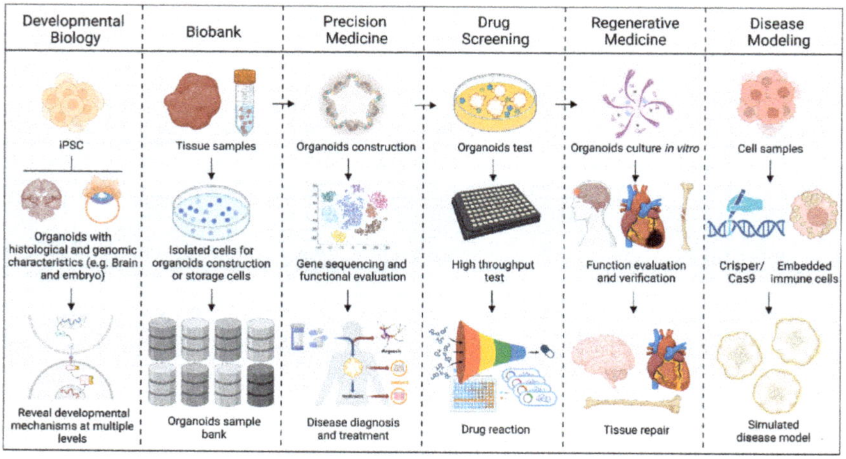

Figure 5. Application of organoids. Reproduced with permission of [49].

In drug discovery, organoids provide platforms to evaluate the efficacy and toxicity of new therapeutic agents. Liver organoids, for example, are valuable tools for studying drug metabolism and identifying novel drug targets [50]. In regenerative medicine, they present the possibility of generating functional tissues for transplantation and other therapeutic purposes [51]. Additionally, organoids are pivotal for exploring human developmental biology, offering critical insights into the mechanisms of various diseases, including genetic and infectious disorders. For example, brain organoids enable the study of human brain development and the underlying causes of neurological disorders [52]. As organoid technology continues to evolve, it promises transformative advancements in these areas, deepening our understanding of human biology and opening the door to innovative therapeutic strategies.

The first example of an organoid was published in 2009 by Sato's group [53]. They demonstrated how single adult intestinal stem cells expressing leucine-rich repeat-containing G protein-coupled receptor 5 could form 3D intestinal organoids within Matrigel, a gelatinous protein mixture rich in extracellular matrix components. It primarily consists of laminin, collagen IV, entactin, and heparan sulfate proteoglycans, which provide a supportive environment resembling in vivo conditions for cell growth and differentiation. These organoids were able to self-organize and differentiate into crypt-villus structures without requiring a mesenchymal niche. This study was the first to establish a 3D organoid culture from a single adult stem cell, paving the way for further development of organoid models in various systems, including endodermal organs (e.g., stomach, liver, pancreas, lung, and kidney) and neuroectodermal tissues (e.g., brain and retina) [47].

Recently, extensive tumor biobanks have been created through 3D cultivation of patient-derived organoids (PDOs), which preserve the genetic and histological diversity of the original tumor tissues [54]. To date, the PDO culture system has been improved and created for various types of cancer, such as breast, colon, gastric, prostate, pancreatic, and renal carcinoma [55]. Weeber et al. developed PDOs from tumor biopsies of 14 patients with metastatic colorectal cancer. Using SOLiD sequencing (sequencing by oligonucleotide ligation and detection), they analyzed 1977 cancer-related genes within both the PDOs and their respective original tumors. The study found a 90% similarity in somatic mutations between PDOs and the corresponding patient biopsies, highlighting the potential of organoids for genomic-based personalized medicine [13].

In a different study, Van De Wetering et al. created a tumor organoid biobank from 20 successive patients with colorectal carcinoma (CRC). Whole exome sequencing demonstrated that this living biobank accurately reflects the somatic copy number variations and range of genetic alterations present in CRC. This study also showed that these organoids are suitable for high-throughput screening (HTS) and facilitate the identification of gene–drug associations [55]. Various clinical trials across different cancer types have been carried out to assess the potential of tumor organoids in personalized cancer treatment. In the clinical study NCT04736043 (https://clinicaltrials.gov/study/NCT04736043?term=NCT04736043&rank=1; accessed on 6 November 2024) the researchers generate organoids from pancreatic cancer tissue obtained through both endoscopic ultrasound-guided fine needle aspiration and endoscopic ultrasound-guided fine needle biopsy during the diagnostic process as well as from tissue obtained after surgery as part of the treatment process. To assess drug responsiveness, they perform cell viability assays on the organoids that were treated with various anticancer drugs, including those used in adjuvant chemotherapy for pancreatic cancer. Additionally, they conduct genomic analysis on each organoid to identify any unique mutations. By examining the relationship between these genomic mutations and the organoid's response to anticancer drugs in patients eligible for surgery, the researchers aim to tailor adjuvant chemotherapy strategies post-surgery, ultimately developing a platform to predict individual patient outcomes. The clinical trial, NCT04279509 (https://clinicaltrials.gov/study/NCT04279509?term=NCT04279509&rank=1; accessed on 6 November 2024), is a single-center study designed to prospectively evaluate whether high-throughput drug screening using PDOs can reliably identify chemotherapeutic agents that lead to objective responses in patients with refractory solid tumors, including head and neck squamous cell carcinoma, colorectal cancer, breast cancer, and epithelial ovarian cancer.

Cancer biomedicine aims to understand both the biological distinctions between tumor types and the variations across individual patients, even within the same pathological classifications. Consequently, there has been a growing focus on analyzing the TME, which has led to several promising findings. PDO technology has played a vital role in advancing personalized or precision medicine in cancer research. However, due to its purely epithelial origin, PDOs lack other cell types that comprise the TME, limiting their ability to replicate the structural and physiological components of the entire tumor accurately. To address this limitation, co-cultures of organoids enriched with specific cell types (such as stromal, inflammatory, immune cells, and pathogens) have been developed, creating models that better mimic the complexity of the TME [56].

In gastrointestinal cancer research, organoid co-cultures have proven valuable for uncovering mechanisms in pathogen-induced carcinogenesis [57]. One well-studied example is the chronic infection caused by Helicobacter pylori (H. pylori), which affects about half of the global population and can lead to gastric cancer in some patients [58]. In gastric organoids, H. pylori infection has been shown to increase both cell proliferation and the activation of stem cells via the oncogenic virulence factor CagA [59]. Additionally, organoids infected with H. pylori exhibit inflammatory responses, including NF-kB pathway activation and IL-8 overexpression [60], along with a notable increase in PD-L1 expression [61]. Similarly, salmonella infection has been identified as a genotoxic agent that acts as an "inflammatory stimulus", potentially contributing to malignant transformation of gallbladder epithelium [62].

Organoid technology represents a breakthrough in biomedical research, but it faces numerous challenges across its construction, data analysis, and application domains. During the construction phase, selecting appropriate matrix materials is critical. Currently, Matrigel dominates the field but presents issues such as poor reproducibility, variability, and potential immunogenicity [63]. The development of synthetic or hybrid hydrogel matrices offers a promising alternative, with advantages like cost-efficiency and closer alignment with the native structure of human organs. However, creating these hydrogels is a complex process due to the diverse chemical compositions and cross-linking mechanisms involved, necessitating iterative optimization [64].

In data analysis, the lack of standardized protocols and real-time monitoring poses significant obstacles. This absence of uniformity introduces variability, which is compounded by manual handling and subjective data interpretation. Furthermore, the limited automation in organoid culture and analysis processes contributes to inconsistencies, undermining the translational potential of organoid technology [65].

The applications of organoids are not without their own challenges, including ethical concerns and economic hurdles. As organoids grow increasingly sophisticated, particularly cerebral organoids, ethical questions about their potential to mimic human consciousness or experience pain become more urgent [66]. Additionally, the high costs of specialized culture media, growth factors, and labor-intensive procedures create barriers to scaling up and clinical translation [67].

In conclusion, while organoid technology provides groundbreaking opportunities to advance our understanding of human biology and disease, it is hindered by challenges in construction, data analysis, and ethical and economic considerations. Addressing these issues systematically is essential to fully harness the potential of this innovative technology.

2.4. *Tumor-on-a-Chip*

Despite both 3D spheroid and organoid models presenting notable advancements compared to 2D cell cultures, there are still unresolved concerns, including physical environmental cues within the TME. Static culture conditions are commonly utilized for spheroids/organoids. However, they are unable to replicate the shear stresses/hydrostatic pressures and interactions between tissues through blood/interstitial flow and vascular perfusion [68]. Therefore, it is crucial to create a reliable evaluation model that can summarize the complex characteristics of the TME and ensure the precise choice of nano-therapeutic agents for patients.

Recent advancements in microfluidic technology and tissue engineering have led to the development of hydrogel-based microfluidic platforms. These platforms are emerging as promising alternatives to traditional experimental models for replicating key TME features and improving the accuracy of patient response predictions. Microfluidic systems are advantageous due to their precise control over microenvironmental variables and the ability to incorporate various cell types in vitro. Notably, these models require significantly fewer cells than patient biopsy samples, enhancing their utility [27]. Additionally, progress in organ-on-a-chip, OoC, techniques has enabled the replication of the micro physiological functions and three-dimensional microstructures of human organs in vivo.

OoC devices are cell cultures created using microfabrication techniques, which merge the benefits of microfluidic technology with 3D cell culture technology to emulate the intricate features of natural organs [69]. Recreating living systems in a lab setting is challenging, so OoCs use "reverse engineering" to isolate the specific functions of target organs [70]. Since their introduction by Prof. Shuler et al., different types of OoC, including liver, lung, heart, kidney, intestine, and multi-organs-on-a-chip, have been created [71]. OoC offers numerous benefits, such as better replication of natural surroundings, ease of use, reduced expenses, and consistency [72]. Cells are grown in tiny chambers inside these small chips, either in 2D layers or 3D suspensions, to mimic organ tissues. The microfluidic parts of OoC platforms replicate the conditions found in living organisms, including flow, pressure, and nutrient levels [73]. Given the increasing prevalence of cancer as a leading cause of death, many researchers are now leveraging OoC technology to test responses to anticancer drugs.

Tumor-on-a-chip (ToC) models are specialized OoC systems that substitute healthy cells with tumor cells to replicate the TME, including the interactions between cancer cells and surrounding healthy tissues (Figure 6). These systems allow for precise control over experimental conditions and enable the study of cancer progression, metastasis, and responses to therapies in a highly controlled setting. ToC models offer numerous advantages that make them highly appealing for both basic and translational research. They enable precise control over cellular, mechanical, and physicochemical conditions, providing a high level of experimental accuracy. Moreover, their compatibility with various analytical techniques, such as transcriptomic analysis and live imaging, significantly deepens the scope of research. ToC models also offer the potential to create human and immunocompetent models, making them highly relevant for personalized medicine. Additionally, the relatively short experimental duration, often just a few days, compared to other 3D cancer models facilitates timely decision-making in clinical settings [74].

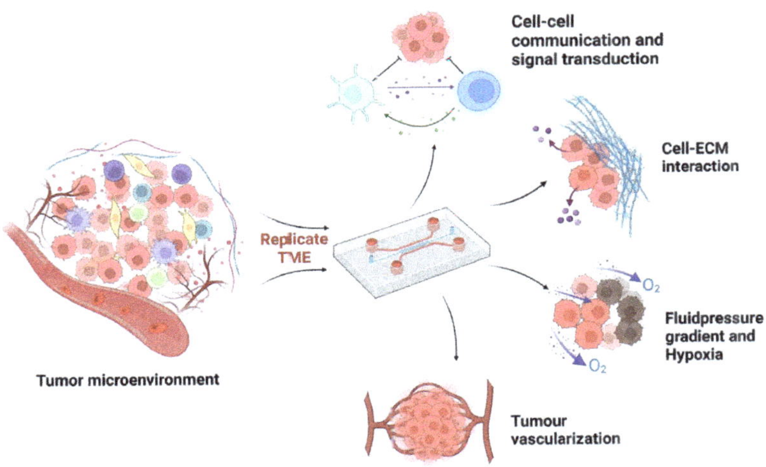

Figure 6. ToC for TME reproduction. Reproduced with permission of [75].

Numerous in vitro ToC models derived from human cells have been created to investigate various features related to tumor progression, including cell growth, blood vessel formation, metastasis, and responses to drugs. Fan et al. developed a 3D brain ToC model using polyethylene glycol diacrylate hydrogel as a medium for drug delivery and biological studies. In details, mepitastatin and irinotecan were administered to cells, and this model demonstrated its effectiveness as a glioma chip model for drug screening and release analysis [76]. In another study, Wang and colleagues created a glioma-focused microfluidic device with four parallel chambers to examine the responses of rat C6 glioma cells to the anticancer drug colchicine. Their findings showed notable changes in cell shape and death rate with increasing colchicine doses or extended treatment, supporting the advancement of glioma-targeted anticancer drugs and the development of glial cell-based biosensors for glioma detection [77]. Carvalho et al. designed a colorectal ToC microfluidic device composed of three primary sections. At the center, a circular hydrogel chamber containing ECM and embedded HCT-116 cells includes an inlet and an outlet. Flanking this central chamber are two perfusion channels. Within this setup, CRC cells and human colonic microvascular endothelial cells in the colon respond to the VEGF growth factor. This vascularized colorectal ToC model was created to explore the process of sprout formation and facilitate applications in drug screening [78].

In 2014, Vidi et al. created a breast-on-a-chip platform that simulates mammary cancer ducts by using a monolayer of breast luminal epithelium on a semicircular acrylic base. Tumor cells cultured in these channels exhibit distinct morphologies and drug sensitivities compared to those grown in traditional flat monoculture systems. These results offer valuable insights for the development and testing of cancer therapies [79].

These platforms provide more precise and sensitive responses to therapeutic effects, making them highly suitable for advancing nanodrug delivery system development. In fact, in 2013, Albanese et al. introduced the first spheroid-based model to investigate nanoparticle diffusion in tumors. This model utilized a polydimethylsiloxane microfluidic device to create controlled flow conditions, housing an immobilized multicellular spheroid. The spheroids, composed of MDA-MB-453 breast carcinoma cells, were embedded in an extracellular matrix that anchored them within the microfluidic chamber. This system demonstrated that active targeting of transferrin receptors enhanced nanoparticle accumulation and retention compared to passive targeting with PEGylated nanoparticles [80].

Nowadays, one of the primary challenges in clinical cancer treatment is tumor metastasis. Metastasis accounts for 90% of deaths related to cancer, yet the cellular and molecular mechanisms that drive this process are still not well understood. To address this, numerous microfluidic models of metastasis have been developed to analyze the cellular and molecular components involved in the metastatic cascade, which is essential for identifying new therapeutic opportunities for early intervention [27].

As a result, Skardal's group showcased the effectiveness of a two-organoid metastasis-on-a-chip platform [81]. Using microfluidics to create circulating flow, this system allowed tumor cells to grow at a primary site, enter circulation, and ultimately settle in downstream liver organoids. In this setup, CRC cells from colon organoids migrated into circulation and later attached to the liver organoids, making it one of the earliest in vitro models to simulate tumor cell metastasis from a three-dimensional primary site to a three-dimensional target site. Recently, the team enhanced the model by expanding the downstream organoids from a single site to four, creating a multisite metastasis-on-a-chip platform to study cancer cells' metastatic preferences [82].

Microfluidic-based ToC platforms are highly effective for replicating the complexities of the TME but face significant challenges in scalability. The intricate networks of microchannels and chambers that characterize these systems are labor-intensive and time-consuming, making it difficult to scale them for larger experiments or high-throughput drug screening. To overcome these limitations, a multifaceted approach is required. Advanced fabrication techniques such as 3D printing and soft lithography provide efficient and scalable methods for creating detailed microfluidic structures. Additionally, modular design principles enable scalability by allowing smaller, standardized components to be assembled into more extensive and complex systems. By employing these strategies, the scalability challenge of microfluidic-based ToC platforms can be effectively mitigated, unlocking their full potential in cancer research and drug development.

While ToC technology is a powerful tool in cancer research, it has inherent limitations. These models are simplified versions of real tumors, often lacking the complexity and heterogeneity found in human tissues. They may focus on specific aspects of tumor biology, potentially overlooking critical interactions. Notable drawbacks include the absence of immune system representation, challenges in modeling metastasis, and issues related to size restrictions and long-term viability. Ethical considerations, the resource-intensive nature of these systems, and variability between models further hinder their widespread adoption. Additionally, integrating systemic effects and ensuring validation against clinical data remain ongoing challenges.

Despite these limitations, ToC models have greatly enhanced our understanding of cancer. Continued research aims to address these constraints, paving the way for even more effective and reliable applications in cancer biology and drug discovery [19].

3. Applications of ML and 3D Models

State-of-the-art tumor modeling methods and immune-profiling instruments produce substantial amounts of high-dimensional information. Consequently, in the absence of AI technologies, even advanced tumor-modeling methods and immune-profiling instruments might not empower clinicians and researchers to recreate the TME, anticipate tumor cell behavior, or formulate diagnosis and treatment plans for improved clinical results [83]. Because of this lack, in vitro tumor modeling and AI have progressed significantly throughout the years (Figure 7).

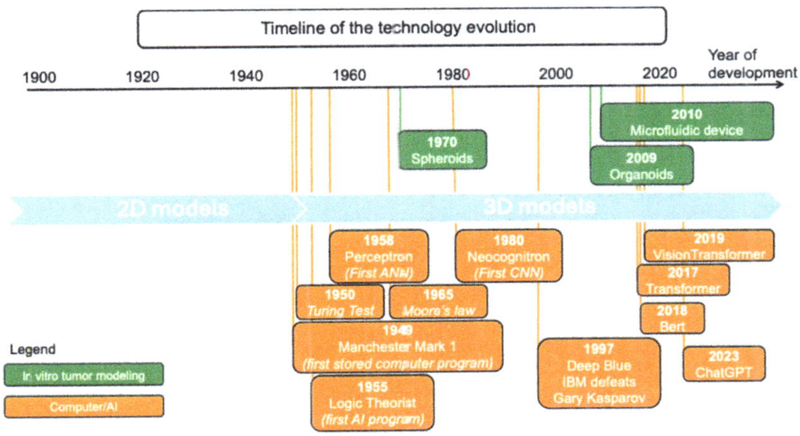

Figure 7. The timeline of the technology evolution for representative techniques used in the context of in vitro tumor modeling and AI. Adapted with permission of [83].

A subcategory of AI is ML that, according to Janiesch et al., is described as "the capacity of systems to learn from problem-specific training data to automate the process of analytical model building and solve associated tasks" [84].

Data used for training can either be labeled or unlabeled before being fed into an ML algorithm, meaning that for a given input, the desired output may or may not be specified. For example, in an image-detection model, the training data can explicitly specify what each image represents, enabling the model to predict the class to which each image belongs. Alternatively, this information can be omitted, and the model will group the data solely based on the input features.

Based on this distinction, ML algorithms can be categorized into three main types: (i) supervised learning, where predictions are made based on prior labeled data; (ii) unsupervised learning, used to uncover hidden patterns using unlabeled data; and (iii) reinforcement learning, which relies on rewards or penalties to guide actions, similar to a video game model [85]. To address some of the latest challenges in healthcare, such as drug development, personalized medicine, and disease diagnosis, ML is being increasingly utilized [86].

A major challenge in this field is the development of anticancer drugs, prompting the use of 3D in vitro tumor models to simulate tumors and their environments for drug testing.

By leveraging ML, researchers can reduce the need for extensive in vitro and animal testing, ultimately saving time and costs by performing in silico drug screening or extracting more meaningful insights from fewer experimental tests.

One of the most common methods for analyzing the development of a 3D in vitro model is by acquiring images, typically using various types of microscopies. ML models are then applied to process the images, primarily for two tasks: feature extraction and solving classification or regression problems. Feature extraction is usually performed by convolutional neural networks (CNNs), a type of deep learning (DL) model specifically designed for processing grid-like data such as images [87]. CNNs are particularly well-suited for tasks like image recognition and classification because, according to Benning et al., "CNN takes advantage of the spatial localization inherent in images and hence identifies and extracts features" [88]. This ability to autonomously discover the relevant features makes them especially useful in fields like computer vision, medical image analysis, and video processing [89]. These extracted features can then be used as input for classification or regression models, which can be the same CNN, to predict desired properties—for example, a drug's effectiveness in killing cancer cells.

An example of AI application is the observation that spheroids undergo characteristic morphological changes when subjected to stress, such as exposure to cytotoxic agents [88]. Images of individual spheroids can then be analyzed and classified using DL models.

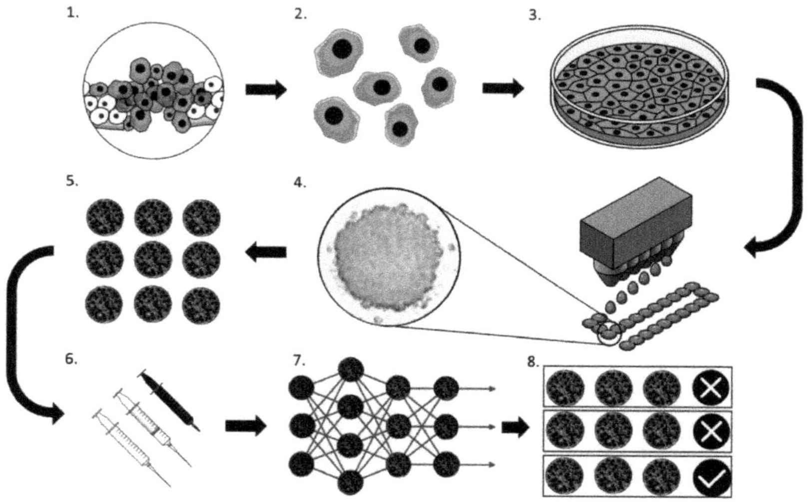

Figure 8. Proposed workflow by Benning et al. to enhance drug screening in solid tumors. (1) A malignant lesion develops in vivo. (2) Primary cells are obtained through biopsy, and (3) cultured to generate a sufficient cell quantity in vitro. (4) Homogeneous spheroids are formed using the hanging drop technique and (5) cultured until they reach full maturation. (6) Various drug combinations and concentrations are administered, and the response of each spheroid is monitored through imaging. (7) The images are processed using a pre-trained CNN, which (8) enables the classification of spheroid sensitivity to specific drugs or concentrations. Reproduced with permission of [88].

In particular, Benning et al. trained a CNN, which takes advantage of the spatial localization inherent in images and hence identifies and extracts features, to estimate the sensitivity of spheroids toward a given drug. This approach can reduce the time- and cost-intensive analysis required to determine spheroid sensitivity to cytotoxic agents, such as flow cytometry, which also demands the production of a large number of spheroids (Figure 8) [88].

A very promising descriptor, that can be extracted by CNNs, is cell motility, which plays a crucial role in life processes and is implicated in various pathologies, such as cancer metastasis [90]. An example of using ML to analyze cell motility in 3D in vitro tumor models is given by Mencattini et al. [91]. They employed a pre-trained DL CNN architecture to extract relevant features from images of cell trajectories, which were acquired with time-lapse microscopy. The sample used consisted of 3D biomimetic gels containing immune cells co-cultured with breast cancer cells in OoC devices, which were treated with an immunotherapy drug. The extracted features were used as input for a classification model, specifically support vector machines to classify trajectories of individual cells based on the presence or absence of the drug.

Another example in the use of CNN is made by Cortesi and colleagues. They introduced ORACLE, a deep-learning neural network designed specifically to identify ovarian cancer cells within complex pre-clinical models [92]. The model was trained on 558 annotated images, featuring both healthy and cancerous cells, which were divided into smaller non-overlapping regions to maximize data input. Each region was labeled with bounding boxes to classify cells as either healthy or cancerous. Leveraging a pre-trained implementation of faster region-based CNNs [93], ORACLE utilizes this widely adopted architecture for object detection. The model's validation demonstrated impressive performance, achieving an F1 score exceeding 0.9 and a ROC-AUC score of 0.99. This high accuracy, along with its ability to analyze co-cultures of patient-derived cells without the need for fluorescent tagging, marks key advantages of the approach. However, challenges remain, such as the dependency on a high-quality training dataset, which may limit generalizability to diverse biological contexts. Additionally, variations in biological behavior across different cell lines could impact the model's broader applicability.

Another study by Li et al. presents a breakthrough approach called "deep cytometry", which uses deep CNNs to classify and sort cells [94]. This method employs time-stretch imaging to capture one-dimensional time-series data from flowing cells. Mode-locked laser pulses convert spatial information into unique waveforms that are recorded by an analog-to-digital converter. By segmenting the waveforms into overlapping portions, the dataset is expanded, enabling the CNN to directly analyze raw data without traditional image processing. The results of this method were outstanding, with an F1 score of 95.71% and accuracy of 95.74% in classifying cancerous and non-cancerous cells within milliseconds. This approach not only improves speed and accuracy but also holds great potential for real-time applications, particularly in label-free early cancer detection.

Oliver et al. developed a 3D microfluidic in vitro device that recapitulates the blood–brain niche in order to understand the development of micro-metastasis and changes to the TME [95]. Fluorescent images of the 3D microfluidic blood–brain niche were captured at different time points using confocal microscopy and reconstructed in three dimensions with confocal tomography. Phenotypic measurements (e.g., cell size, shape descriptors, distance from the endothelial barrier) were extracted from these images. Then, these measurements were used as features for an ML model to predict the metastatic potential of cancer cells. This was used to identify intrinsic phenotypic differences between brain-metastatic and non-brain metastatic cancer cells [96].

A different approach is provided by Yoo et al., exploring the use of spheroid size (SoS) and hypoxic regions (SoH) for efficient drug efficacy evaluation through ML. They trained an ML model to predict changes in SoS and SoH over time using images of untreated spheroids. Then, HepG2 spheroids were treated with sorafenib, a molecule known for its potential to inhibit tumor growth and suppress angiogenesis [97]. The drug's efficacy was assessed by comparing the experimental changes in SoS and SoH of the treated spheroids with the ML-predicted changes in SoS and SoH based on the untreated spheroids. Experiments revealed that for spheroids treated with sorafenib, cellular growth was inhibited three days post-treatment, and this study demonstrates that SoS and SoH can be accurately predicted by ML models. This approach holds significant promise for biomedical research, by reducing the time, costs and resources associated with preliminary experiments [98].

While image acquisition is a common method to analyze the development of 3D in vitro models, an effective alternative approach is transcriptomic profiling. As demonstrated by Kong et al., organoids can be analyzed by extracting RNA and obtaining gene expression data [99]. Pharmacogenomic data are derived from 3D organoid culture models, including gene expression profiles and drug response information. This latest pharmacogenomic data is analyzed to identify relevant features associated with drug response, particularly by examining pathways that are proximal to drug targets within a protein–protein interaction network. The selected features, which represent the gene expression levels of these identified pathways, are then used to train an ML model. This model learns to associate specific expression patterns with drug responses observed in the organoid models. Once trained, the ML model can be applied to predict drug responses for cancer patients based on their transcriptomic profiles, enabling predictions about the likelihood of a patient responding to a particular treatment.

Another study by Tang et al. combines 3D bioprinting and ML to enhance the assessment of glioma treatment responses [100]. The authors developed an ML workflow, GlioML, which incorporates nine classic algorithms and a weighted ensemble model for reliable drug response predictions based on gene expression profiles (Figure 9).

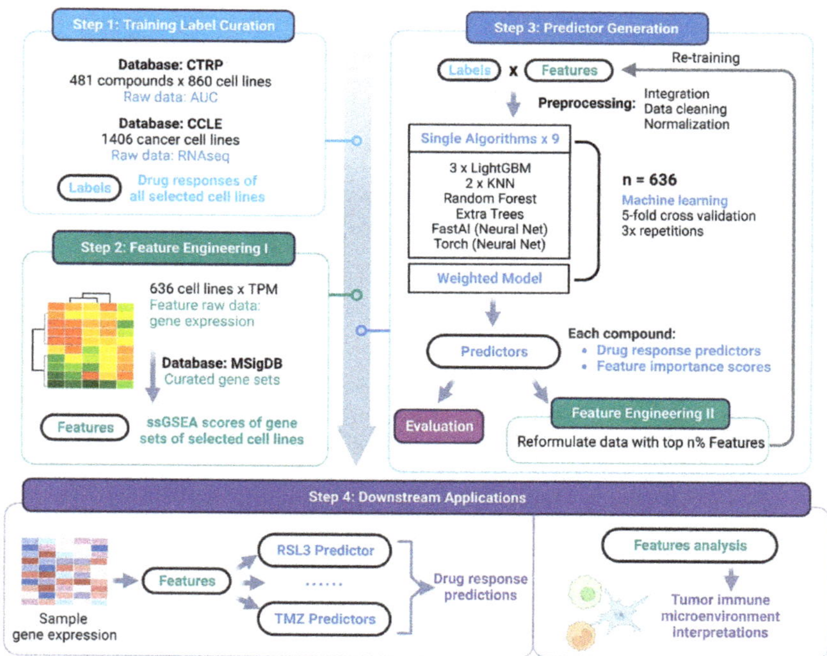

Figure 9. Representation of the four main stages of the GlioML workflow: (1) data preprocessing; (2) feature engineering; (3) training module; and (4) downstream applications, enabling drug response predictions and microenvironment analysis for enhanced therapeutic insights. Reproduced with permission of [100].

These profiles were collected from patient-derived glioma tissue samples and glioblastoma stem cells. The results highlighted the effectiveness of neural networks and gradientboosting algorithms, which outperformed models like K-nearest neighbors in producing the best predictors. Additionally, the integration of 3D bioprinting enabled the creation of biomimetic tissues that replicated the genomic characteristics and drug responses of native tumors, enhancing the clinical relevance of GlioML's predictions. However, challenges persist, such as the potential for overfitting when using narrower datasets and the resource-intensive nature of implementing multiple algorithms. Nonetheless, this pioneering approach offers new avenues for improving glioma treatment strategies by incorporating patient-specific drug response insights and addressing the complexities of clinical research.

Sammut et al. presented an ML framework designed to integrate diverse datasets—clinical, digital pathology, genomic, and transcriptomic features—to predict a pathologically complete response in breast cancer patients undergoing neoadjuvant treatment [101]. Using an ensemble learning approach, the researchers developed six predictive models based on varying combinations of features, from clinical data alone to a full integration of DNA, RNA, digital pathology, and treatment data. These data were extracted from the biopsies of 180 women with early and locally advanced breast cancer. The framework demonstrated improved predictive power across multiple cancer types, although challenges remain, such as interpretability and the need for high-quality multi-omic data. Despite these hurdles, integrating clinicopathological and molecular data through ML can significantly enhance predictions of therapy responses, paving the way for more personalized oncology treatments.

In the realm of ovarian cancer, a study from Boehm et al. explored multimodal ML approaches to improve risk stratification for patients with high-grade serous ovarian cancer [102]. By integrating clinical, genomic, histopathological, and radiological data from routine diagnostic procedures, the study developed risk stratification models. These models, which used radiomic features from contrast-enhanced computed tomography and histopathological data from pre-treatment tissue samples, were validated using a late-fusion multimodal statistical framework. The integration of these diverse data types significantly enhanced predictive accuracy. However, challenges emerged in generalizing the models across different institutions, as varying imaging protocols introduced variability. Moreover, reliance on data not originally intended for computational modelling, such as incomplete genomic samples, added complexity. Despite these challenges, the study highlights the potential of integrating quantitative imaging features with clinical and genomic data to improve cancer prognosis. Further exploration of multimodal ML strategies could lead to enhanced cancer prognostics and more effective treatment planning [103,104].

ML models have become transformative tools in drug discovery, enabling the efficient prediction of molecular properties, protein–ligand interactions, and clinical outcomes. These models, particularly DL architectures, excel in processing high-dimensional datasets, extracting complex patterns, and offering predictive insights beyond simple correlative associations [105,106]. Predictive ML models have demonstrated utility in lead identification and optimization, such as predicting absorption, distribution, metabolism, excretion, and toxicity (ADMET) profiles, thus reducing reliance on expensive and time-consuming experimental methods [86].

However, the integration of ML into drug discovery presents several challenges. One critical issue is the dependence on high-quality, domain-specific datasets. Many public and proprietary datasets suffer from noise and inconsistencies, which can lead to biased or overfitted models with limited generalizability [107]. Furthermore, the interpretability of ML predictions remains a significant limitation, especially for models using DL. The "black-box" nature of these approaches can impede the extraction of mechanistic insights, which are crucial for drug design and regulatory approval [18]. Another challenge lies in transferring models trained on existing chemical space to novel or underexplored regions, as these often lack representative training examples. This issue is compounded by the need for rigorous validation strategies, including external validation on independent datasets and experimental confirmation, to ensure clinical relevance. Finally, the ethical and logistical aspects of implementing ML-driven predictions in drug discovery pipelines must also be considered. For example, biases inherent in datasets may inadvertently lead to inequitable drug development or flawed prioritization of candidate molecules [108].

Given the growing complexity of tumor biology, ML applied to 3D tumor models offers a significant advancement over traditional methods, including RNA-based drug response profiling. Unlike transcriptomic snapshots, which provide static molecular readouts, ML approaches in 3D systems capture dynamic cell–cell interactions, migration behaviors, and microenvironmental remodeling, factors that are not discernible through RNA profiling alone. Deep learning models, particularly convolutional neural networks, autonomously extract biologically meaningful features from microscopy images of 3D cultures, offering a scalable and label-free alternative to traditional transcriptomic analyses. Additionally, ML frameworks can integrate RNA sequencing data with imaging, metabolic, and proteomic profiles, allowing for a more comprehensive view of drug response mechanisms that extends beyond differential gene expression analysis. The ability of ML to train on phenotypic datasets further enhances predictive accuracy, enabling the identification of novel drug response markers that might otherwise be overlooked by transcriptomics alone.

4. Conclusions and Perspectives

In conclusion, the integration of advanced 3D in vitro tumor models with AI and ML represents a transformative shift in cancer research and drug discovery. These innovative approaches provide unprecedented opportunities to mimic the tumor microenvironment with high precision, enabling better insights into tumor biology, progression, and therapeutic responses. Compared to traditional 2D cultures and animal models, 3D systems such as spheroids, organoids, and ToC platforms offer superior physiological relevance, making them key tools for preclinical testing and personalized medicine.

Looking forward, the combination of 3D models and AI-driven analytics has the potential to revolutionize cancer therapy by facilitating high-throughput drug screening, optimizing treatment regimens, and identifying novel therapeutic targets. However, challenges remain, including the standardization of protocols, scalability of 3D systems, and ensuring robust clinical validation of these models. Future research should focus on addressing these limitations through advancements in synthetic matrices, real-time monitoring technologies, and modular microfluidic designs. Additionally, further integration of AI techniques, such as DL and multi-omics data analysis, will enhance the predictive accuracy and efficiency of these models, paving the way for personalized, patient-specific cancer therapies. By overcoming these challenges, the synergy between 3D in vitro models and AI can redefine the landscape of oncology, offering a path toward more effective and less invasive cancer treatments, ultimately improving patient outcomes and reducing reliance on animal testing.

Therefore, the integration of ML and AI into tumor spheroids, organoids, and OoC models represents a transformative approach in cancer research, offering significant enhancements in predictive power and efficiency. By leveraging ML and AI, researchers can extract more complex, multidimensional data from these 3D culture systems, thereby improving their accuracy in mimicking human tumor biology and reducing reliance on animal testing.

Author Contributions: Conceptualization: G.V., E.M. and C.F.; provision, collection, and assembly of data: C.M., B.C., L.L., M.T. and M.D.P.; review of the literature: all authors.; writing—original draft preparation: all authors; writing—review and editing: all authors; final approval of the manuscript: all authors. All authors have read and agreed to the published version of the manuscript.

Funding: This review paper received no external funding.

Data Availability Statement: No new data were created or analyzed in this review paper. Data sharing is not applicable to this article.

Conflicts of Interest: The authors declare no conflicts of interest.

References

1. Siegel, R.L.; Giaquinto, A.N.; Jemal, A. Cancer statistics, 2024. *CA. Cancer J. Clin.* **2024**, *74*, 12–49. [CrossRef]
2. Aboulkheyr Es, H.; Montazeri, L.; Aref, A.R.; Vosough, M.; Baharvand, H. Personalized Cancer Medicine: An Organoid Approach. *Trends Biotechnol.* **2018**, *36*, 358–371. [CrossRef] [PubMed]
3. Vitale, I.; Shema, E.; Loi, S.; Galluzzi, L. Intratumoral heterogeneity in cancer progression and response to immunotherapy. *Nat. Med.* **2021**, *27*, 212–224. [CrossRef]
4. Aldea, M.; Andre, F.; Marabelle, A.; Dogan, S.; Barlesi, F.; Soria, J.-C. Overcoming Resistance to Tumor-Targeted and Immune-Targeted Therapies. *Cancer Discov.* **2021**, *11*, 874–899. [CrossRef] [PubMed]
5. Manduca, N.; Maccafeo, E.; De Maria, R.; Sistigu, A.; Musella, M. 3D cancer models: One step closer to in vitro human studies. *Front. Immunol.* **2023**, *14*, 1175503. [CrossRef] [PubMed]
6. Russell, W.M.S.; Burch, R.L.; Hume, C.W. The Principles of Humane Experimental Technique. *Med. J. Aust.* **1960**, *1*, 500. [CrossRef]
7. Vitale, A.; Manciocco, A.; Alleva, E. The 3R principle and the use of non-human primates in the study of neurodegenerative diseases: The case of Parkinson's disease. *Neurosci. Biobehav. Rev.* **2009**, *33*, 33–47. [CrossRef] [PubMed]
8. Cacciamali, A.; Villa, R.; Dotti, S. 3D Cell Cultures: Evolution of an Ancient Tool for New Applications. *Front. Physiol.* **2022**, *13*, 836480. [CrossRef] [PubMed]
9. Schachtschneider, K.M.; Schwind, R.M.; Newson, J.; Kinachtchouk, N.; Rizko, M.; Mendoza-Elias, N.; Grippo, P.; Principe, D.R.; Park, A.; Overgaard, N.H.; et al. The Oncopig Cancer Model: An Innovative Large Animal Translational Oncology Platform. *Front. Oncol.* **2017**, *7*, 190. [CrossRef] [PubMed]
10. Van Norman, G.A. Limitations of Animal Studies for Predicting Toxicity in Clinical Trials. *JACC Basic Transl. Sci.* **2019**, *4*, 845–854. [CrossRef]
11. Shanks, N.; Greek, R.; Greek, J. Are animal models predictive for humans? *Philos. Ethics Humanit. Med.* **2009**, *4*, 2. [CrossRef] [PubMed]
12. Sachdeva, U.M.; Shimonosono, M.; Flashner, S.; Cruz-Acuña, R.; Gabre, J.T.; Nakagawa, H. Understanding the cellular origin and progression of esophageal cancer using esophageal organoids. *Cancer Lett.* **2021**, *509*, 39–52. [CrossRef] [PubMed]
13. Weeber, F.; Van De Wetering, M.; Hoogstraat, M.; Dijkstra, K.K.; Krijgsman, O.; Kuilman, T.; Gadellaa-van Hooijdonk, C.G.M.; Van Der Velden, D.L.; Peeper, D.S.; Cuppen, E.P.J.G.; et al. Preserved genetic diversity in organoids cultured from biopsies of human colorectal cancer metastases. *Proc. Natl. Acad. Sci. USA* **2015**, *112*, 13308–13311. [CrossRef]
14. Schütte, M.; Risch, T.; Abdavi-Azar, N.; Boehnke, K.; Schumacher, D.; Keil, M.; Yildiriman, R.; Jandrasits, C.; Borodina, T.; Amstislavskiy, V.; et al. Molecular dissection of colorectal cancer in pre-clinical models identifies biomarkers predicting sensitivity to EGFR inhibitors. *Nat. Commun.* **2017**, *8*, 14262. [CrossRef]
15. Salinas-Vera, Y.M.; Valdés, J.; Pérez-Navarro, Y.; Mandujano-Lazaro, G.; Marchat, L.A.; Ramos-Payán, R.; Nuñez-Olvera, S.I.; Pérez-Plascencia, C.; López-Camarillo, C. Three-Dimensional 3D Culture Models in Gynecological and Breast Cancer Research. *Front. Oncol.* **2022**, *12*, 826113. [CrossRef]
16. Chen, Z.; Ma, N.; Sun, X.; Li, Q.; Zeng, Y.; Chen, F.; Sun, S.; Xu, J.; Zhang, J.; Ye, H.; et al. Automated evaluation of tumor spheroid behavior in 3D culture using deep learning-based recognition. *Biomaterials* **2021**, *272*, 120770. [CrossRef] [PubMed]
17. Iqbal, M.J.; Javed, Z.; Sadia, H.; Qureshi, I.A.; Irshad, A.; Ahmed, R.; Malik, K.; Raza, S.; Abbas, A.; Pezzani, R.; et al. Clinical applications of artificial intelligence and machine learning in cancer diagnosis: Looking into the future. *Cancer Cell Int.* **2021**, *21*, 270. [CrossRef] [PubMed]
18. Jiménez Pérez, M.; Grande, R.G. Application of artificial intelligence in the diagnosis and treatment of hepatocellular carcinoma: A review. *World J. Gastroenterol.* **2020**, *26*, 5617–5628. [CrossRef]
19. Hwangbo, H.; Chae, S.; Kim, W.; Jo, S.; Kim, G.H. Tumor-on-a-chip models combined with mini-tissues or organoids for engineering tumor tissues. *Theranostics* **2024**, *14*, 33–55. [CrossRef]
20. Huh, D.; Hamilton, G.A.; Ingber, D.E. From 3D cell culture to organs-on-chips. *Trends Cell Biol.* **2011**, *21*, 745–754. [CrossRef] [PubMed]

21. Growth of Multicell Spheroids in Tissue Culture as a Model of Nodular Carcinomas2. *JNCI J. Natl. Cancer Inst.* **1971**. [CrossRe
22. Huh, D.; Matthews, B.D.; Mammoto, A.; Montoya-Zavala, M.; Hsin, H.Y.; Ingber, D.E. Reconstituting organ-level lung functior on a chip. *Science* **2010**, *328*, 1662–1668. [CrossRef]
23. Shi, R.; Tang, Y.; Miao, H. Metabolism in tumor microenvironment: Implications for cancer immunotherapy. *MedComm* **2020**, 47–68. [CrossRef] [PubMed]
24. Anderson, N.M.; Simon, M.C. The tumor microenvironment. *Curr. Biol.* **2020**, *30*, R921–R925. [CrossRef] [PubMed]
25. Ziółkowska-Suchanek, I. Mimicking Tumor Hypoxia in Non-Small Cell Lung Cancer Employing Three-Dimensional In Vitı Models. *Cells* **2021**, *10*, 141. [CrossRef]
26. Piñeiro Fernández, J.; Luddy, K.A.; Harmon, C.; O'Farrelly, C. Hepatic Tumor Microenvironments and Effects on NK Ce Phenotype and Function. *Int. J. Mol. Sci.* **2019**, *20*, 4131. [CrossRef]
27. Li, W.; Zhou, Z.; Zhou, X.; Khoo, B.L.; Gunawan, R.; Chin, Y.R.; Zhang, L.; Yi, C.; Guan, X.; Yang, M. 3D Biomimetic Mode to Reconstitute Tumor Microenvironment In Vitro: Spheroids, Organoids, and Tumor-on-a-Chip. *Adv. Healthc. Mater.* **202** *12*, 2202609. [CrossRef]
28. Hirschhaeuser, F.; Menne, H.; Dittfeld, C.; West, J.; Mueller-Klieser, W.; Kunz-Schughart, L.A. Multicellular tumor spheroids: A underestimated tool is catching up again. *J. Biotechnol.* **2010**, *148*, 3–15. [CrossRef]
29. Hanahan, D.; Coussens, L.M. Accessories to the Crime: Functions of Cells Recruited to the Tumor Microenvironment. *Cancer Ce* **2012**, *21*, 309–322. [CrossRef] [PubMed]
30. Costa, E.C.; Moreira, A.F.; de Melo-Diogo, D.; Gaspar, V.M.; Carvalho, M.P.; Correia, I.J. 3D tumor spheroids: An overview on th tools and techniques used for their analysis. *Biotechnol. Adv.* **2016**, *34*, 1427–1441. [CrossRef] [PubMed]
31. Nath, S.; Devi, G.R. Three-dimensional culture systems in cancer research: Focus on tumor spheroid model. *Pharmacol. Ther.* **201** *163*, 94–108. [CrossRef]
32. Mayer, B.; Klement, G.; Kaneko, M.; Man, S.; Jothy, S.; Rak, J.; Kerbel, R.S. Multicellular gastric cancer spheroids recapitulat growth pattern and differentiation phenotype of human gastric carcinomas. *Gastroenterology* **2001**, *121*, 839–852. [CrossRe [PubMed]
33. Breslin, S.; O'Driscoll, L. Three-dimensional cell culture: The missing link in drug discovery. *Drug Discov. Today* **2013**, *18*, 240–24 [CrossRef] [PubMed]
34. Kelm, J.M.; Timmins, N.E.; Brown, C.J.; Fussenegger, M.; Nielsen, L.K. Method for generation of homogeneous multicellula tumor spheroids applicable to a wide variety of cell types. *Biotechnol. Bioeng.* **2003**, *83*, 173–180. [CrossRef]
35. Roy, M.; Alix, C.; Bouakaz, A.; Serrière, S.; Escoffre, J.-M. Tumor Spheroids as Model to Design Acoustically Mediated Dru Therapies: A Review. *Pharmaceutics* **2023**, *15*, 806. [CrossRef]
36. Benien, P.; Swami, A. 3D Tumor Models: History, Advances and Future Perspectives. *Future Oncol.* **2014**, *10*, 1311–1327. [CrossRe
37. Kim, J.B. Three-dimensional tissue culture models in cancer biology. *Semin. Cancer Biol.* **2005**, *15*, 365–377. [CrossRef] [PubMed
38. Brancato, V.; Comunanza, V.; Imparato, G.; Corà, D.; Urciuolo, F.; Noghero, A.; Bussolino, F.; Netti, P.A. Bioengineered tumora microtissues recapitulate desmoplastic reaction of pancreatic cancer. *Acta Biomater.* **2017**, *49*, 152–166. [CrossRef] [PubMed]
39. Kim, J.H.; Verwilst, P.; Won, M.; Lee, J.; Sessler, J.L.; Han, J.; Kim, J.S. A Small Molecule Strategy for Targeting Cancer Stem Cell in Hypoxic Microenvironments and Preventing Tumorigenesis. *J. Am. Chem. Soc.* **2021**, *143*, 14115–14124. [CrossRef] [PubMed
40. Ishiguro, T.; Ohata, H.; Sato, A.; Yamawaki, K.; Enomoto, T.; Okamoto, K. Tumor-derived spheroids: Relevance to cancer sten cells and clinical applications. *Cancer Sci.* **2017**, *108*, 283–289. [CrossRef]
41. Lapidot, T.; Sirard, C.; Vormoor, J.; Murdoch, B.; Hoang, T.; Caceres-Cortes, J.; Minden, M.; Paterson, B.; Caligiuri, M.A.; Dick, J.I A cell initiating human acute myeloid leukaemia after transplantation into SCID mice. *Nature* **1994**, *367*, 645–648. [CrossRef [PubMed]
42. Singh, S.K.; Clarke, I.D.; Terasaki, M.; Bonn, V.E.; Hawkins, C.; Squire, J.; Dirks, P.B. Identification of a cancer stem cell in huma brain tumors. *Cancer Res.* **2003**, *63*, 5821–5828.
43. Dontu, G.; Abdallah, W.M.; Foley, J.M.; Jackson, K.W.; Clarke, M.F.; Kawamura, M.J.; Wicha, M.S. In vitro propagation an transcriptional profiling of human mammary stem/progenitor cells. *Genes Dev.* **2003**, *17*, 1253–1270. [CrossRef] [PubMed]
44. Ricci-Vitiani, L.; Lombardi, D.G.; Pilozzi, E.; Biffoni, M.; Todaro, M.; Peschle, C.; De Maria, R. Identification and expansion o human colon-cancer-initiating cells. *Nature* **2007**, *445*, 111–115. [CrossRef] [PubMed]
45. Masuda, M.; Uno, Y.; Ohbayashi, N.; Ohata, H.; Mimata, A.; Kukimoto-Niino, M.; Moriyama, H.; Kashimoto, S.; Inoue, T.; Gotc N.; et al. TNIK inhibition abrogates colorectal cancer stemness. *Nat. Commun.* **2016**, *7*, 12586. [CrossRef]
46. Tuveson, D.; Clevers, H. Cancer modeling meets human organoid technology. *Science* **2019**, *364*, 952–955. [CrossRef] [PubMed]
47. Corrò, C.; Novellasdemunt, L.; Li, V.S.W. A brief history of organoids. *Am. J. Physiol. Cell Physiol.* **2020**, *319*, C151–C16 [CrossRef] [PubMed]
48. Yuan, J.; Li, X.; Yu, S. Cancer organoid co-culture model system: Novel approach to guide precision medicine. *Front. Immuno* **2023**, *13*, 1061388. [CrossRef]

Bai, L.; Wu, Y.; Li, G.; Zhang, W.; Zhang, H.; Su, J. AI-enabled organoids: Construction, analysis, and application. *Bioact. Mater.* **2024**, *31*, 525–548. [CrossRef]

Mun, S.J.; Ryu, J.-S.; Lee, M.-O.; Son, Y.S.; Oh, S.J.; Cho, H.-S.; Son, M.-Y.; Kim, D.-S.; Kim, S.J.; Yoo, H.J.; et al. Generation of expandable human pluripotent stem cell-derived hepatocyte-like liver organoids. *J. Hepatol.* **2019**, *71*, 970–985. [CrossRef] [PubMed]

Arjmand, B.; Rabbani, Z.; Soveyzi, F.; Tayanloo-Beik, A.; Rezaei-Tavirani, M.; Biglar, M.; Adibi, H.; Larijani, B. Advancement of Organoid Technology in Regenerative Medicine. *Regen. Eng. Transl. Med.* **2023**, *9*, 83–96. [CrossRef] [PubMed]

Sun, N.; Meng, X.; Liu, Y.; Song, D.; Jiang, C.; Cai, J. Applications of brain organoids in neurodevelopment and neurological diseases. *J. Biomed. Sci.* **2021**, *28*, 30. [CrossRef]

Sato, T.; Vries, R.G.; Snippert, H.J.; van de Wetering, M.; Barker, N.; Stange, D.E.; van Es, J.H.; Abo, A.; Kujala, P.; Peters, P.J.; et al. Single Lgr5 stem cells build crypt-villus structures in vitro without a mesenchymal niche. *Nature* **2009**, *459*, 262–265. [CrossRef]

Sachs, N.; de Ligt, J.; Kopper, O.; Gogola, E.; Bounova, G.; Weeber, F.; Balgobind, A.V.; Wind, K.; Gracanin, A.; Begthel, H.; et al. A Living Biobank of Breast Cancer Organoids Captures Disease Heterogeneity. *Cell* **2018**, *172*, 373–386.e10. [CrossRef] [PubMed]

van de Wetering, M.; Francies, H.E.; Francis, J.M.; Bounova, G.; Iorio, F.; Pronk, A.; van Houdt, W.; van Gorp, J.; Taylor-Weiner, A.; Kester, L.; et al. Prospective derivation of a living organoid biobank of colorectal cancer patients. *Cell* **2015**, *161*, 933–945. [CrossRef]

Obreque, J.; Vergara-Gómez, L.; Venegas, N.; Weber, H.; Owen, G.I.; Pérez-Moreno, P.; Leal, P.; Roa, J.C.; Bizama, C. Advances towards the use of gastrointestinal tumor patient-derived organoids as a therapeutic decision-making tool. *Biol. Res.* **2023**, *56*, 63. [CrossRef] [PubMed]

Dutta, D.; Clevers, H. Organoid culture systems to study host-pathogen interactions. *Curr. Opin. Immunol.* **2017**, *48*, 15–22. [CrossRef]

Alipour, M. Molecular Mechanism of Helicobacter pylori-Induced Gastric Cancer. *J. Gastrointest. Cancer* **2021**, *52*, 23–30. [CrossRef] [PubMed]

McCracken, K.W.; Catá, E.M.; Crawford, C.M.; Sinagoga, K.L.; Schumacher, M.; Rockich, B.E.; Tsai, Y.-H.; Mayhew, C.N.; Spence, J.R.; Zavros, Y.; et al. Modelling human development and disease in pluripotent stem-cell-derived gastric organoids. *Nature* **2014**, *516*, 400–404. [CrossRef]

Bartfeld, S.; Bayram, T.; van de Wetering, M.; Huch, M.; Begthel, H.; Kujala, P.; Vries, R.; Peters, P.J.; Clevers, H. In vitro expansion of human gastric epithelial stem cells and their responses to bacterial infection. *Gastroenterology* **2015**, *148*, 126–136.e6. [CrossRef]

Holokai, L.; Chakrabarti, J.; Broda, T.; Chang, J.; Hawkins, J.A.; Sundaram, N.; Wroblewski, L.E.; Peek, R.M.; Wang, J.; Helmrath, M.; et al. Increased Programmed Death-Ligand 1 is an Early Epithelial Cell Response to Helicobacter pylori Infection. *PLoS Pathog.* **2019**, *15*, e1007468. [CrossRef]

Iyer, N.V.; Kotch, L.E.; Agani, F.; Leung, S.W.; Laughner, E.; Wenger, R.H.; Gassmann, M.; Gearhart, J.D.; Lawler, A.M.; Yu, A.Y.; et al. Cellular and developmental control of O_2 homeostasis by hypoxia-inducible factor 1α. *Genes Dev.* **1998**, *12*, 149–162. [CrossRef] [PubMed]

Blondel, D.; Lutolf, M.P. Bioinspired Hydrogels for 3D Organoid Culture. *Chimia* **2019**, *73*, 81. [CrossRef] [PubMed]

Revete, A.; Aparicio, A.; Cisterna, B.A.; Revete, J.; Luis, L.; Ibarra, E.; Segura González, E.A.; Molino, J.; Reginensi, D. Advancements in the Use of Hydrogels for Regenerative Medicine: Properties and Biomedical Applications. *Int. J. Biomater.* **2022**, *2022*, 1–16. [CrossRef]

Peng, W.; Datta, P.; Wu, Y.; Dey, M.; Ayan, B.; Dababneh, A.; Ozbolat, I.T. Challenges in Bio-fabrication of Organoid Cultures. In *Cell Biology and Translational Medicine, Volume 3*; Turksen, K., Ed.; Advances in Experimental Medicine and Biology; Springer International Publishing: Cham, Switzerland, 2018; Volume 1107, pp. 53–71, ISBN 978-3-030-04184-7.

Yadav, A.; Seth, B.; Chaturvedi, R.K. Brain Organoids: Tiny Mirrors of Human Neurodevelopment and Neurological Disorders. *Neuroscientist* **2021**, *27*, 388–426. [CrossRef] [PubMed]

Kozlowski, M.T.; Crook, C.J.; Ku, H.T. Towards organoid culture without Matrigel. *Commun. Biol.* **2021**, *4*, 1387. [CrossRef] [PubMed]

Guan, P.-P.; Yu, X.; Guo, J.-J.; Wang, Y.; Wang, T.; Li, J.-Y.; Konstantopoulos, K.; Wang, Z.-Y.; Wang, P. By activating matrix metalloproteinase-7, shear stress promotes chondrosarcoma cell motility, invasion and lung colonization. *Oncotarget* **2015**, *6*, 9140–9159. [CrossRef]

Esch, E.W.; Bahinski, A.; Huh, D. Organs-on-chips at the frontiers of drug discovery. *Nat. Rev. Drug Discov.* **2015**, *14*, 248–260. [CrossRef]

Tian, C.; Zheng, S.; Liu, X.; Kamei, K. Tumor-on-a-chip model for advancement of anti-cancer nano drug delivery system. *J. Nanobiotechnol.* **2022**, *20*, 338. [CrossRef]

Herland, A.; Maoz, B.M.; Das, D.; Somayaji, M.R.; Prantil-Baun, R.; Novak, R.; Cronce, M.; Huffstater, T.; Jeanty, S.S.F.; Ingram, M.; et al. Quantitative prediction of human pharmacokinetic responses to drugs via fluidically coupled vascularized organ chips. *Nat. Biomed. Eng.* **2020**, *4*, 421–436. [CrossRef]

72. Whitesides, G.M. The origins and the future of microfluidics. *Nature* **2006**, *442*, 368–373. [CrossRef] [PubMed]
73. Fetah, K.L.; DiPardo, B.J.; Kongadzem, E.; Tomlinson, J.S.; Elzagheid, A.; Elmusrati, M.; Khademhosseini, A.; Ashammakhi Cancer Modeling-on-a-Chip with Future Artificial Intelligence Integration. *Small* **2019**, *15*, 1901985. [CrossRef] [PubMed]
74. Bouquerel, C.; Dubrova, A.; Hofer, I.; Phan, D.T.T.; Bernheim, M.; Ladaigue, S.; Cavaniol, C.; Maddalo, D.; Cabel, L.; Mech Grigoriou, E.; et al. Bridging the gap between tumor-on-chip and clinics: A systematic review of 15 years of studies. *Lab. C* **2023**, *23*, 3906–3935. [CrossRef] [PubMed]
75. Xu, H.; Wen, J.; Yang, J.; Zhou, S.; Li, Y.; Xu, K.; Li, W.; Li, S. Tumor-microenvironment-on-a-chip: The construction and applicati *Cell Commun. Signal.* **2024**, *22*, 515. [CrossRef]
76. Fan, Y.; Nguyen, D.T.; Akay, Y.; Xu, F.; Akay, M. Engineering a Brain Cancer Chip for High-throughput Drug Screening. *Sci. R* **2016**, *6*, 25062. [CrossRef] [PubMed]
77. Liu, W.; Sun, P.; Yang, L.; Wang, J.; Li, L.; Wang, J. Assay of glioma cell responses to an anticancer drug in a cell-based microflui device. *Microfluid. Nanofluid.* **2010**, *9*, 717–725. [CrossRef]
78. Carvalho, M.R.; Barata, D.; Teixeira, L.M.; Giselbrecht, S.; Reis, R.L.; Oliveira, J.M.; Truckenmüller, R.; Habibovic, P. Colorec tumor-on-a-chip system: A 3D tool for precision onco-nanomedicine. *Sci. Adv.* **2019**, *5*, eaaw1317. [CrossRef]
79. Vidi, P.-A.; Maleki, T.; Ochoa, M.; Wang, L.; Clark, S.M.; Leary, J.F.; Lelièvre, S.A. Disease-on-a-chip: Mimicry of tumor growth mammary ducts. *Lab. Chip* **2014**, *14*, 172–177. [CrossRef] [PubMed]
80. Albanese, A.; Lam, A.K.; Sykes, E.A.; Rocheleau, J.V.; Chan, W.C.W. Tumour-on-a-chip provides an optical window i nanoparticle tissue transport. *Nat. Commun.* **2013**, *4*, 2718. [CrossRef] [PubMed]
81. Skardal, A.; Devarasetty, M.; Forsythe, S.; Atala, A.; Soker, S. A reductionist metastasis-on-a-chip platform for in vitro tum progression modeling and drug screening. *Biotechnol. Bioeng.* **2016**, *113*, 2020–2032. [CrossRef]
82. Aleman, J.; Skardal, A. A multi-site metastasis-on-a-chip microphysiological system for assessing metastatic preference of can cells. *Biotechnol. Bioeng.* **2019**, *116*, 936–944. [CrossRef] [PubMed]
83. Lee, R.Y.; Wu, Y.; Goh, D.; Tan, V.; Ng, C.W.; Lim, J.C.T.; Lau, M.C.; Yeong, J.P.S. Application of Artificial Intelligence to In Vi Tumor Modeling and Characterization of the Tumor Microenvironment. *Adv. Healthc. Mater.* **2023**, *12*, 2202457. [CrossR [PubMed]
84. Janiesch, C.; Zschech, P.; Heinrich, K. Machine learning and deep learning. *Electron. Mark.* **2021**, *31*, 685–695. [CrossRef]
85. Badillo, S.; Banfai, B.; Birzele, F.; Davydov, I.I.; Hutchinson, L.; Kam-Thong, T.; Siebourg-Polster, J.; Steiert, B.; Zhang, J.D. Introduction to Machine Learning. *Clin. Pharmacol. Ther.* **2020**, *107*, 871–885. [CrossRef]
86. Vamathevan, J.; Clark, D.; Czodrowski, P.; Dunham, I.; Ferran, E.; Lee, G.; Li, B.; Madabhushi, A.; Shah, P.; Spitzer, M.; et Applications of machine learning in drug discovery and development. *Nat. Rev. Drug Discov.* **2019**, *18*, 463–477. [CrossR [PubMed]
87. Li, Z.; Liu, F.; Yang, W.; Peng, S.; Zhou, J. A Survey of Convolutional Neural Networks: Analysis, Applications, and Prospe *IEEE Trans. Neural Netw. Learn. Syst.* **2022**, *33*, 6999–7019. [CrossRef] [PubMed]
88. Benning, L.; Peintner, A.; Finkenzeller, G.; Peintner, L. Automated spheroid generation, drug application and efficacy screeni using a deep learning classification: A feasibility study. *Sci. Rep.* **2020**, *10*, 11071. [CrossRef] [PubMed]
89. Kshatri, S.S.; Singh, D. Convolutional Neural Network in Medical Image Analysis: A Review. *Arch. Comput. Methods Eng.* **20** *30*, 2793–2810. [CrossRef]
90. D'Orazio, M.; Corsi, F.; Mencattini, A.; Di Giuseppe, D.; Colomba Comes, M.; Casti, P.; Filippi, J.; Di Natale, C.; Ghibelli, Martinelli, E. Deciphering Cancer Cell Behavior From Motility and Shape Features: Peer Prediction and Dynamic Selection Support Cancer Diagnosis and Therapy. *Front. Oncol.* **2020**, *10*, 580698. [CrossRef]
91. Mencattini, A.; Di Giuseppe, D.; Comes, M.C.; Casti, P.; Corsi, F.; Bertani, F.R.; Ghibelli, L.; Businaro, L.; Di Natale, C.; Parri M.C.; et al. Discovering the hidden messages within cell trajectories using a deep learning approach for in vitro evaluation cancer drug treatments. *Sci. Rep.* **2020**, *10*, 7653. [CrossRef]
92. Cortesi, M.; Liu, D.; Powell, E.; Barlow, E.; Warton, K.; Ford, C.E. Accurate Identification of Cancer Cells in Complex Pre-Clini Models Using a Deep-Learning Neural Network: A Transfection-Free Approach. *Adv. Biol.* **2024**, *8*, 2400034. [CrossRef]
93. Ren, S.; He, K.; Girshick, R.; Sun, J. Faster R-CNN: Towards Real-Time Object Detection with Region Proposal Networks. *IE Trans. Pattern Anal. Mach. Intell.* **2017**, *39*, 1137–1149. [CrossRef]
94. Li, Y.; Mahjoubfar, A.; Chen, C.L.; Niazi, K.R.; Pei, L.; Jalali, B. Deep Cytometry: Deep learning with Real-time Inference in C Sorting and Flow Cytometry. *Sci. Rep.* **2019**, *9*, 11088. [CrossRef]
95. Oliver, C.R.; Westerhof, T.M.; Castro, M.G.; Merajver, S.D. Quantifying the Brain Metastatic Tumor Micro-Environment using Organ-On-A Chip 3D Model, Machine Learning, and Confocal Tomography. *J. Vis. Exp.* **2020**, 61654. [CrossRef]
96. Oliver, C.R.; Altemus, M.A.; Westerhof, T.M.; Cheriyan, H.; Cheng, X.; Dziubinski, M.; Wu, Z.; Yates, J.; Morikawa, A.; He J.; et al. A platform for artificial intelligence based identification of the extravasation potential of cancer cells into the bra metastatic niche. *Lab. Chip* **2019**, *19*, 1162–1173. [CrossRef] [PubMed]

7. Rodríguez-Hernández, M.A.; Chapresto-Garzón, R.; Cadenas, M.; Navarro-Villarán, E.; Negrete, M.; Gómez-Bravo, M.A.; Victor, V.M.; Padillo, F.J.; Muntané, J. Differential effectiveness of tyrosine kinase inhibitors in 2D/3D culture according to cell differentiation, p53 status and mitochondrial respiration in liver cancer cells. *Cell Death Dis.* **2020**, *11*, 339. [CrossRef]
3. Yoo, J.; Choi, J.W.; Kim, E.; Park, E.-J.; Baek, A.; Kim, J.; Jeong, M.S.; Cho, Y.; Lee, T.G.; Heo, M.B. Evaluating cell growth and hypoxic regions of 3D spheroids via a machine learning approach. *Mach. Learn. Sci. Technol.* **2024**, *5*, 035063. [CrossRef]
9. Kong, J.; Lee, H.; Kim, D.; Han, S.K.; Ha, D.; Shin, K.; Kim, S. Network-based machine learning in colorectal and bladder organoid models predicts anti-cancer drug efficacy in patients. *Nat. Commun.* **2020**, *11*, 5485. [CrossRef] [PubMed]
00. Tang, M.; Jiang, S.; Huang, X.; Ji, C.; Gu, Y.; Qi, Y.; Xiang, Y.; Yao, E.; Zhang, N.; Berman, E.; et al. Integration of 3D bioprinting and multi-algorithm machine learning identified glioma susceptibilities and microenvironment characteristics. *Cell Discov.* **2024**, *10*, 39. [CrossRef]
01. Sammut, S.-J.; Crispin-Ortuzar, M.; Chin, S.-F.; Provenzano, E.; Bardwell, H.A.; Ma, W.; Cope, W.; Dariush, A.; Dawson, S.-J.; Abraham, J.E.; et al. Multi-omic machine learning predictor of breast cancer therapy response. *Nature* **2022**, *601*, 623–629. [CrossRef] [PubMed]
02. Boehm, K.M.; Aherne, E.A.; Ellenson, L.; Nikolovski, I.; Alghamdi, M.; Vázquez-García, I.; Zamarin, D.; Long Roche, K.; Liu, Y.; Patel, D.; et al. Multimodal data integration using machine learning improves risk stratification of high-grade serous ovarian cancer. *Nat. Cancer* **2022**, *3*, 723–733. [CrossRef] [PubMed]
03. Adam, G.; Rampášek, L.; Safikhani, Z.; Smirnov, P.; Haibe-Kains, B.; Goldenberg, A. Machine learning approaches to drug response prediction: Challenges and recent progress. *NPJ Precis. Oncol.* **2020**, *4*, 19. [CrossRef] [PubMed]
04. Huang, C.; Clayton, E.A.; Matyunina, L.V.; McDonald, L.D.; Benigno, B.B.; Vannberg, F.; McDonald, J.F. Machine learning predicts individual cancer patient responses to therapeutic drugs with high accuracy. *Sci. Rep.* **2018**, *8*, 16444. [CrossRef] [PubMed]
05. Zhavoronkov, A.; Vanhaelen, Q.; Oprea, T.I. Will Artificial Intelligence for Drug Discovery Impact Clinical Pharmacology? *Clin. Pharmacol. Ther.* **2020**, *107*, 780–785. [CrossRef] [PubMed]
06. Stokes, J.M.; Yang, K.; Swanson, K.; Jin, W.; Cubillos-Ruiz, A.; Donghia, N.M.; MacNair, C.R.; French, S.; Carfrae, L.A.; Bloom-Ackermann, Z.; et al. A Deep Learning Approach to Antibiotic Discovery. *Cell* **2020**, *180*, 688–702.e13. [CrossRef] [PubMed]
07. Bonner, S.; Barrett, I.P.; Ye, C.; Swiers, R.; Engkvist, O.; Hoyt, C.T.; Hamilton, W.L. Understanding the performance of knowledge graph embeddings in drug discovery. *Artif. Intell. Life Sci.* **2022**, *2*, 100036. [CrossRef]
08. Artrith, N.; Butler, K.T.; Coudert, F.-X.; Han, S.; Isayev, O.; Jain, A.; Walsh, A. Best practices in machine learning for chemistry. *Nat. Chem.* **2021**, *13*, 505–508. [CrossRef] [PubMed]

cancers

MDPI

Correction

Correction: Momoli et al. The Evolution of Anticancer 3D In Vitro Models: The Potential Role of Machine Learning and AI in the Next Generation of Animal-Free Experiments. *Cancers* 2025, *17*, 700

Carolina Momoli †, Beatrice Costa †, Lorenzo Lenti, Matilde Tubertini, Marco Daniele Parenti, Elisa Martella *, Greta Varchi * and Claudia Ferroni

Institute for the Organic Synthesis and Photoreactivity—Italian National Research Council, 40129 Bologna, Italy; carolina.momoli@isof.cnr.it (C.M.); beatrice.costa@isof.cnr.it (B.C.); lorenzo.lenti@isof.cnr.it (L.L.); matilde.tubertini@isof.cnr.it (M.T.); marcodaniele.parenti@isof.cnr.it (M.D.P.); claudia.ferroni@isof.cnr.it (C.F.)

* Correspondence: elisa.martella@isof.cnr.it (E.M.); greta.varchi@isof.cnr.it (G.V.)

† These authors contributed equally to this work.

Addition of Authorship

Beatrice Costa was not included as a first co-author in the original publication [1]. The authors state that the scientific conclusions are unaffected. This correction was approved by the Academic Editor. The original publication has also been updated.

Reference

1. Momoli, C.; Costa, B.; Lenti, L.; Tubertini, M.; Parenti, M.D.; Martella, E.; Varchi, G.; Ferroni, C. The Evolution of Anticancer 3D In Vitro Models: The Potential Role of Machine Learning and AI in the Next Generation of Animal-Free Experiments. *Cancers* **2025**, *17*, 700. [CrossRef] [PubMed]

cancers

MDPI

view

ual-Action Therapeutics: DNA Alkylation and Antimicrobial eptides for Cancer Therapy

lia María Curieses Andrés [1], José Manuel Pérez de la Lastra [2,*], Elena Bustamante Munguira [1], lia Andrés Juan [3] and Eduardo Pérez-Lebeña [4]

1 Hospital Clínico Universitario de Valladolid, Avenida de Ramón y Cajal, 3, 47003 Valladolid, Spain; cmcuriesesa@saludcastillayleon.es (C.M.C.A.); ebustamante@saludcastillayleon.es (E.B.M.)
2 Institute of Natural Products and Agrobiology, CSIC-Spanish Research Council, Avda. Astrofísico Fco. Sánchez, 3, 38206 La Laguna, Spain
3 Cinquima Institute and Department of Organic Chemistry, Faculty of Sciences, Valladolid University, Paseo de Belén, 7, 47011 Valladolid, Spain; celia.andres.juan@uva.es
4 Sistemas de Biotecnología y Recursos Naturales, 47625 Valladolid, Spain; info@glize.eu
* Correspondence: jm.perezdelalastra@csic.es

Simple Summary: Conventional cancer treatments, based on chemotherapy and radiotherapy, are often effective but suffer from serious side effects and a potential risk of resistance. Dual therapies, combining DNA alkylating agents and antimicrobial peptides, are generating great interest. Within chemotherapies, a frequently used mechanism is DNA alkylation, inducing DNA damage and subsequent cell death. Antimicrobial peptides, in turn, have demonstrated their efficacy as anticancer agents due to their ability to selectively alter cancer cell membranes. In this review, our aim has been to explore the synergistic potential of these two therapeutic modalities when used together.

Abstract: Cancer remains one of the most difficult diseases to treat, requiring continuous research into innovative therapeutic strategies. Conventional treatments such as chemotherapy and radiotherapy are effective to a certain extent but often have significant side effects and carry the risk of resistance. In recent years, the concept of dual-acting therapeutics has attracted considerable attention, particularly the combination of DNA alkylating agents and antimicrobial peptides. DNA alkylation, a well-known mechanism in cancer therapy, involves the attachment of alkyl groups to DNA, leading to DNA damage and subsequent cell death. Antimicrobial peptides, on the other hand, have been shown to be effective anticancer agents due to their ability to selectively disrupt cancer cell membranes and modulate immune responses. This review aims to explore the synergistic potential of these two therapeutic modalities. It examines their mechanisms of action, current research findings, and the promise they offer to improve the efficacy and specificity of cancer treatments. By combining the cytotoxic power of DNA alkylation with the unique properties of antimicrobial peptides, dual-action therapeutics may offer a new and more effective approach to fighting cancer.

Keywords: chemotherapy; alkylating agents; DNA alkylation; cancer; DNA damage and mutations

1. Introduction

Chemotherapy uses drugs to kill cancer cells. Chemical compounds are used to selectively shrink or destroy a tumor or at least limit its growth. Compared to surgery and radiotherapy, chemotherapy has the advantage that the drugs are distributed in most tissues of the body after oral or intravenous administration and can, therefore, also destroy tumor cells that metastasize or are located in protected areas such as the brain [1].

Today, many types of chemotherapy or chemotherapeutic drugs are used to fight cancer, which can be administered as combination treatments or individually or together with other treatments. These drugs vary greatly in their chemical composition, how they are prescribed and administered, how effective they are in treating certain types of cancer, and what side effects they may have. Cancer encompasses a large group of diseases characterized by the development of abnormal cells that divide and grow uncontrollably in any part of the body [2].

The chemical compounds used in chemotherapy are cytotoxic, meaning they are designed to kill cancer cells, which have the property of multiplying faster than normal cells, but they also act on normal tissue, meaning they are able to stop the replication or growth of cancer cells and cause their death. A specific type of drug is used, with the number of treatments and the length of treatment depending on the type of cancer and the patient's particular circumstances [3].

Cancer drugs do not all work in the same way, and it is also important to consider the side effects they can cause depending on the dose being administered. This leads to side effects that vary from patient to patient. The most common are bone marrow suppression, hair loss, fatigue, nausea, and vomiting. These side effects reflect the fact that the cells of the bone marrow, gastrointestinal tract, and hair follicles divide faster than most normal tissues [4].

Cancer treatment usually involves more than one approach, and the strategy chosen depends on the type of cancer, the presence of a biomarker, and the stage of tumor progression. Some cancer patients require only one type of treatment, while others may need several, administered at different times and in specific combinations [5].

A comparison between targeted therapy and chemotherapy shows that both use drugs to attack and destroy cancer cells. However, targeted therapy is designed to act specifically on proteins with mutations that are unique to cancer cells, reducing the potential damage to normal tissue. Targeted therapy is often used in combination with other treatments. Currently, it is not an ideal method for eliminating cancer, as it still requires the use of strong chemicals that can also cause side effects such as hair and skin problems or high blood pressure [6].

2. Types of Chemotherapy Drugs

The anticancer drugs currently in use and under development act via different mechanisms to interfere with a number of processes, such as tumor cell growth, motility, survival, or angiogenesis. Chemotherapeutic drugs can be categorized into groups according to their mode of action, chemical structure, and interactions with other drugs. Some drugs act in more than one way and may belong to more than one group.

DNA-damaging agents, such as alkylating agents, platinum derivatives, anthracyclines, and camptothecins, are some of the most commonly used drugs in chemoattractants. Knowledge of ligand–DNA interactions is a key step in the development of new drugs. It is necessary to determine their DNA sequence specificity, the mode of binding to DNA, and the kinetic, dynamic, and structural parameters of binding [7].

The drugs used to fight cancer are called antineoplastic or cytotoxic. They have different mechanisms of action based on agents that are able to mimic properties of biomolecules to disrupt growth or destroy cells and are characterized by being non-selective, i.e., they act on both cancer cells and healthy cells [2].

To classify anticancer drugs, the following characteristics can be considered: (i) chemical structure, (ii) function, and (iii) interaction with other drugs [8]. The treatments used are based on the action of the pharmacological groups shown in Figure 1.

Different antineoplastic drugs can act on one or more phases of the cell cycle or on the mechanisms controlling cell proliferation. The same drug may have more than one mode of action on the tumor cell, although usually one of them predominates. The classification currently used is based on the target or site of action of the antineoplastic [9,10] and is shown in Table 1.

Within the antineoplastic drugs, alkylating agents are commonly used. These treatments are able to intervene in one or more phases of the cell cycle or in the mechanisms that control cell proliferation, and at least two drugs are used to act on different therapeutic targets and avoid the emergence of resistance.

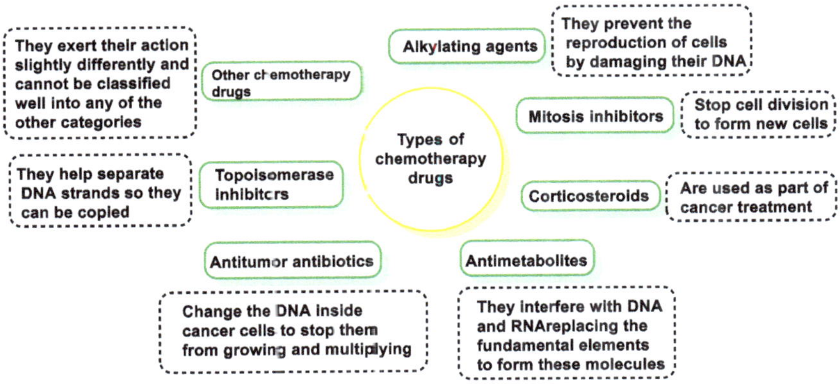

Figure 1. Classification of oncological drugs and mechanism of action of each drug type.

Table 1. Classification of antineoplastic drugs according to the target and mechanisms of action.

Category	Mechanism of Action	Examples
Alkylating Agents	Act on the entire reproduction process; most effective during the DNA synthesis phase (S).	Derivatives of Nitrogen Mustards: cyclophosphamide, chlorambucil, ifosfamide. Tetrazenes–Triazenes: dacarbazine. Derivatives of Platinum: cisplatin, carboplatin, oxaliplatin.
Medications that Affect the Entire Process of Reproduction	Interfere with various stages of the cell cycle and reproduction process.	Nitrosureas: carmustine (bcnu), lomustine (ccnu), estramustine.
Antitumor Antibiotics	Interfere with DNA duplication and alter the membrane surrounding the cells.	Anthracyclines: doxorubicin, daunorubicin, epirrubicin, idarubicin, mitoxantrone. Others: bleomycin, mitomycin c, dactinomycin.
Antimetabolites	Act on the S phase (DNA synthesis), interfere with DNA and RNA.	Folic Acid Analogues: methotrexate. Pyrimidine Analogues: fluorouracil, cytarabine, gemcitabine, capecitabine. Purine Analogues: fludarabine, mercaptopurine, cladribine.
Campotecin Derivatives	Inhibit topoisomerase I, leading to DNA damage.	Irinotecan, topotecan.
Mitotic Inhibitors	Act during the M phase (mitosis), inhibit or stop mitosis, or inhibit enzymes needed for cell reproduction.	Vinca Alkaloids: vincristine, vinblastine, vindesine, vinorelbine. Taxoids: paclitaxel, docetaxel. Epipodophyllotoxins: etoposide, teniposide.
Other Mechanisms of Action	Affect tumors with hormone dependence or stimulate the immune system to attack cancer cells.	Hormones: diethylbestrol. Antiandrogens: flutamide, nilutamide, cyproterone. Antiestrogens: tamoxifen, anastrazole. Gonadotrophin Agonists: leuprolide, goserelin, buserelin, triptorelin. Progestagens: megestrol, medroxyprogesterone.
Immunotherapy	Stimulates the immune system's natural defenses to destroy abnormal cells.	Interferons: interferon alpha 2a, alpha 2b. Monoclonal Antibodies: rituximab, cetuximab, trastuzumab.

Unfortunately, due to the lack of specificity of cytotoxic treatments, there are a variety of adverse effects, the most common of which are rapid death of healthy cells, bone marrow depression, anemia, nausea, vomiting, alopecia, impaired wound healing, growth failure in children, necrosis, sterility, and possible long-term carcinogenesis [11].

Anthracycline derivatives such as doxorubicin (DOX) are another of the most widely used compounds and are the main component of cancer treatment regimens currently used in clinical practice. However, the precise mechanisms of their action are not fully understood [12].

The anticancer activity of DOX may be attributed to the drug's ability to intercalate into DNA, inhibit topoisomerase II, alter mitochondrial function, and enhance free radical generation and oxidative damage. The precise mechanisms of action of DOX are complex and still relatively unknown [13–15].

Like other anthracyclines, DOX intercalates DNA through the formation of hydrogen bonds with guanines in adjacent GC base pairs [16,17].

DOX induces DNA damage through three main mechanisms: DNA adduct formation, single-strand break (SSB) induction, and double-strand break (DSB) induction, in which DNA strands remain attached to trapped topoisomerase enzymes through DNA–protein cross-linking (DPC) and DOX intercalation into the DNA molecule. One possible mechanism postulated to explain the effect of DOX in malignant cells is that DOX intercalation into DNA unwinds the molecule and results in positive supercoiling of the DNA helix [18,19]. There is also evidence that DOX increases the turnover of nucleosomes surrounding active gene promoters due to its intercalation and induces changes in DNA topology [20]. It is possible that the DNA unwinding that occurs as a result of DOX intercalation provides a significant amount of positive torsional stress that destabilizes nucleosomes [21].

Topoisomerase II inhibitors are among the most effective antitumor drugs used in the treatment of human tumors [22].

Lemke K. et al., 2005, have described the antitumor drug 5-dimethylaminopropylamino-8-hydroxytriazoloacridinone, C-1305, which belongs to the group of acridine derivatives, binds to DNA by intercalation, and is an inhibitor of DNA topoisomerase II [23]. C-1305 induces structural perturbations in regions of DNA with three adjacent guanine residues. C-1305 binds to DNA by intercalation and induces unusual structural changes in regions of DNA with guanine triplets.

A study of direct C-1305-mediated microtubule stabilization as the central mediator leading to apoptosis was performed in 2020. Furthermore, C-1305 was found to promote G2 cell cycle arrest by modulating gene expression, indicating that C-1305 is the first microtubule-stabilizing agent identified together with topoisomerase II inhibitory activity [24].

3. Binding of a Ligand to DNA

The binding of a ligand to DNA can be classified into covalent and non-covalent binding (Figure 2).

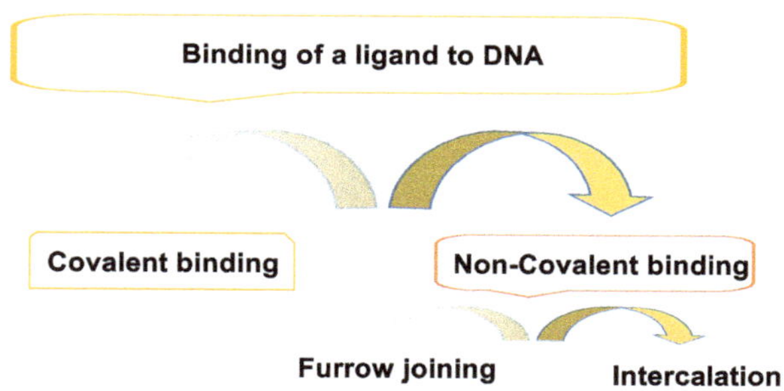

Figure 2. Modes of ligand binding to DNA. DNA can bind to small molecules or drugs through covalent or non-covalent interactions. Covalent binding in DNA can be irreversible and can lead to inhibition of all DNA processes that subsequently lead to cell death. Covalent interactions lead to permanent changes in the structure of nucleic acids. Non-covalent interaction of molecules with DNA can be due to electrostatic interaction, intercalation, and groove binding.

3.1. Non-Covalent Bond

Non-covalent bonds are hydrogen bonds, electrostatic interactions, or non-covalent bonds with ethidium bromide and quinacrine. Non-covalent binding can occur in different ways, e.g., groove binding and intercalation. Agents that interact with non-covalent DNA (groove, intercalators, and external binders) are generally considered less cytotoxic than agents that produce covalent DNA adducts and other types of DNA damage. The non-covalent mode of binding is reversible and is generally preferred over covalent adduct formation in terms of drug metabolism and toxic side effects. In addition, drugs that interact with non-covalent DNA can alter DNA conformation, torsionally stress DNA, disrupt protein–DNA interactions, and potentially lead to DNA strand breaks [25].

3.2. Covalent Bonding

The first cancer chemotherapeutics were those that formed a covalent bond by alkylating the nitrogenous bases of DNA. This binding is irreversible and leads to a complete inhibition of DNA processes and subsequent cell death [26].

The alkylation of adenine and guanine at the nitrogen atom and of guanine at the oxygen atom can easily be caused by alkylating chemicals. If these errors are not corrected by DNA repair processes, cell mutations can occur. In medicine, DNA alkylation is used beneficially in the treatment of cancer. Alkylating chemicals impair DNA replication and can, therefore, cause cell death. This effect is particularly pronounced in rapidly dividing cells, such as cancer cells [27].

4. DNA Alkylating Agents

Alkylating agents prevent the proliferation of cells by damaging their DNA. These drugs work at all stages of the cell cycle and are used to treat many different types of cancer, including lung, breast, and ovarian cancer, as well as leukemia, lymphoma, Hodgkin's disease, multiple myeloma, and sarcoma.

As these drugs damage DNA, they can affect the bone marrow cells that form new blood cells. The risk of leukemia from alkylating agents is dose-dependent, i.e., the risk is lower at lower doses but increases as the total amount of the drug is increased. The risk of leukemia after administration of alkylating agents is highest about 5 to 10 years after treatment.

The characteristic toxicities of alkylating agents are (i) hematopoietic toxicity, (ii) gastrointestinal toxicity, and (iii) gonadal and central nervous system (CNS) toxicity. However, each of these agents has a characteristic set of toxicities that are determined by the reactivity, metabolism, and distribution of the drug.

Despite their toxicity, DNA alkylating drugs remain the cornerstone of cancer therapy. DNA-targeting alkylating agents were among the first anticancer drugs and are still the most commonly used in chemotherapy today. Alkylating agents used as anticancer drugs are able to react with DNA and proteins and disrupt cell function at several stages of the cell cycle, either killing the cell or preventing it from growing. The National Cancer Institute defines an alkylating agent as a type of drug that is used to treat cancer by interfering with the cell's deoxyribonucleic acid (DNA) and inhibiting cancer growth. Although alkylating agents can be used to treat most types of cancer, they are usually most useful in treating slow-growing cancers [28]. Alkylating agents are used to treat the types of cancer shown in Figure 3.

Based on the number of alkyl groups in their molecule, alkylating agents are classified as monofunctional, bifunctional, and polyfunctional; the last two are the most effective antitumor agents. Alkylating compounds cause mono-alkylation or can intercalate into DNA. In addition, bis-alkylating compounds can form bridges within a single DNA strand or between two complementary DNA strands (Figure 4), as well as cross-links between DNA and associated proteins [29].

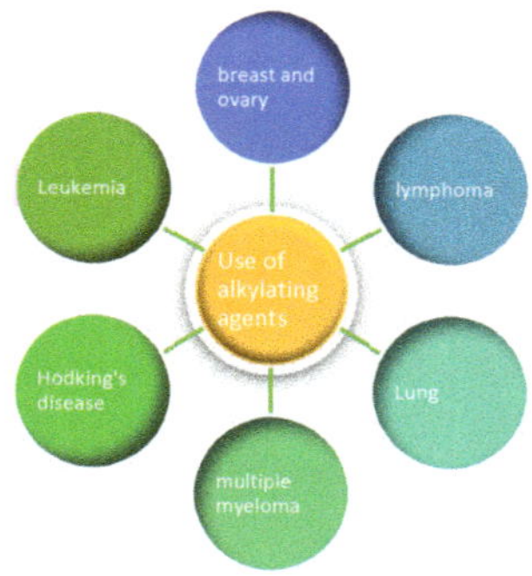

Figure 3. Types of cancers treated with alkylating agents.

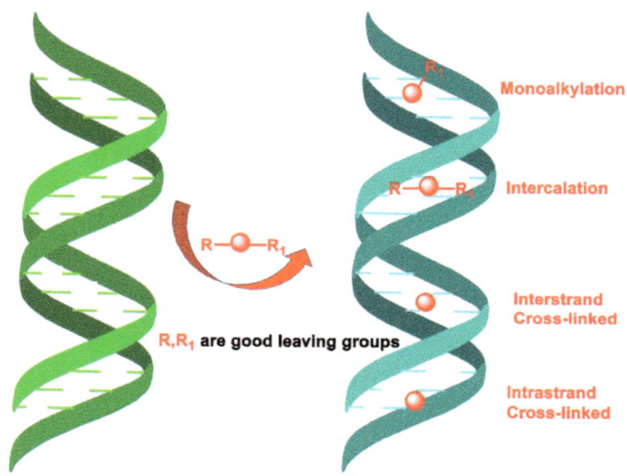

Figure 4. Different types of links produced on DNA by bis-alkylating agents.

The structure and dynamics of DNA are strongly affected by the alkylation of its bases, leading to different types of effects: (i) prevention of DNA replication, (II) DNA fragmentation, (iii) induces nucleotide mismatches by disrupting the normal hydrogen bonds between bases, (iv) bis-alkylation causes intra-catenary cross-linking, (v) cross-linking between two complementary DNA strands preventing their separation during DNA replication or transcription, (vi) cross-linking between DNA and associated proteins. Bifunctional alkylating compounds are more cytotoxic than monofunctional ones. Many anticancer drugs interact with DNA by intercalation, whereby compounds containing aromatic or heteroaromatic rings are inserted between neighboring base pairs perpendicular to the helix axis without altering the overall stacking pattern due to Watson–Crick hydrogen bonding [30,31].

Based on the chemical structure of the reactive groups in antineoplastic drugs, they are categorized into the following groups: nitrogen mustards, nitrosoureas, alkyl sulphonates, triazines, and ethylenimines, and this group of covalent bonds also includes platinum compounds (Figure 5). Alkylating chemotherapies have been used for years and can be effective against many types of cancer. However, because they attack all cells, the side effects can be severe. The goal in using these older alkylating neoplastic agents is to carefully titrate the dose to ensure that the cancer cells are attacked the most while minimizing damage to non-cancerous cells.

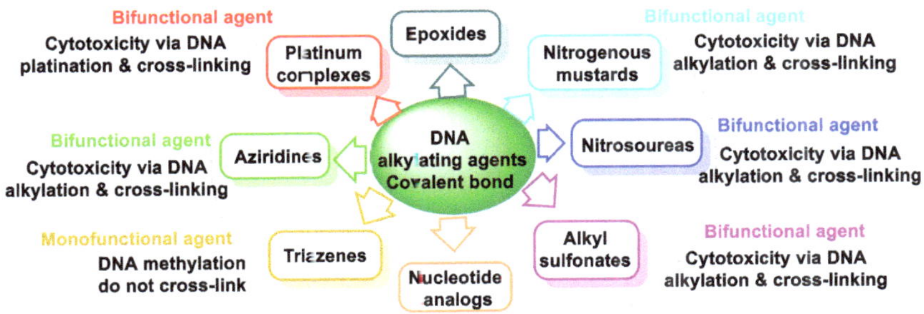

Figure 5. Classification of antineoplastics–alkylating agents according to the reactive groups. They may act as monofunctional agents or may form bifunctional derivatives by forming cross-links (inter- or intra-chain) in DNA or between DNA and proteins.

An alkylating agent is defined as a compound capable of covalently bonding an alkyl group to a biomolecule under physiological conditions (aqueous solution, 37 °C, pH 7.4); DNA alkylating agents interact with resting and proliferating cells at any stage of the cell cycle but are most cytotoxic in late G1 and S phase, as there is insufficient time to repair DNA damage. Alkylating agents, especially nitrogen-based alkylating agents, are commonly used in the treatment of hematologic and solid malignancies. They exert their antineoplastic effect in all phases of the cell cycle and prevent tumor cells from proliferating.

All nitrogen and oxygen atoms of these bases are nucleophilic, with the exception of the nitrogen atoms involved in nucleoside binding (N9 or N1 in purines or pyrimidines), therapeutically useful drugs always behaving as carbon electrophiles. Alkylating agents are electrophilic and bind covalently to electron-rich functional groups, such as the N-7, N-1, and O-6 positions of guanine [32]. Other nucleobases that can be alkylated and the atomic positions where alkylation preferentially occurs are the N-1, N-3, and N-7 positions of adenine, the N-3 position of cytosine, and the O-4 position of thymidine (Figure 6).

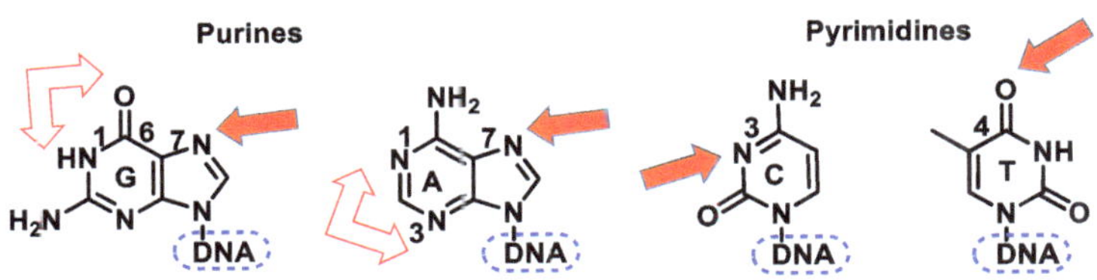

Figure 6. Arrows indicate DNA base alkylation sites.

Nitrogen-based alkylating agents, which are frequently used to treat hematological and solid malignancies, often in combination with other chemotherapeutic agents, exert their antineoplastic effect at all stages of the cell cycle and prevent the replication of tumor cells. During this alkylation process, an alkyl residue is introduced into the DNA bases, a change that leads to fragmentation of the DNA by DNA repair enzymes that attempt to replace the alkylated base [33].

5. Nitrogen Mustards

One of the largest and most diverse groups of drugs are those that affect the structure and function of DNA. Part of the reason that DNA-interfering drugs are the largest group of agents in clinical use today is that nitrogen mustard was the first type of chemotherapeutic agent to be discovered. They have been used for more than 60 years [34]. Nitrogen mustard acts on the kidneys, heart, bladder, central nervous system, and gonads.

After ingestion, the alkylating agent is metabolized to a highly reactive aziridinium derivative that alkylates DNA and inhibits DNA reduplication. There are several generations of nitrogen-based alkylating drugs. The mustard compounds generated from the nitro group contain the bis(2-chloroethyl) group, which is the major chemical constituent, and all have a common mechanism of action, namely the in situ generation of a highly electrophilic aziridinium cation (Figure 7) by an intramolecular nucleophilic substitution reaction [35].

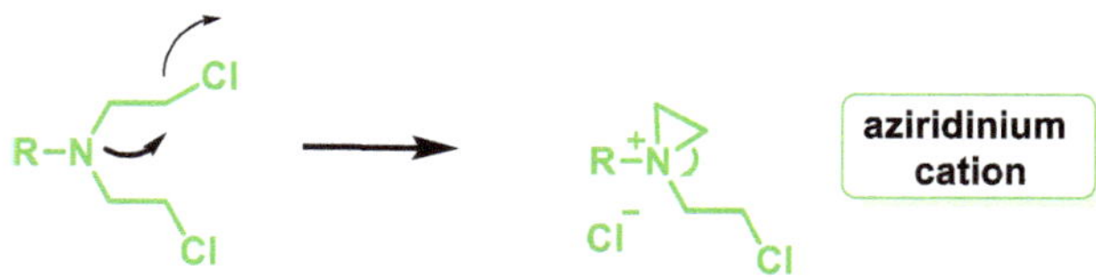

Figure 7. In situ generation of an aziridinium cation.

5.1. *Aliphatic Mustard*

Mechlorethamine (Figure 8) was the first antitumor agent developed during the First World War as part of the mustard gas study. Due to its short half-life, it must be prepared and administered immediately. It can only be administered intravenously. At physiological pH, mechlorethamine is converted via the free base into a relatively stable aziridinium ion, which in turn reacts with available nucleophilic centers. Mechlorethamine has been replaced by new derivatives to improve selectivity and specificity in the fight against cancer cells [36].

Figure 8. Mechlorethamine structure.

5.2. *Phosphamide Nitrogen Mustard*

The oxazaphosphorins have complex pharmacokinetics, and metabolism is crucial for their cytotoxic activity. There are several classes, such as cyclophosphamide, iphosphamide, maphosphamide, trophosphamide, bromophosphamide, and gluphosphamide (Figure 9).

Cyclophosphamide is the most commonly used alkylating agent. It is a prodrug that is converted to its active form after metabolism by cytochrome P-450. Hydroxylation of the carbon at position 4 is the first metabolic reaction of cyclophosphamide and iphosphamide. This reaction occurs in the liver microsomes via the enzyme systems of the 2A6, 2B6, 2C, 3A4, and 3A5 subfamilies of the cytochrome P450 (CYP) enzyme complex. The main metabolite of this reaction is 4-hydroxycyclophosphamide or 4-hydroxyphosphamide, which is in tautomeric equilibrium with its keto form, aldophosphamide. From aldophosphamide, a β-elimination reaction produces phosphoramide mustard with an alkylating effect and acrolein, a urotoxic compound responsible for hemorrhagic cystitis [37] (Figure 10).

Its main toxic effects are myelosuppression for 10–14 days, hair loss, nausea, and vomiting. It can also cause hemorrhagic cystitis due to the effect of some of its metabolites, such as acrolein, on the bladder epithelium; high fluid intake for 24 to 48 h and Sodium-2-mercapto-sulphonate (mesna) to chelate acrolein and avoid this complication. It is a drug widely used in oncology and is part of polychemotherapy regimens, as its efficacy has been demonstrated in various neoplasms (leukemias, lymphomas, breast cancer, ovarian cancer, and sarcomas). It is also part of the main induction regimens prior to bone marrow transplantation.

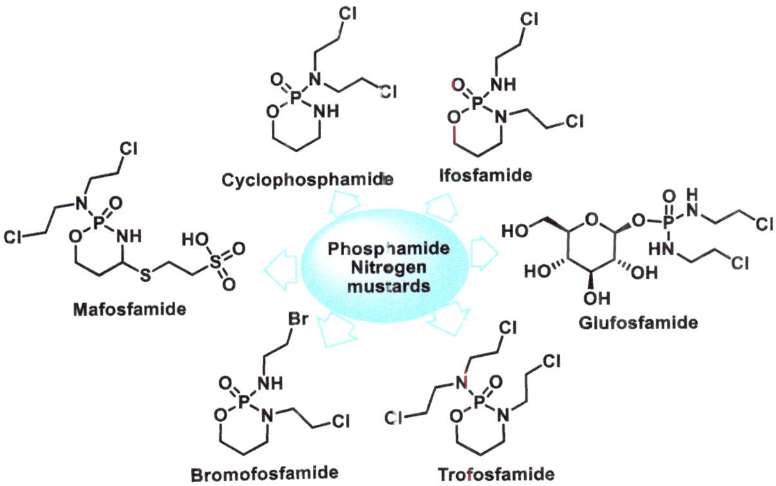

Figure 9. Phosphamide nitrogen mustard.

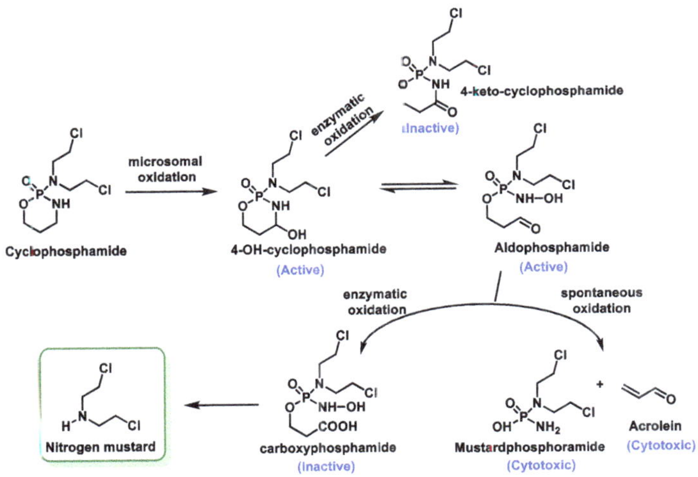

Figure 10. Biotransformation mechanism of cyclophosphamide.

Isophosphamide is a drug similar to cyclophosphamide but requires higher doses to achieve the same antitumor effect. It is only administered intravenously, always with sufficient fluid intake and mesna as a prophylactic measure [38]. Iphosphamide is a prodrug that must be converted into a cytotoxic metabolite (4-hydroxy derivatives) by the action of CYP3A4 isoforms [39]. Iphosphamide generates seven-atom DNA cross-links (compared to five-atom cross-links for cyclophosphamide).

Acrolein, a metabolite of cyclophosphamide and iphosphamide, is a highly toxic aldehyde. As a bifunctional electrophile, acrolein reacts with protein nucleophiles and DNA via Michael addition and Schiff base formation [40] (Figure 11).

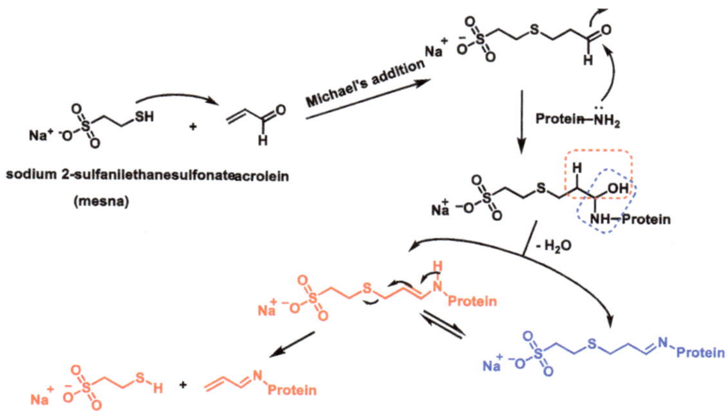

Figure 11. Mechanism of acrolein toxicity.

Maphosphamide is a cyclophosphamide analog, an alkylating agent, and an immunomodulator that induces a pro-inflammatory response and increased endothelial cell permeability in vitro [41]. Maphosphamide breaks down spontaneously into 4-OH-CP and mesna (Figure 12). The released 4-OH-CP is in equilibrium with its tautomer aldophosphamide, which can be degraded by β-elimination to generate cytotoxic phosphoramide mustard and acrolein. Thus, maphosphamide is directly active in cells and intracellular fluids. Although the drug is similar to CP, it does not need to be activated by the hepatic enzyme system [42].

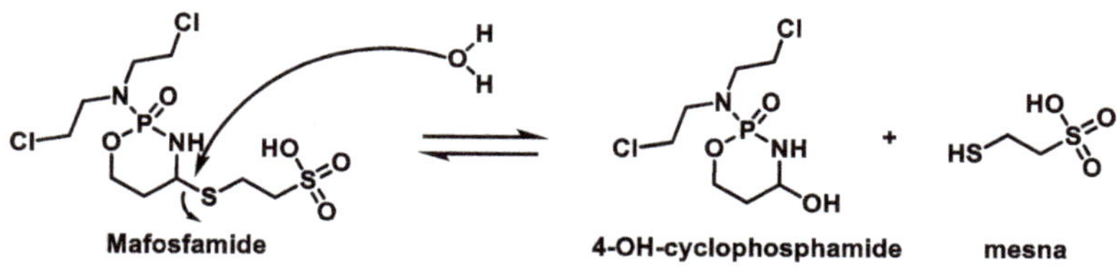

Figure 12. Spontaneous decomposition of Maphosphamide into 4-HO-CP and mesna.

Maphosphamide has significant in vitro activity against leukemia and solid tumor cell lines such as U-937, ML-1, MOLT-4, rhabdomyosarcoma, and MCF-7, and also in vivo against several transplantable mouse tumors, including P388 and L1210 leukemia, B16 melanoma, Lewis lung carcinoma, colon tumor 38, and cyclophosphamide-resistant P388 leukemia.

Trophosphamide was approved as one of the anticancer drugs in 1973. Like phosphamide and cyclophosphamide, they are non-specific cell cycle alkylating agents that are bioactivated by hepatic cytochrome P450 enzymes through either 4-hydroxylation or N-dechloroalkylation; both are NADPH-dependent [43] (Figure 13).

Bromophosphamide, an analog of iphosphamide with a bromine substituent, causes less urotoxicity. It acts by cross-linking DNA strands or deoxyribose–protein cross-links and alkali-labile sites in a concentration-dependent manner in HeLa cells [44].

(S)-(-)-bromophosphamide (Figure 14) is an alkylating anticancer drug introduced into clinical practice in the 1980s and used to treat soft tissue sarcomas and a variety of pediatric tumors [45,46].

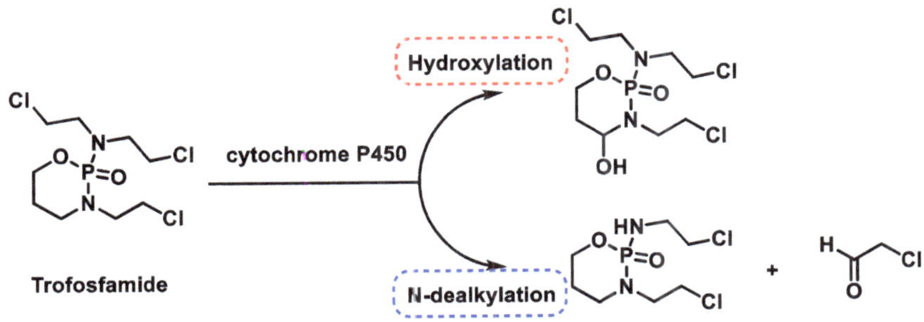

Figure 13. Activation of trophosphamide by cytochrome P450 into oxazaphosphorine mustards. N-dealkylation by cytochrome P450s results in the formation of inactive metabolites and chloracetaldehyde.

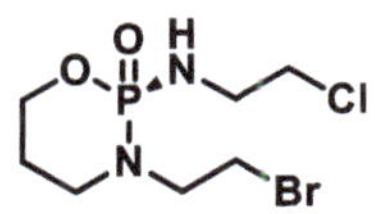

(S)-3-(2-bromoethyl)-2-((2-chloroethyl)amino)-1,3,2-oxazaphosphinane 2-oxide

Figure 14. Chemical structure of (S)-(-)-Bromophosphamide Gluphosphamide (β-D-glucoseisophosphoramide-mustard).

Gluphosphamide, also known as glucophosphamide (β-D-glucose isophosphoramide mustard), is a glycosidic conjugate between β-D-glucose and the active alkylating moiety of the antineoplastic drug iphosphamide. In gluphosphamide, the active part of iphosphamide is linked to a glucose molecule. Inside the cells, the bond between the glucose and the alkylating agent must be broken in order to release the active substance. Phase III trials are currently underway with this drug, comparing gluphosphamide with 5-FU, both in combination with gemcitabine, for the second-line treatment of metastatic pancreatic cancer [47].

5.3. Aromatic Nitrogen Mustards

The main aromatic nitrogen mustards are shown in Figure 15.

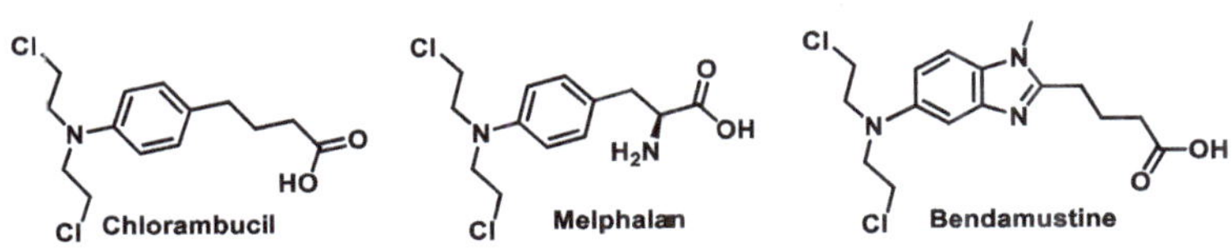

Figure 15. Aromatic nitrogen mustards.

Melphalan is a phenylalanine derivative of nitrogen mustard and was approved by the FDA in 1964 [48]. It can be administered orally or intravenously. Melphalan is 90% bound to plasma proteins and is actively transported into cells primarily via the L-type amino acid transporter system. The oral bioavailability is 50–80%, and the half-life can be up to 8 h. It is partially excreted in the feces and partially in the urine. Melphalan was developed for the treatment of melanoma because it selectively targets tumor cells that actively use tyrosine. However, it has shown little effect in this neoplasm. Currently, its main indication is multiple myeloma. It has been shown to be effective at conventional doses in ovarian cancer and lymphoma and at high doses in breast cancer and acute myeloid leukemia.

Chlorambucil was approved by the FDA in 1957 [48]. It is stable in aqueous solution so that it is almost completely absorbed after administration and is rapidly and completely excreted from the blood. It is metabolized in the liver and has the alkylating metabolite PAAM (phenylacetic acid mustard). Adverse effects are uncommon, with the exception of bone marrow suppression. However, it can cause a severe generalized rash, which may develop into Stevens–Johnson syndrome or toxic epidermal necrolysis. In the event of a rash, further treatment with chlorambucil is contraindicated. This drug is indicated for the palliative treatment of chronic lymphocytic leukemia, malignant lymphoma, lymphosarcoma, follicular giant lymphoma, and Hodgkin's disease [49].

Bendamustine: N-methylbenzimidazole replaced the benzene ring of chlorambucil to form bendamustine. The drug was approved by the FDA in 2008, although it was discovered in 1960. The incorporation of the benzimidazole ring gives the molecule the properties of a purine analog [50].

5.4. Steroid-Coupled Nitrogen Mustards

Alkylating agents have high toxicity and low specificity for alkylating agents, but when coupled with a steroid, they are more effective in improving the therapeutic profile. Steroid-coupled nitrogen mustards utilize the action of both the alkylating agent and a steroid for selective uptake or a combined anticancer effect. The steroid acts as a biological carrier, allowing the nitrogen mustard component to be easily transported and increasing the uptake of the mustard through the lipid bilayer of cells [51]. The result is a more effective, more selective, and less toxic antineoplastic treatment. Saha et al., 2018, have shown that steroid-bound mustard enhances the damaging effect on a specific DNA sequence and achieves better selectivity and lower toxicity compared to nitrogen mustard itself [51,52]. Some of the most important drugs in this class are listed below:

Estramustine phosphate sodium: Estramustine has a β-estradiol moiety attached to a nitrogen mustard moiety (mechlorethamine) via a carbamate bridge. Chemically, it is estra-1,3,5(10)-triene-3,17β-diol-3-bis(2-chloroethyl)-carbamate. Due to the stability of the carbamate bridge, which is not cleaved to release the nitrogen mustard, the compound has no alkylating activity.

In 1986, estramustine sodium phosphate was synthesized by condensation of estradiol with bis(2-chloroethyl) carbamyl chloride, followed by phosphatidylation with phosphorus oxychloride and finally treatment with NaOH in EtOH [53] (Figure 16).

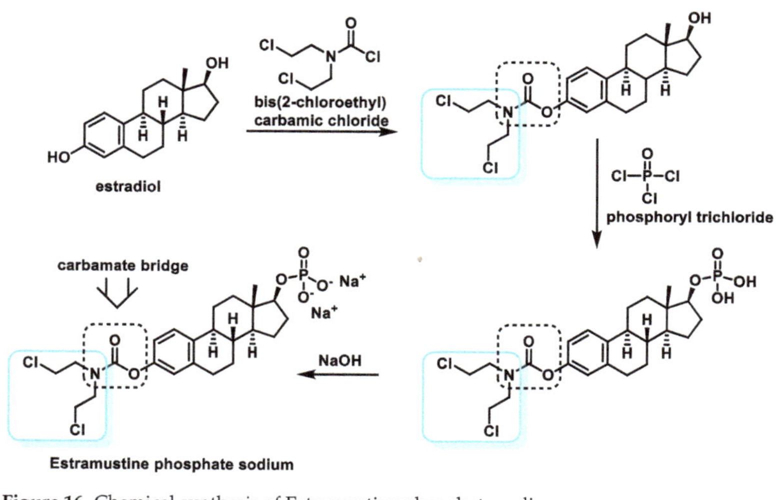

Figure 16. Chemical synthesis of Estramustine phosphate sodium.

The compound is dephosphorylated during absorption and broken down into estradiol, estrone, and estramustine as the main metabolites in the blood. The estradiol component increases the uptake of the alkylating agent into the target cells. Estramustine phosphate is a prodrug marketed for the treatment of advanced prostate cancer.

Prednimustine: The drug was produced from prednisolone ester (corticosteroid) and chlorambucil (Figure 17). It has both alkylating and corticosteroid properties. Acute leukemia, breast cancer, and lymphocytic leukemia are diseases for which this drug is commonly used [54,55].

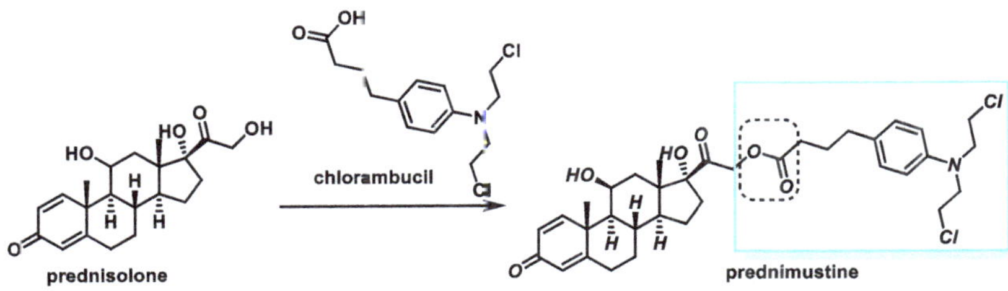

Figure 17. Chemical synthesis of Prednimustine.

Homoazasteroid (lactam steroids): This type of drug is formed by attaching a lactam residue to the esterified form of the alkylating steroid drug. They are effective in various neoplastic tumors and leukemias both in vivo and in vitro. Lactandrate and lactestoxate (Figure 18) showed excellent anticancer results in colon carcinomas. However, the latter showed better results than lactandrate [56]. In 2016, Trafalis et al. reported positive results with four modified forms of lactamsteroids [57].

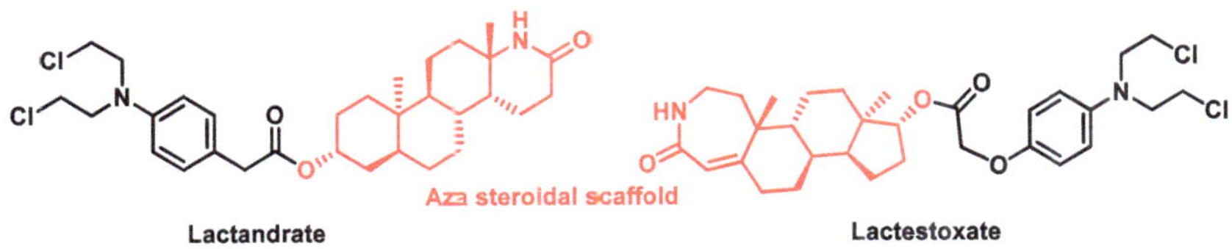

Figure 18. Chemical structure of Lactandrate and Lactestoxate.

5.5. Peptide-Coupled Nitrogen Mustards

Melphalan flufenamide, also known as melphalan flufenamide, is a third-generation drug consisting of the ethyl ester of melphalan and para-fluoro-L-phenylalanine (Figure 19).

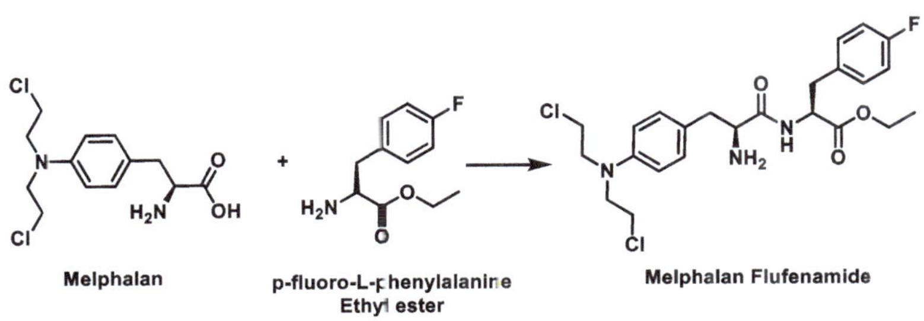

Figure 19. Chemical structure of Melflufen.

Being highly lipophilic, it is readily taken up by rare cells (such as myeloma cells) across the membrane barrier, leading to activation by hydrolytic cleavage of the peptide bond using peptidases and esterases overexpressed in multiple myeloma cells to release a toxic cargo into the cells, damaging the DNA and killing the cancer cells [58]. The alkylating component produced during hydrolysis is melphalan, which enters the cell nucleus and causes DNA alteration (Figure 20).

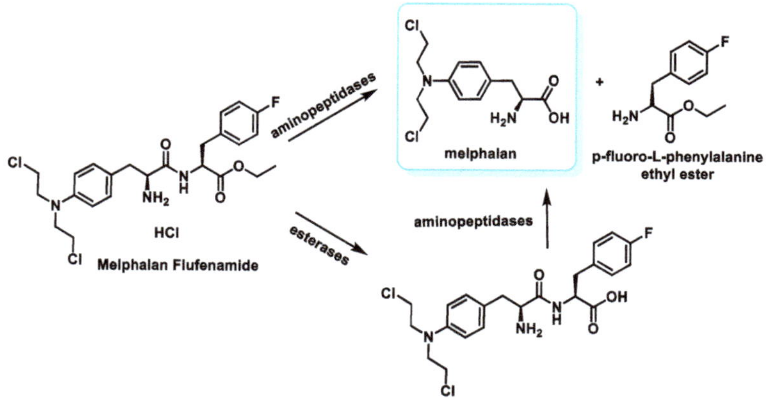

Figure 20. Metabolism of Melflufen hydrochloride.

It has a high affinity for solid tumors and multiple myeloma, and phase I/II trials have led to its successful use over time [51]. The drug is effective in third-line treatment in patients who have not previously received alkylating chemotherapeutic agents and in patients who have received melphalan in previous treatments and who are also unsuitable for transplantation [59].

In comparison between melphalan and melflufen, melflufen has a superior cytotoxic effect, higher efficacy at a lower dose, and is effective in treating multiple myeloma cells in which melphalan was not actively involved [60].

Melflufen received accelerated approval in the HORIZON trial on 26 February 2021 [61]. In March 2021, the FDA approved melflufen in combination with dexamethasone for the treatment of adult patients with multiple myeloma. In October 2021, it was withdrawn from the market in the USA after too many deaths occurred in a confirmatory phase 3 study [61].

5.6. Mode of Action of Nitrogen Mustards

Nitrogen mustard reacts with DNA nucleobases preferentially at the N-7 position of guanine. Their alkylation leads to important changes in the chemical properties of guanine. Since a positive charge is generated on the imidazole ring, the formation of the enolic tautomer is favored, which leads to guanine preferentially pairing with thymine and not with cytosine. In purification processes, hydrolysis leads to local destabilization of the DNA, which would favor strand cleavage. Another consequence is the opening of the imidazole ring, which would also trigger purification (Figure 21). The mechanism of action involves the formation of DNA cross-links, which leads to an inhibition of DNA synthesis and function.

The reason for the high reactivity of nitrogen mustard as an alkylating agent under mild conditions is the formation of an aziridinium cation, which acts as an alkylating agent and is formed by intramolecular nucleophilic substitution. The aziridinium cation is highly reactive due to the positive charge of the N and the tension of the three-membered ring. Since nitrogen mustards are bifunctional alkylating agents, one of their cytotoxicity mechanisms is related to the ability of monoalkylated DNA species to form a covalent intercatalytic DNA strand and intracatalytic cross-links or DNA–protein complexes that lead to disruption of replication or transcription (Figure 22).

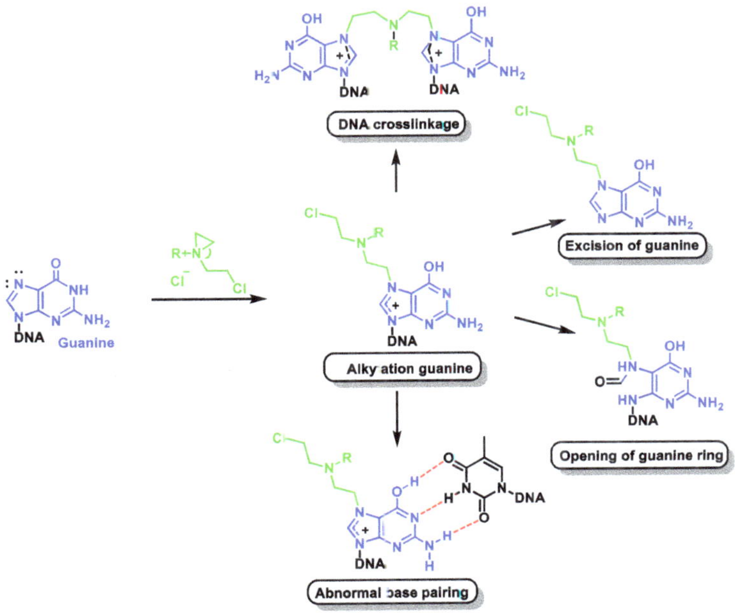

Figure 21. Action of alkylating agents on guanine.

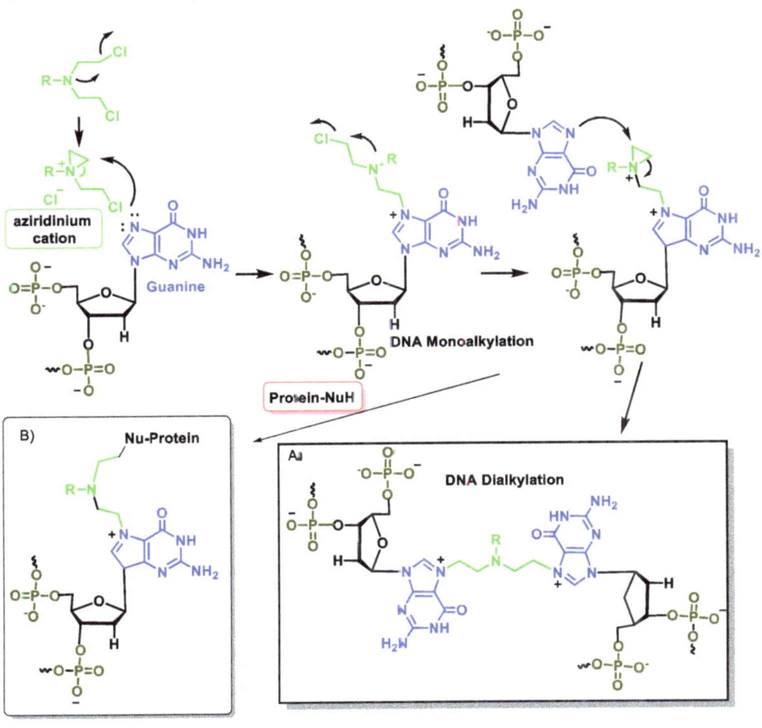

Figure 22. (A) DNA monoalkylation and dialkylation mechanism. (B) DNA–protein complexes.

During alkylation, the normal pairing of the DNA bases between adenine–thymine and guanine–cytosine (Watson–Crick base pairs) is altered. The three hydrogen bonds connecting guanine and cytosine, for example, require the existence of a carbonyl at the C-6 position of the purine derivative. Alkylation at the N-7 position creates a positive charge in this center adjacent to the partial positive charge at C-6, which means that the tautomeric equilibrium shifts to the more stable 6-hydroxy form [62]. This change in the tautomeric form turns the hydrogen bond acceptor groups into hydrogen bond donors and vice versa. This leads to a weakening of the hydrogen bond with cytosine, as only two bonds can be formed, whereas the pairing of the 6-hydroxy species with thymine leads to a more stable complex (three hydrogen bonds). The normal pairing is changed to guanine–non-thymine, which leads to mutations (Figure 23).

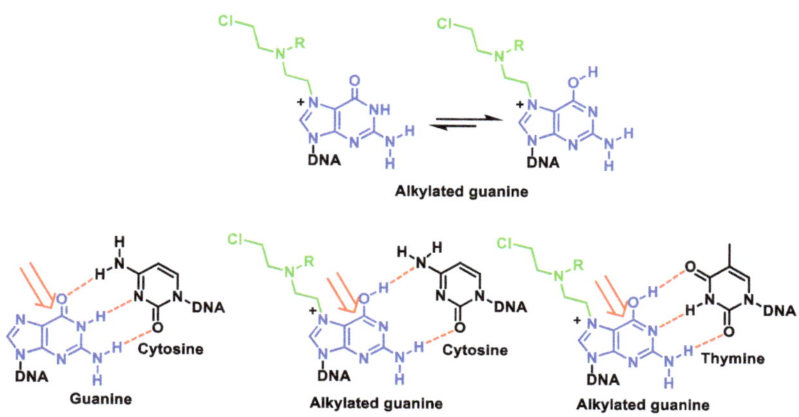

Figure 23. Hydrogen bonding interactions in guanine–cytosine and guanine–thymine pairs.

Another consequence of guanine alkylation is an increase in the electrophilicity of the positions adjacent to or conjugated with the positive charge on N-7, which leads to various hydrolytic reactions that alter the DNA structure. This increased electrophilicity allows cleavage of the heteroside bond (Figure 24) and leads to purification of the DNA.

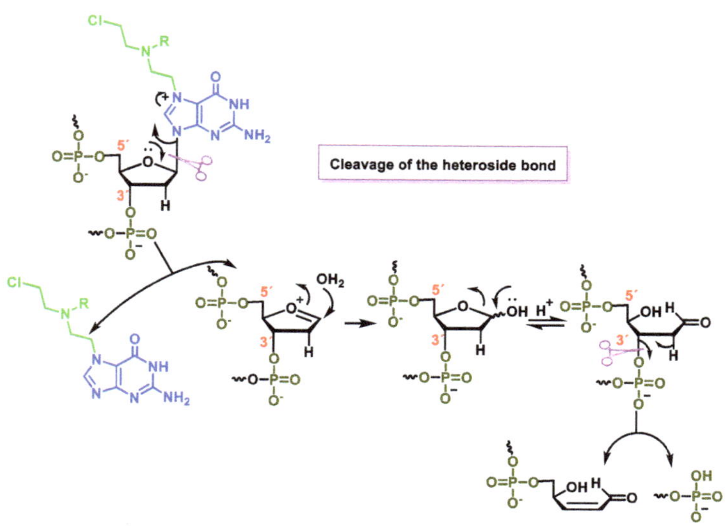

Figure 24. Cleavage of the heteroside bond.

When attacked by water, a cyclic hemiketal forms in equilibrium with the open form, which has a good leaving group (oxygen phosphate) at the β-position with respect to the carbonyl group, facilitating its elimination and leading to DNA fragmentation.

C-8 of the guanine ring becomes more electrophilic as it is close to the positive charge generated in the N-7 alkylation reaction. The addition of water to C-8 provides the hydroxylated intermediate, which opens the purine ring, which then evolves into the imine that develops after hydrolysis to cleave the DNA strand [63] (Figure 25).

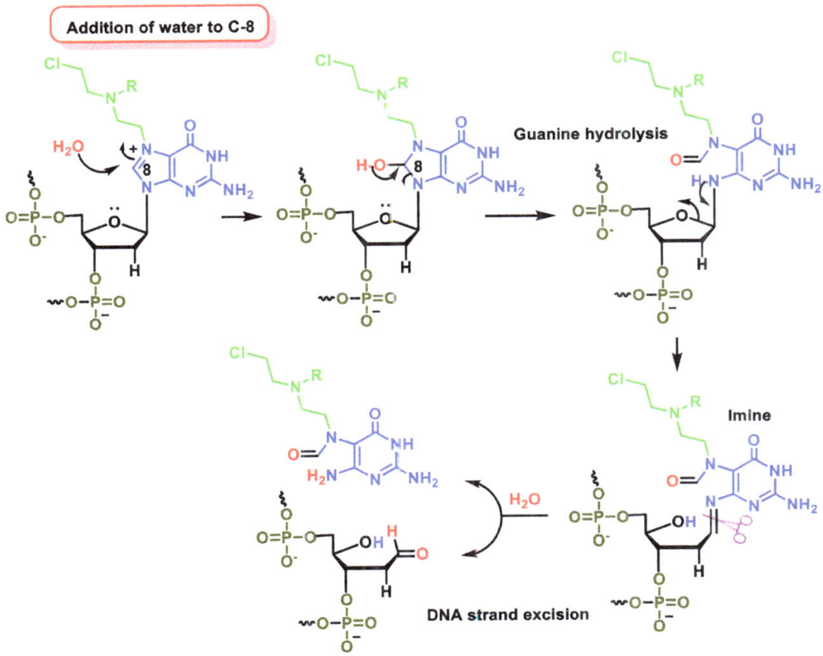

Figure 25. DNA fragmentation that takes place following guanine monoalkylation.

6. Aziridines

Several aziridine derivatives have been described as antitumor agents [64]. It has been found that two aziridine units are required for these compounds to show good activity. It was also found that the presence of a third or fourth aziridine is not required, suggesting that the cytotoxicity is due to a cross-linking mechanism, as observed with nitrogen mustard. Aziridine alkylating agents are characterized by an aziridine ring that is structurally similar to the iminium found in nitrogen mustard. Since they have no charge on the nitrogen, they are less reactive than the nitrogen mustards. It should be noted that the reactivity of the aziridine group increases with protonation and is, therefore, enhanced at low pH. It is assumed that the ring opening of the aziridine ring is responsible for the alkylating activity of the molecule. The mechanism of action of these drugs is less understood. The first compounds of this family to be used are shown in Figure 26.

Thiotepa (tris(1-aziridinyl) phosphine sulfide) is an alkylating agent approved for the treatment of breast cancer, ovarian cancer, and bladder cancer [65–67]. Thiotepa is oxidatively desulfurized by hepatic cytochrome P450 to produce TEPA, a less cytotoxic form of the molecule [68,69]. Both compounds have been found in the plasma of patients undergoing this treatment. Thiotepa is currently still used for the treatment of bladder carcinomas. Position 7 of the guanine is the preferred alkylation site of thiotepa (Figure 27).

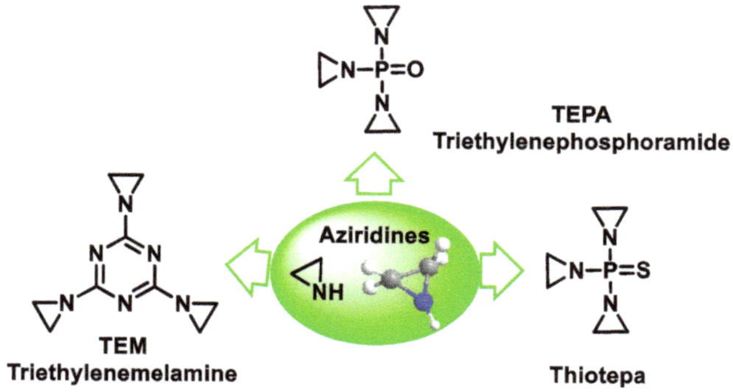

Figure 26. The first compounds derived from aziridines introduced into therapeutics.

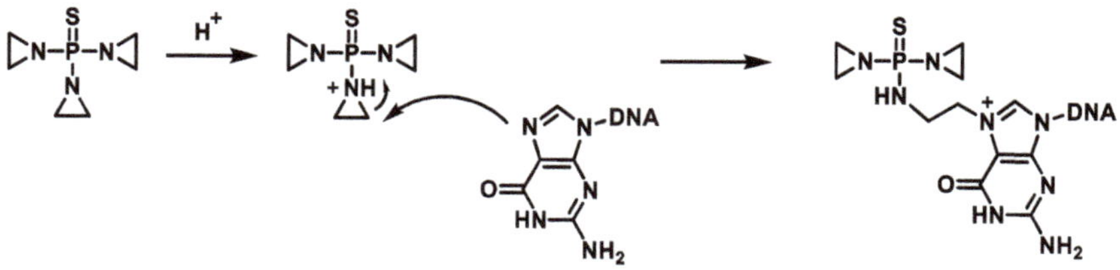

Figure 27. Monoalkylation of guanine by thiotepa.

In vivo and in vitro studies show that DNA alkylation by ThioTEPA can occur via two pathways, but it is not yet clear which pathway is the exact mechanism of action [70] (Figure 28).

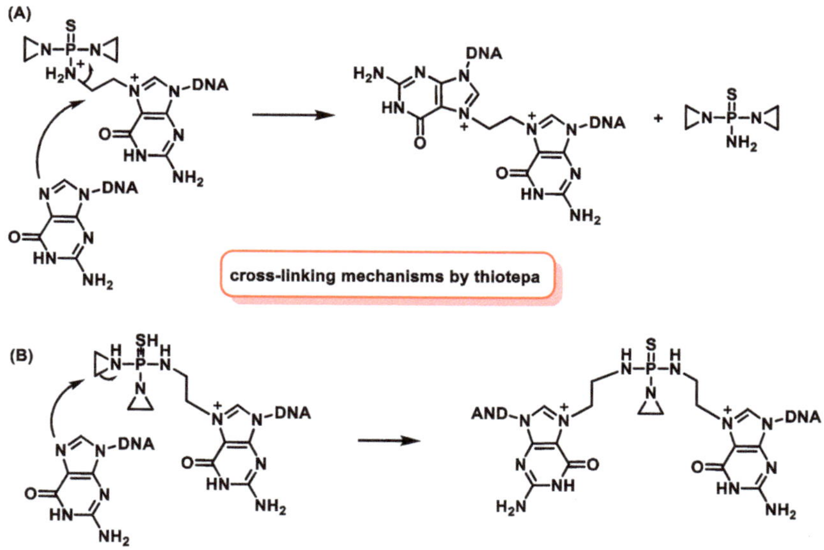

Figure 28. (**A**) alkylation and cross-linking by sequential reaction of a single aziridine group. (**B**) Cross-linking produced by sequential alkylating reactions of two aziridine groups of thiotepa.

6.1. Aziridines Linked to a Benzoquinone System

Alkylating agents containing a quinone moiety require reduction of the quinone structure to activate their alkylating substituents, which are used in the clinic for cancer treatment [71,72].

6.1.1. Natural Aziridinyylbenzoquinone

Mitomycin C (mutamycin) [73] is a natural antitumor quinone from *Streptomyces caespitosus* [74] that contains a quinone unit and an aziridine (Figure 29). It has been used since the 1960s for the treatment of breast cancer and tumors of the gastrointestinal tract and is considered a cytotoxic agent [75]. Mitomycin C was approved by the FDA in 2020 for adult patients with low-grade urothelial carcinoma of the upper tract.

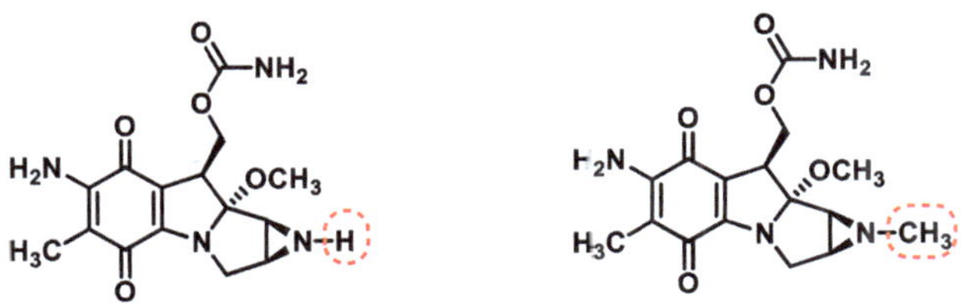

Figure 29. Chemical structure of mitomycins.

Porfyromycin is the N-methyl derivative of mitomycin C and is also a natural product. Porfyromycin has been tested in phase III clinical trials for the treatment of head and neck cancer in combination with radiotherapy with acceptable toxicity and stimulating effect [76].

Mitomycin C and porfyromycin can be considered as prototypes of reductively activated alkylating agents leading to cytotoxic species [77]. Figure 30 includes two reduction steps with one electron transfer to semiquinone and subsequently to hydroquinone. Both species can initiate the reaction cascade leading to DNA alkylation, but the available evidence points to hydroquinone as the active species [78]. The next step would be the formation of the imine derivative by methanol elimination of the hydroquinone. The next step is the formation of an indole derivative, protonation of the aziridine nitrogen of this derivative, and subsequent elimination with simultaneous opening of the aziridine ring yielding methylquinone methylide. A Michael reaction with nucleophilic DNA groups takes place on this intermediate product, in which the N-2 amino group of the guanine or the N-7 position is involved [79]. With mitomycin, removal of the carbamate group generates an electrophilic iminium, which undergoes a second alkylation by attacking a 2-amino group of guanine, leading to DNA cross-linking [80,81] (Figure 30).

Both inter- and intra-strand cross-linking by mitomycin C have been observed, with the former predominating.

6.1.2. Synthetic Aziridinyylbenzoquinone

Antitumor compounds with two or three aziridine rings linked to a benzoquinone system can act as DNA bisalkylators and cross-linking agents and cross the blood–brain barrier due to their high ionization and reductively activated lipophilicity (Figure 31).

Diaziquone (AZQ) Diaziquone (AZQ) [82] was granted orphan drug status by the FDA in the early 1980s but showed no improvement over existing drugs. AZQ is a synthetic, lipophilic aziridine that can cross cell membranes and the blood–brain barrier. This drug has been used in CNS cancer and has shown therapeutic effect in these tumors. AZQ has also been used in the treatment of other solid tumors and leukemia.

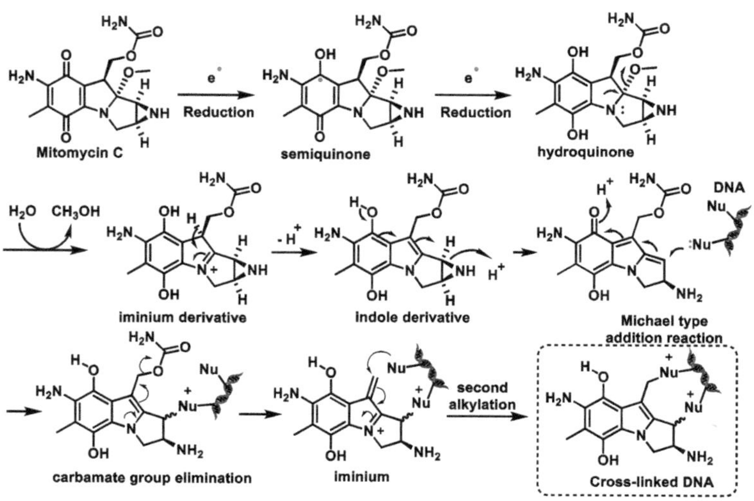

Figure 30. Bioreductive alkylation of DNA by mitomycin C.

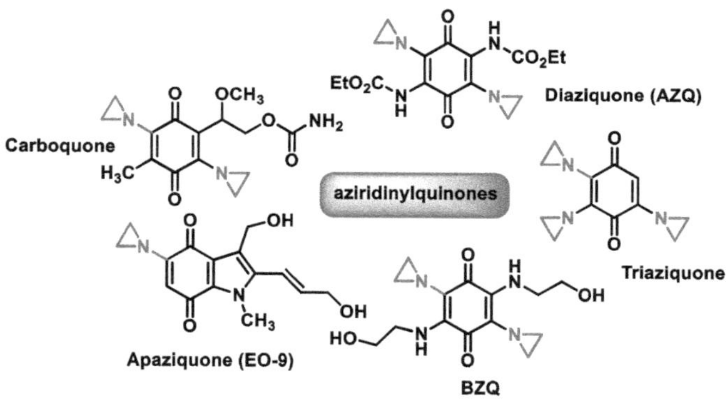

Figure 31. Chemical structure of Aziridinylquinones.

Diazichon (AZQ) and carboquinone (carbazilquinone) have been tested in the treatment of various cancers but show low efficacy and toxicity. Triazichon (Oncovedex) was used clinically for the treatment of cancer in 1960. It is highly toxic to the walls of blood vessels and bone marrow, so it was replaced by other, more important and effective agents [83].

Apaziquone (EO9), a bioreductive drug with only one azidirine moiety in its structure, showed no activity in phase II clinical trials when administered intravenously, but in subsequent studies, it proved to be active and well tolerated in patients with superficial transitional cell carcinoma of the bladder after intravesical administration [84]. Approval was denied in 2016 after poor results in phase III clinical trials.

Aziridinylquinone BZQ was also the subject of several clinical trials in humans [72]. Mitomycin C and diaziquone (AZQ) are currently used in chemotherapy.

6.1.3. Aziridines β-D-Galactopyranosides

Calderón-Montaño et al., 2021, investigated various β-D-galactopyranoside-aziridine derivatives as anticancer agents. The best performing was 2-methyl-2,3-[N-(4-methylbenzenesulphonyl)imino] propyl 2,3-di-O-benzyl-4,6-O-(S)-benzylidene-β-D-galactopyranoside (AzGalp) (Figure 32), which has been shown to induce DNA damage [64].

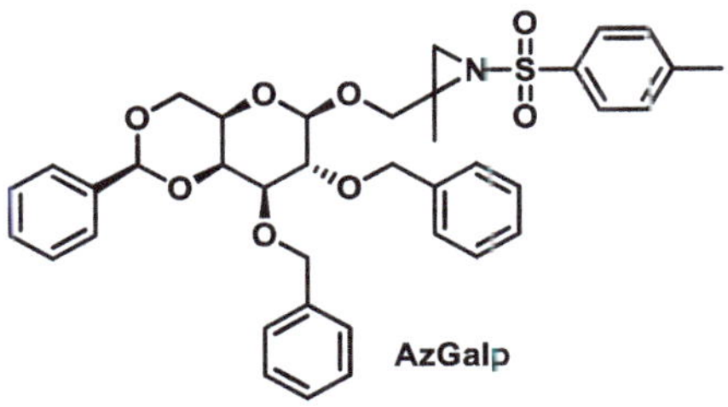

Figure 32. Chemical structure of AzGalp.

6.2. Mechanism of DNA Alkylation by Aziridinylbenzoquinones

The mechanism of DNA alkylation by aziridinylbenzoquinone derivatives involves bioreductive processes [64,85–87] due to the action of one- or two-electron reductases. Aziridinylbenzoquinones enter the cell in an inactive, oxidized form and are reduced to hydroquinone by enzymes in the cell, increasing the basicity of the aziridine groups and enhancing protonation until they become cytotoxic (Figure 33). This DNA damage increases when electron-donating groups are present on the benzoquinone ring, while the presence of electron-withdrawing functions or bulky groups at the C-6 position results in lower potency or inactive compounds [88].

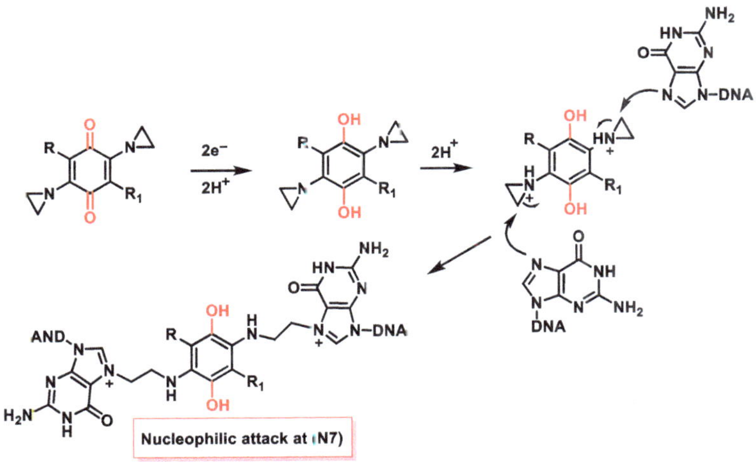

Figure 33. Reduction mechanism and structure of cross-links in DNA by aziridinylbenzoquinone derivatives.

One of these enzymes is DT-diaphorase (DTD, NQO1), a two-electron reductase that is located in the cell nucleus and whose level is elevated in many tumors [89]. The reduction is mainly due to the hypoxic conditions that often prevail in tumor cells [90]. Research is mainly aimed at finding new drugs that target this enzyme and have lower toxicity in normal tissue.

6.3. Inactivation of Aziridinylbenzoquinones

In addition to alkylation, other reactions can also occur with aziridinylquinone derivatives. Other possible transformations that lead to the inactivation of these derivatives are shown in Figure 34—(A) Formation of quinone derivatives by loss of aziridine; (B) 1,5-sigmatropic hydrogen shift [91], which is then converted to ethylaminoquinone by tautomerism or to aminoquinone by a second 1,5-sigmatropic shift, followed by hydrolysis of the imine derivative in which acetaldehyde is lost; (C) formation of a quinone derivative by loss of aziridine; (D) formation of a quinone derivative by loss of aziridine, which then undergoes a second 1,5-sigmatropic hydrogen shift, followed by hydrolysis of the imine derivative, which undergoes loss of acetaldehyde.

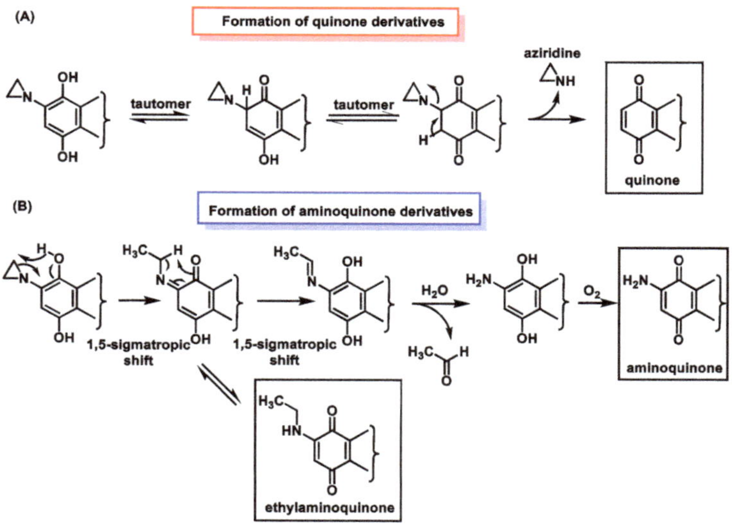

Figure 34. Inactivation of aziridinium benzoquinones (**A**) with loss of the aziridine ring and (**B**) initiated by a 1,5-sigmatropic shift.

The metabolic one-electron reduction of aziridinylquinones yields semiquinones. Their protonated derivatives undergo a 1,5-sigmatropic shift leading to inactive ethylaminoquinones and aminoquinones (Figure 35), as in the two-electron reduction [92].

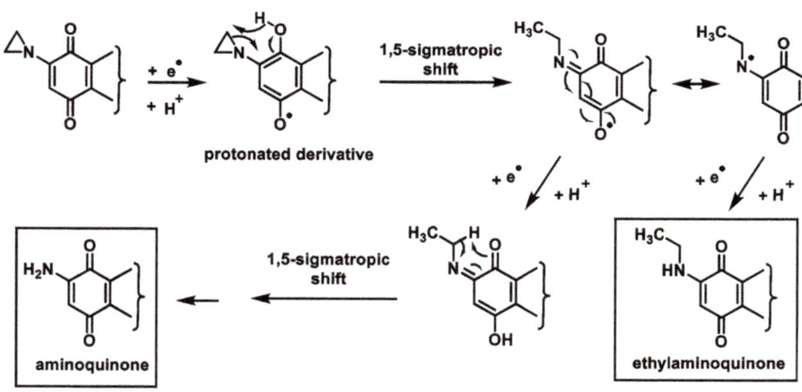

Figure 35. Inactivation of hydroquinone forms of aziridinylbenzoquinones by one-electron reduction.

7. Epoxides

The high reactivity of the epoxide ring towards nucleophilic groups in biomolecules is the basis for the use of ethylene oxide in side chains of compounds intended for the alkylation of DNA.

Treosulphan and mitobronitol are two prodrugs that undergo intramolecular double nucleophilic displacement to yield the diepoxide derivative. This compound is able to crosslink DNA by alkylating the guanine bases of DNA [93] (Figure 36).

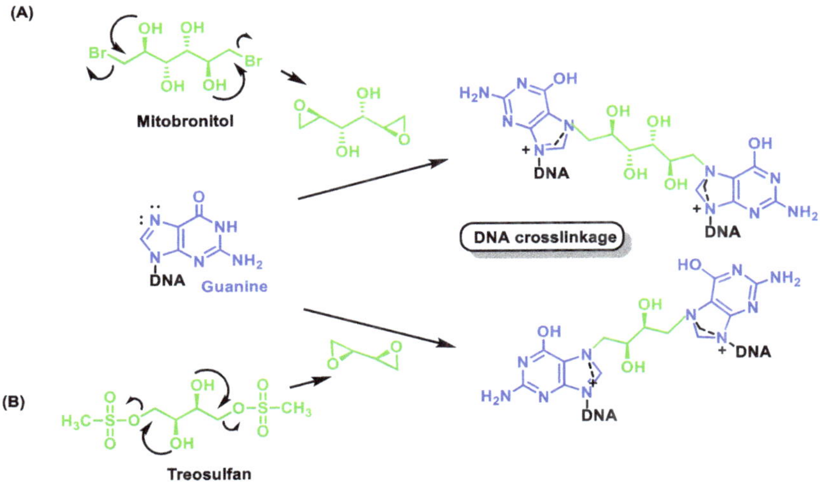

Figure 36. (**A**) DNA alkylation by mitobronitol and (**B**) DNA alkylation by treosulphan.

Treosulphan is mainly used for the treatment of ovarian cancer. In addition, it has shown preclinical and clinical activity in some other solid tumors and hematological malignancies and is used for bone marrow ablation prior to stem cell transplantation and for the treatment of malignant melanoma and breast cancer. Mitobronitol is the biomolecular analog of mannitol and is used for myelosuppression prior to allogeneic bone marrow transplantation in accelerated chronic granulocytic leukemia [94].

8. Nitrosoureas

Antineoplastic agents derived from nitrosoureas are alkylating agents. The most important property is their ability to reach the brain, and they are used specifically to treat brain tumors [95]. All other alkylating agents lack this property and cannot reach the brain. Nitrosoureas can penetrate the brain because they are able to cross the so-called blood–brain barrier, a special area that prevents most drugs from reaching the brain. Nitrosourea derivatives such as carmustine and lomustine are effective in brain tumors, Hodgkin's disease, and other lymphomas. Side effects of nitrosourea derivatives include nausea, vomiting, phlebitis, and suppression of hematopoiesis. Some of the nitrosourea-derived drugs that can treat certain brain tumors are shown in Figure 37.

1-Methyl-1-nitrosourea, the most important compound of the nitrosourea group, was found to be effective against intracerebrally implanted murine leukemia. However, it was found that the introduction of a 2-chloroethyl chain at the nitrogen-containing nitroso group led to a significant increase in the activity of the nitrosoureas. These 2-chloroethyl derivatives are lipophilic and can cross the blood–brain barrier and enable the treatment of brain tumors. In view of this activity, the nitrosoureas lomustine (CCNU), semustine, carmustine (BCNU), nimustine (ACNU), tauromustine, and fotemustine were synthesized, all of which are water-soluble but have significant toxicity problems.

Carmustine was tested in clinical trials in 1964 and approved by the FDA in 1977 for the treatment of various types of brain tumors (including glioma, glioblastoma multiforme, medulloblastoma, and astrocytoma), multiple myeloma, and lymphomas (Hodgkin's and non-Hodgkin's lymphoma).

A new formulation of carmustine with reduced systemic toxicity has been developed for the local treatment of brain tumors. The formulation, in the form of a slow-release "wafer" (Gliadel Wafer®; Polifeprosan 20 with carmustine), is implanted into the resection cavity remaining after surgical removal of the tumor. It was approved by the FDA in 1997 as an adjunct to surgery to prolong the survival of patients with recurrent GBM for whom surgical resection is indicated.

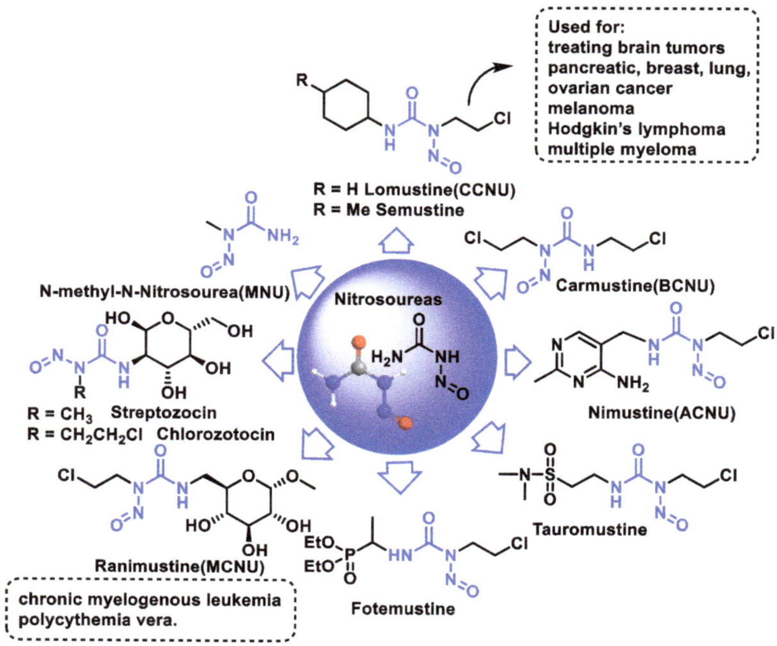

Figure 37. Chemical structure of the main nitrosoureas.

BCNU is a prodrug that breaks down to produce chloroethylalkylation chloroethyl residues that can form cross-links between DNA strands [96]. Carbamoylation of lysine residues in nucleoproteins via isocyanate intermediates may also play a role in BCNU's anticancer mode of action [97].

Nimustine is used in combination with teniposide as second- or third-line chemotherapy for recurrent glioblastoma.

Streptozotocin (a glucosamine nitrosourea), a natural hydrophilic nitrosourea, was isolated from Streptomyces achromogenes (1967). This compound proved to be more effective and less toxic, was approved by the FDA in 1982 for the treatment of pancreatic islet cell cancer and marketed as Zanosar®, although its use is generally limited to patients whose cancer cannot be removed by surgery, and several analogs have been produced, such as chlorozotocin and ranimustine.

Streptozotocin differs from other nitrosoureas in that it does not cross the blood–brain barrier due to its high hydrophilicity and also has relatively low myelosuppression, as it penetrates less into the cells of the bone marrow.

Chlorozotocin has been shown in a phase II study to be as active against metastatic melanoma as other clinically used chloroethylnitrosoureas without causing bone marrow toxicity [98]. Chlorozotocin is probably carcinogenic to humans.

Mechanism of Action of Nitrosoureas

The mechanism of action of nitrosoureas has been extensively studied. Nitrosoureas decompose spontaneously to form two electrophiles: an isocyanate and a diazene hydroxide. The diazene hydroxide produces a diazonium salt that alkylates the DNA, and this appears to be the main reaction responsible for the antitumor activity [99]. The isocyanate reacts with the amino groups of proteins to cause carbamoylation of proteins, which leads to inhibition of various DNA repair mechanisms. This fragmentation pathway was proposed based on studies of thermal decomposition of nitrosoureas under anhydrous conditions, but in aqueous solution the reaction proved to be more complex [71] (Figure 38).

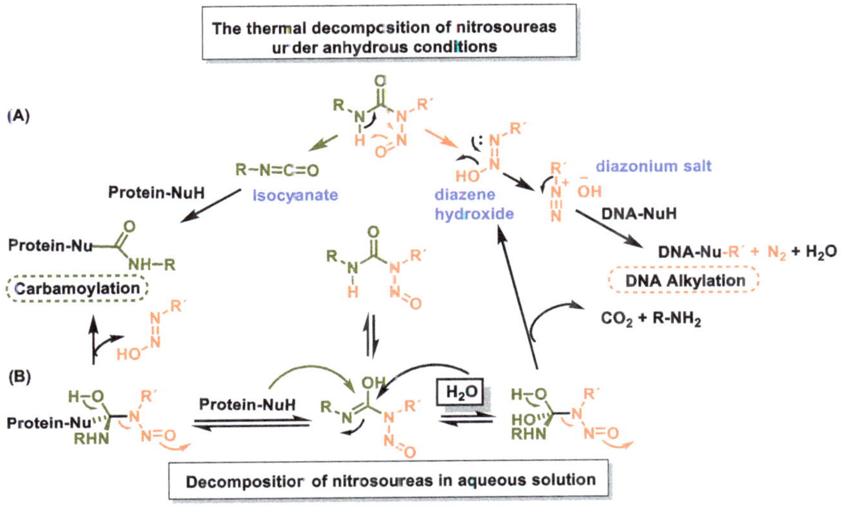

Figure 38. (A) Decomposition of nitrosoureas under anhydrous conditions. (B) In aqueous solution with the products of the alkylation of DNA and proteins.

Almost all nitrosoureas contain a chloroethyl chain in the nitroside, which enables them to act as DNA cross-linkers. This monoalkylated product forms a 5-linked cyclic intermediate and then reacts with the N-3 atom of the cytosine unit on the complementary DNA strand to form the cross-linked end product (Figure 39).

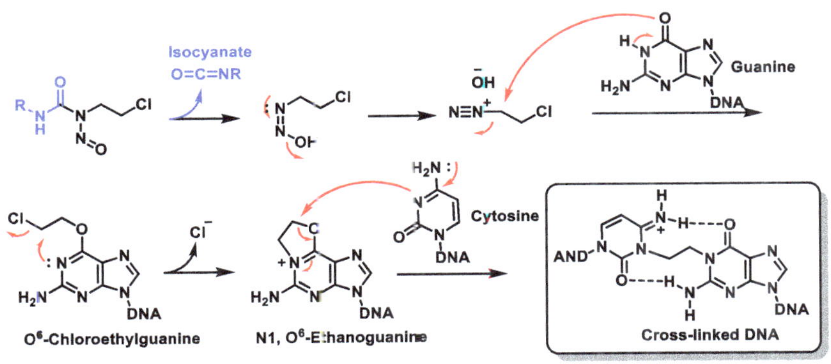

Figure 39. DNA cross-linking by nitrosoureas.

An alternative mechanism would be the direct alkylation of DNA by the nitrosourea itself and not by the diazonium salt. Most drugs from the nitrosourea group are no longer in use and have been replaced by other alkylating agents.

9. Methanesulphonates

The mesylate is a good leaving group due to the delocalization of the negative charge between its three oxygen atoms (mesylate anion) (Figure 40).

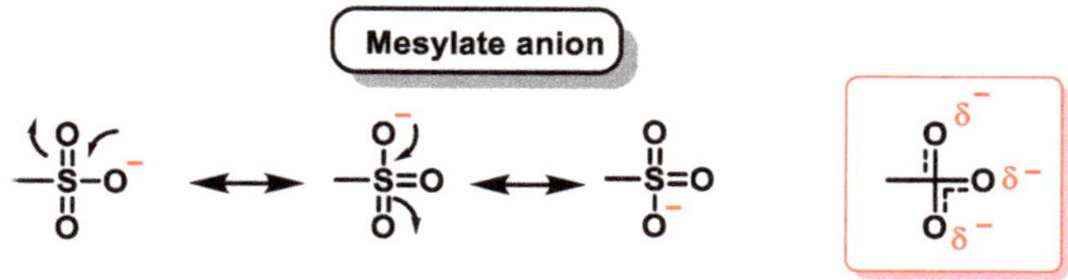

Figure 40. Mesylate ion resonant forms and average structure.

Several compounds containing two methanesulphonate groups separated by a hydrocarbon–carbon chain was tested as antitumor agents, and the compound with four carbons (busulphan) was found to have the optimal activity. The other methanesulphonate derivatives tested include piposulphan, improsulphan, hepsulphan, and treosulphan, a di-epoxide prodrug, which are shown in Figure 41.

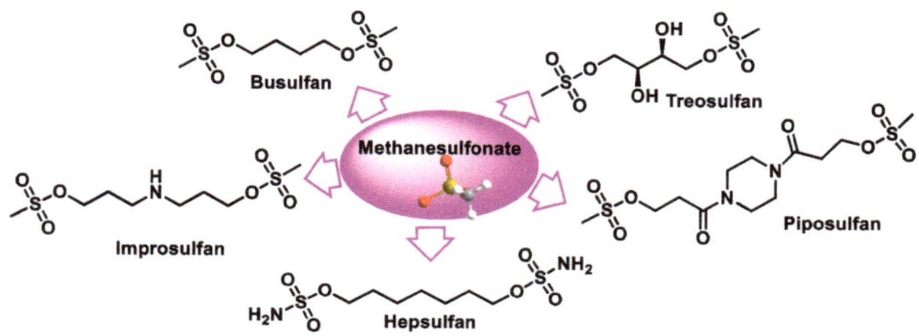

Figure 41. Methanesulphonate used as antitumor agents.

Busulphan, 4-methylsulphonyloxybutylmethanesulphonate, is an antineoplastic agent from the class of bifunctional alkylating agents that has been used to treat various types of cancer since 1959. Busulphan is used to treat chronic myeloid leukemia and certain blood disorders such as polycythemia vera and myeloid metaplasia and is also used in some conditioning programs prior to bone marrow transplants [100–103]. In contrast to nitrogen mustard and nitrosoureas, busulphan mainly affects myeloid cells. In high doses, busulphan is used in combination with cyclophosphamide as part of a preparatory regimen for bone marrow transplants to treat childhood leukemias, lymphomas, and some solid tumors. Busulphan is toxic to lung tissue and this toxicity may limit the dose that can be used.

Close monitoring of blood counts is required in patients treated with busulphan. Hematologic toxicity, including severe neutropenia, leukopenia, thrombocytopenia, and anemia, is the dose-limiting factor for busulphan.

In aqueous solutions, busulphan hydrolyzes to release methanesulphate groups and generates carbonium cations, which are reactive and can alkylate DNA and other proteins, leading to cytotoxicity. Busulphan reacts more readily with the thiol groups of amino acids, with cysteine, which involves a double nucleophilic attack of busulphan on the amino acid (Figure 42), and also with proteins as nitrogen mustard. In the final step of this reaction, elimination of the tetrahydrothiophene takes place and a C-C double bond is formed. The alkylation reaction depends on the concentrations of both busulphan and the target compound. Busulphan binds to DNA at the N-7 position of guanine. The alkylation of guanine with busulphan occurs via an SN2 mechanism [104]. The mechanism involves the attack of the nucleophile on the carbon containing the leaving group (mesylate). This carbon has a clearly positive polarity due to the electro-negativity of the sulphonate residue. Simultaneously with the attack on the nucleophile (N-7), the carbon-mesylate bond is cleaved, forming the monoalkylation product. The formation of cross-links between DNA strands has been demonstrated for busulphan cross-links within the DNA strand, mainly in the guanine–adenine sequence [102,105] (Figure 42).

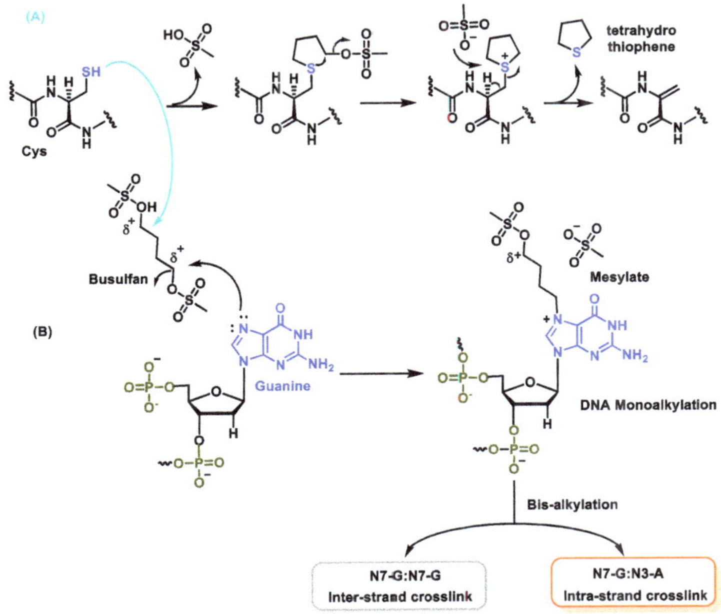

Figure 42. (**A**) Alkylation of cysteine residues by busulphan and (**B**) mechanism of DNA alkylation by busulphan.

Hepsulpham induced higher levels of DNA strand cross-linking than busulpham in three isolated samples from patients with chronic myeloid leukemia in blast crisis. However, busulphan caused a low number of DNA strand breaks in human cells.

Treosulphan in combination with fludarabine is a drug that is given to people before they undergo a bone marrow transplant from a donor, called an allogeneic hematopoietic stem cell transplant. It is used as a 'conditioning treatment' to cleanse the patient's bone marrow and make room for the transplanted bone marrow cells, which can then produce healthy blood cells. It is used in both adult and pediatric patients older than one month with malignant neoplasms and benign diseases.

10. Triazenes

Triazenes are molecules that contain three contiguous nitrogen atoms in a linear arrangement. They have anticancer potential in many types of tumors such as leukemia, melanoma, and brain tumors as well as other therapeutic applications. The three triazenes used therapeutically are the anticancer drugs dacarbazine and temozolomide, and diaminazole acetate, which is used in the treatment of trypanosomiasis in animals (Figure 43).

Dacarbazine (DAC) (5-(3,3-dimethyl-1-triazenyl) imidazole-4-carboxamide) belongs to the triazene group of alkylating agents used in cancer therapy [106–109]. DAC has a triazene group linked to the imidazole ring, which, in turn, is linked to the carboxamide group. It is an electrophilic agent that acts specifically in the S phase of the cell cycle. Dacarbazine was approved in the 1970s in the USA and France for the treatment of metastatic malignant melanoma, Hodgkin's lymphoma, sarcoma [110], and pancreatic islet cell carcinoma and is used in combination with other cancer drugs in more extreme cases. It has been classified in the group of antitumor agents for which there is no safe exposure range [111].

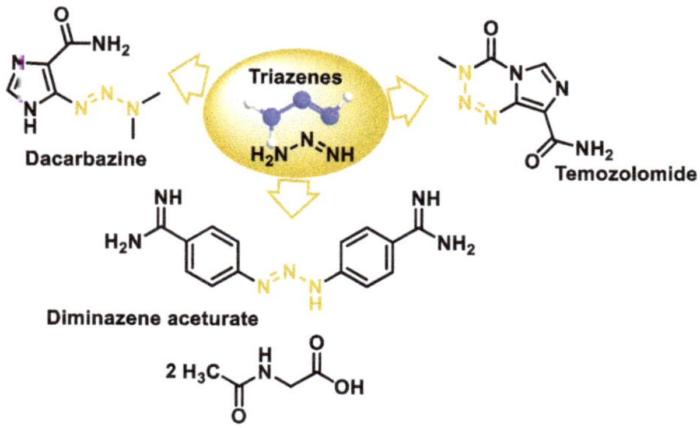

Figure 43. Current therapeutic use triazenes.

It is a prodrug that is activated in the liver and its cytotoxic activity is due to the formation of methyldiazonium during its metabolism, which methylates DNA [112]. Methyldiazonium has a very short half-life of about 0.4 s in aqueous solution. A mechanism for this process is summarized in Figure 44, where activation of dacarbazine by metabolic oxidative demethylation with formation of formaldehyde and (E)-5-(3-methyltriaz-1-en-1-yl)-1H-imidazole-4-carboxamide from the tautomer generates the reagent methyldiazonium. The formation of this cation is considered to be the main mechanism of its antitumor effect (Figure 44).

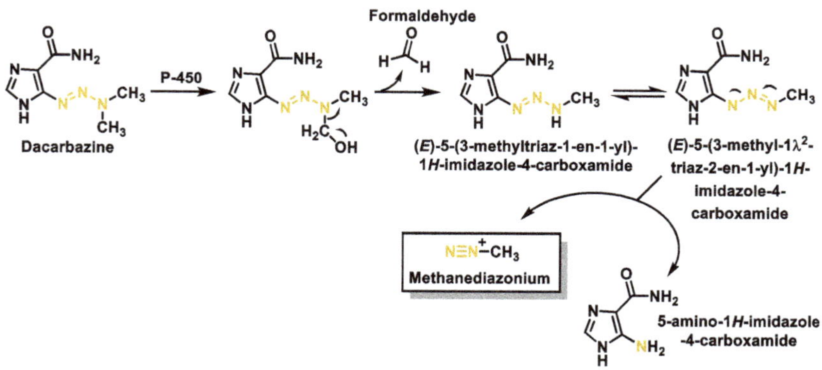

Figure 44. Generation of methanediazonium from dacarbazine.

The most important methylation reaction takes place at the N-7 guanine atom and is relatively non-toxic. Methylation of the O-6 guanine also occurs and is considered to be the primary cytotoxic mechanism [113] (Figure 45).

Disadvantages of dacarbazine are: (i) toxicity, (ii) excessive hydrophilicity leading to slow and incomplete oral absorption, necessitating intravenous administration, and (iii) its high photosensitivity with a very short half-life (about 30 min). This antineoplastic drug is excreted in the urine and its half-life in the body is 5 h. Dacarbazine is unstable in aqueous solution and is spontaneously degraded in the light, decomposing to 5-diazomidazole-4-carboxamide and di-methylamine and also undergoing a structural rearrangement to 2-azahipoxanthine [114,115] (Figure 46).

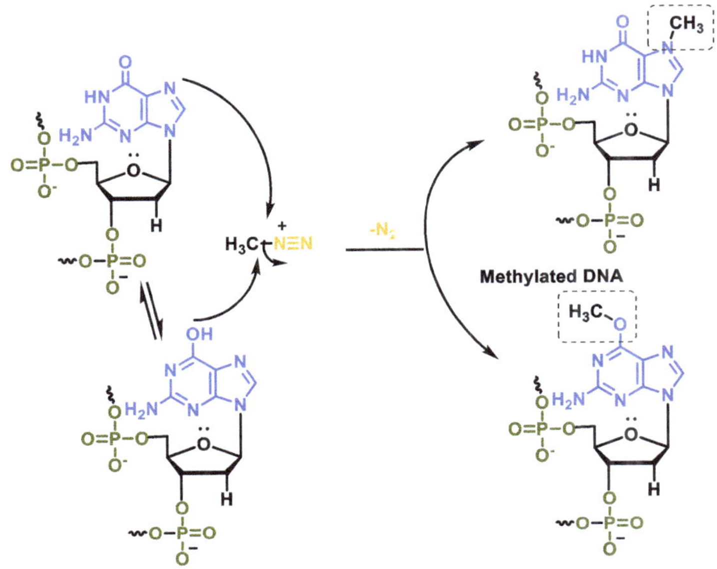

Figure 45. Mechanism of DNA methylation by Dacarbazine.

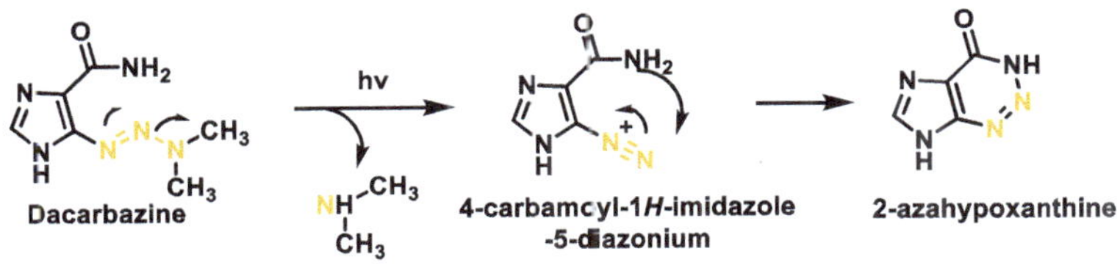

Figure 46. Dacarbazine photodegradation products.

5-Diazomidazole-4-carboxamide can attack nucleophilic groups of DNA. However, the degradation products of dacarbazine probably do not contribute significantly to its cytotoxicity, although they may be involved in the painful local burning sensation on intravenous injection and in systemic problems associated with the drug.

Temozolamide (TMZ) was synthesized in 1980 as part of a series of new imidazotetrazinones to address the problems caused by the use of dacarbazine and has replaced dacarbazine in the clinic. It is a cytotoxic alkylating prodrug used for the treatment of glioblastoma multiforme [116,117] {Stevens, 1984 #12198}. Temozolomide, a methylating agent, was approved by the FDA in the USA in 1999 for the treatment of refractory anaplastic astrocytoma. The European Agency for the Evaluation of Medicinal Products approved the drug in 1998 for (i) progressive or recurrent glioblastoma multiforme after standard surgery and (ii) in 1999 for anaplastic astrocytoma, which is also resistant to treatment. Importantly, temozolomide is one of the few drugs approved specifically for a brain tumor and the first that can be administered orally

The chemotherapeutic activity is achieved by converting temozolamide to the same intermediate (5-methyltriazenoimidazole-4-carboxamide) as dacarbazine, but in the case of temozolomide, the bioactivation process involves a non-enzymatic hydrolysis reaction followed by decarboxylation (Figure 47).

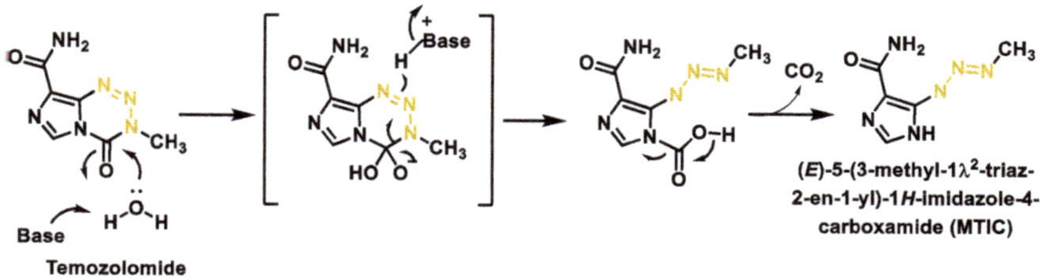

Figure 47. Temozolomide hydrolysis.

The methyldiazonium ion formed during the degradation of MTIC mainly methylates the guanine residues in the DNA molecule, leading to the formation of O6- and N7-methylguanine. This aberrant DNA can no longer be corrected by the cellular machinery and ultimately leads to cell death through apoptosis. In addition, unlike dacarbazine, this drug can be administered both parenterally and orally. The main problem with this drug is bone marrow toxicity.

11. Fate and Effects of Anticancer Drugs in the Environment

Cytotoxic drugs, i.e., cancer drugs used in chemotherapy, are administered in European countries in the order of tons per year. After administration, these compounds are excreted by the chemotherapy patient either in their original form or as a product of the metabolization of the original compound into wastewater and subsequently discharged into rivers, seas, and oceans and can enter the environment via wastewater treatment plants, where they can have harmful effects. Due to their physicochemical properties, they can also be adsorbed in sewage sludge. The incidence of cancer has increased in recent years and is expected to continue to rise, as is the consumption of cytotoxic drugs. Therefore, their control in the environment is of utmost importance [118].

Assessing the risk posed by cytotoxic drugs is of particular interest as they are cytotoxic and, in many cases, carcinogenic. There is currently little data on the occurrence and effects of cytotoxic drugs in the environment.

The release of cytotoxic drugs used in chemotherapy has received less attention than other pharmaceuticals, although they may have cytotoxic, genotoxic, mutagenic, carcinogenic, endocrine, and/or teratogenic effects on various organisms.

The main sources of these effluents are hospital effluents, effluents from drugs excreted during outpatient treatment, and drug manufacturers. The immediate consequence is the release into the aquatic and terrestrial environment, even in minute quantities. The removal efficiency of wastewater treatment varies and depends on both the physicochemical properties of these by-products and the treatment process used [119].

In the 1990s, major efforts were made to investigate the occurrence, phase, and potential risks of human pharmaceuticals in the environment. Several large-scale collaborative programs have been conducted in Europe and the US, such as 'Poseidon' (EU, 2001–2004), Norman (EU, 2005–2008), PILLS (EU, 2007–2012), and Emerging Contaminants in the Environment (US Geological Survey, 2007–2011). The most recent PHARMAS (EU, 2011–2013) and CytoThreat (EU, 2011–2013) specifically target antibiotics and cancer drugs.

There is a need to sensitize both the scientific community and the general public to the potential of anticancer drugs as pollutants of increasing concern in aquatic and terrestrial ecosystems [120].

Antineoplastic drugs (ANPs) have raised concerns about their safe remediation. An excellent review paper by Abhilash Kumar Tripathi and co-authors provides an overview of the environmental sources of ANP agents, focusing on the remediation methods currently in use [121].

Other necessary considerations include the solid cytotoxic wastes generated during the preparation, reconstitution, and administration of these drugs to chemotherapy patients. There are no suitable treatment technologies for this waste in most parts of Spain, so it has to be transported to France and/or Germany for incineration, resulting in the emission of toxic and greenhouse gasses.

12. Antimicrobial Peptides

The search for more effective and less toxic cancer treatments has driven research into numerous innovative therapeutic strategies. Among these, antimicrobial peptides (AMPs) have emerged as a promising class of agents with potent anticancer properties [122]. Originally identified for their role in innate immunity and their ability to fight microbial infections, AMPs have shown a broad spectrum of biological activities that also have anticancer effects. These peptides exert their anticancer effects primarily through the selective destruction of cancer cell membranes, induction of apoptosis, and modulation of immune responses, making them highly versatile and potent candidates for cancer therapy [123]. Their unique mechanism of attacking cancer cells while sparing normal cells offers a significant advantage over conventional chemotherapeutic agents, which often result in significant collateral damage to healthy tissue [124,125].

Antimicrobial peptides (AMPs) are a critical component of the innate immune system and serve as one of the first lines of defense against a broad spectrum of pathogens, including bacteria, viruses, fungi, and parasites [126]. These small, naturally occurring peptides are ubiquitous in a variety of species, from microorganisms to humans, underscoring their evolutionary importance [126]. The discovery of AMPs can be traced back to the early 20th century, when scientists first discovered the antimicrobial properties of lysozyme, an enzyme found in egg white and human tears [127]. But it was not until the late 1980s that the field of AMPs gained prominence after magainin were isolated and characterized from the skin of the African clawed frog *Xenopus laevis* [128]. This groundbreaking work highlighted the strong antimicrobial activity of these peptides and initiated extensive research into their potential therapeutic applications.

AMPs are characterized by their amphipathic nature, which enables them to interact with and destroy microbial membranes [129]. Structurally, they typically consist of 12–50 amino acids and can adopt various conformations, including alpha helices, beta sheets, and extended or looped structures [129,130]. The ability of AMPs to attack and destroy the lipid bilayers of microbial membranes, while mammalian cells are generally spared, is attributed to differences in membrane composition [122]. This selectivity is crucial for their role in host defense and represents a significant advantage over conventional antibiotics, which often cause problems with resistance and collateral damage to host tissue [129].

Beyond their antimicrobial properties, AMPs have attracted attention for their diverse role in modulating the immune response, promoting wound healing and even their anticancer activity. In the context of cancer therapy, AMPs have been shown to selectively target cancer cells through mechanisms similar to those used to fight microbes [122]. Cancer cells often have particular membrane characteristics, such as a higher proportion of negatively charged phospholipids, which make them susceptible to AMP-induced membrane disruption. In addition, AMPs can induce apoptosis in cancer cells and modulate the tumor microenvironment to enhance antitumor immunity. These properties make AMPs promising candidates for novel cancer therapies, either as stand-alone treatments or in combination with other therapeutics [122].

The potential of AMPs in cancer therapy has generated considerable interest in the scientific community, particularly with regard to their use in combination with other therapeutic modalities [131]. One of these promising combinations is with DNA alkylating agents [132–135]. DNA alkylation involves the attachment of alkyl groups to DNA molecules, resulting in DNA damage and inhibition of cancer cell proliferation. Alkylating agents, such as cyclophosphamide and cisplatin, have been cornerstones of cancer treatment for decades [60]. However, their use is often limited by the development of resistance and severe side effects [136]. The combination of DNA alkylating agents with AMPs offers a novel approach to overcome these limitations by increasing the efficacy and specificity of the attack on cancer cells [137,138]. The reason for this combination lies in the complementary mechanisms of action of AMPs and alkylating agents. While alkylating agents have a direct genotoxic effect on cancer cells, AMPs can enhance this effect by compromising the integrity of cancer cell membranes, enabling better intracellular uptake of the drug and promoting apoptosis. In addition, AMPs can modulate the tumor microenvironment by enhancing the immune response against cancer cells, potentially attenuating the immunosuppressive effects often associated with alkylating chemotherapy. This synergistic interaction promises not only better therapeutic outcomes but also a reduction in the required doses of alkylating agents and thus a minimization of their undesirable side effects [2,139–141].

13. Mechanisms of Action of AMPs in Cancer Cells

Antimicrobial peptides (AMPs) exhibit a versatile mode of action against cancer cells by utilizing their innate properties to attack and destroy malignant cells. One of the most important mechanisms by which AMPs exert their anticancer effect is by destroying the membranes of cancer cells [122]. Cancer cells often have special membrane characteristics, including a higher content of negatively charged phospholipids and an altered lipid composition compared to normal cells. This makes them particularly susceptible to the amphipathic and cationic nature of AMPs. Upon interaction with the cancer cell membrane, AMPs can penetrate the lipid bilayer and form pores or degrade the membrane. This disruption impairs the integrity of the cell membrane and leads to cell lysis and cell death [142].

In addition to the direct disruption of the membrane, AMPs can also trigger apoptosis in cancer cells. Apoptosis, or programmed cell death, is a tightly regulated process that eliminates damaged or unwanted cells without causing inflammation. AMPs can trigger apoptosis via different pathways, including the mitochondrial pathway and the death receptor pathway. In the mitochondrial pathway, AMPs can cause the release of cytochrome c from the mitochondria into the cytosol, which subsequently activates caspases, the enzymes responsible for carrying out apoptosis. Alternatively, AMPs can interact with cell surface receptors to activate the death receptor pathway, leading to the formation of the death-inducing signaling complex (DISK) and activation of caspase-8. These pathways culminate in the cleavage of important cellular proteins and the eventual destruction of the cell.

AMPs can not only trigger apoptosis but also modulate the immune response, creating a hostile environment for cancer cells [143]. AMPs are known to have immunomodulatory properties. These include the ability to attract immune cells to the tumor site, promote the maturation and activation of dendritic cells, and enhance the presentation of tumor antigens. This immunomodulation can stimulate both the innate and adaptive immune response, leading to a stronger attack on the cancer cells [144]. By altering the tumor microenvironment, AMPs can help overcome the immunosuppressive barriers that often protect tumors from surveillance and destruction by the immune system [122,145].

In addition, some AMPs have been found to inhibit angiogenesis, the process by which tumors establish their own blood supply to sustain growth and metastasis [122,146]. Angiogenesis is a crucial step in tumor progression, and inhibiting it can supply the tumor with essential nutrients and oxygen [147]. AMPs can interfere with angiogenic signaling pathways by reducing the expression of pro-angiogenic factors such as vascular endothelial growth factor (VEGF) and disrupting the formation of new blood vessels [148,149]. This anti-angiogenic effect not only inhibits tumor growth but also limits the potential for metastasis [150].

14. Advantages of AMPs and Their Synergy with Alkylating Agents

One of the main problems with alkylating agents is their lack of selectivity, which results in damage to both cancer cells and healthy cells. This non-specificity leads to a number of serious side effects, including myelosuppression (bone marrow suppression), gastrointestinal toxicity, and an increased risk of secondary malignancies [151]. These side effects may limit the dose of alkylating agents that can be safely administered to patients, reducing their overall efficacy [136]. The membranes of cancer cells often contain higher levels of negatively charged phospholipids and other unique lipid components that attract the cationic and amphipathic nature of AMPs [122,152]. This selectivity minimizes collateral damage to healthy cells and reduces the side effects commonly associated with conventional chemotherapy [153] (Figure 48).

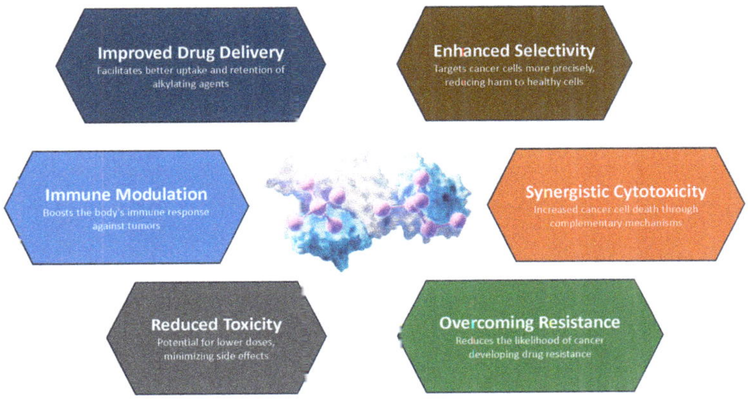

Figure 48. Advantages of antimicrobial peptides in combination with alkylating agents for cancer therapy.

AMPs also exhibit a broad spectrum of anticancer mechanisms, making it difficult for cancer cells to develop resistance. Their ability to disrupt cell membranes, induce apoptosis, modulate immune responses, and inhibit angiogenesis provides a versatile approach to destroying cancer cells. By attracting immune cells to the tumor site and promoting the activation and maturation of dendritic cells, AMPs can stimulate both the innate and adaptive immune systems. This immune modulation can create a more hostile environment for cancer cells, further enhancing the anticancer effects of AMPs and also helps to prevent the emergence of resistant cancer cell populations, which is a major problem with single-target chemotherapeutics [154].

The natural origin of many AMPs and their relatively simple structure also facilitate their synthesis and modification, enabling the development of a wide range of AMP derivatives with improved stability, potency, and selectivity. Advances in peptide engineering and delivery systems have further enhanced the therapeutic potential of AMPs, making them more suitable for clinical use [142]. These properties make AMPs promising candidates for the development of novel cancer therapies, either as stand-alone treatments or in combination with existing chemotherapeutics to increase efficacy and reduce toxicity.

The combination of AMPs with alkylating agents can improve the efficacy of cancer therapy by several mechanisms (Table 2): (i) by enhancing DNA damage, since alkylating agents act by attaching alkyl groups to DNA and AMPs can enhance this effect by interfering with DNA repair mechanisms, making cancer cells more susceptible to damage by alkylating agents; (ii) by modulating the tumor microenvironment, affecting angiogenesis (formation of new blood vessels) and immune responses, making tumor cells more susceptible to the cytotoxic effects of alkylating agents; (iii) overcoming drug resistance by disrupting cell signaling pathways that cancer cells use to escape the effects of DNA damage; and (iv) activating signaling pathways that lead to apoptosis (programmed cell death) and complement the cytotoxic effects of alkylating agents. This dual approach may lead to more effective elimination of cancer cells.

Table 2. Synergistic mechanisms of AMPs and alkylating agents in cancer therapy.

Mechanism	Description	Reference
Membrane Disruption	AMPs disrupt cancer cell membranes, leading to cell lysis, facilitating the entry of alkylating agents.	[155]
DNA Damage	Alkylating agents cause DNA strand breaks, facilitated by AMPs.	[156]
Immune Modulation–Induction of Apoptosis	AMPs trigger apoptosis, which can be amplified by alkylating agents.	[157,158]
Anti-Angiogenesis	AMPs inhibit angiogenesis, which complements the cytotoxic effects of alkylating agents.	[159]
Overcoming Drug Resistance	AMPs reduce the potential for drug resistance to alkylating agents.	[160]

15. Technological Advances in the Administration of AMPs and Alkylating Drugs

The development of effective delivery systems is critical to the therapeutic success of antimicrobial peptides (AMPs) and alkylating agents in cancer treatment (Figure 49). One of the most important advances in AMP delivery is the use of nanoparticle-based systems [161]. Nanoparticles can be engineered to encapsulate AMPs, protect them from enzymatic degradation, and improve their stability in the bloodstream [162,163]. These nanoscale carriers can be functionalized with targeted ligands that recognize and bind to specific markers on the cancer cells to ensure that the AMPs are delivered precisely to the tumor. This targeted delivery not only increases the therapeutic efficacy of the AMP, but also minimizes off-target effects and reduces systemic toxicity. Various nanoparticle platforms, including liposomes, polymeric nanoparticles, and metallic nanoparticles, have been evaluated for their ability to effectively deliver AMPs to tumors [164].

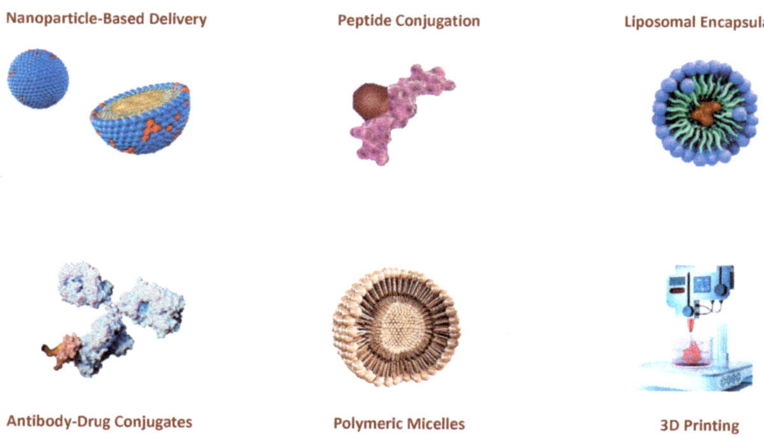

Figure 49. Technological advances in the administration of AMPs and alkylating drugs for cancer therapy.

Similarly, advances in the delivery of alkylating agents have focused on improving their selectivity and reducing their toxicity. Liposomal encapsulation is one of these promising strategies. Liposomes are spherical vesicles consisting of lipid bilayers that can encapsulate hydrophilic and hydrophobic drugs. Encapsulation of alkylating agents in liposomes may help to protect them from premature degradation and reduce their systemic toxicity by enhancing their accumulation in tumor tissue through the effect of enhanced permeability and retention (EPR). This effect takes advantage of the leaky vasculature and poor lymphatic drainage of tumors to preferentially deliver liposomal encapsulated drugs to the tumor [165–167].

Another significant advance is the development of antibody–drug conjugates (ADCs), which combine the targeting ability of monoclonal antibodies with the potent cytotoxic activity of alkylating agents. ADCs are designed to bind specifically to antigens expressed on the surface of cancer cells, delivering the alkylating agent directly to the tumor and sparing normal tissue. This targeted approach maximizes the therapeutic index of alkylating agents and increases their efficacy while minimizing adverse effects [168,169].

Polymeric micelles have also been used to enhance the delivery of AMP and alkylating agents. These self-assembling nanocarriers can encapsulate hydrophobic actives in their core and form a hydrophilic shell that improves the solubility and stability of the encapsulated actives. Polymeric micelles can be engineered to release their payload in response to specific stimuli, such as changes in pH or temperature, ensuring controlled and sustained drug release at the tumor site [164,170–172].

In addition to nanoparticle-based systems, peptide conjugation strategies have been developed to improve the delivery of AMPs. Conjugation of AMPs with cell-penetrating peptides (CPPs) can facilitate their translocation across cell membranes, improving their intracellular release and therapeutic efficacy. CPPs can be designed to recognize and bind to specific receptors on cancer cells, ensuring targeted delivery of AMPs and enhancing their anticancer effects [173,174]. For example, a combination of the cell-penetrating peptide octoarginine–oxaliplatin was developed to efficiently and rapidly transport oxaliplatin into colon cancer cells. The in vivo study in mice showed that the compound successfully suppressed tumor development and had a significantly strong anticancer effect [175]. Similarly, the complex of the cell-penetrating peptide Tansportan-10 and cisplatin produced a stronger effect on the cancer cell lines (HeLa, OS143B) compared to that observed after separate treatment [176]. Conjugation of cell-penetrating peptides with cobalt complexes also showed superior efficacy against the HepG2 liver cancer cell line compared to the original Co(III) complex [177]. The combination of cell-penetrating peptides with nanotechnology is a promising strategy. Encapsulation of the cell-penetrating peptide EIP103, which specifically targets the nucleus, in PEG-PE micelles showed significant inhibitory effects in H446 and A549 cells and demonstrated a promising therapy to improve the clinical treatment of lung cancer [178].

Advances in 3D printing technology have opened new avenues for the local delivery of AMPs and alkylating agents; 3D-printed scaffolds and hydrogels can be designed to release these agents in a controlled manner and ensure sustained therapeutic concentration at the tumor site. This local delivery approach can minimize systemic exposure and reduce the risk of adverse effects [179–181]. These advances demonstrate the importance of continuing drug delivery research and development to fully realize the therapeutic potential of AMPs and alkylating agents.

16. Can AI Tools Accelerate Current Research into Alkylating Agents?

The question of whether artificial intelligence (AI) can accelerate research into alkylating agents in cancer treatment is a hot topic that is increasingly being answered in the affirmative.

Alkylating agents are a class of chemotherapeutic agents that damage the DNA of cancer cells. However, their use is limited by significant side effects and the emergence of tumor resistance. This is where AI can make a decisive difference.

AI can accelerate research by discovering new drugs by analyzing large amounts of data. AI can analyze huge databases of molecules to identify new chemical structures with potential antitumor activity and through molecular modeling that simulates the interaction between molecules and proteins [182]. AI can predict the efficacy and toxicity of new drugs, reducing the time and cost of preclinical trials [183].

It can also help optimize treatments, such as personalizing therapies by analyzing each patient's genomic data and identifying the most appropriate treatments that minimize side effects. It can also be used to predict response to treatment by analyzing medical imaging and clinical data so that therapy can be better tailored [184]. Another potential area of application is understanding resistance mechanisms and identifying potential biomarkers that predict the occurrence of resistance to alkylating agents [185].

In the negative perception, AI relies on large amounts of high-quality data. Therefore, more comprehensive and standardized databases need to be developed, and tools need to be developed to understand how the models arrive at their conclusions [186].

AI therefore has the potential to revolutionize alkylating agent research and ultimately improve outcomes for cancer patients. However, it is important to address the existing challenges and encourage collaboration between researchers from different disciplines to achieve the most out of this technology [187].

17. Conclusions

In the last thirty years, more and more drugs have been used to treat tumors, as genetic and environmental factors have led to an increased incidence of cancer in the human population. The structure and dynamics of DNA are strongly influenced by the alkylation of its bases. Alkylation prevents DNA replication and RNA transcription of the affected DNA molecule. It also leads to DNA fragmentation through hydrolytic reactions and through the action of repair enzymes that attempt to remove the alkylated bases. Alkylation also leads to mismatching of nucleotides by disrupting the normal hydrogen bonds between the bases.

Current research on alkylating agents should focus on the development of more selective alkylating agents with fewer side effects and adequate bioavailability. New derivatives of nitrogen mustard continue to be investigated for improved efficacy and reduced toxicity with combinations of nitrogen mustard with monoclonal antibodies and small molecule targeted agents.

The increasing use of these drugs and their degradation products has resulted in their appearance in water bodies around the world, impacting ecosystems, polluting the environment, and having negative effects on biota.

The combination of antimicrobial peptides (AMPs) and alkylating agents is a promising strategy towards more effective and targeted cancer therapies. AMPs offer significant advantages over conventional chemotherapeutics with their versatile mechanisms of action, including membrane disruption, apoptosis induction, immune modulation, and anti-angiogenic properties. These peptides selectively target cancer cells while sparing healthy tissue, minimizing systemic toxicity and reducing the risk of resistance development. Meanwhile, alkylating agents remain central to cancer treatment due to their ability to cause DNA damage and inhibit cell proliferation, albeit with significant limitations such as non-specific toxicity and the potential for resistance.

The combination of AMPs with alkylating agents capitalizes on the complementary strengths of these agents and may improve therapeutic efficacy through synergistic interactions. Technological advances in drug delivery, such as nanoparticle-based systems, antibody–drug conjugates, and 3D printing technologies, have further optimized the delivery of AMPs and alkylating agents to tumor sites, improving their bioavailability and therapeutic outcomes.

Combining AMPs with alkylating agents can also improve the efficacy of cancer therapy by enhancing DNA damage, modulating the tumor microenvironment, overcoming drug resistance, and activating signaling pathways that lead to apoptosis. These efforts may yield a more effective elimination of malignant cells.

While challenges such as resistance mechanisms, off-target effects, and optimal dosing strategies remain unresolved, ongoing research and clinical trials continue to explore the full potential of this dual mode of action. Future directions include the development of novel AMP derivatives, personalized medicine approaches, and combination therapies tailored to the individual patient profile. Ultimately, the convergence of AMPs and alkylating agents represents a promising paradigm shift in cancer therapy, offering hope for more effective treatments and better outcomes for patients with this difficult disease.

Author Contributions: Conceptualization, C.M.C.A. and C.A.J.; investigation, E.B.M., C.M.C.A. and C.A.J.; writing—review and editing, E.B.M., C.M.C.A., C.A.J., J.M.P.d.l.L. and E.P.-L.; supervision, E.B.M., C.M.C.A., C.A.J. and J.M.P.d.l.L. All authors have read and agreed to the published version of the manuscript.

Funding: This research was funded by the State Plan for Scientific, Technical Research and Innovation 2021–2023 from the Spanish Ministry of Science and Innovation (project PLEC2022-009507) and by the COCREA innovation project of the Spanish National Research Council (CSIC) to promote solutions to global challenges.

Conflicts of Interest: The authors declare no conflicts of interest. The funders had no role in the design of the study; in the collection, analyses or interpretation of data; in the writing of the manuscript; or in the decision to publish the results.

ferences

Amjad, M.T.; Chidharla, A.; Kasi, A. *Cancer Chemotherapy*; StatPearls Publishing: Treasure Island, FL, USA, 2023.
Anand, U.; Dey, A.; Chandel, A.K.S.; Sanyal, R.; Mishra, A.; Pandey, D.K.; De Falco, V.; Upadhyay, A.; Kandimalla, R.; Chaudhary, A.; et al. Cancer chemotherapy and beyond: Current status, drug candidates, associated risks and progress in targeted therapeutics. *Genes Dis.* **2023**, *10*, 1367–1401. [CrossRef] [PubMed]
Wiltse, J.; Dellarco, V.L. US Environmental Protection Agency guidelines for carcinogen risk assessment: Past and future. *Mutat. Res./Rev. Genet. Toxicol.* **1996**, *365*, 3–15. [CrossRef] [PubMed]
De, Surya Kanta. An Overview of Cancer. In *Fundamentals of Cancer Detection, Treatment, and Prevention*; John Wiley & Sons: Hoboken, NJ, USA, 2022; pp. 1–20.
Passaro, A.; Al Bakir, M.; Hamilton, E.G.; Diehn, M.; André, F.; Roy-Chowdhuri, S.; Mountzios, G.; Wistuba, I.I.; Swanton, C.; Peters, S. Cancer biomarkers: Emerging trends and clinical implications for personalized treatment. *Cell* **2024**, *187*, 1617–1635 [CrossRef] [PubMed]
Lee, Y.T.; Tan, Y.J.; Oon, C.E. Molecular targeted therapy: Treating cancer with specificity. *Eur. J. Pharmacol.* **2018**, *834*, 188–196. [CrossRef] [PubMed]
Misiak, M.; Mantegazza, F.; Beretta, L.G. Methods for Elucidation of DNA-Anticancer Drug Interactions and their Applications in the Development of New Drugs. *Curr. Pharm. Des.* **2016**, *22*, 6596–6611. [CrossRef]
Thurston, D.E.; Pysz, I. *Chemistry and Pharmacology of Anticancer Drugs*; CRC Press: Boca Raton, FL, USA, 2021.
Lagunas-Rangel, F.A.; Liu, W.; Schiöth, H.B. Can Exposure to Environmental Pollutants Be Associated with Less Effective Chemotherapy in Cancer Patients? *Int. J. Environ. Res. Public Health* **2022**, *19*, 2064. [CrossRef] [PubMed]
Hoofnagle, J.H.; Serrano, J.; Knoben, J.E.; Navarro, V.J. *LiverTox: Clinical and Research Information on Drug-Induced Liver Injury*; National Institute of Diabetes and Digestive and Kidney Diseases: Bethesda, MD, USA, 2012.
Holland, J.F. *Holland-Frei Cancer Medicine 8*; PMPH-USA: Shelton, CT, USA, 2010; Volume 8.
Kciuk, M.; Gielecińska, A.; Mujwar, S.; Kołat, D.; Kałuzińska-Kołat, Ż.; Celik, I.; Kontek, R. Doxorubicin—An Agent with Multiple Mechanisms of Anticancer Activity. *Cells* **2023**, *12*, 659. [CrossRef]
van der Zanden, S.Y.; Qiao, X.; Neefjes, J. New insights into the activities and toxicities of the old anticancer drug doxorubicin. *FEBS J.* **2021**, *288*, 6095–6111. [CrossRef]
Carvalho, C.; Santos, R.X.; Cardoso, S.; Correia, S.; Oliveira, P.J.; Santos, M.S.; Moreira, P.I. Doxorubicin: The good, the bad and the ugly effect. *Curr. Med. Chem.* **2009**, *16*, 3267–3285. [CrossRef]
Tacar, O.; Sriamornsak, P.; Dass, C.R. Doxorubicin: An update on anticancer molecular action, toxicity and novel drug delivery systems. *J. Pharm. Pharmacol.* **2013**, *65*, 157–170. [CrossRef] [PubMed]
Lipscomb, L.A.; Peek, M.E.; Zhou, F.X.; Bertrand, J.A.; VanDerveer, D.; Williams, L.D. Water ring structure at DNA interfaces: Hydration and dynamics of DNA-anthracycline complexes. *Biochemistry* **1994**, *33*, 3649–3659. [CrossRef] [PubMed]
Howerton, S.B.; Nagpal, A.; Dean Williams, L. Surprising roles of electrostatic interactions in DNA–ligand complexes. *Biopolymers* **2003**, *69*, 87–99. [CrossRef] [PubMed]
Zlatanova, J.; Victor, J.M. How are nucleosomes disrupted during transcription elongation? *HFSP J.* **2009**, *3*, 373–378. [CrossRef]
Yang, F.; Kemp, C.J.; Henikoff, S. Doxorubicin enhances nucleosome turnover around promoters. *Curr. Biol.* **2013**, *23*, 782–787. [CrossRef] [PubMed]
Salerno, D.; Brogioli, D.; Cassina, V.; Turchi, D.; Beretta, G.L.; Seruggia, D.; Ziano, R.; Zunino, F.; Mantegazza, F. Magnetic tweezers measurements of the nanomechanical properties of DNA in the presence of drugs. *Nucleic Acids Res.* **2010**, *38*, 7089–7099. [CrossRef]
Yang, F.; Teves, S.S.; Kemp, C.J.; Henikoff, S. Doxorubicin, DNA torsion, and chromatin dynamics. *Biochim. Biophys. Acta* **2014**, *1845*, 84–89. [CrossRef]

22. Hande, K.R. Clinical applications of anticancer drugs targeted to topoisomerase II. *Biochim. Biophys. Acta* **1998**, *1400*, 173–1
[CrossRef]
23. Lemke, K.; Wojciechowski, M.; Laine, W.; Bailly, C.; Colson, P.; Baginski, M.; Larsen, A.K.; Skladanowski, A. Induction of uni
structural changes in guanine-rich DNA regions by the triazoloacridone C-1305, a topoisomerase II inhibitor with antitun
activities. *Nucleic Acids Res.* **2005**, *33*, 6034–6047. [CrossRef]
24. Króliczewski, J.; Bartoszewska, S.; Dudkowska, M.; Janiszewska, D.; Biernatowska, A.; Crossman, D.K.; Krzymiński, K.; Wyso
M.; Romanowska, A.; Baginski, M.; et al. Utilizing Genome-Wide mRNA Profiling to Identify the Cytotoxic Chemotherape
Mechanism of Triazoloacridone C-1305 as Direct Microtubule Stabilization. *Cancers* **2020**, *12*, 864. [CrossRef]
25. Kellett, A.; Molphy, Z.; Slator, C.; McKee, V.; Farrell, N.P. Molecular methods for assessment of non-covalent metallodrug-D
interactions. *Chem. Soc. Rev.* **2019**, *48*, 971–988. [CrossRef]
26. Bauer, G.B.; Povirk, L.F. Specificity and kinetics of interstrand and intrastrand bifunctional alkylation by nitrogen mustards a
G-G-C sequence. *Nucleic Acids Res.* **1997**, *25*, 1211–1218. [CrossRef]
27. Andrés, C.M.C.; Lastra, J.M.P.d.l.; Juan, C.A.; Plou, F.J.; Pérez-Lebeña, E. Chemical Insights into Oxidative and Nitrat
Modifications of DNA. *Int. J. Mol. Sci.* **2023**, *24*, 15240. [CrossRef] [PubMed]
28. David, E.G.; Luminita, M.V. Intercalative Binding of Small Molecules to Nucleic Acids. *Curr. Org. Chem.* **2000**, *4*, 915–9
[CrossRef]
29. Ralhan, R.; Kaur, J. Alkylating agents and cancer therapy. *Expert Opin. Ther. Pat.* **2007**, *17*, 1061–1075. [CrossRef]
30. Guerra, C.F.; Bickelhaupt, F.M. CHAPTER 4-Watson–Crick hydrogen bonds: Nature and role in DNA replication. In *Mod
Methods for Theoretical Physical Chemistry of Biopolymers*; Starikov, E.B., Lewis, J.P., Tanaka, S., Eds.; Elsevier Science: Amsterda
The Netherlands, 2006; pp. 79–97.
31. Langkjær, N.; Wengel, J.; Pasternak, A. Watson-Crick hydrogen bonding of unlocked nucleic acids. *Bioorg. Med. Chem. Lett.* **20**
25, 5064–5066. [CrossRef] [PubMed]
32. Nelson, S.M.; Ferguson, L.R.; Denny, W.A. DNA and the chromosome-varied targets for chemotherapy. *Cell Chromosome* **2004**
2. [CrossRef] [PubMed]
33. Paul, A.; Bhattacharya, S. Chemistry and biology of DNA-binding small molecules. *Curr. Sci.* **2012**, *102*, 212–231.
34. Highley, M.S.; Landuyt, B.; Prenen, H.; Harper, P.G.; De Bruijn, E.A. The Nitrogen Mustards. *Pharmacol. Rev.* **2022**, *74*, 552–5
[CrossRef]
35. Chen, Y.; Jia, Y.; Song, W.; Zhang, L. Therapeutic Potential of Nitrogen Mustard Based Hybrid Molecules. *Front. Pharmacol.* **20**
9, 1453. [CrossRef]
36. Li, J.J.; Li, J.J. Cancer Drugs: From Nitrogen Mustards to Gleevec. In *Laughing Gas, Viagra, and Lipitor: The Human Stories Behind
Drugs We Use*; Oxford University Press: Oxford, UK, 2006.
37. Ogino, M.H.; Tadi, P. Cyclophosphamide. In *StatPearls*; StatPearls Publishing LLC.: Treasure Island, FL, USA, 2024.
38. Idle, J.R.; Beyoğlu, D. Ifosfamide-History, efficacy, toxicity and encephalopathy. *Pharmacol. Ther.* **2023**, *243*, 108366. [CrossRef
39. Scripture, C.D.; Sparreboom, A.; Figg, W.D. Modulation of cytochrome P450 activity: Implications for cancer therapy. *Lan
Oncol.* **2005**, *6*, 780–789. [CrossRef]
40. Burcham, P.C.; Fontaine, F.R.; Kaminskas, L.M.; Petersen, D.R.; Pyke, S.M. Protein adduct-trapping by hydrazinophthalaz
drugs: Mechanisms of cytoprotection against acrolein-mediated toxicity. *Mol. Pharmacol.* **2004**, *65*, 655–664. [CrossRef]
41. Martinez-Sanchez, J.; Pascual-Diaz, R.; Palomo, M.; Moreno-Castaño, A.B.; Ventosa, H.; Salas, M.Q.; Rovira, M.; Escolar,
Carreras, E.; Diaz-Ricart, M. Mafosfamide, a cyclophosphamide analog, causes a proinflammatory response and increas
permeability on endothelial cells in vitro. *Bone Marrow Transplant.* **2023**, *58*, 407–413. [CrossRef] [PubMed]
42. Moon, K.-Y.; Kwon, C.-H. N3-methyl-mafosfamide as a chemically stable, alternative prodrug of mafosfamide. *Bioorganic M
Chem. Lett.* **1998**, *8*, 1673–1678. [CrossRef] [PubMed]
43. May-Manke, A.; Kroemer, H.; Hempel, G.; Bohnenstengel, F.; Hohenlöchter, B.; Blaschke, G.; Boos, J. Investigation of the ma
human hepatic cytochrome P450 involved in 4-hydroxylation and N-dechloroethylation of trofosfamide. *Cancer Chemot
Pharmacol.* **1999**, *44*, 327–334. [CrossRef] [PubMed]
44. Liang, J.; Huang, M.; Duan, W.; Yu, X.Q.; Zhou, S. Design of new oxazaphosphorine anticancer drugs. *Curr. Pharm. Des.* **2007**,
963–978. [CrossRef]
45. O'Byrne, K.; Steward, W.P. The role of chemotherapy in the treatment of adult soft tissue sarcomas. *Oncology* **1999**, *56*, 13–
[CrossRef]
46. Advani, S.H. The role of ifosfamide in paediatric cancer. *Aust. N. Z. J. Med.* **1998**, *28*, 410–413. [CrossRef]
47. Dayyani, F.; Macarulla, T.; Johnson, A.; Wainberg, Z.A. Second-line treatment options for patients with metastatic pancrea
ductal adenocarcinoma: A systematic literature review. *Cancer Treat. Rev.* **2023**, *113*, 102502. [CrossRef]
48. Diethelm-Varela, B.; Ai, Y.; Liang, D.; Xue, F. Nitrogen Mustards as Anticancer Chemotherapies: Historic Perspective, Curre
Developments and Future Trends. *Curr. Top. Med. Chem.* **2019**, *19*, 691–712. [CrossRef]
49. Summerfield, G.P.; Taylor, P.R.; Mounter, P.J.; Proctor, S.J. High-dose chlorambucil for the treatment of chronic lymphocy
leukaemia and low-grade non-Hodgkin's lymphoma. *Br. J. Haematol.* **2002**, *116*, 781–786. [CrossRef] [PubMed]
50. Lalic, H.; Aurer, I.; Batinic, D.; Visnjic, D.; Smoljo, T.; Babic, A. Bendamustine: A review of pharmacology, clinical use a
immunological effects (Review). *Oncol. Rep.* **2022**, *47*, 114. [CrossRef] [PubMed]

1. Singh, R.K.; Kumar, S.; Prasad, D.N.; Bhardwaj, T.R. Therapeutic journey of nitrogen mustard as alkylating anticancer agents: Historic to future perspectives. *Eur. J. Med. Chem.* **2018**, *151*, 401–433. [CrossRef] [PubMed]
2. Saha, P.; Debnath, C.; Bérubé, G. Steroid-linked nitrogen mustards as potential anticancer therapeutics: A review. *J. Steroid Biochem. Mol. Biol.* **2013**, *137*, 271–300. [CrossRef] [PubMed]
3. Scholar, E. Estramustine. In *xPharm: The Comprehensive Pharmacology Reference*; Enna, S.J., Bylund, D.B., Eds.; Elsevier: New York, NY, USA, 2007; pp. 1–5.
4. Bastholt, L.; Johansson, C.J.; Pfeiffer, P.; Svensson, L.; Johansson, S.A.; Gunnarsson, P.O.; Mouridsen, H. A pharmacokinetic study of prednimustine as compared with prednisolone plus chlorambucil in cancer patients. *Cancer Chemother. Pharmacol.* **1991**, *28*, 205–210. [CrossRef] [PubMed]
5. Zolottsev, V.A.; Latysheva, A.S.; Pokrovsky, V.S.; Khan, I.I.; Misharin, A.Y. Promising applications of steroid conjugates for cancer research and treatment. *Eur. J. Med. Chem.* **2021**, *210*, 113089. [CrossRef]
6. Trafalis, D.T.P. Hybrid aza-steroid alkylators in the treatment of colon cancer. *Cancer Lett.* **2006**, *243*, 202–210. [CrossRef]
7. Trafalis, D.; Geromichalou, E.; Dalezis, P.; Nikoleousakos, N.; Sarli, V. Synthesis and evaluation of new steroidal lactam conjugates with aniline mustards as potential antileukemic therapeutics. *Steroids* **2016**, *115*, 1–8. [CrossRef]
8. Wickström, M.; Nygren, P.; Larsson, R.; Harmenberg, J.; Lindberg, J.; Sjöberg, P.; Jerling, M.; Lehmann, F.; Richardson, P.; Anderson, K.; et al. Melflufen—A peptidase-potentiated alkylating agent in clinical trials. *Oncotarget* **2017**, *8*, 66641–66655. [CrossRef]
9. Mateos, M.V.; Bladé, J.; Bringhen, S.; Ocio, E.M.; Efebera, Y.; Pour, L.; Gay, F.; Sonneveld, P.; Gullbo, J.; Richardson, P.G. Melflufen: A Peptide-Drug Conjugate for the Treatment of Multiple Myeloma. *J. Clin. Med.* **2020**, *9*, 3120. [CrossRef]
10. Lehmann, F.; Wennerberg, J. Evolution of Nitrogen-Based Alkylating Anticancer Agents. *Processes* **2021**, *9*, 377. [CrossRef]
11. Olivier, T.; Prasad, V. The approval and withdrawal of melphalan flufenamide (melflufen): Implications for the state of the FDA. *Transl. Oncol.* **2022**, *18*, 101374. [CrossRef] [PubMed]
12. Persmark, M.; Guengerich, F.P. Spectroscopic and thermodynamic characterization of the interaction of N7-guanyl thioether derivatives of d(TGCTG*CAAG) with potential complements. *Biochemistry* **1994**, *33*, 14368. [CrossRef]
13. Ryan, B.J.; Yang, H.; Bacurio, J.H.T.; Smith, M.R.; Basu, A.K.; Greenberg, M.M.; Freudenthal, B.D. Structural Dynamics of a Common Mutagenic Oxidative DNA Lesion in Duplex DNA and during DNA Replication. *J. Am. Chem. Soc.* **2022**, *144*, 8054–8065. [CrossRef]
14. Burgos-Morón, E.; Pastor, N.; Orta, M.L.; Jiménez-Alonso, J.J.; Palo-Nieto, C.; Vega-Holm, M.; Vega-Pérez, J.M.; Iglesias-Guerra, F.; Mateos, S.; López-Lázaro, M.; et al. In Vitro Anticancer Activity and Mechanism of Action of an Aziridinyl Galactopyranoside. *Biomedicines* **2022**, *10*, 41. [CrossRef] [PubMed]
15. Eder, J.P.; Elias, A.; Shea, T.C.; Schryber, S.M.; Teicher, B.A.; Hunt, M.; Burke, J.; Siegel, R.; Schnipper, L.; Frei, E., 3rd. A phase I-II study of cyclophosphamide, thiotepa, and carboplatin with autologous bone marrow transplantation in solid tumor patients. *J. Clin. Oncol.* **1990**, *8*, 1239–1245. [CrossRef]
16. Langer, C.J.; Nash, S.; Catalano, R.; Rosenblum, N.G.; Hogan, W.M.; Comis, R.L.; O'Dwyer, P.J. Phase II trial of thio-TEPA in relapsed and refractory ovarian carcinoma. *Gynecol. Oncol.* **1991**, *43*, 242–246. [CrossRef] [PubMed]
17. Teicher, B.A.; Herman, T.S.; Holden, S.A.; Epelbaum, R.; Liu, S.-d.; Frei, E., III. Lonidamine as a modulator of alkylating agent activity in vitro and in vivo. *Cancer Res.* **1991**, *51*, 780–784.
18. Miller, B.; Tenenholz, T.; Egorin, M.J.; Sosnovsky, G.; Rao, N.U.M.; Gutierrez, P.L. Cellular pharmacology of N, N′, N″-triethylene thiophosphoramide. *Cancer Lett.* **1988**, *41*, 157–168. [CrossRef]
19. MILLER, D.G. ALKYLATING AGENTS AND HUMAN SPERMATOGENESIS. *Obstet. Gynecol. Surv.* **1972**, *27*, 150–151. [CrossRef]
20. Van Maanen, M.; Smeets, C.; Beijnen, J. Chemistry, pharmacology and pharmacokinetics of N, N′, N″-triethylenethiophosphoramide (ThioTEPA). *Cancer Treat. Rev.* **2000**, *26*, 257–268. [CrossRef] [PubMed]
21. Beall, H.D.; Winski, S. Mechanisms of action of quinone-containing alkylating agents. I: NQO1-directed drug development. *Front. Biosci.* **2000**, *5*, D639–D648. [PubMed]
22. Begleiter, A. Clinical applications of quinone-containing alkylating agents. *Front. Biosci.* **2000**, *5*, 153–171. [CrossRef] [PubMed]
23. Bass, P.D.; Gubler, D.A.; Judd, T.C.; Williams, R.M. Mitomycinoid alkaloids: Mechanism of action, biosynthesis, total syntheses, and synthetic approaches. *Chem. Rev.* **2013**, *113*, 6816–6863. [CrossRef] [PubMed]
24. Puyo, S.; Montaudon, D.; Pourquier, P. From old alkylating agents to new minor groove binders. *Crit. Rev. Oncol./Hematol.* **2014**, *89*, 43–61. [CrossRef] [PubMed]
25. DeVita, V.T.; Lawrence, T.S.; Rosenberg, S.A. *DeVita, Hellman, and Rosenberg's Cancer: Principles & Practice of Oncology*; Lippincott Williams & Wilkins: Philadelphia, PA, USA, 2008; Volume 2.
26. Haffty, B.G.; Son, Y.H.; Wilson, L.D.; Papac, R.; Fischer, D.; Rockwell, S.; Sartorelli, A.C.; Ross, D.; Sasaki, C.T.; Fischer, J.J. Bioreductive alkylating agent porfiromycin in combination with radiation therapy for the management of squamous cell carcinoma of the head and neck. *Radiat. Oncol. Investig. Clin. Basic Res.* **1997**, *5*, 235–245. [CrossRef]
27. Wolkenberg, S.E.; Boger, D.L. Mechanisms of in situ activation for DNA-targeting antitumor agents. *Chem. Rev.* **2002**, *102*, 2477–2496. [CrossRef]
28. Paz, M.M.; Pritsos, C.A. Chapter Seven-The Molecular Toxicology of Mitomycin C. In *Advances in Molecular Toxicology*; Fishbein, J.C., Ed.; Elsevier: Amsterdam, The Netherlands, 2012; Volume 6, pp. 243–299.

79. Andrés, C.M.C.; Pérez de la Lastra, J.M.; Bustamante Munguira, E.; Andrés Juan, C.; Pérez-Lebeña, E. Michael Acceptors a Anti-Cancer Compounds: Coincidence or Causality? *Int. J. Mol. Sci.* **2024**, *25*, 6099. [CrossRef]
80. Tomasz, M.; Palom, Y. The mitomycin bioreductive antitumor agents: Cross-linking and alkylation of DNA as the molecular bas of their activity. *Pharmacol. Ther.* **1997**, *76*, 73–87. [CrossRef]
81. Tomasz, M.; Chawla, A.K.; Lipman, R. Mechanism of monofunctional and bifunctional alkylation of DNA by mitomycin C *Biochemistry* **1988**, *27*, 3182–3187. [CrossRef]
82. Eagan, R.T.; Dinapoli, R.P.; Cascino, T.L.; Scheithauer, B.; O'Neill, B.P.; O'Fallon, J.R. Comprehensive phase II evaluation c aziridinylbenzoquinone (AZQ, diaziquone) in recurrent human primary brain tumors. *J. Neuro-Oncol.* **1987**, *5*, 309–314. [CrossRe [PubMed]
83. Obe, G.; Beek, B. Trenimon: Biochemical, physiological and genetic effects on cells and organisms. *Mutat. Res.* **1979**, *65*, 21–7 [CrossRef] [PubMed]
84. Phillips, R.M.; Hendriks, H.R.; Peters, G.J. EO9 (Apaziquone): From the clinic to the laboratory and back again. *Br. J. Pharmaco* **2013**, *168*, 11–18. [CrossRef] [PubMed]
85. Pierce, S.E.; Guziec, L.J.; Guziec, F.S., Jr.; Brodbelt, J.S. Characterization of Aziridinylbenzoquinone DNA Cross-Links by Liqui Chromatography— Infrared Multiphoton Dissociation— Mass Spectrometry. *Chem. Res. Toxicol.* **2010**, *23*, 1097–1104. [CrossRef] [PubMed]
86. Hargreaves, R.; Hartley, J.A.; Butler, J. Mechanisms of action of quinone-containing alkylating agents: DNA alkylation b aziridinylquinones. *Front. Biosci.* **2000**, *5*, E172–E180. [CrossRef]
87. Naylor, M.; Thomson, P. Recent advances in bioreductive drug targeting. *Mini Rev. Med. Chem.* **2001**, *1*, 17–29. [CrossRef]
88. Fourie, J.; Guziec, F., Jr.; Guziec, L.; Monterrosa, C.; Fiterman, D.J.; Begleiter, A. Structure-activity study with bioreductiv benzoquinone alkylating agents: Effects on DT-diaphorase-mediated DNA crosslink and strand break formation in relation t mechanisms of cytotoxicity. *Cancer Chemother. Pharmacol.* **2004**, *53*, 191–203. [CrossRef]
89. Danson, S.; Ward, T.H.; Butler, J.; Ranson, M. DT-diaphorase: A target for new anticancer drugs. *Cancer Treat. Rev.* **2004**, *30* 437–449. [CrossRef]
90. Denny, W.A. Prodrug strategies in cancer therapy. *Eur. J. Med. Chem.* **2001**, *36*, 577–595. [CrossRef]
91. Zhou, R.; Skibo, E.B. Chemistry of the pyrrolo [1, 2-a] benzimidazole antitumor agents: Influence of the 7-substituent on th ability to alkylate DNA and inhibit topoisomerase II. *J. Med. Chem.* **1996**, *39*, 4321–4331. [CrossRef] [PubMed]
92. Li, B.; Gutierrez, P.L.; Amstad, P.; Blough, N.V. Hydroxyl radical production by mouse epidermal cell lines in the presence o quinone anti-cancer compounds. *Chem. Res. Toxicol.* **1999**, *12*, 1042–1049. [CrossRef] [PubMed]
93. Hartley, J.; O'hare, C.; Baumgart, J. DNA alkylation and interstrand cross-linking by treosulfan. *Br. J. Cancer* **1999**, *79*, 264–26 [CrossRef] [PubMed]
94. Szebeni, J.; Barna, K.; Uher, F.; Milosevits, J.; Pálóczi, K.; Gaál, D.; Petrányi, G.; Kelemen, E. Comparison of the lymphoid toxicitie of mitobronitol and busulphan in mice: Reduced B cell toxicity and improved thymic recovery as possible contributors to th reduced risk for complications following BMT with mitobronitol preconditioning. *Leukemia* **1997**, *11*, 1769–1774. [CrossRef [PubMed]
95. Schepartz, S.A. Early history and development of the nitrosoureas. *Cancer Treat. Rep.* **1976**, *60*, 647. [PubMed]
96. Kohn, K.W. Interstrand cross-linking of DNA by 1, 3-bis (2-chloroethyl)-1-nitrosourea and other 1-(2-haloethyl)-1-nitrosoureas *Cancer Res.* **1977**, *37*, 1450–1454. [PubMed]
97. Panasci, L.C.; Green, D.; Nagourney, R.; Fox, P.; Schein, P.S. A structure-activity analysis of chemical and biological parameters o chloroethylnitrosoureas in mice. *Cancer Res.* **1977**, *37*, 2615–2618. [PubMed]
98. Samson, M.; Baker, L.; Cummings, G.; Talley, R.; McDonald, B.; Bhathena, D. Clinical trial of chlorozotocin, DTIC, an dactinomycin in metastatic malignant melanoma. *Cancer Treat. Rep.* **1982**, *66*, 371–373.
99. Montgomery, J.A.; James, R.; McCaleb, G.S.; Johnston, T.P. The modes of decomposition of 1, 3-bis (2-chloroethyl)-1-nitrosoure and related compounds. *J. Med. Chem.* **1967**, *10*, 668–674. [CrossRef] [PubMed]
100. Buggia, I.; Locatelli, F.; Regazzi, M.B.; Zecca, M. Busulfan. *Ann. Pharmacother.* **1994**, *28*, 1055–1062. [CrossRef]
101. Krivoy, N.; Hoffer, E.; Lurie, Y.; Bentur, Y.; Rowe, J.M. Busulfan use in hematopoietic stem cell transplantation: Pharmacolog dose adjustment, safety and efficacy in adults and children. *Curr. Drug Saf.* **2008**, *3*, 60–66. [CrossRef]
102. Iwamoto, T.; Hiraku, Y.; Oikawa, S.; Mizutani, H.; Kojima, M.; Kawanishi, S. DNA intrastrand cross-link at the 5′-GA-3′ sequenc formed by busulfan and its role in the cytotoxic effect. *Cancer Sci.* **2004**, *95*, 454–458. [CrossRef] [PubMed]
103. Myers, A.L.; Kawedia, J.D.; Champlin, R.E.; Kramer, M.A.; Nieto, Y.; Ghose, R.; Andersson, B.S. Clarifying busulfan metabolism and drug interactions to support new therapeutic drug monitoring strategies: A comprehensive review. *Expert Opin. Drug Metab Toxicol.* **2017**, *13*, 901–923. [CrossRef] [PubMed]
104. Farmer, P. Metabolism and reactions of alkylating agents. *Pharmacol. Ther.* **1987**, *35*, 301–358. [CrossRef] [PubMed]
105. Bedford, P.; Fox, B.W. DNA-DNA interstrand crosslinking by dimethanesulphonic acid esters: Correlation with cytotoxicity and antitumour activity in the yoshida lymphosarcoma model and relationship to chain length. *Biochem. Pharmacol.* **1983**, *32* 2297–2301. [CrossRef]
106. Serrone, L.; Zeuli, M.; Sega, F.M.; Cognetti, F. Dacarbazine-based chemotherapy for metastatic melanoma: Thirty-year experience overview. *J. Exp. Clin. Cancer Res.* **2000**, *19*, 21–34.

7. Reid, J.M.; Kuffel, M.J.; Miller, J.K.; Rios, R.; Ames, M.M. Metabolic activation of dacarbazine by human cytochromes P450: The role of CYP1A1, CYP1A2, and CYP2E1. *Clin. Cancer Res.* **1999**, *5*, 2192–2197.
8. Nussbaumer, S.; Bonnabry, P.; Veuthey, J.L.; Fleury-Souverain, S. Analysis of anticancer drugs: A review. *Talanta* **2011**, *85*, 2265–2289. [CrossRef]
9. Moody, C.L.; Wheelhouse, R.T. The medicinal chemistry of imidazotetrazine prodrugs. *Pharmaceuticals* **2014**, *7*, 797–838. [CrossRef] [PubMed]
0. García-del-Muro, X.; López-Pousa, A.; Maurel, J.; Martín, J.; Martínez-Trufero, J.; Casado, A.; Gómez-España, A.; Fra, J.; Cruz, J.; Poveda, A. Randomized phase II study comparing gemcitabine plus dacarbazine versus dacarbazine alone in patients with previously treated soft tissue sarcoma: A Spanish Group for Research on Sarcomas study. *J. Clin. Oncol.* **2011**, *29*, 2528–2533. [CrossRef]
1. Bouraoui, S.; Brahem, A.; Tabka, F.; Mrizek, N.; Saad, A.; Elghezal, H. Assessment of chromosomal aberrations, micronuclei and proliferation rate index in peripheral lymphocytes from Tunisian nurses handling cytotoxic drugs. *Environ. Toxicol. Pharmacol.* **2011**, *31*, 250–257. [CrossRef] [PubMed]
2. Meer, L.; Janzer, R.C.; Kleihues, P.; Kolar, G.F. In vivo metabolism and reaction with DNA of the cytostatic agent, 5-(3,3-dimethyl-1-triazeno)imidazole-4-carboxamide (DTIC). *Biochem. Pharmacol.* **1986**, *35*, 3243–3247. [CrossRef] [PubMed]
3. Kyrtopoulos, S.A.; Souliotis, V.L.; Valavanis, C.; Boussiotis, V.A.; Pangalis, G.A. Accumulation of O6-methylguanine in human DNA after therapeutic exposure to methylating agents and its relationship with biological effects. *Environ. Health Perspect.* **1993**, *99*, 143–147. [CrossRef] [PubMed]
4. Horton, J.K.; Stevens, M.F. A new light on the photo-decomposition of the antitumour drug DTIC. *J. Pharm. Pharmacol.* **1981**, *33*, 808–811. [CrossRef] [PubMed]
5. Iwamoto, T.; Hiraku, Y.; Okuda, M.; Kawanishi, S. Mechanism of UVA-dependent DNA damage induced by an antitumor drug dacarbazine in relation to its photogenotoxicity. *Pharm. Res.* **2008**, *25*, 598–604. [CrossRef] [PubMed]
6. Newlands, E.S.; Stevens, M.F.; Wedge, S.R.; Wheelhouse, R.T.; Brock, C. Temozolomide: A review of its discovery, chemical properties, pre-clinical development and clinical trials. *Cancer Treat. Rev.* **1997**, *23*, 35–61. [CrossRef] [PubMed]
7. Baker, S.D.; Wirth, M.; Statkevich, P.; Reidenberg, P.; Alton, K.; Sartorius, S.E.; Dugan, M.; Cutler, D.; Batra, V.; Grochow, L.B.; et al. Absorption, metabolism, and excretion of 14C-temozolomide following oral administration to patients with advanced cancer. *Clin. Cancer Res.* **1999**, *5*, 309–317. [PubMed]
8. Turci, R.; Sottani, C.; Spagnoli, G.; Minoia, C. Biological and environmental monitoring of hospital personnel exposed to antineoplastic agents: A review of analytical methods. *J. Chromatogr. B Analyt. Technol. Biomed. Life Sci.* **2003**, *789*, 169–209. [CrossRef]
9. Ioannou-Ttofa, L.; Fatta-Kassinos, D. Cytostatic drug residues in wastewater treatment plants: Sources, removal efficiencies and current challenges. In *Fate and Effects of Anticancer Drugs in the Environment*; Springer: Cham, Switzerland, 2020; pp. 103–138.
0. Li, D.; Chen, H.; Liu, H.; Schlenk, D.; Mu, J.; Lacorte, S.; Ying, G.-G.; Xie, L. Anticancer drugs in the aquatic ecosystem: Environmental occurrence, ecotoxicological effect and risk assessment. *Environ. Int.* **2021**, *153*, 106543. [CrossRef]
1. Tripathi, A.K.; David, A.; Govil, T.; Rauniyar, S.; Rathinam, N.K.; Goh, K.M.; Sani, R.K. Environmental Remediation of Antineoplastic Drugs: Present Status, Challenges, and Future Directions. *Processes* **2020**, *8*, 747. [CrossRef]
2. Tornesello, A.L.; Borrelli, A.; Buonaguro, L.; Buonaguro, F.M.; Tornesello, M.L. Antimicrobial peptides as anticancer agents: Functional properties and biological activities. *Molecules* **2020**, *25*, 2850. [CrossRef]
3. Mader, J.S.; Hoskin, D.W. Cationic antimicrobial peptides as novel cytotoxic agents for cancer treatment. *Expert Opin. Investig. Drugs* **2006**, *15*, 933–946. [CrossRef] [PubMed]
4. Roudi, R.; Syn, N.L.; Roudbary, M. Antimicrobial peptides as biologic and immunotherapeutic agents against cancer: A comprehensive overview. *Front. Immunol.* **2017**, *8*, 1320. [CrossRef] [PubMed]
5. Deslouches, B.; Di, Y.P. Antimicrobial peptides with selective antitumor mechanisms: Prospect for anticancer applications. *Oncotarget* **2017**, *8*, 46635. [CrossRef] [PubMed]
6. Erdem Büyükkiraz, M.; Kesmen, Z. Antimicrobial peptides (AMPs): A promising class of antimicrobial compounds. *J. Appl. Microbiol.* **2022**, *132*, 1573–1596.
7. Wang, G.S. *Antimicrobial Peptides: Discovery, Design and Novel Therapeutic Strategies*; Cabi: Wallingford, UK, 2017.
8. Narayana, J.L.; Chen, J.-Y. Antimicrobial peptides: Possible anti-infective agents. *Peptides* **2015**, *72*, 88–94. [CrossRef] [PubMed]
9. Li, S.; Wang, Y.; Xue, Z.; Jia, Y.; Li, R.; He, C.; Chen, H. The structure-mechanism relationship and mode of actions of antimicrobial peptides: A review. *Trends Food Sci. Technol.* **2021**, *109*, 103–115.
0. Decker, A.P.; Mechesso, A.F.; Wang, G. Expanding the landscape of amino acid-rich antimicrobial peptides: Definition, deployment in nature, implications for peptide design and therapeutic potential. *Int. J. Mol. Sci.* **2022**, *23*, 12874. [CrossRef] [PubMed]
1. Luong, H.X.; Thanh, T.T.; Tran, T.H. Antimicrobial peptides–Advances in development of therapeutic applications. *Life Sci.* **2020**, *260*, 118407.
2. Ramos-Martín, F.; Herrera-León, C.; d'Amelio, N. Molecular basis of the anticancer, apoptotic and antibacterial activities of Bombyx mori Cecropin A. *Arch. Biochem. Biophys.* **2022**, *715*, 109095. [CrossRef] [PubMed]
3. de Almeida Gomes, I.; da Lima, A.B.; da Silva Brito, D.M.; Almeida Lima, A.; de Oliveira, F.L.; Espino Zelaya, E.A.; Magalhães Rebello Alencar, L.; Castelo Branco de Souza Collares Maia, D.; Amaral de Moraes, M.E.; Pantoja Mesquita, F. Recalculating the Route: Repositioning Antimicrobial Peptides for Cancer Treatment. *Chem. Biodivers.* **2024**, *21*, e202301840. [CrossRef]

134. Tolos, A.M.; Moisa, C.; Dochia, M.; Popa, C.; Copolovici, L.; Copolovici, D.M. Anticancer potential of antimicrobial peptic Focus on buforins. *Polymers* **2024**, *16*, 728. [CrossRef] [PubMed]
135. Kanaujia, K.A.; Mishra, N.; Rajinikanth, P.; Saraf, S.A. Antimicrobial peptides as antimicrobials for wound care managem A comprehensive review. *J. Drug Deliv. Sci. Technol.* **2024**, *95*, 105570. [CrossRef]
136. van den Boogaard, W.M.; Komninos, D.S.; Vermeij, W.P. Chemotherapy side-effects: Not all DNA damage is equal. *Cancers* **2(** *14*, 627. [CrossRef] [PubMed]
137. Toribio, M.; Campos, E.L.; Monteiro, G.J.; Henriques, J.C.G.; Gonzaga, W.J.C.; Lorenzon, E.N.; Miranda, C.S.S. The combinatior peptides and cisplatin for the treatment of oral cancer. *Oral Surg. Oral Med. Oral Pathol. Oral Radiol.* **2024**, *137*, e308. [CrossR
138. Khine, H.E.E.; Ecoy, G.A.U.; Roytrakul, S.; Phaonakrop, N.; Pornputtapong, N.; Prompetchara, E.; Chanvorachote, P.; Chaoth C. Chemosensitizing activity of peptide from Lentinus squarrosulus (Mont.) on cisplatin-induced apoptosis in human lu cancer cells. *Sci. Rep.* **2021**, *11*, 4060. [CrossRef] [PubMed]
139. Bhattacharjee, R.; Dey, T.; Kumar, L.; Kar, S.; Sarkar, R.; Ghorai, M.; Malik, S.; Jha, N.K.; Vellingiri, B.; Kesari, K.K. Cellu landscaping of cisplatin resistance in cervical cancer. *Biomed. Pharmacother.* **2022**, *153*, 113345. [CrossRef]
140. Aghamiri, S.; Zandsalimi, F.; Raee, P.; Abdollahifar, M.-A.; Tan, S.C.; Low, T.Y.; Najafi, S.; Ashrafizadeh, M.; Zarrabi, Ghanbarian, H. Antimicrobial peptides as potential therapeutics for breast cancer. *Pharmacol. Res.* **2021**, *171*, 105777. [CrossR
141. Teng, Q.-X.; Luo, X.; Lei, Z.-N.; Wang, J.-Q.; Wurpel, J.; Qin, Z.; Yang, D.-H. The multidrug resistance-reversing activity of a no antimicrobial peptide. *Cancers* **2020**, *12*, 1963. [CrossRef]
142. Jafari, A.; Babajani, A.; Sarrami Forooshani, R.; Yazdani, M.; Rezaei-Tavirani, M. Clinical applications and anticancer effects antimicrobial peptides: From bench to bedside. *Front. Oncol.* **2022**, *12*, 819563. [CrossRef]
143. Seyfi, R.; Kahaki, F.A.; Ebrahimi, T.; Montazersaheb, S.; Eyvazi, S.; Babaeipour, V.; Tarhriz, V. Antimicrobial peptides (AM Roles, functions and mechanism of action. *Int. J. Pept. Res. Ther.* **2020**, *26*, 1451–1463. [CrossRef]
144. Aria, H.; Rezaei, M. Immunogenic cell death inducer peptides: A new approach for cancer therapy, current status and fut perspectives. *Biomed. Pharmacother.* **2023**, *161*, 114503. [CrossRef] [PubMed]
145. Tripathi, A.K.; Vishwanatha, J.K. Role of anti-cancer peptides as immunomodulatory agents: Potential and design strate *Pharmaceutics* **2022**, *14*, 2686. [CrossRef] [PubMed]
146. Kardani, K.; Bolhassani, A. Antimicrobial/anticancer peptides: Bioactive molecules and therapeutic agents. *Immunotherapy* **20** *13*, 669–684. [CrossRef] [PubMed]
147. Al-Ostoot, F.H.; Salah, S.; Khamees, H.A.; Khanum, S.A. Tumor angiogenesis: Current challenges and therapeutic opportunit *Cancer Treat. Res. Commun.* **2021**, *28*, 100422. [CrossRef]
148. Hu, Q.; Chen, C.; Lin, Z.; Zhang, L.; Guan, S.; Zhuang, X.; Dong, G.; Shen, J. The antimicrobial peptide Esculentin-1a (1–21) N stimulates wound healing by promoting angiogenesis through the PI3K/AKT pathway. *Biol. Pharm. Bull.* **2023**, *46*, 382–3 [CrossRef]
149. Kang, M.J.; Roh, K.-H.; Lee, J.S.; Lee, J.H.; Park, S.; Lim, D.W. Vascular Endothelial Growth Factor Receptor 1 Targeting Fusi Polypeptides with Stimuli-Responsiveness for Anti-angiogenesis. *ACS Appl. Mater. Interfaces* **2023**, *15*, 32201–32214. [CrossR
150. Del Genio, V.; Bellavita, R.; Falanga, A.; Hervé-Aubert, K.; Chourpa, I.; Galdiero, S. Peptides to overcome the limitations current anticancer and antimicrobial nanotherapies. *Pharmaceutics* **2022**, *14*, 1235. [CrossRef]
151. Basak, D.; Arrighi, S.; Darwiche, Y.; Deb, S. Comparison of anticancer drug toxicities: Paradigm shift in adverse effect profile. *l* **2021**, *12*, 48. [CrossRef]
152. Szlasa, W.; Zendran, I.; Zalesińska, A.; Tarek, M.; Kulbacka, J. Lipid composition of the cancer cell membrane. *J. Bioen Biomembr.* **2020**, *52*, 321–342. [CrossRef]
153. Vedadghavami, A.; Zhang, C.; Bajpayee, A.G. Overcoming negatively charged tissue barriers: Drug delivery using catio peptides and proteins. *Nano Today* **2020**, *34*, 100898. [CrossRef]
154. Divyashree, M.; Mani, M.K.; Reddy, D.; Kumavath, R.; Ghosh, P.; Azevedo, V.; Barh, D. Clinical applications of antimicrob peptides (AMPs): Where do we stand now? *Protein Pept. Lett.* **2020**, *27*, 120–134. [CrossRef] [PubMed]
155. Chernov, A.N.; Kim, A.V.; Skliar, S.S.; Fedorov, E.V.; Tsapieva, A.N.; Filatenkova, T.A.; Chutko, A.L.; Matsko, M.V.; Galimova, E Shamova, O.V. Expression of molecular markers and synergistic anticancer effects of chemotherapy with antimicrobial peptic on glioblastoma cells. *Cancer Chemother. Pharmacol.* **2024**, *93*, 455–469. [CrossRef] [PubMed]
156. Jana, A.; Narula, P.; Chugh, A.; Kulshreshtha, R. Efficient delivery of anti-miR-210 using Tachyplesin, a cell penetrating pepti for glioblastoma treatment. *Int. J. Pharm.* **2019**, *572*, 118789. [CrossRef] [PubMed]
157. Wu, J.; Chen, X.; Zhang, J.; Chen, J.; Wang, Y.; Wei, T.; Ma, J.; Li, Y.; Mo, T.; He, Z.; et al. Tachyplesin induces apoptosis in non-sm cell lung cancer cells and enhances the chemosensitivity of A549/DDP cells to cisplatin by activating Fas and necroptosis pathw *Chem. Biol. Drug Des.* **2021**, *97*, 809–820. [CrossRef]
158. Lin, M.-C.; Lin, S.-B.; Chen, J.-C.; Hui, C.-F.; Chen, J.-Y. Shrimp anti-lipopolysaccharide factor peptide enhances the antitum activity of cisplatin in vitro and inhibits HeLa cells growth in nude mice. *Peptides* **2010**, *31*, 1019–1025. [CrossRef]
159. Chen, L.; Wang, Q.; Jiang, Y.; Xu, L.; Wei, N.; Lu, C.; Chang, C.; Song, D.; Wang, Y.; Wu, L.; et al. A novel anti-angiogene peptide in combination with cisplatin self-assembling into tube-like nanomedicine for oral treatment of gastric cancer. *Chem. E J.* **2024**, *496*, 154169. [CrossRef]
160. Luo, X.; Teng, Q.-X.; Dong, J.-Y.; Yang, D.-H.; Wang, M.; Dessie, W.; Qin, J.-J.; Lei, Z.-N.; Wang, J.-Q.; Qin, Z. Antimicrobial pepti reverses ABCB1-mediated chemotherapeutic drug resistance. *Front. Pharmacol.* **2020**, *11*, 1208. [CrossRef]

61. Teixeira, M.C.; Carbone, C.; Sousa, M.C.; Espina, M.; Garcia, M.L.; Sanchez-Lopez, E.; Souto, E.B. Nanomedicines for the delivery of antimicrobial peptides (AMPs). *Nanomaterials* **2020**, *10*, 560. [CrossRef]
62. Parchebafi, A.; Tamanaee, F.; Ehteram, H.; Ahmad, E.; Nikzad, H.; Haddad Kashani, H. The dual interaction of antimicrobial peptides on bacteria and cancer cells; mechanism of action and therapeutic strategies of nanostructures. *Microb. Cell Factories* **2022**, *21*, 118. [CrossRef]
63. Gera, S.; Kankuri, E.; Kogermann, K. Antimicrobial peptides–unleashing their therapeutic potential using nanotechnology. *Pharmacol. Ther.* **2022**, *232*, 107990. [CrossRef]
64. Tang, Z.; Ma, Q.; Chen, X.; Chen, T.; Ying, Y.; Xi, X.; Wang, L.; Ma, C.; Shaw, C.; Zhou, M. Recent advances and challenges in nanodelivery systems for antimicrobial peptides (AMPs). *Antibiotics* **2021**, *10*, 990. [CrossRef]
65. Ghafari, M.; Haghiralsadat, F.; Khanamani Falahati-pour, S.; Zavar Reza, J. Development of a novel liposomal nanoparticle formulation of cisplatin to breast cancer therapy. *J. Cell. Biochem.* **2020**, *121*, 3584–3592. [CrossRef]
66. Tsvetkova, D.; Ivanova, S. Application of approved cisplatin derivatives in combination therapy against different cancer diseases. *Molecules* **2022**, *27*, 2466. [CrossRef] [PubMed]
67. Renault-Mahieux, M.; Vieillard, V.; Seguin, J.; Espeau, P.; Le, D.T.; Lai-Kuen, R.; Mignet, N.; Paul, M.; Andrieux, K. Co-encapsulation of fisetin and cisplatin into liposomes for glioma therapy: From formulation to cell evaluation. *Pharmaceutics* **2021**, *13*, 970. [CrossRef] [PubMed]
68. Tan, X.; Wang, Y.; Wu, Z.; Zhou, Q.; Tang, Y.; Liu, Z.; Yuan, G.; Luo, S.; Zou, Y.; Guo, S. The role of Her-2 in penile squamous cell carcinoma progression and cisplatin chemoresistance and potential for antibody-drug conjugate-based therapy. *Eur. J. Cancer* **2023**, *194*, 113360. [CrossRef]
69. Fuentes-Antrás, J.; Genta, S.; Vijenthira, A.; Siu, L.L. Antibody–drug conjugates: In search of partners of choice. *Trends Cancer* **2023**, *9*, 339–354. [CrossRef] [PubMed]
70. Işık, G. Development and Characterization of PEG-B-PCL Micelles Carrying Anticancer Agents. Ph.D. Thesis, Middle East Technical University, Ankara, Turkey, 2022.
71. Xia, W.; Tao, Z.; Zhu, B.; Zhang, W.; Liu, C.; Chen, S.; Song, M. Targeted delivery of drugs and genes using polymer nanocarriers for cancer therapy. *Int. J. Mol. Sci.* **2021**, *22*, 9118. [CrossRef] [PubMed]
72. Eskandari, R.; Asoodeh, A.; Mousavi, S.-D.; Firouzi, Z. The effect of a novel drug delivery system using encapsulated antimicrobial peptide protonectin (IL-12) into nano micelle PEG-PCL on A549 adenocarcinoma lung cell line. *J. Polym. Res.* **2021**, *28*, 341. [CrossRef]
73. Gessner, I.; Neundorf, I. Nanoparticles modified with cell-penetrating peptides: Conjugation mechanisms, physicochemical properties, and application in cancer diagnosis and therapy. *Int. J. Mol. Sci.* **2020**, *21*, 2536. [CrossRef]
74. Zhou, M.; Zou, X.; Cheng, K.; Zhong, S.; Su, Y.; Wu, T.; Tao, Y.; Cong, L.; Yan, B.; Jiang, Y. The role of cell-penetrating peptides in potential anti-cancer therapy. *Clin. Transl. Med.* **2022**, *12*, e822. [CrossRef]
75. Singh, T.; Kang, D.H.; Kim, T.W.; Kong, H.J.; Ryu, J.S.; Jeon, S.; Ahn, T.S.; Jeong, D.; Baek, M.J.; Im, J. Intracellular delivery of oxaliplatin conjugate via cell penetrating peptide for the treatment of colorectal carcinoma in vitro and in vivo. *Int. J. Pharm.* **2021**, *606*, 120904. [CrossRef]
76. Izabela, R.; Jarosław, R.; Magdalena, A.; Piotr, R.; Ivan, K. Transportan 10 improves the anticancer activity of cisplatin. *Naunyn-Schmiedeberg's Arch. Pharmacol.* **2016**, *389*, 485–497. [CrossRef] [PubMed]
77. Śmiłowicz, D.; Metzler-Nolte, N. Bioconjugates of Co(III) complexes with Schiff base ligands and cell penetrating peptides: Solid phase synthesis, characterization and antiproliferative activity. *J. Inorg. Biochem.* **2020**, *206*, 111041. [CrossRef]
78. Jiang, M.; Fang, X.; Ma, L.; Liu, M.; Chen, M.; Liu, J.; Yang, Y.; Wang, C. A nucleus-targeting peptide antagonist towards EZH2 displays therapeutic efficacy for lung cancer. *Int. J. Pharm.* **2022**, *622*, 121894. [CrossRef] [PubMed]
79. Safhi, A.Y. Three-dimensional (3D) printing in cancer therapy and diagnostics: Current status and future perspectives. *Pharmaceuticals* **2022**, *15*, 678. [CrossRef] [PubMed]
80. Zhu, H.; Zheng, J.; Oh, X.Y.; Chan, C.Y.; Low, B.Q.L.; Tor, J.Q.; Jiang, W.; Ye, E.; Loh, X.J.; Li, Z. Nanoarchitecture-integrated hydrogel systems toward therapeutic applications. *ACS Nano* **2023**, *17*, 7953–7978. [CrossRef] [PubMed]
81. Binaymotlagh, R.; Chronopoulou, L.; Haghighi, F.H.; Fratoddi, I.; Palocci, C. Peptide-based hydrogels: New materials for biosensing and biomedical applications. *Materials* **2022**, *15*, 5871. [CrossRef]
82. Chinnaiyan, K.; Mugundhan Laakshmi, S.; Narayanasamy, D.; Mohan, M. Revolutionizing Healthcare and Drug Discovery: The Impact of Artificial Intelligence on Pharmaceutical Development. *Curr. Drug Ther.* **2024**, *19*, 1–16. [CrossRef]
83. Saraf, S.; De, A.; Tripathy, B. Effective Use of Computational Biology and Artificial Intelligence in the Domain of Medical Oncology. In *Computational Intelligence for Oncology and Neurological Disorders*; CRC Press: Boca Raton, FL, USA, 2024; pp. 228–252.
84. Dlamini, Z.; Francies, F.Z.; Hull, R.; Marima, R. Artificial intelligence (AI) and big data in cancer and precision oncology. *Comput. Struct. Biotechnol. J.* **2020**, *18*, 2300–2311. [CrossRef]
85. Riaz, I.B.; Khan, M.A.; Haddad, T.C. Potential application of artificial intelligence in cancer therapy. *Curr. Opin. Oncol.* **2024**, *36*, 437–448. [CrossRef]
86. Kim, J.; Kusko, R.; Zeskind, B.; Zhang, J.; Escalante-Chong, R. A primer on applying AI synergistically with domain expertise to oncology. *Biochim. Biophys. Acta (BBA)-Rev. Cancer* **2021**, *1876*, 188548. [CrossRef] [PubMed]
87. Mukherjee, D.; Roy, D.; Thakur, S. Transforming Cancer Care: The Impact of AI-Driven Strategies. *Curr. Cancer Drug Targets* **2024**, *24*, 1–4. [CrossRef] [PubMed]

cancers

MDPI

iew

reen Tea Components: In Vitro and In Vivo Evidence for Their nticancer Potential in Colon Cancer

derica Randisi †, Gianpaolo Perletti †, Emanuela Marras and Marzia Bruna Gariboldi *

Department of Biotechnology and Life Sciences (DBSV), University of Insubria, 21100 Varese, Italy; frandisi1@studenti.uninsubria.it (F.R.); gianpaolo.perletti@uninsubria.it (G.P.); emanuela.marras@uninsubria.it (E.M.)
* Correspondence: marzia.gariboldi@uninsubria.it; Tel.: +39-0331339418
† These authors contributed equally to this work.

Simple Summary: Traditionally used in Asian countries, green tea is nowadays one of the most widely used beverages worldwide. Polyphenols are the primary bioactive compounds found in green tea, with catechins representing the principal constituents. Among these, epigallocatechin-3-gallate (EGCG) is the most abundant and extensively investigated catechin. The pharmacological potential of green tea polyphenols has been well-documented, encompassing antibacterial, anti-inflammatory, antioxidant, antiobesity, antiviral, and anticancer activities. Notably, anticancer properties have been observed against several malignancies, including colorectal cancer. This review aims to comprehensively examine the anticancer effects of green tea constituents in colorectal cancer through in vitro (cell-based) and in vivo (animal-based) studies, with a focus on elucidating their underlying mechanisms of action. Additionally, particular emphasis is placed on clinical studies, including meta-analyses and key investigations involving human participants.

Abstract: Green tea consumption has been implicated in various biological activities, with particular emphasis on its anticancer properties. The antineoplastic effects of green tea are primarily attributed to its rich polyphenol content, among which, epigallocatechin-3-gallate (EGCG) is recognized as the most bioactive and potent catechin, responsible for the majority of its anticancer activity. This review provides a detailed examination of the in vitro and in vivo effects of green tea components, focusing on their potential therapeutic implications in colorectal cancer. The molecular mechanisms of action and bioactive constituents of green tea are systematically discussed, alongside an evaluation of experimental evidence supporting their efficacy. Furthermore, insights into the relationship between green tea dietary intake and colorectal cancer risk are analyzed, with a particular emphasis on clinical data and findings from meta-analyses involving patients diagnosed with colon cancer. The aggregated evidence underscores the necessity for well-designed randomized controlled trials and longitudinal cohort studies to substantiate the role of green tea as a chemopreventive agent. Additionally, future investigations should prioritize determining the optimal dosages, the appropriate durations of consumption, and the potential modulatory effects of dietary or lifestyle factors on green tea's anticancer efficacy.

Keywords: green tea; colorectal cancer; catechins; epigallocatechin-3-gallate; in vitro studies; in vitro studies; human studies

https://doi.org/10.3390/cancers17040623

1. Introduction

Colorectal cancer (CRC) represents a major global health concern, ranking as the third most frequently diagnosed malignancy and the second leading cause of cancer-related mortality worldwide [1–4]. Current projections estimate a substantial rise in CRC burden, with the annual incidence predicted to reach 3.29 million new cases and 1.66 million deaths by 2045 (Global Cancer Observatory: Cancer Tomorrow, accessed on 5 January 2025, available online at https://gco.iarc.fr/tomorrow). Furthermore, the incidence of CRC is strongly correlated with the Human Development Index (HDI), as countries with a high HDI currently exhibit rates approximately four times higher than those observed in low-HDI regions [2,3]. However, longitudinal trends indicate a marked disparity in the growth of colorectal cancer incidence between areas with differing HDI levels. Between 2022 and 2045, colorectal cancer incidence is projected to increase by 111.8% in low-HDI countries, compared to a 67.9% increase in very high-HDI countries (Global Cancer Observatory: Cancer Tomorrow, accessed on 5 January 2025, available online at https://gco.iarc.fr/tomorrow). This anticipated rise in low-HDI regions is closely linked to increased exposure to colorectal cancer risk factors, driven by the adoption of a Westernized lifestyle, a shift often accompanying socioeconomic and industrial transitions [2,4].

Colorectal cancer can be classified into three primary subtypes based on the origin of the underlying genetic alterations: sporadic, inherited, and familial [5]. Sporadic colorectal cancer constitutes approximately 70% of all cases and is characterized by mutations affecting a range of genes. The initial driver mutation often arises in the adenomatous polyposis coli (*APC*) gene, a critical tumor suppressor, promoting the formation of benign adenomas or polyps. Subsequent mutations in other genes, such as *Kristen rat sarcoma* viral oncogene homolog (*KRAS*), *p53*, and *Deleted in Colorectal Cancer* (*DCC*), further drive tumor progression [5,6]. Inherited colorectal cancer constitutes approximately 5% of cases and encompasses both polyposis and non-polyposis syndromes. Polyposis includes familial adenomatous polyposis (FAP), while non-polyposis forms are predominantly linked to deficiencies in DNA mismatch repair proteins, including MSH2, MLH1, MSH6, PMS1, and PMS2, often classified under Lynch syndrome [7,8]. Familial colorectal cancer constitutes about 25% of cases but does not typically result from known germline mutations. Instead, it arises from complex interactions between genetic predispositions (e.g., gene–gene interactions, epigenetic modifications), environmental factors, or their combined effects [7]. Studies indicate that chromosomal abnormalities and translocations affecting key signaling pathways, including *Wnt*, *mitogen-activated protein kinases/phosphatidylinositole 3 kinase* (*MAPK/PI3K*), and (*tumor growth factor-β*) *TGF-β*, along with disruptions in cellular functions such as *p53* and cell-cycle regulation, are also involved in the pathogenesis of familial colorectal cancer [5,9].

According to the World Health Organization (WHO), several factors contribute to an increased risk of developing colorectal cancer. These include aging, a family history of the pathology, hereditary conditions such as familial adenomatous polyposis (FAP), and a personal history of colorectal cancer or adenomatous polyps. In addition to these non-modifiable risk factors, various lifestyle factors play a significant role. Among these, excess body weight is particularly notable as a major risk factor [2,10]. This relationship is believed to be mediated by the proinflammatory cytokines produced by adipose tissue, chronic hyperinsulinemia, and elevated levels of insulin-like growth factor 1 (IGF-I), all of which may facilitate carcinogenesis [11]. Dietary habits have also been extensively studied in the context of colorectal cancer risk. High consumption of processed and red meats has been linked to impaired gut immune responses and increased gut inflammation, both of which may promote tumorigenesis. Conversely, a diet abundant in fiber, fruits, vegetables, dairy products, fish, beta-carotene, vitamins C, E, and D, and folate is associated with a protective effect against colorectal cancer development [12–14]. Lifestyle factors beyond diet also influence colorectal cancer risk. Regular physical activity and the use of aspirin or non-steroidal anti-inflammatory drugs (NSAIDs) have been shown to confer a protective effect [15,16]. In contrast, alcohol consumption and tobacco use are consistently associated with increased colorectal cancer risk, underscoring the critical importance of lifestyle modifications in colorectal cancer prevention [17,18].

The treatment of colorectal cancer is directed by tumor type, stage, and the patient's overall medical history [19]. Early detection is critical as it enables the implementation of more effective therapeutic strategies, often leading to improved prognoses. For resectable colorectal cancer, surgical excision remains the cornerstone of treatment. In cases of non-resectable colorectal cancer, standard therapeutic modalities include chemotherapy, radiotherapy, and immunotherapy [20]. However, these conventional approaches are associated with significant limitations, including non-specific cytotoxic effects on healthy tissues. Moreover, more than 50% of patients experience disease relapse due to the emergence of acquired multidrug resistance [21]. To address these challenges, recent advancements in colorectal cancer therapeutics have focused on strategies to overcome drug resistance and enhance tumor responsiveness to treatment. These include the development of immune checkpoint inhibitors (ICIs), chimeric antigen receptor (CAR) T cell therapy, T cell receptor (TCR) modifications, cytokine-based therapies, and monoclonal antibodies [19]. In addition, innovative approaches such as probiotics, RNA-based therapies, oncolytic viral therapies, and natural product-based treatments have demonstrated promising potential in preclinical and early clinical studies [22–25]. Emerging evidence also highlights the role of dietary factors in colorectal cancer prevention and treatment. Specifically, green tea and its bioactive constituents have attracted interest for their potential anticancer effects, underscoring the need for further research into dietary interventions as complementary strategies in colorectal cancer management [26].

Green tea, constituting approximately 20% of global tea production, is one of the most commonly consumed beverages globally. Derived from the *Camellia sinensis* plant, tea is cultivated in over 30 countries, with China, India, Kenya, Sri Lanka, Taiwan, and Japan recognized as leading producers. Unlike white tea, which undergoes minimal processing and is derived from young leaves and buds, or black tea, which involves full fermentation, green tea is produced from more mature leaves and is distinguished by the lack of fermentation during its processing [27,28]. The chemical composition of green tea varies based on environmental conditions such as soil quality, climate, light exposure, geographic factors, microbial interactions, and temperature [29]. Green tea is available in several varieties, each distinguished by unique flavor profiles and antioxidant properties. The method of tea extraction significantly influences its antioxidant potential.

Green tea consumption is associated with diverse biological activities, including antiobesity, antioxidant, and anticancer effects. The antineoplastic properties of green tea are predominantly attributed to its polyphenol content, particularly epigallocatechin-3-gallate (EGCG), the most potent catechin present in green tea [30]. EGCG has been demonstrated to exert chemopreventive effects by targeting multiple stages of carcinogenesis, such as initiation, promotion, and progression. This compound modulates critical cellular signaling pathways that regulate essential processes, including cell proliferation, apoptosis, and angiogenesis [31]. In addition to its role as a chemopreventive agent, EGCG exhibits significant antioxidant activity. It interacts with endogenous antioxidants such as glutathione and generates reactive oxygen species (ROS) to activate pro-oxidative pathways, leading to selective induction of cancer cell death. This dual antioxidant and pro-oxidant functionality underscores the therapeutic potential of EGCG in cancer prevention and treatment [26].

This article aims to provide a comprehensive overview of green tea's in vitro and in vivo effects, focusing on its potential therapeutic benefits in colorectal cancer. Particular attention was paid to the analysis of clinical trials and meta-analyses related to the use of green tea or green tea catechins in humans.

2. Methods

PubMed, Google Scholar, and Web of Science searches were performed to collect the published data using the keywords of green tea, phytochemical, chemical composition, EGCG, pharmacology, tea polyphenols, antioxidants, colon cancer, colorectal cancer, and toxicology. Several websites and related articles were also incorporated. Additionally, some selected articles were manually searched. The inclusion criteria for this review encompassed systematic reviews and experimental studies on green tea, with no restrictions on the publication timeframe.

An in-depth literature search was also conducted to identify relevant human studies investigating the association between green tea consumption (or green tea extracts) and colorectal cancer risk. Searches were performed across several electronic databases, including Medline, PubMed, and Embase, covering articles in English, published between 1 January 1994 and 30 November 2024. A combination of medical subject headings (MeSH) terms was used as a search strategy to ensure comprehensive retrieval of relevant literature. Key search terms included "cancer*", "colorectal", "carcinoma*", "Green Tea", "Tea", and "Catechin*". A flowchart of the search and screening process is shown in Figure 1.

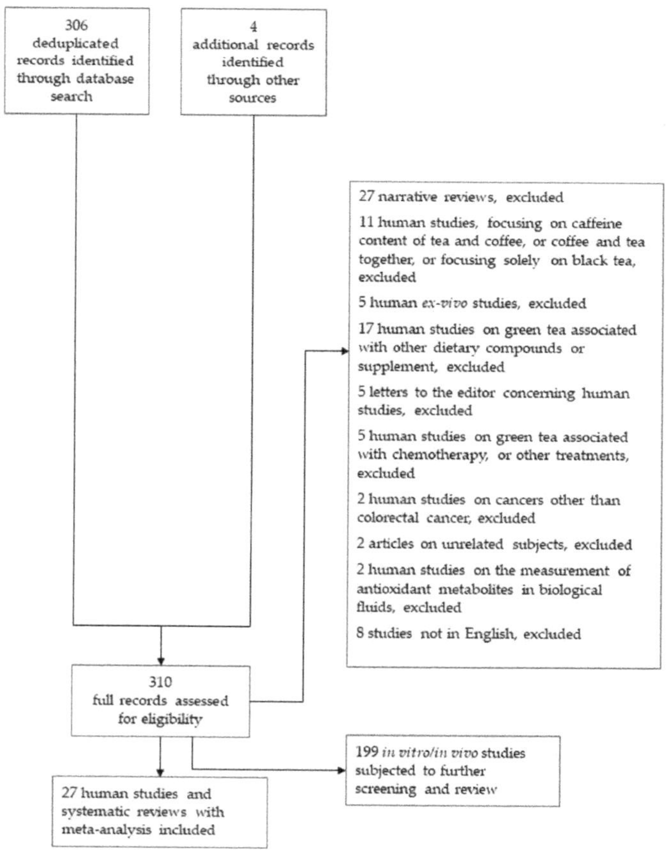

Figure 1. Flowchart of the search and screening process.

3. Green Tea Constituents

Since the early 20th century, over 500 chemical compounds have been isolated from tea, many of which are associated with beneficial effects on human health [32]. In particular, these bioactive molecules have demonstrated a range of pharmacological properties, including antihypertensive, antihyperlipidemic, antiarteriosclerotic, antidiabetic, anticancer, and antioxidative effects, as well as neuroprotective and blood glucose-lowering activities [33–37].

The primary constituents of green tea include polyphenols, alkaloids, carbohydrates, amino acids, aromatic compounds, and various mineral elements (Figure 2) [38,39].

Green tea polyphenols, including catechins, flavonoids, anthocyanins, and phenolic acids, constitute the primary bioactive components of green tea leaves and account for 20–30% of the total dry leaf mass [40]. Among these, catechins are the most abundant and are widely regarded as the primary contributors to the biological activities observed in green tea studies [32,41]. Catechins identified in green tea include catechin (C), epicatechin (EC), epigallocatechin (EGC), epicatechin gallate (ECG), gallocatechin (GC), catechin gallate (CG), gallocatechin gallate (GCG), and epigallocatechin gallate (EGCG). Notably, EGCG comprises approximately 60% of the total catechin content, followed by EGC (about 20%), ECG (about 14%), and EC (about 6%) [42,43]. Notably, black tea also contains specific polyphenolic flavonoids exerting antioxidant activities, such as theaflavin-3-gallate.

In addition to catechins, green tea is rich in flavonoids, such as myricetin, quercetin, and behenyl glycosides. Although anthocyanins, a family of water-soluble pigments, are present in low concentrations, they contribute to the bitter taste of tea, significantly influencing its quality [32,44]. The primary alkaloids in green tea are purine alkaloids, with caffeine being the most prevalent, accounting for 2–5% of the total content. Trace amounts of theophylline and theobromine are also present. Together, these three alkaloids constitute the principal components responsible for tea's invigorating effects [32,45].

Among the primary factors influencing the quality of tea, its amino acid composition plays a pivotal role. Tea typically contains between 1% and 4% amino acids, with a total of 26 distinct amino acids identified to date. These include 20 proteinogenic amino acids and 6 non-proteinogenic amino acids. Prominent amino acids within tea include theanine, glutamic acid, arginine, serine, and aspartic acid. Notably, theanine and γ-aminobutyric acid (GABA) are bioactive compounds in tea that have garnered attention for their neuroprotective properties. Theanine accounts for approximately 50% of the total amino acids present in tea, while GABA is typically found in comparatively lower concentrations.

The aroma of green tea predominantly originates from volatile aromatic compounds, which, although comprising only 0.005% to 0.020% of its chemical composition, exhibit remarkable complexity. Several studies have investigated the volatile constituents of green tea, continually identifying and characterizing novel compounds. In addition to volatiles, water-soluble organic acids significantly contribute to the aroma and flavor of green tea infusions. To date, over 40 organic acids have been identified in tea, including both free organic acids within the infusion and more than 30 associated with aromatic components. Volatile acids such as acetic acid, butyric acid, and hexenoic acid are classified among the aromatic substances responsible for the characteristic scent profile of green tea. Furthermore, green tea contains phenolic acids, including gallic acid, chlorogenic acid, caffeic acid, p-coumaric acid, ellagic acid, quinic acid, tea gallate, and proanthocyanidins. Although these compounds are present in relatively low concentrations, they may contribute to green tea's sensory attributes and potential health benefits. Research focused on the phenolic acid content and its implications within green tea remains limited, warranting further scientific exploration [32,44].

The slight sweetness of tea infusion is attributed to the presence of small quantities of monosaccharides and disaccharides, including glucose, fructose, galactose, and sucrose. However, the majority of carbohydrates in tea are polysaccharides such as cellulose, starch, and pectin, which are predominantly water-insoluble.

The inorganic constituents of tea, commonly referred to as "ash", consist primarily of various mineral elements and their oxides. Ash content serves as a critical parameter in the quality evaluation of tea, particularly for export standards. The most abundant mineral elements include phosphorus (P) and potassium (K), followed by calcium (Ca), magnesium (Mg), iron (Fe), manganese (Mn), aluminum (Al), sulfur (S), and silicon (Si). Trace elements such as zinc (Zn), copper (Cu), and fluoride (F) are also present. These mineral elements not only play essential roles in the physiological functions of tea plants but are also recognized for their potential health benefits in humans, prompting significant scientific investigation.

In addition to its complex chemical composition and micronutrient content, green tea contains a variety of vitamins, including vitamin B, vitamin C, and vitamin E. Moreover, it is a source of active enzymes such as glucosidases and lipoxidases, as well as chlorophyll, a natural edible pigment. These components collectively contribute to the nutritional and sensory attributes of green tea and underscore its value as a functional beverage [32,46,47].

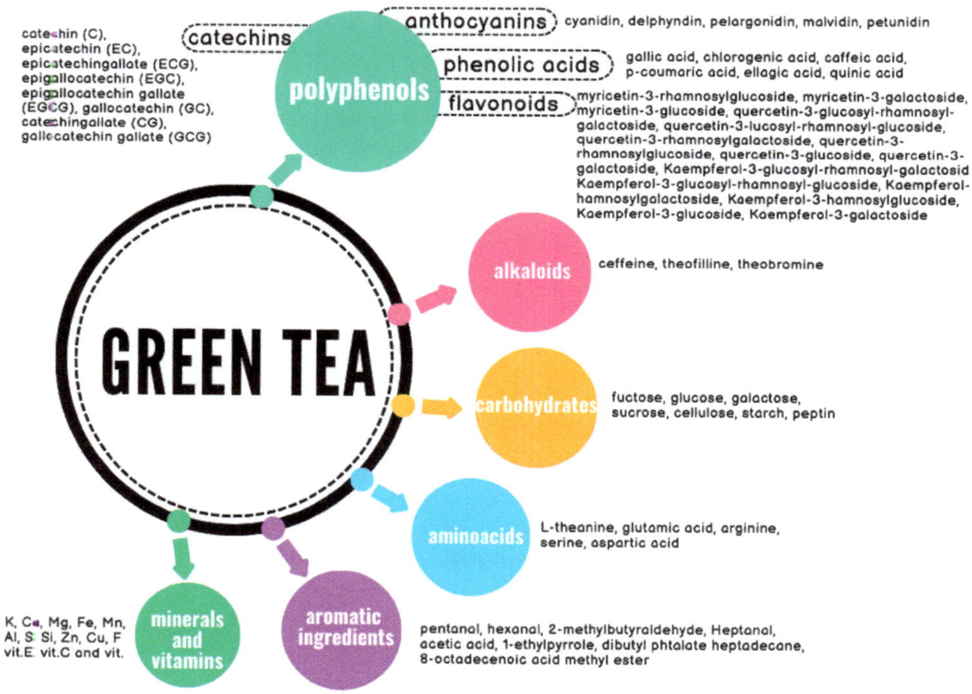

Figure 2. Summary of the main green tea constituents.

4. Green Tea Activities

The majority of the documented biological effects of green tea, summarized in Figure 3, have been primarily attributed to its catechin content. Nevertheless, additional components of green tea, such as theanine and caffeine, have also been implicated in several physiological activities, including antioxidant effects [48]. As previously mentioned, among the catechins, EC, EGC, ECG, and EGCG are the most abundant and biologically active. These catechins contribute to a wide spectrum of health-promoting effects, including antioxidant, antitumor, anti-inflammatory, antimicrobial, antiviral, antidiabetic, antiobesity, and hypotensive properties [49–51]. The synergistic interaction between catechins and other bioactive compounds in green tea underscores its potential as a multifunctional dietary agent with significant therapeutic applications. Further research is warranted to elucidate the specific contributions and mechanisms of these components in mediating green tea's health benefits.

Green tea's anti-inflammatory effects are primarily attributed to its capacity to inhibit protein denaturation and enhance the production of anti-inflammatory cytokines. These properties hold therapeutic potential in various diseases characterized by chronic inflammation, including arthritis, cardiovascular diseases, diabetes, obesity, and cancer [51,52]. Moreover, green tea's combined anti-inflammatory and antioxidant activities contribute to its neuroprotective effects, with evidence suggesting that its bioactive metabolites can cross the blood-brain barrier [53,54]. These properties also underlie green tea's antibacterial efficacy and its role in the management of cardiovascular diseases (CVD) [55].

Regular consumption of green tea has been demonstrated to offer cardioprotective benefits by preventing atherosclerosis, reducing total cholesterol levels, and improving lipid profiles through a more favorable ratio among low-density lipoprotein (LDL) and high-density lipoprotein (HDL) [51,52]. This multifaceted functionality underscores green tea's potential as a dietary intervention for both the prevention and management of inflammatory and metabolic disorders.

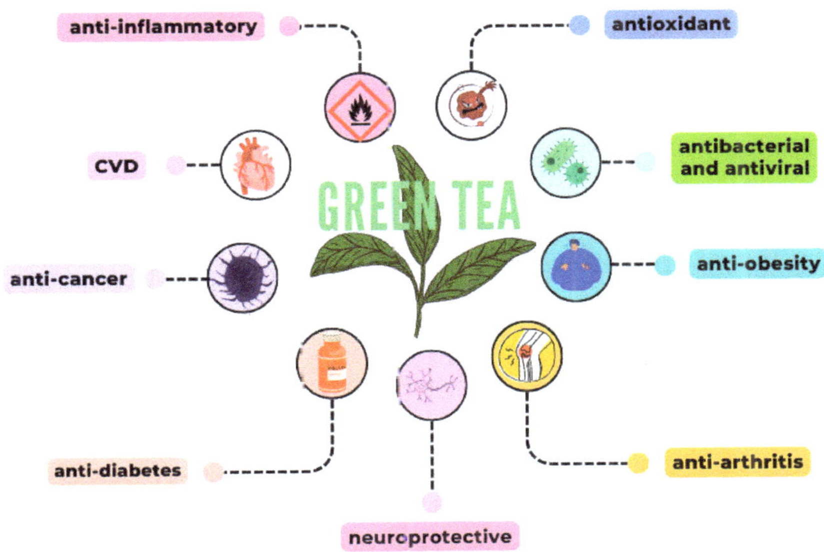

Figure 3. Green tea activities.

5. Anticancer Effects of Green Tea

Among the various types of tea, green tea has been the most extensively studied for its cancer-preventive and therapeutic properties [41,51,56]. Its anticancer effects are predominantly attributed to catechins, with EGCG exhibiting the highest activity, followed by ECG, EGC, and EC [51]. The biological effects of green tea catechins on cancer cells are mediated through multiple mechanisms, including the induction of apoptosis, cell cycle arrest, inhibition of angiogenesis, and modulation of antioxidant and pro-oxidant activities (Figure 4) [32,35,41]. Neutralization of reactive oxygen species (ROS) is a critical mechanism by which green tea catechins inhibit cancer pathogenesis. These compounds demonstrate potent antioxidant activity by scavenging free radicals, chelating transition metals, and upregulating the expression and activity of antioxidant enzymes such as superoxide dismutase (SOD), catalase (CAT), glutathione-S-transferase (GST), glutathione peroxidase (GSP), and glutathione reductase (GSR) [30,57]. Additionally, green tea catechins help prevent cellular and tissue oxidative damage by inhibiting the activity of pro-oxidant enzymes, including cyclooxygenase (COX) and xanthine oxidase [58].

Interestingly, recent evidence suggests that the anticancer effects of green tea may be more pronounced under conditions of excessive oxidative stress, highlighting the potential context-dependent nature of its therapeutic benefits [32,35,36,41,51]. In this regard, green tea polyphenols exhibit pro-oxidant activity through self-oxidation processes that generate hydroxyl radicals, hydrogen peroxide (H_2O_2), and quinonoid intermediates [35,41,59]. This pro-oxidant behavior contributes to their anticancer effects by upregulating *AMP-activated protein kinase* (*AMPK*), which subsequently downregulates *mTOR* and then *nuclear factor of the κ-chain in B-cells* (*NF-κB*) signaling pathways. These pathways are critical for the antineoplastic effects of green tea catechins. Reactive oxygen species (ROS) generated by green tea are integral to the induction of apoptosis, which inhibits cancer cell growth [30]. The pro-apoptotic activity of green tea polyphenols has been linked to their capacity to activate cell death receptors, regulate caspase-8, -3, and -9, and modulate the expression of key apoptosis-regulatory proteins. Specifically, green tea reduces c-Myc, Bcl-2, and Bcl-xL while upregulating Bax, BAD, Apaf-1, Bak, and PUMA protein levels [30,35,60]. Additionally, EGCG has been shown to influence several signaling pathways involved in the regulation of apoptotic cell death, including the *PI3K/AKT/mTOR* (*phosphatidylinositol 3 kinase/protein kinase B/molecular target of rapamycin*), *ERK1/2* (*extracellular signal-regulated kinase 1/2*), and *MAPK* pathways [61].

Autophagy, another critical mechanism in green tea's anticancer effects is activated by polyphenols, likely through the transcription factor Nrf-2 (nuclear factor erythroid 2-related factor 2). This activation has been shown to inhibit the growth of several cancer cell lines [42,62,63].

Green tea polyphenols also inhibit cancer progression by arresting the cell cycle. This effect is predominantly associated with G0/G1 phase arrest, potentially mediated by the regulation of *cyclin D1*, *cdk4*, and *WAF1/p21* expression [32,64]. However, G_2/M phase arrest has also been reported [65]. Furthermore, inhibition of the Wnt/β-catenin signaling pathway has been implicated in EGCG-mediated control of cancer cell proliferation [66].

Green tea catechins also repress tumor angiogenesis. This is achieved by inhibiting the activation of *NF-κB*, *STAT3*, *HIF-1*, and the expression of *vascular endothelial growth factor* (*VEGF*) [30,51]. In addition to their antiangiogenic activity, green tea catechins regulate key processes such as epigenetic modification [67], tumor cell migration, and invasion. These effects are facilitated by modulating the expression of *matrix metalloproteinase-9* (*MMP-9*) and its endogenous inhibitor TIMP-1.

Further insights into the molecular mechanisms of catechins, particularly EGCG, reveal their interaction with (67)-laminin receptors, leading to downregulation of *receptor tyrosine kinase* (*RTK*)-mediated signaling pathways [30,51,68]. EGCG also inhibits the expression of *indoleamine 2,3-dioxygenase* (*IDO*), thereby reversing *IDO*-induced immune tolerance and restoring effective antitumor immunity [35]. The chemopreventive properties of green tea extend to its antiobesity effects, which ameliorate metabolic syndrome, a condition linked to increased carcinogenesis risk [69]. Furthermore, emerging evidence highlights EGCG as a potent epigenetic modulator that rectifies abnormal epigenetic events implicated at various stages of carcinogenesis [41].

Green tea's strong anti-inflammatory effects further contribute to its anticancer activity. These effects involve the suppression of pro-inflammatory cytokine production, inhibition of *cyclooxygenase-2* (*COX-2*) expression, and attenuation of inflammatory signaling pathways such as *STAT3* and *NF-κB* [30,70,71].

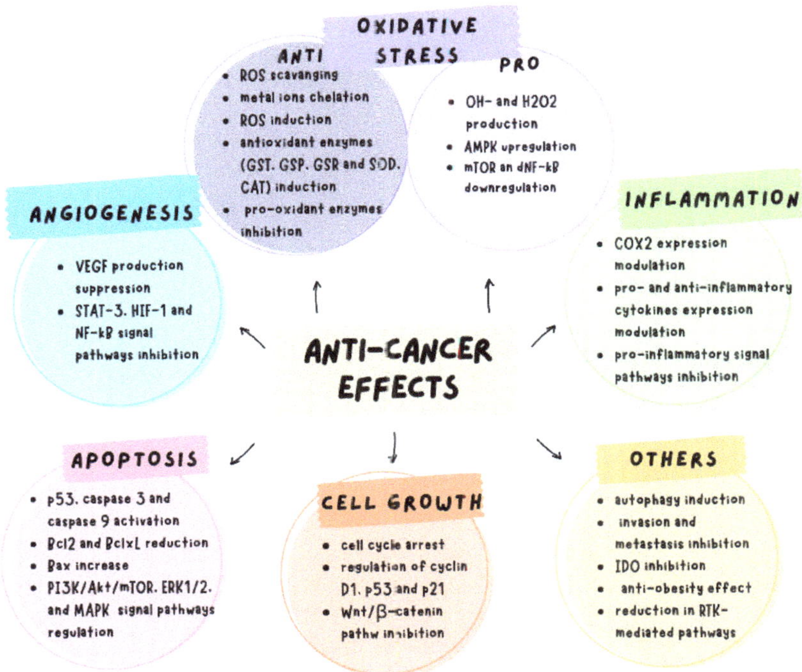

Figure 4. Summary of the anticancer effects of green tea.

6. Effects of Green Tea on Colorectal Cancer

As highlighted previously, green tea and its bioactive components have been extensively investigated for their potential antineoplastic effects, particularly in colorectal cancer. While genetic predispositions significantly contribute to the etiology of colorectal cancer, lifestyle factors, particularly dietary habits, are increasingly recognized as modifiable determinants of cancer risk. Green tea, primarily due to its rich composition of polyphenolic compounds, has demonstrated protective effects against the development of multiple cancer types, including esophageal, lung, prostate, stomach, breast, pancreatic, intestinal, and bladder cancers [41,72,73]. Notably, epidemiological evidence has established a correlation between the dietary intake of green tea and a reduced risk of colorectal cancer [74–77].

Over the years, several in vitro and in vivo studies utilizing animal models have investigated the anticancer effects of green tea components, with a primary focus on catechins, in the context of colorectal cancer. A comprehensive summary of these principal studies, including the active principles evaluated and the proposed anticancer mechanisms, is presented in Table 1. Furthermore, a simplified schematic representation of the mechanisms through which green tea exerts its anticancer effects on colorectal cancer is illustrated in Figure 5, providing an integrative overview of its multifaceted action.

6.1. In Vitro Studies

Most in vitro studies investigating the anticancer properties against colorectal carcinoma have predominantly focused on catechins, with a particular emphasis on epigallocatechin-3-gallate (EGCG). Notably, two independent studies conducted by Luo et al. demonstrated that EGCG treatment significantly induces apoptosis and suppresses cellular proliferation in SW480, SW620, and LS411N colon cancer cell lines. These effects were mediated through the activation of caspase-3 and poly (ADP-ribose) polymerase (PARP), alongside the downregulation of phosphorylated signal transducer and activator of transcription 3 (p-STAT3). Additionally, EGCG was observed to reduce the expression of the antiapoptotic proteins Bcl-2, Bim, and MCL-1, further elucidating its potential therapeutic mechanisms [78,79]. In the same studies, EGCG has also been reported to exhibit antimigratory effects, potentially mediated by its ability to upregulate *E-cadherin* expression and downregulate Vimentin protein levels [79]. This phenomenon was also observed in a study by Zhang et al., wherein SW480 colon cancer cells co-cultured with neutrophils isolated from colorectal cancer patients demonstrated inhibited migration following EGCG treatment [80]. Beyond the findings reported by Luo et al., other investigators have highlighted the pro-apoptotic activity of EGCG in various colon cancer cell lines [61,81–85]. Several mechanisms have been suggested to explain the pro-apoptotic properties of EGCG. These include the activation of caspase-3, -8, and -9, the upregulation of pro-apoptotic proteins, the downregulation of antiapoptotic proteins, and the suppression of key signaling pathways. Concerning this last issue, EGCG-induced apoptosis and proliferation reduction in DLD1, CaCo-2, HCT116, and HT29 colorectal cancer cell lines, among others, have been associated with the inhibition of different signaling cascades, including the previously mentioned *STAT3* and *AKT/mTOR*, *ERK1/2*, *Wnt/β-catenin*, *MAPK*, and *NF-κB* pathways [61,79,82,84–86].

The ability of green tea to inhibit colorectal cancer cell proliferation has also been linked to cell cycle dysregulation. Several studies have demonstrated that EGCG induces cell cycle arrest, primarily at the G0/G1 phase, through the downregulation of *Cyclin D1* expression [81,85,87]. However, G2/M phase arrest has also been observed in some contexts [81,88].

In a recent study, Lin et al. reported that another green tea component, namely, gallic acid (GA), also inhibited proliferation and induced apoptosis in HCT116 and HT29 colon cancer cell lines. These effects were predominantly attributed to an increased ratio of cleaved caspase-3/pro-caspase-3 and cleaved caspase-9/pro-caspase-9, as well as a reduction in phosphorylated epidermal growth factor receptor (p-EGFR), AKT, and STAT3 levels [89].

The antiangiogenic potential of EGCG has been highlighted by Jung et al., who demonstrated that treatment with EGCG reduced *vascular endothelial growth factor* (*VEGF*) expression in HT29 cells [82]. Additionally, recent investigations have revealed that EGCG suppresses cancer stem cell markers and diminishes sphere formation capacity in DLD-1, SW480, and HCT116 cell lines [83,84]. Further, EGCG exerts epigenetic regulatory effects, evidenced by reduced DNA methylation in HCT116, HT29, and SW48 cell lines [87].

6.2. *In Vivo Studies (Animal Models)*

Several studies have shown that green tea polyphenols, mainly EGCG, effectively reduce colon cancer cell growth in animal models. These effects are mediated by mechanisms similar to those observed in vitro [78,81,82,90–92]. For example, Jin et al. reported that EGCG treatment induced apoptotic cell death and cell cycle alterations in BALB/c nude mice bearing orthotopically established HT29 tumors [81]. Similarly, Luo et al. observed a significant reduction in tumor volume in mice implanted with SW620 tumors following EGCG administration. This antitumor activity was attributed to the downregulation of phosphorylated STAT3 and its associated downstream signaling pathways [78].

Shimizu et al. demonstrated that drinking water supplemented with EGCG effectively suppressed the growth of SW837 human colorectal cancer xenografts in BALB/c nude mice. This effect was mediated through the inhibition of *hypoxia-inducible factor-1* (*HIF-1*) expression, which subsequently disrupted the *VEGF/VEGFR* signaling axis [91]. Similar findings were reported by Jung et al. in an HT29 xenograft model, where EGCG treatment led to *VEGF* downregulation, correlating with the inhibition of *extracellular signal-regulated kinase 1/2* (*ERK1/2*) [82].

In addition, a recent study by Shojaei-Zarghani et al. highlighted the chemopreventive properties of theanine, a component of green tea. Theanine was shown to inhibit tumorigenesis by suppressing the *AKT/mTOR* and *Janus kinase 2* (*JAK2*)/*STAT3* pathways [92]. Furthermore, the previously mentioned study by Li et al. confirmed the in vivo anticancer efficacy of gallic acid (GA) in a xenograft tumor model. GA treatment inhibited tumor growth, induced apoptosis, and reduced the phosphorylation levels of EGFR, STAT3, and AKT [89].

Table 1. Summary of the key findings of in vitro (cell-based) and in vivo (animal-based) studies on the use of green tea or its components, mainly EGCG, in colon cancer and the potential mechanisms involved in their effects.

Type of Model	Treatment	Potential Mechanisms	Effects
In vitro			
SW480 and colon cancer cells from patients	0–50 μM EGCG for 48 h	*STAT3/CXCL8* signaling pathway	Significant inhibition of SW480 cells migration and invasion [80]
SW480, SW620 and LS411N cells	10, 20 and 40 μg/mL EGCG	Activation of caspase-3 and PARP, downregulation of *STAT3*, and decrease in Bcl-2 protein levels	Significant suppression of cell proliferation and induction of apoptosis [78]
HT29 cells	EGCG	Inhibition of *ERK1/2* signaling and *VEGF* expression	Significant induction of apoptosis, and angiogenesis inhibition [82]
LoVo, SW480, HCT8, HT29 cells lines	0–35 μg/mL EGCG	Induction of cell cycle alterations and inhibition of *Notch* signaling	Significant inhibition of cell proliferation and apoptosis induction [81]
Caco-2 cells	0–80 μM EGCG	Induction of G2/M cell cycle arrest and activation of laminin receptor-mediated myosin phosphatase	Significant inhibition of cell proliferation [88]
HCT116, HT29, SW480, IEC6 cell lines	0–150 μM EGCG	RXRa activity restoration and *RXRa* promoter methylation reduction	Reversal of gene silencing in colon carcinogenesis modulation [87]
HCT116, HEK293, SW480 cell lines	0–80 μM EGCG	*Wnt/β-catenin* signaling suppression	β-catenin phosphorylation and proteasomal degradation promotion [86]
HCT116, HT29, Caco-2, SW480, and SW837 cells	0–50 μM EGCG or polyphenon E	Induction of G1 cell cycle arrest and reduction of AKT, EGFR, and activation of *HER2*	Significant cell growth inhibition and apoptosis induction [85]
SW480 cells	25 and 50 μM EGCG	*EGFR* downregulation	Significant cell growth inhibition [93]
HT29 cells	0–50 μM EGCG	*AKT*, *ERK1/2*, and *p38 MAPK* signaling pathways modulation	Significant induction of apoptotic cell death [61]

Table 1. *Cont.*

Type of Model	Treatment	Potential Mechanisms	Effects
HCT116 spheroids	50 µM EGCG for 1 week	*CD133* and *NANOG* and *ABCC1* and *ABCG2* gene expression decrease	Significant decrease in sphere formation, and apoptosis and cell cycle alterations induction [84]
DLD-1 and SW480 spheroids	0–60 µM EGCG	Colorectal CSC properties and *Wnt/β-catenin* pathway inhibition	Significant reduction of cell proliferation and induction of apoptosis in colorectal CSCs [83]
SW480, SW620 and LS411N cell lines	0–100 µg/mL EGCG	Activation of caspase-3 and PARP, decrease in Bcl-2, MCL-1, and vimentin levels, and increase in E-cadherin levels; downregulation of STAT3 and p-STAT3	Significant inhibition of cell proliferation and migration [79]
HCT116 and HT29 cell lines	30–60 µM GA	Activation of caspase-3 and caspase-9, decrease in STAT3, EGFR, and AKT phosphorylation	Cell proliferation inhibition, apoptosis induction [89]
In vivo (animal models)			
SW620 xenograft in BALB/c mice	50 and 100 mg/Kg EGCG i.p. for 4 weeks	STAT3 deregulation	Suppression of tumor volume and weight [78]
azoxymethane-induced premalignant lesions in mice	0.01% and 0.1% EGCG in drinking water for 7 weeks	*β-catenin*, *COX2*, and *cyclin D1* expression decrease	Significant inhibition of premalignant lesions development [90]
xenograft of SW837 cells in BALB/c mice	0.01% and 0.1% EGCG in drinking water for 7 weeks	Suppression of *VEGF/VEGFR* signaling activation	Xenograft tumor growth inhibition [91]
xenograft of HT29 cells in BALB/c mice	5, 10, and 20 mg/Kg/die EGCG intragastrically for 14 days	Inhibition of *Notch* signaling	Significant xenograft tumor volume reduction, and apoptosis induction [81]
xenograft of HT29 cells in BALB/c mice	1.5 mg/day EGCG	VEGF decrease	Significant tumor growth and microvessels density reduction; apoptosis induction [82]
1,2-dimethylhydrazine-induced colon cancer in Wistar rats	400 mg/Kg Theanine for 2 weeks	Ki-67, AKT/mTOR, and JAK2/STAT3 reduction; Smad2 tumor suppressor increase	Precancerous and cancerous lesions, tumor volume reduction [92]
xenograft of HCT116 and HT29 cells in BALB/c mice	80 mg/Kg/day GA	Activation of caspase-3 and caspase-9; decrease in STAT3, EGFR, and AKT phosphorylation	Xenograft volume reduction, apoptosis induction [89]

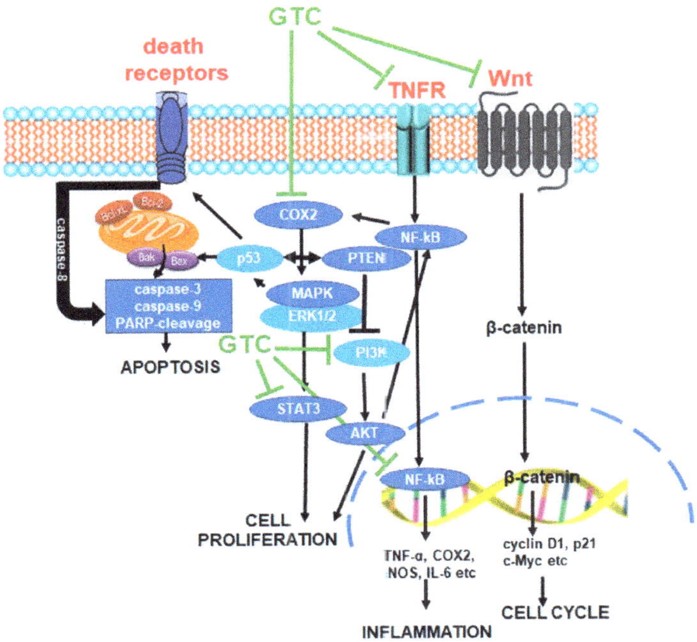

Figure 5. Main mechanisms of action of green tea constituents (GTC) in colorectal cancer.

7. Human Studies

This chapter summarizes and synthesizes the findings from studies investigating the relationship between green tea consumption and colorectal cancer risk. In particular, meta-analyses and major studies according to specific geographic regions are reported.

7.1. Meta-Analyses

A 2023 systematic review and meta-analysis by Huang et al. assessed the association between tea consumption and colorectal cancer risk across multiple population-based studies. The analysis included data from 15 studies with over 2 million participants [94]. The pooled relative risk (RR) for individuals consuming tea regularly (compared to non-consumers) was 0.76 (95% CI, 0.49–1; $p = 0.02$), indicating a non-significant lowering of the risk of developing CRC. American subgroup analysis confirmed this evidence (RR, 0.32; 95% CI, 0.11–0.91), whereas data from the UK (RR, 1.454; 95% CI, 1.031–2.050) and Italian subgroups (RR, 1.15; 95% CI, 0.087–1.23) showed non-significant risk ratios. Thus, green tea consumption may be associated with a reduction in colorectal cancer risk, although the effect may vary by population.

In a previous meta-analysis by Zhu and colleagues, data from 20 prospective cohort studies involving 2,068,137 participants and 21,437 colorectal cancer cases were analyzed. The overall relative risk for colorectal cancer comparing the highest vs. lowest tea consumption was 0.97 (95% CI, 0.94–1.01; $p = 0.09$), indicating a non-significant association between tea consumption and colorectal cancer risk across all studies [95]. Stratified analysis by geographic region, tea type, age, or smoking showed no significant differences between subgroups. Gender-specific meta-analysis indicated no significant association between tea consumption and colorectal cancer risk in men and a borderline-significant inverse relationship in women (RR with no adjustment for coffee intake, 0.9; 95% CI, 0.82–1.00; $p < 0.05$), suggesting a trend towards protective effect. Overall, conclusions of the study do not associate tea consumption to colorectal cancer risk reduction for both genders combined, though a marginally significant inverse association may exist in women.

A 2017 dose–response meta-analysis from China evaluated the relationship between tea consumption and colorectal cancer risk by pooling 29 studies. The summary odds ratio (OR) suggested a mild but non-significant protective effect of tea on colorectal cancer risk, with moderate heterogeneity observed (OR, 0.93; 95% CI, 0.87–1.00; $I^2 = 43.4\%$) [76]. Stratified analysis according to study design revealed that green tea had a protective effect in patients with rectal cancer (OR, 0.91; 95% CI, 0.85–0.99). A significant inverse relationship between a 1-cup/day increase in tea consumption and colorectal cancer risk was shown among green tea drinkers (OR = 0.87; 95% CI, 0.76–0.98), and particularly in women (OR, 0.86; 95% CI, 0.78–0.94). In contrast with this evidence, a 2014 dose–response meta-analysis of prospective studies, which evaluated the effect of tea consumption on five major cancers, including breast and colorectal cancers, found no overall inverse association between tea consumption and cancer risk. The study concluded that tea consumption does not appear to have a protective role in preventing five major cancers, including colorectal cancer [96].

A systematic meta-analysis by Wang et al. included six prospective cohort studies, involving 352,275 participants and 1675 cases of colorectal cancer from Eastern Asian countries (Shanghai, Japan, Singapore) [97]. The analysis compared the highest versus lowest levels of green tea consumption and assessed the effect of increasing consumption by one cup per day. The overall relative risk for the highest compared to the lowest green tea consumption was 0.90 (95% CI, 0.72–1.08), suggesting no significant protective effect of green tea against colorectal cancer. Interestingly, considerable heterogeneity, found across the cohort studies for overall risk of colorectal cancer ($I^2 = 59.6\%$; $p = 0.011$), disappeared after stratification by geographic region, suggesting the presence of ethnic variations in the response of Eastern Asians to tea consumption. Subgroup analysis conducted according to geographic region showed a significant inverse association in the Shanghai population (women and men), where green tea consumption was linked to a 30% reduced risk of colorectal (RR, 0.70; 95% CI, 0.55–0.85) and colon cancers (RR, 0.69; 95% CI, 0.48–0.98). Similar to the meta-analyses summarized above, the protective effect against colorectal cancer was more pronounced in women (one study, RR, 0.63; 95% CI, 0.45–0.88). In contrast, a 36% higher risk of colorectal cancer was found in Singaporean men (one study, RR, 1.36; 95% CI, 1.06–1.74) [97].

By pooling 13 cohort studies, a 2010 USA meta-analysis examined the association between the risk of colorectal cancer and the consumption of coffee, tea, and sugar-sweetened carbonated soft drinks. The analysis included 731,441 participants followed for 6 to 20 years, with 5604 colon cancer cases assessed. No significant association was found between colon cancer risk and coffee consumption or carbonated drinks. Tea consumption showed a modest, significant association with colon cancer risk, with a pooled RR of 1.28 (95% CI, 1.02–1.61; $p = 0.01$) for those drinking more than four 8 oz cups daily. These associations did not differ according to the tumor site, body mass index, physical activity, alcohol use, smoking status, or sex. The study concluded that a modest positive association with colorectal cancer in the case of higher amounts of tea consumed was observed, although data on the type of tea in most studies were not available [98].

A 2012 meta-analysis of case–control studies included 13 studies involving 12,636 cases and 38,419 controls, extracted from studies performed in Asia, North and South America, and Europe [99]. Pooled analysis revealed a non-significant reduction in colorectal cancer risk associated with high green tea intake (OR, 0.87; 95% CI, 0.48–1.58; $p = 0.65$). Analysis was characterized by very high heterogeneity ($I^2 = 90.9$; $p < 0.001$). Stratification by continent did not result in significant odds ratios. Thus, the data presented in this meta-analysis were insufficient to definitively establish green tea as a protective factor against human colorectal cancer.

In conclusion, while the overall evidence from meta-analyses focusing on tea consumption and CRC risk remains uncertain, a consistent trend suggests a protective effect, particularly in women. Further research, particularly focusing on gender differences and geographic variations, may confirm and refine these findings.

Table 2 summarizes the meta-analysis evidence on the relationship between green tea consumption and colorectal cancer, obtained in the studies considered.

Table 2. Meta-analysis evidence on the relationship between green tea consumption and colorectal cancer (RR, risk ratio; OR, odds ratio; 95% CI, 95% confidence interval).

Authors, Year	Number of Pooled Studies	Comparisons	Major Findings	Reference
Huang et al., 2023	15	Regular tea consumption vs. non-consumers	Risk of CRC not significantly decreased in the general population; significant 68% decrease in American subgroup	[94]
Zhu et al., 2020	20	Highest vs. lowest tea consumption	Risk of CRC not significantly decreased in the general population; significant, 10% decrease in women	[95]
Chen et al., 2017	29	Increase in tea consumption (one cup increase)	Odds of CRC not significantly decreased in the general population; significant 9% decreased odds found against rectal cancer; significant, 14% decrease in CRC odds in women stratum	[83]
Yu et al., 2014	15	Risk estimate per 3 cups increase	No significant decrease in CRC risk	[96]
Wang et al., 2012	6	Highest vs. lowest tea consumption	Risk of CRC not significantly decreased in the general population; significant, 30% decrease in risk in Shanghai subgroup; 36% increased CRC risk in Singapore subgroup	[97]
Wang et al., 2012	13		No significant decrease in CRC risk	[99]
Zhang et al., 2010	13	Consumption increment of more than 4 cups/day of tea (type of tea not disclosed)	Significant increase in CRC risk (28%)	[98]

7.2. Major Studies According to Specific Geographic Regions

7.2.1. Asian Studies

Japan

A case–control analysis performed in the frame of the Fukuoka colorectal cancer study investigated the association between dietary intake of polyphenols and colorectal cancer risk in 816 colorectal cancer cases and 815 community-based controls. Polyphenol intake was assessed from 97 food items using the Phenol-Explorer database, specifically focusing on tea and coffee polyphenols [100]. There was no correlation found between tea polyphenols and colorectal, colon, or rectal cancer (OR adjusted for age, sex, residential area, familial history, smoking, alcohol, BMI, profession, and physical activity, 1.37; 95% CI, 0.99–1.90; $p = 0.09$).

In a population-based prospective cohort study, the association between green tea consumption and colorectal cancer risk was evaluated in 13,957 men and 16,374 women from Japan over a 12-year follow-up period. The study found no overall significant association between green tea consumption and colorectal cancer risk in both men and women. However, there was a weak inverse correlation between men with increased green tea consumption and their colon cancer risk. Compared to men consuming no or less than one cup of green tea per day, the relative risks for colon cancer were 1.32, 0.76, and 0.78 for those consuming 1, 2–3, and 4 or more cups per day, respectively ($p = 0.045$ for a linear trend). While this trend suggests a potential protective effect of higher green tea consumption on male colon cancer risk, no significant association was found [101].

A pilot study by Shimizu et al. (2008) investigated the potential of green tea extracts in preventing metachronous colorectal adenomas. In this trial, participants who had previously undergone polypectomy for colorectal adenomas were randomly assigned to receive either green tea extract or a placebo for 12 months. At end-point colonoscopy, at least one colorectal adenoma was diagnosed in 31% (20 of 65) of the patients in the control group. Supplementation with 1.5 g GTE appeared to prevent metachronous adenomas to 15% (9 of 60 patients; relative risk, 0.49; 95% confidence interval, 0.24–0.99; $p < 0.05$). These findings suggest that green tea extracts may have a protective effect against the recurrence of colorectal adenomas, though the study called for larger trials to confirm the efficacy and explore the mechanisms behind this effect [102].

In a study conducted in 2005, Suzuki and coworkers pooled two prospective cohort studies conducted in northern Japan, including a total of 65,915 subjects. The analysis showed no significant association between green tea consumption and colorectal cancer risk [103]. In particular, the hazard ratio (HR) for colon cancer for those drinking five or more cups of green tea per day, compared to fewer than one cup, was 0.97 (95% CI, 0.70–1.35). Similarly, for rectal cancer, the HR was 0.85 (95% CI, 0.58–1.23).

Thus, studies focusing on the Japanese population found contradictory evidence concerning the relationship between green tea consumption and the risk of colorectal cancer.

Korea

A case–control study conducted in Korea examined the association between green tea intake and colorectal cancer risk, considering the influence of lifestyle factors. The study included 922 cases and 1820 controls. The results showed that participants in the highest tertile of green tea intake showed significantly lower odds of colorectal cancer compared to those in the lowest tertile (OR, 0.48; 95% CI, 0.39–0.59; $p < 0.001$) [104]. Regardless of lifestyle changes, a high green tea intake was consistently associated with a lower risk of colorectal cancer when green tea consumption and lifestyle factors interactions were considered.

The ability of green tea extract (GTE) supplements to prevent metachronous colorectal adenomas and cancer in 176 participants who had undergone complete removal of colorectal adenomas by endoscopic polypectomy was evaluated in a randomized controlled trial also conducted in Korea. Participants were randomly assigned to receive either 0.9 g of GTE per day for 12 months or no GTE supplementation. The incidence of metachronous adenomas was significantly lower in the GTE group (23.6%) compared to the control group (42.3%), with a relative risk (RR) of 0.56 (95% CI, 0.34–0.92; $p = 0.018$). Additionally, the GTE group experienced fewer relapsed adenomas (0.3 ± 0.6) compared to the control group (0.7 ± 1.1), with a statistically significant difference ($p = 0.010$) [105]. Similarly to the results of the Japanese trial reported by Shimizu and coworkers [102], GTE supplementation appears to have a favorable chemopreventive effect on metachronous colorectal adenomas in Korean patients.

Singapore

The Singapore study, which was the main outlier in the Wang et al. (2012) meta-analysis described in the previous chapter, was a prospective cohort study examining the relationship between green or black tea consumption and colorectal cancer risk in 60,000 men and women. After an average follow-up of 8.9 years, 845 colorectal cancer cases were identified [74]. Although green tea drinkers had a statistically non-significant increased colorectal cancer risk compared to non-drinkers in the overall population, a significant risk increase was calculated in men (RR, 1.31; 95% CI, 1.08–1.58), while women showed no significant association. Among men, green tea was particularly strongly associated to colorectal cancer for advanced stages of the disease, with an RR of 1.53 (95% CI, 1.19–1.97) and showed a dose-dependent relationship (p for trend = 0.0002). Black tea intake showed no significant association with colorectal cancer risk for either gender (RR, 0.92; 95% CI, 0.79–1.07). Thus, in contrast with other studies performed on various Asian populations, these findings suggest that green tea consumption may be linked to a slightly higher risk of colorectal cancer and that such an effect may be related to the genetics of the specific population tested.

Taiwan

In a cross-sectional study by Chen et al., the association between cumulative tea consumption and colorectal adenomas was examined in a Taiwanese population of 1688 adults who underwent colonoscopy screening [106]. The variable "cup of tea" used in this study was obtained by multiplying the daily cups with the years of tea consumption. A negative association was found between tea consumption and the likelihood of high-risk adenomas in the group in which the largest amount of tea was consumed (≥42 cup-years) compared to non-consumers (OR, 0.57; 95% CI, 0.36–0.89; p = 0.02). Similar results were obtained for low-risk adenomas (OR, 0.66; 95% CI, 0.48–0.90; p = 0.01), even for consumers of lower amounts of tea (e.g., 12–41.9 cup-years; OR, 0.64; 95% CI, 0.46–0.90; p = 0.01). These results suggest that cumulative tea consumption may be associated with a significantly reduced risk of colorectal adenomas in the Taiwanese population and are in agreement with the results of the GTE Korean study by Shin et al. [105].

China

Two key Shanghai-based studies included in the 2012 meta-analysis by Wang et al., cited in the previous chapter, reported a protective effect of green tea against colorectal cancer. These studies were the main contributors to the significant protective effect observed in the meta-analysis, highlighting green tea's potential in reducing colorectal cancer risk, particularly in specific subgroups.

The first study, conducted by Ji et al. (1997), examined the association between green tea intake and the risk of colon, rectal, and pancreatic cancers [77]. This study included 931 colon, 884 rectal, and 451 pancreatic cancer cases diagnosed between 1990 and 1993, with 1552 age- and gender-matched controls. Multivariate analyses, adjusted for age, income, education, and smoking, showed an inverse relationship between green tea consumption and cancer risk, particularly for rectal cancer. Among men, those in the 200–299 g/month consumption range had an odds ratio of 0.66 (95% CI, 0.43–0.99) for rectal cancer. No significant associations were found for colon cancer. A more pronounced trend was observed in women, who showed significant protective activity against rectal cancer for both the lower (1–200 g/month, OR, 0.51; 95% CI, 0.33–0.79) and the highest consumption categories (≥200 g/month, OR, 0.57, 95% CI, 0.34–0.97). Non-significant results were found for colon cancer. These findings suggest that green tea consumption may have a preventive activity against the risk of rectal cancer, especially in Chinese women.

Focusing on a cohort of 60,567 Chinese men, the second Shanghai study found that regular green tea consumption (namely, drinking green tea at least three times per week for more than six months) was associated with a reduced risk of colorectal cancer (HR, 0.77; 95%CI, 0.59–1.01), which was more pronounced and statistically significant among non-smokers (multivariable-adjusted HR, 0.54; 95% CI, 0.34–0.86). Interestingly, the more green tea that was consumed, the more the risk decreased. Specifically, a 12% reduction in risk was associated with each 2 g increment in daily intake of dry green tea leaves (HR, 0.88; 95% CI, 0.78–0.99). No significant association was found between green tea consumption and colorectal cancer risk among smokers (HR, 0.94; 95% CI, 0.66–1.34) [75]. This study suggested that regular green tea consumption may reduce the incidence of colorectal cancer, particularly among non-smokers.

A prospective cohort study, which included 69,710 Chinese women aged 40 to 70 years, evaluated the association between green tea consumption and colorectal cancer risk. Baseline in-person interviews were used to assess tea consumption and were reassessed 2–3 years later. Over 6 years of follow-up, 256 new cases of colorectal cancer were identified. Women who reported moderate consumption of green tea (1–4 g/day) showed an age-adjusted relative risk of colorectal cancer of 0.64 (95% CI, 0.44–0.94) compared to non-regular tea drinkers. Women drinking higher amounts of tea (>/=5 g/day) had a lower relative risk of colorectal cancer (0.51; 95% CI, 0.29–0.89; p/trend = 0.002). Both colon and rectal cancers showed this inverse association [107].

Negative results were reported by Li et al., who conducted a large prospective cohort study within the China Kadoorie Biobank to assess the relationship between tea consumption and cancer risk among 455,981 Chinese adults aged 30–79 years. Data, adjusted for confounders such as alcohol consumption and smoking, showed that regular green tea drinkers did not show a decreased hazard of developing cancer overall (HR, 1.03; 95% CI, 0.93–1.13). This negative finding included all kinds of assessed cancers, including colorectal cancer (daily consumers of tea leaves >4.0 g/day vs. less-than-weekly consumers: HR, 1.08; CI, 0.81–1.45) [108].

Studies conducted in China explored the relationship between genetic susceptibility to CRC and environmental factors, such as diet and, in particular, tea consumption.

An interesting population-based case–control study, including 421 colorectal cancer patients and 845 controls, was conducted in China to assess the association between XPC polymorphisms (Lys939Gln, Ala499Val, and PAT) and colorectal cancer risk [109]. The XPC gene plays a crucial role in repairing bulky, helix-distorting DNA lesions, and several polymorphisms in this gene may influence susceptibility to colorectal cancer [110]. Importantly, the study investigated the impact of green tea consumption on the odds of having colorectal cancer in carriers of specific XPC polymorphisms. For the Lys939Gln polymorphism, the CC genotype showed significantly higher odds for colorectal cancer (OR, 1.5; 95% CI, 1.0–2.2) compared to the AA genotype. Interestingly, those who never drank tea had higher odds ratios (ORs) for the associations between colorectal cancer risk and the AC + CC genotype (OR, 3.0; 95% CI, 1.9–4.7), as well as the AA genotype (OR, 2.4; 95% CI, 1.5–3.9), than tea drinkers with the AA genotype. In addition, the PAT +/+ genotype was also associated with a significantly greater risk of colorectal cancer (OR, 1.5; 95% CI, 1.0–2.3) compared to the PAT −/− genotype. Once more, comparing tea drinkers with the PAT −/− and −/+ genotypes, the genotype impact was more noticeable among never tea drinkers with the PAT +/+ genotype (OR, 3.3; 95% CI, 1.8–6.1), as well as the PAT −/− and −/+ genotypes (OR, 2.5; 95% CI, 1.7–3.6). While no significant association was found between the Ala499Val polymorphism and overall colorectal cancer, the authors observed significantly increased odds among never-tea drinkers carrying either the CC genotype (OR, 2.8; 95% CI, 1.8–4.4) or the CT + TT genotype (OR, 2.2; 95% CI, 1.4–3.3) as compared with tea drinkers carrying the CT + TT genotype [110].

Yu et al. (2012) investigated the influence of gene mutations in the cytosolic phospholipase A2 PLA2G4A gene on the protective effect of tea consumption against colorectal cancer. From a cohort of 1200 individuals, it was found that in carriers of the PLA2G4A rs6666834 polymorphism, tea consumption exhibited a significant protective effect. Compared with non-tea drinkers with the TT/CT genotype of rs6666834, tea drinkers with TT/CT or CC showed significantly lower odds of colorectal cancer (OR, 0.6; 95% CI, 0.36–1.00 for TT/CT [p = 0.047]; 0.38, 0.19-0.74 for CC, [p = 0.0049]). This suggests that the analyzed genetic variants may influence the cancer preventive properties of tea, underscoring the complex interplay between diet, genetic factors, and cancer susceptibility [111].

These results suggest that genetic variations may predispose individuals to colorectal cancer, but green tea consumption appears to modulate this risk, providing a protective effect for carriers of certain genotypes. In addition, the interplay between genetics and diet could inform personalized prevention strategies for colorectal cancer.

Except for the Singapore study, which showed the opposite trend, most trials conducted in Asia consistently support the conclusion that green tea consumption may be associated with a reduced risk of colorectal cancer. These studies, particularly those from regions like Shanghai, suggest that green tea may confer a protective effect against colorectal cancer, highlighting a potential link between its regular consumption and lower cancer risk. In addition, studies from Korea and Taiwan have demonstrated a protective effect of green tea or its extracts against colorectal adenomas.

7.2.2. European, American, and Australian Studies

Europe

Data from the European Prospective Investigation into Cancer and Nutrition (EPIC) cohort, a comprehensive epidemiological study exploring various dietary and lifestyle factors related to cancer risk, were used in two different studies. In one of these, the association between the dietary intake of total and individual polyphenol classes and colorectal cancer (CRC) risk was evaluated in 476,160 European men and women. Polyphenol intake was assessed using country-specific dietary questionnaires. Over a follow-up period of 14 years, 5991 colorectal cancer cases were recorded [112]. The analysis found that a doubling of total dietary polyphenol intake was not associated with colorectal cancer risk in either women (hazard ratio [HR] log2, 1.06; 95% CI, 0.99–1.14) or men (HR log2, 0.97; 95% CI, 0.90–1.05). In a second study by Dik and coworkers, the association between coffee and tea consumption and colorectal cancer risk was investigated using data from 477,071 participants of the EPIC cohort. Habitual coffee and tea consumption was assessed using dietary questionnaires [113]. Over a median follow-up of 11.6 years, 4234 participants developed CRC. Concerning tea consumption, no significant association was found between total consumption and CRC risk (HR, 0.97; 95% CI, 0.86–1.09). Risk estimates for tea consumption and colorectal cancer risk were not modified by sex, diabetes, BMI, and intake of red or processed meat. Furthermore, when considering potential effect modifications by genotypes related to caffeine metabolism (CYP1A2 and NAT2), no significant differences in colorectal cancer risk were observed between high tea consumers with slow enzyme activity and non-/low consumers with fast enzyme activity. These findings suggest that caffeine metabolism does not significantly influence the relationship between coffee and tea consumption and colorectal cancer risk. Overall, both studies from the EPIC cohort found no significant association between polyphenol or tea consumption and colorectal cancer risk, suggesting that these dietary factors may not play a major role in influencing colorectal cancer risk in the European population investigated.

Non-significant results also emerged from a randomized, double-blind trial performed in Germany. The effectiveness of a green tea extract (GTE), standardized to 150 mg of epigallocatechin gallate (EGCG) twice daily, was assessed for preventing colorectal adenomas in 1001 participants with a history of colon adenomas. After a 4-week safety run-in period, participants were randomized to receive either GTE or a placebo for 3 years. There were no significant variations in the adverse events between the two groups, indicating that GTE has a good safety profile. Furthermore, there was no significant reduction of adenoma incidence with GTE in the modified intention-to-treat population. Specifically, the adenoma rate in the placebo group was 55.7%, and in the GTE group, it was 51.1% (adjusted relative risk, 0.905, $p = 0.16$). Similarly, in the per-protocol population, the adenoma rate difference was also not significant (adjusted relative risk 0.883; $p = 0.117$). The study concluded that GTE did not significantly reduce the incidence of colorectal adenomas compared to placebo [114].

Over a median follow-up of 8 years, 111 cases of colon cancer and 83 cases of rectal cancer were diagnosed in Finland. Using proportional hazard regression models, the authors examined the association between coffee and tea consumption and cancer incidence, adjusting for potential confounders. Coffee consumption was not significantly associated with either colon or rectal cancer. However, a positive association was observed between tea consumption and colon cancer risk. Compared to non-tea drinkers, those who consumed more than 1 cup/day had an RR of 2.09 (95% CI, 1.34–3.26). On the other hand, rectal cancer incidence was not significantly improved by tea consumption. Thus, the hypothesis that coffee or tea can protect against colorectal cancer in Finnish men was not supported by this study [115].

In a large Swedish cohort study, Terry and coworkers examined 61,463 women over an average 9.6 years of follow-up and identified 460 incident cases of colorectal cancer. No overall associations were found between tea consumption and colorectal cancer in age- or multivariate-adjusted models. For the highest tea consumption category (two or more cups/day), the multivariate-adjusted relative risk compared to non-consumers was 0.97 (95% CI, 0.63–1.48). No significant association was found for proximal, distal, or combined colon cancer. However, a non-significant positive association with rectal cancer was observed, which became statistically significant in women aged over 65 years (RR, 2.26; 95% CI, 1.03–4.96) [116]. In conclusion, cohort studies conducted in Scandinavian populations yielded mixed results regarding the relationship between tea consumption and colorectal cancer risk. The Finnish study found a positive association between tea consumption and colon cancer risk, while the Swedish study reported no significant overall association, although a marginally significant positive association with rectal cancer risk was observed in women over 65. These findings contrast with data from studies in China, which consistently appear to show a protective effect of tea consumption against colorectal cancer. This suggests that the impact of tea on colorectal cancer risk may vary by population, age, and other factors.

USA

In a study by Shi et al. (2023), the association between flavonoid intake and survival after colorectal cancer diagnosis was assessed in two large U.S. cohorts: the Nurses' Health Study and the Health Professionals Follow-Up Study. The analysis included 3695 CRC survivors and evaluated the impact of flavonoid consumption on overall survival and CRC-specific survival. The results indicated that higher flavonoid intake was associated significant improvements in both overall survival and colorectal cancer-specific survival. Specifically, for every standard deviation increase in total flavonoid intake, the hazard ratio for overall mortality was 0.89 (95% CI, 0.83–0.95), and for colorectal cancer-specific mortality, the HR was 0.84 (95% CI, 0.71–0.98). Flavonoid subclasses, mainly the flavan-3-ol tea component, are particularly associated with improved survival outcomes. The multivariable HRs for CRC mortality per 1 daily cup of tea was 0.86 (0.75–0.99; $p = 0.03$) [117].

A US-based study aimed at examining the relationship between tea consumption and colon cancer risk using data from the NHANES I Epidemiologic Follow-up Study (NHEFS). The researchers applied Cox proportional hazard models to test whether frequent tea consumption might have a protective effect against colon cancer. The study included 7656 tea users and 4514 non-tea users. After adjusting for potential confounders, the relative risks of colon cancer for tea drinkers were significantly lower compared to non-tea drinkers either for those drinking up to 1.5 cups per day (RR, 0.57; 95% CI, 0.42–0.78) or for subjects consuming more than 1.5 cups per day (RR, 0.59; 95% CI, 0.35–1.00). Notably, the protective effect was more pronounced in men, with an RR of 0.41 (95% CI, 0.25–0.66) for men consuming up to 1.5 cups per day and an RR of 0.30 (95% CI, 0.09–0.98) for those consuming more than 1.5 cups per day. In conclusion, this study suggests an inverse relationship between habitual tea consumption and colon cancer risk, particularly among men [118].

A smaller population-based case–control study in Midwestern US included 685 colon cancer cases and 655 rectal cancer cases, aged 40–85 years, identified through the Iowa SEER Cancer Registry. Controls (n = 2434) were frequency-matched by sex and age. Total tea consumption was categorized as none, low ($\leq$3 cups/day), medium (3.1–5 cups/day), and high (>5 cups/day). The study found no association between total tea consumption and either colon or rectal cancer after adjusting for age, sex, education, physical activity, smoking history, and intake of coffee, fiber, and fruits and vegetables [119].

In conclusion, studies conducted in the USA suggest varying associations between tea consumption and colorectal cancer risk. While a study in the Nurses' Health Study and the Health Professionals Follow-Up Study found that higher flavonoid intake, particularly from tea, was associated with improved survival outcomes after colorectal cancer diagnosis, and the NHANES study showed a marked reduction in colorectal cancer risk, other studies, such as the one based on the Iowa SEER Cancer Registry, did not find a significant protective effect of tea consumption on colorectal cancer risk. These findings highlight the complex and context-dependent nature of the relationship between tea intake and colorectal cancer outcomes in U.S. populations.

Australia

A case–control study conducted in Western Australia (2005–2007) analyzed the associations between tea, coffee, and milk consumption and colorectal cancer risk in 854 incident cases and 948 controls. No significant positive or inverse associations were found between black tea, green tea, decaffeinated coffee, or milk and colorectal cancer risk [120]. However, the highest rate of green tea consumption in this study was rather undefined (one or more cups a week), and subgroup analysis in this intake cohort was not performed.

8. Discussion

Green tea, traditionally consumed for centuries as a tonic beverage, has only recently become the focus of extensive scientific research aimed at elucidating its potential health benefits, particularly in disease prevention, including cancer. According to epidemiological studies, green tea consumption may lower the risk of developing several malignancies, including colorectal cancer [101,104,120]. Current evidence highlights that green tea catechins, particularly epigallocatechin-3-gallate (EGCG), possess both therapeutic and preventive potential. These compounds inhibit cancer initiation, development, and progression by modulating essential cellular processes such as proliferation, differentiation, apoptosis, angiogenesis, and metastasis [35,41,51].

Studies utilizing tumor cell lines and animal models have provided significant insights into the molecular mechanisms underlying the anticancer properties of green tea catechins. Extensive research has demonstrated that these catechins regulate critical biological pathways involved in colorectal carcinogenesis. Notably, EGCG effectively induces apoptotic cell death and enforces cell cycle arrest in cancer cells while exerting minimal effects on normal cells, emphasizing its selective activity [32,35,41,56]. Furthermore, multiple animal studies have corroborated these findings, revealing that green tea treatment markedly reduces the incidence and growth of colorectal tumors, thereby supporting its potential role as a chemopreventive and therapeutic agent [81,82,85,90].

Evidence gathered from individual studies and meta-analyses consistently indicates that green tea consumption may exert a protective effect against colorectal cancer in specific populations, such as China (Shanghai), Taiwan, and Korea, whereas it appears to be ineffective, or even detrimental (e.g., Singapore), in a number of other ethnic groups. Moreover, the magnitude of this effect appears to vary substantially across studies. This variability is likely attributable to differences in study designs (e.g., case–control versus cohort studies), methods of exposure assessment (e.g., self-reported green tea intake), and the heterogeneity of the study populations. These factors underscore the inherent complexity of investigating dietary interventions in cancer prevention. Geographically, regular green tea consumption has been most strongly associated with a reduced risk of colorectal cancer in East Asian populations, where the intake is typically higher. Moreover, research into gene–environment interactions, such as the influence of genetic polymorphisms related to DNA repair mechanisms, suggests that certain subpopulations may derive enhanced benefits from consistent green tea consumption. However, similar protective associations have not been consistently observed in European populations, highlighting potential regional differences in dietary patterns, genetic predisposition, or environmental factors.

Despite experimental evidence supporting the association between green tea catechin intake and a reduced risk of colorectal cancer, their clinical applications remain limited, primarily due to poor bioavailability. The chemopreventive and anticancer effects of EGCG are strongly influenced by its absorption efficiency and interactions with target tissues. Thus, EGCG's limited absorption and instability within the gastrointestinal tract significantly impede its ability to effectively reach therapeutic targets, leading to discrepancies between its demonstrated in vitro and in vivo biological effects. Catechins, including EGCG, are rapidly metabolized under physiological conditions, often resulting in the formation of degradation products or pro-oxidant molecules, irrespective of the administration method [41,121,122]. While administering higher doses of catechins might compensate for these bioavailability challenges, excessive intake is associated with dose-dependent toxicological effects. Consequently, achieving an effective therapeutic balance remains a challenge for clinical application [123]. Significant attempts are being made to overcome these limitations and enhance EGCG's bioavailability and improve its cellular uptake. Among these strategies, encapsulation of EGCG within hydrophobic nanocarriers has shown considerable promise, particularly in colorectal cancer models. Such nanocarrier systems provide increased stability and targeted delivery of EGCG, thereby overcoming traditional barriers to its clinical use [124–127].

Although not specifically discussed in this review, studies investigating the co-administration of EGCG with other nutraceuticals, such as curcumin, or conventional chemotherapeutic agents, including irinotecan, have yielded promising results [128,129]. These findings suggest that EGCG may serve as a valuable adjuvant in traditional chemotherapy, potentially enhancing the efficacy of chemotherapeutic agents while mitigating their associated side effects. This dual action could improve therapeutic outcomes and contribute to an overall enhancement in the quality of life for patients undergoing cancer treatment.

9. Conclusions

Despite the promising findings obtained both in vitro and in vivo, the evidence regarding the use of green tea components, mainly green tea catechins, in colon cancer prevention and treatment remains inconclusive. Some studies have reported no significant associations or, in certain cases, positive associations between green tea consumption and colorectal cancer risk. To establish its efficacy as a chemopreventive agent, rigorous randomized controlled trials and long-term cohort studies are essential. Particular emphasis should be placed on assessing potential sex-specific effects, especially on women, as these may differ due to hormonal and metabolic factors.

Additionally, future research should aim to determine the optimal dosages and durations of green tea catechin intake required for maximum protective benefits while minimizing risks. Investigations into the interactions between green tea and other dietary or lifestyle factors that influence cancer risk would further clarify its role in colorectal cancer prevention and treatment. It would be of significant interest to further investigate the molecular mechanisms underlying the synergistic interactions between EGCG and conventional chemotherapeutic/chemopreventive agents or nutraceutical compounds. Expanding the scope of research to include additional conventional pharmacological agents could provide valuable insights into optimizing their clinical applications. The ability of EGCG to enhance therapeutic efficacy may enable the use of lower drug dosages, thereby minimizing associated adverse effects and improving the overall therapeutic index.

Author Contributions: Conceptualization, writing—original draft preparation, writing—review and editing, F.R., G.P., E.M., M.B.G.; supervision: M.B.G. All authors have read and agreed to the published version of the manuscript.

Funding: This research received no external funding.

Conflicts of Interest: The authors declare no conflicts of interest.

References

1. Hossain, M.S.; Karuniawati, H.; Jairoun, A.A.; Urbi, Z.; Ooi, D.J.; John, A.; Lim, Y.C.; Kaderi Kibria, K.M.; Mohiuddin, A.K.M.; Ming, L.C.; et al. Colorectal Cancer: A Review of Carcinogenesis, Global Epidemiology, Current Challenges, Risk Factors, Preventive and Treatment Strategies. *Cancers* **2022**, *14*, 1732. [CrossRef] [PubMed]
2. Saraiva, M.R.; Rosa, I.; Claro, I. Early-Onset Colorectal Cancer: A Review of Current Knowledge. *World J. Gastroenterol.* **2023**, *29*, 1289–1303. [CrossRef] [PubMed]
3. Arnold, M.; Sierra, M.S.; Laversanne, M.; Soerjomataram, I.; Jemal, A.; Bray, F. Global Patterns and Trends in Colorectal Cancer Incidence and Mortality. *Gut* **2017**, *66*, 683–691. [CrossRef] [PubMed]
4. Sawicki, T.; Ruszkowska, M.; Danielewicz, A.; Niedźwiedzka, E.; Arłukowicz, T.; Przybyłowicz, K.E. A Review of Colorectal Cancer in Terms of Epidemiology, Risk Factors, Development, Symptoms and Diagnosis. *Cancers* **2021**, *13*, 2025. [CrossRef] [PubMed]
5. Mármol, I.; Sánchez-de-Diego, C.; Dieste, A.P.; Cerrada, E.; Yoldi, M.J.R. Colorectal Carcinoma: A General Overview and Future Perspectives in Colorectal Cancer. *Int. J. Mol. Sci.* **2017**, *18*, 197. [CrossRef] [PubMed]
6. Fearon, E.R.; Vogelstein, B. A Genetic Model for Colorectal Tumorigenesis. *Cell* **1990**, *61*, 759–767. [CrossRef]
7. Wells, K.; Wise, P.E. Hereditary Colorectal Cancer Syndromes. *Surg. Clin. N. Am.* **2017**, *97*, 605–625. [CrossRef] [PubMed]
8. Umar, A.; Boland, C.R.; Terdiman, J.P.; Syngal, S.; de la Chapelle, A.; Rüschoff, J.; Fishel, R.; Lindor, N.M.; Burgart, L.J.; Hamelin, R.; et al. Revised Bethesda Guidelines for Hereditary Nonpolyposis Colorectal Cancer (Lynch Syndrome) and Microsatellite Instability. *J. Natl. Cancer Inst.* **2004**, *96*, 261–268. [CrossRef]
9. Willett, C.G.; Chang, D.T.; Czito, B.G.; Meyer, J.; Wo, J.; Cancer Genome Atlas Network. Comprehensive Molecular Characterization of Human Colon and Rectal Cancer. *Nature* **2012**, *487*, 330–337.

10. Karahalios, A.; English, D.R.; Simpson, J.A. Weight Change and Risk of Colorectal Cancer: A Systematic Review and Meta Analysis. *Am. J. Epidemiol.* **2015**, *181*, 832–845. [CrossRef] [PubMed]
11. Van Kruijsdijk, R.C.M.; Van Der Wall, E.; Visseren, F.L.J. Obesity and Cancer: The Role of Dysfunctional Adipose Tissue. *Cance Epidemiol. Biomark. Prev.* **2009**, *18*, 2569–2578. [CrossRef] [PubMed]
12. Vieira, A.R.; Abar, L.; Chan, D.S.M.; Vingeliene, S.; Polemiti, E.; Stevens, C.; Greenwood, D.; Norat, T. Foods and Beverages an Colorectal Cancer Risk: A Systematic Review and Meta-Analysis of Cohort Studies, an Update of the Evidence of the WCRF-AICR Continuous Update Project. *Ann. Oncol.* **2017**, *28*, 1788–1802. [CrossRef] [PubMed]
13. Rosato, V.; Bosetti, C.; Levi, F.; Polesel, J.; Zucchetto, A.; Negri, E.; La Vecchia, C. Risk Factors for Young-Onset Colorectal Cance *Cancer Causes Control* **2013**, *24*, 335–341. [CrossRef]
14. Ma, Y.; Zhang, P.; Wang, F.; Yang, J.; Liu, Z.; Qin, H. Association between Vitamin D and Risk of Colorectal Cancer: A Systemati Review of Prospective Studies. *J. Clin. Oncol.* **2011**, *29*, 3775–3782. [CrossRef]
15. Wolin, K.Y.; Yan, Y.; Colditz, G.A.; Lee, I.M. Physical Activity and Colon Cancer Prevention: A Meta-Analysis. *Br. J. Cancer* **200** *100*, 611–616. [CrossRef] [PubMed]
16. Rostom, A.; Dubé, C.; Lewin, G.; Tsertsvadze, A.; Barrowman, N.; Code, C.; Sampson, M.; Moher, D. Nonsteroidal Ant Inflammatory Drugs and Cyclooxygenase-2 Inhibitors for Primary Prevention of Colorectal Cancer: A Systematic Review Prepared for the U.S. Preventive Services Task Force. *Ann. Intern. Med.* **2007**, *146*, 376–389. [CrossRef] [PubMed]
17. McNabb, S.; Harrison, T.A.; Albanes, D.; Berndt, S.I.; Brenner, H.; Caan, B.J.; Campbell, P.T.; Cao, Y.; Chang-Claude, J.; Char A.; et al. Meta-Analysis of 16 Studies of the Association of Alcohol with Colorectal Cancer. *Int. J. Cancer* **2020**, *146*, 861–87 [CrossRef] [PubMed]
18. Botteri, E.; Iodice, S.; Bagnardi, V.; Raimondi, S.; Lowenfels, A.B.; Maisonneuve, P. Smoking and Colorectal Cancer: A Meta Analysis. *JAMA* **2008**, *300*, 2765–2778. [CrossRef] [PubMed]
19. Kumar, A.; Gautam, V.; Sandhu, A.; Rawat, K.; Sharma, A.; Saha, L. Current and Emerging Therapeutic Approaches for Colorecta Cancer: A Comprehensive Review. *World J. Gastrointest. Surg.* **2023**, *15*, 495–519. [CrossRef] [PubMed]
20. Schuell, B.; Gruenberger, T.; Kornek, G.V.; Dworan, N.; Depisch, D.; Lang, F.; Schneeweiss, B.; Scheithauer, W. Side Effects durin Chemotherapy Predict Tumour Response in Advanced Colorectal Cancer. *Br. J. Cancer* **2005**, *93*, 744–748. [CrossRef] [PubMed]
21. Zhou, J.; Ji, Q.; Li, Q. Resistance to Anti-EGFR Therapies in Metastatic Colorectal Cancer: Underlying Mechanisms and Reversa Strategies. *J. Exp. Clin. Cancer Res.* **2021**, *40*, 328. [CrossRef] [PubMed]
22. Ding, S.; Hu, C.; Fang, J.; Liu, G. The Protective Role of Probiotics against Colorectal Cancer. *Oxid. Med. Cell. Longev.* **2020**, *2020*, 8884583. [CrossRef] [PubMed]
23. Liu, J.; Guo, B. RNA-Based Therapeutics for Colorectal Cancer: Updates and Future Directions. *Pharmacol. Res.* **2020**, *152*, 10455 [CrossRef] [PubMed]
24. Fukuhara, H.; Ino, Y.; Todo, T. Oncolytic Virus Therapy: A New Era of Cancer Treatment at Dawn. *Cancer Sci.* **2016**, *107*, 1373–137 [CrossRef] [PubMed]
25. Rajamanickam, S.; Agarwal, R. Natural Products and Colon Cancer: Current Status and Future Prospects. *Drug Dev. Res.* **200** *69*, 460–471. [CrossRef]
26. Filippini, T.; Malavolti, M.; Borrelli, F.; Izzo, A.A.; Fairweather-Tait, S.J.; Horneber, M.; Vinceti, M. Green Tea (Camellia Sinensis for the Prevention of Cancer. *Cochrane Database Syst. Rev.* **2020**, *2020*, CD005004.
27. Botten, D.; Fugallo, G.; Fraternali, F.; Molteni, C. Structural Properties of Green Tea Catechins. *J. Phys. Chem. B* **2015**, *11* 12860–12867. [CrossRef]
28. Jigisha, A.; Nishant, R.; Navin, K.; Pankaj, G. Green Tea: A Magical Herb with Miraculous Outcomes. *Int. Res. J. Pharm.* **2012**, 139–148.
29. Yanagimoto, K.; Ochi, H.; Lee, K.G.; Shibamoto, T. Antioxidative Activities of Volatile Extracts from Green Tea, Oolong Tea, and Black Tea. *J. Agric. Food Chem.* **2003**, *51*, 7396–7401. [CrossRef]
30. Alam, M.; Ali, S.; Ashraf, G.M.; Bilgrami, A.L.; Yadav, D.K.; Hassan, M.I. Epigallocatechin 3-Gallate: From Green Tea to Cance Therapeutics. *Food Chem.* **2022**, *379*, 132135. [CrossRef] [PubMed]
31. Almatroodi, S.A.; Almatroudi, A.; Khan, A.A.; Alhumaydh, F.A.; Alsahl, M.A.; Rahmani, A.H. Potential Therapeutic Targets o Epigallocatechin Gallate (EGCG), the Most Abundant Catechin in Green Tea, and Its Role in the Therapy of Various Types o Cancer. *Molecules* **2020**, *25*, 3146. [CrossRef] [PubMed]
32. Zhao, T.; Li, C.; Wang, S.; Song, X. Green Tea (Camellia Sinensis): A Review of Its Phytochemistry, Pharmacology, and Toxicology *Molecules* **2022**, *27*, 3909. [CrossRef] [PubMed]
33. Ríos, J.L.; Francini, F.; Schinella, G.R. Natural Products for the Treatment of Type 2 Diabetes Mellitus. *Planta Med.* **2015**, *81* 975–994. [CrossRef] [PubMed]
34. Yoneda, Y.; Kuramoto, N.; Kawada, K. The Role of Glutamine in Neurogenesis Promoted by the Green Tea Amino Acid Theanine in Neural Progenitor Cells for Brain Health. *Neurochem. Int.* **2019**, *129*, 104505. [CrossRef] [PubMed]

Shirakami, Y.; Shimizu, M. Possible Mechanisms of Green Tea and Its Constituents against Cancer. *Molecules* **2018**, *23*, 2284. [CrossRef] [PubMed]

Lambert, J.D.; Elias, R.J. The Antioxidant and Pro-Oxidant Activities of Green Tea Polyphenols: A Role in Cancer Prevention. *Arch. Biochem. Biophys.* **2010**, *501*, 65–72. [CrossRef] [PubMed]

Teramoto, M.; Yamagishi, K.; Muraki, I.; Tamakoshi, A.; Iso, H. Coffee and Green Tea Consumption and Cardiovascular Disease Mortality Among People with and Without Hypertension. *J. Am. Heart Assoc.* **2023**, *12*, e026477. [CrossRef]

Ahmad, N.; Mukhtar, H. Green Tea Polyphenols and Cancer: Biologic Mechanisms and Practical Implications. *Nutr. Rev.* **1999**, *57*, 78–83. [CrossRef] [PubMed]

Graham, H.N. Green Tea Composition, Consumption, and Polyphenol Chemistry. *Prev. Med.* **1992**, *21*, 334–350. [CrossRef] [PubMed]

Tallei, T.E.; Fatimawali; Niode, N.J.; Idroes, R.; Zidan, B.M R.M.; Mitra, S.; Celik, I.; Nainu, F.; Ağagündüz, D.; Emran, T.B.; et al. A Comprehensive Review of the Potential Use of Green Tea Polyphenols in the Management of COVID-19. *Evid.-Based Complement. Altern. Med.* **2021**, *2021*, 7170736. [CrossRef] [PubMed]

Farhan, M. Green Tea Catechins: Nature's Way of Preventing and Treating Cancer. *Int. J. Mol. Sci.* **2022**, *23*, 10713. [CrossRef] [PubMed]

Brimson, J.M.; Prasanth, M.I.; Malar, D.S.; Sharika, R.; Sivamaruthi, B.S.; Kesika, P.; Chaiyasut, C.; Tencomnao, T.; Prasansuklab, A. Role of Herbal Teas in Regulating Cellular Homeostasis and Autophagy and Their Implications in Regulating Overall Health. *Nutrients* **2021**, *13*, 2162. [CrossRef] [PubMed]

Li, F.; Wang, Y.; Li, D.; Chen, Y.; Qiao, X.; Fardous, R.; Lewandowski, A.; Liu, J.; Chan, T.H.; Dou, Q.P. Perspectives on the Recent Developments with Green Tea Polyphenols in Drug Discovery. *Expert Opin. Drug Discov.* **2018**, *13*, 643–660.

Musial, C.; Kuban-Jankowska, A.; Gorska-Ponikowska, M. Beneficial Properties of Green Tea Catechins. *Int. J. Mol. Sci.* **2020**, *21*, 1744. [CrossRef] [PubMed]

Del Rio, D.; Stewart, A.J.; Mullen, W.; Burns, J.; Lean, M.E.J.; Brighenti, F.; Crozier, A. HPLC-MSn Analysis of Phenolic Compounds and Purine Alkaloids in Green and Black Tea. *J. Agric. Food Chem.* **2004**, *52*, 2807–2815. [CrossRef] [PubMed]

Khan, N.; Mukhtar, H. Tea Polyphenols in Promotion of Human Health. *Nutrients* **2019**, *11*, 39. [CrossRef] [PubMed]

Luo, Q.; Zhang, J.R.; Li, H.B.; Wu, D.T.; Geng, F.; Corke, H.; Wei, X.L.; Gan, R.Y. Green Extraction of Antioxidant Polyphenols from Green Tea (Camellia Sinensis). *Antioxidants* **2020**, *9*, 785. [CrossRef] [PubMed]

Yokozawa, T.; Dong, E. Influence of Green Tea and Its Three Major Components upon Low-Density Lipoprotein Oxidation. *Exp. Toxicol. Pathol.* **1997**, *49*, 329–335. [CrossRef]

Hamilton-Miller, J.M.T. Anti-Cariogenic Properties of Tea (Camellia Sinensis). *J. Med. Microbiol.* **2001**, *50*, 299–302. [CrossRef]

Chacko, S.M.; Thambi, P.T.; Kuttan, R.; Nishigaki, I. Beneficial Effects of Green Tea: A Literature Review. *Chin. Med.* **2010**, *5*, 13. [CrossRef]

Oh, J.W.; Muthu, M.; Pushparaj, S.S.C.; Gopal, J. Anticancer Therapeutic Effects of Green Tea Catechins (GTCs) When Integrated with Antioxidant Natural Components. *Molecules* **2023**, *28*, 2151. [CrossRef]

Chatterjee, P.; Chandra, S.; Dey, P.; Bhattacharya, S. Evaluation of Anti-Inflammatory Effects of Green Tea and Black Tea: A Comparative in Vitro Study. *J. Adv. Pharm. Technol. Res.* **2012**, *3*, 136–138. [CrossRef] [PubMed]

Unno, K.; Pervin, M.; Nakagawa, A.; Iguchi, K.; Hara, A.; Takagaki, A.; Nanjo, F.; Minami, A.; Nakamura, Y. Blood–Brain Barrier Permeability of Green Tea Catechin Metabolites and Their Neuritogenic Activity in Human Neuroblastoma SH-SY5Y Cells. *Mol. Nutr. Food Res.* **2017**, *61*, 1700294. [CrossRef] [PubMed]

Prasanth, M.I.; Sivamaruthi, B.S.; Chaiyasut, C.; Tencomnao, T. A Review of the Role of Green Tea (Camellia Sinensis) in Antiphotoaging, Stress Resistance, Neuroprotection, and Autophagy. *Nutrients* **2019**, *11*, 474. [CrossRef]

Wolfram, S. Effects of Green Tea and Egcg on Cardiovascular and Metabolic Health. *J. Am. Coll. Nutr.* **2007**, *26*, 373S–388S. [CrossRef]

Cheng, Z.; Zhang, Z.; Han, Y.; Wang, J.; Wang, Y.; Chen, X.; Shao, Y.; Cheng, Y.; Zhou, W.; Lu, X.; et al. A Review on Anti-Cancer Effect of Green Tea Catechins. *J. Funct. Foods* **2020**, *74*, 104172. [CrossRef]

Zhou, P.; Yu, J.F.; Zhao, C.G.; Sui, F.X.; Teng, X.; Wu, Y. Bin Therapeutic Potential of EGCG on Acute Renal Damage in a Rat Model of Obstructive Nephropathy. *Mol. Med. Rep.* **2013**, *7*, 1096–1102. [CrossRef] [PubMed]

Zhu, S.; Li, Y.; Li, Z.; Ma, C.; Lou, Z.; Yokoyama, W.; Wang, H. Lipase-Catalyzed Synthesis of Acetylated EGCG and Antioxidant Properties of the Acetylated Derivatives. *Food Res. Int.* **2014**, *56*, 279–286. [CrossRef]

Nakagawa, H.; Hasumi, K.; Woo, J.T.; Nagai, K.; Wachi, M. Generation of Hydrogen Peroxide Primarily Contributes to the Induction of Fe(II)-Dependent Apoptosis in Jurkat Cells by (−)-Epigallocatechin Gallate. *Carcinogenesis* **2004**, *25*, 1567–1574. [CrossRef]

Shankar, S.; Suthakar, G.; Srivastava, R.K. Epigallocatechin-3-Gallate Inhibits Cell Cycle and Induces Apoptosis in Pancreatic Cancer. *Front. Biosci.* **2007**, *12*, 5039–5051. [CrossRef] [PubMed]

61. Cerezo-Guisado, M.I.; Zur, R.; Lorenzo, M.J.; Risco, A.; Martín-Serrano, M.A.; Alvarez-Barrientos, A.; Cuenda, A.; Centeno Implication of Akt, ERK1/2 and Alternative P38MAPK Signalling Pathways in Human Colon Cancer Cell Apoptosis Induced Green Tea EGCG. *Food Chem. Toxicol.* **2015**, *84*, 125–132. [CrossRef] [PubMed]
62. Walker, A.; Singh, A.; Tully, E.; Woo, J.; Le, A.; Nguyen, T.; Biswal, S.; Sharma, D.; Gabrielson, E. Nrf2 Signaling and Autopha Are Complementary in Protecting Breast Cancer Cells during Glucose Deprivation. *Free Radic. Biol. Med.* **2018**, *120*, 407–4 [CrossRef]
63. Enkhbat, T.; Nishi, M.; Yoshikawa, K.; Jun, H.; Tokunaga, T.; Takasu, C.; Kashihara, H.; Ishikawa, D.; Tominaga, M.; Shimada, Epigallocatechin-3-Gallate Enhances Radiation Sensitivity in Colorectal Cancer Cells through Nrf2 Activation and Autopha *Anticancer Res.* **2018**, *38*, 6247–6252. [CrossRef] [PubMed]
64. Gupta, S.; Ahmad, N.; Nieminen, A.L.; Mukhtar, H. Growth Inhibition, Cell-Cycle Dysregulation, and Induction of Apopto by Green Tea Constituent (−)-Epigallocatechin-3-Gallate in Androgen-Sensitive and Androgen-Insensitive Human Prost Carcinoma Cells. *Toxicol. Appl. Pharmacol.* **2000**, *164*, 82–90. [CrossRef]
65. Zan, L.; Chen, Q.; Zhang, L.; Li, X. Epigallocatechin Gallate (EGCG) Suppresses Growth and Tumorigenicity in Breast Can Cells by Downregulation of MiR-25. *Bioengineered* **2019**, *10*, 374–382. [CrossRef]
66. Yang, C.; Du, W.; Yang, D. Inhibition of Green Tea Polyphenol EGCG((−)-Epigallocatechin-3-Gallate) on the Proliferation Gastric Cancer Cells by Suppressing Canonical Wnt/β-Catenin Signalling Pathway. *Int. J. Food Sci. Nutr.* **2016**, *67*, 818–8 [CrossRef]
67. Li, F.; Qasim, S.; Li, D.; Dou, Q.P. Updated Review on Green Tea Polyphenol Epigallocatechin-3-Gallate as a Cancer Epigene Regulator. *Semin. Cancer Biol.* **2022**, *83*, 335–352. [CrossRef]
68. Adachi, S.; Shimizu, M.; Shirakami, Y.; Yamauchi, J.; Natsume, H.; Matsushima-Nishiwaki, R.; To, S.; Weinstein, I.B.; Moriwa H.; Kozawa, O. (−)-Epigallocatechin Gallate Downregulates EGF Receptor via Phosphorylation at Ser1046/1047 by P38 MAPK Colon Cancer Cells. *Carcinogenesis* **2009**, *30*, 1544–1552. [CrossRef]
69. Shimizu, M.; Shirakami, Y.; Sakai, H.; Kubota, M.; Kochi, T.; Ideta, T.; Miyazaki, T.; Moriwaki, H. Chemopreventive Potential Green Tea Catechins in Hepatocellular Carcinoma. *Int. J. Mol. Sci.* **2015**, *16*, 6124–6139. [CrossRef]
70. Misra, S.; Ikbal, A.M.A.; Bhattacharjee, D.; Hore, M.; Mishra, S.; Karmakar, S.; Ghosh, A.; Srinivas, R.; Das, A.; Agarwal, et al. Validation of Antioxidant, Antiproliferative, and in Vitro Anti-Rheumatoid Arthritis Activities of Epigallo-Catechin-Ri Bioactive Fraction from Camellia Sinensis Var. Assamica, Assam Variety White Tea, and Its Comparative Evaluation with Gre Tea Fract. *J. Food Biochem.* **2022**, *46*, e14487. [CrossRef] [PubMed]
71. Yang, E.J.; Lee, J.; Lee, S.Y.; Kim, E.K.; Moon, Y.M.; Jung, Y.O.; Park, S.H.; Cho, M. La EGCG Attenuates Autoimmune Arthritis Inhibition of STAT3 and HIF-1α with Th17/Treg Control. *PLoS ONE* **2014**, *9*, e86062. [CrossRef]
72. Bernatoniene, J.; Kopustinskiene, D.M. The Role of Catechins in Cellular Responses to Oxidative Stress. *Molecules* **2018**, *23*, 9 [CrossRef] [PubMed]
73. Perletti, G.; Magri, V.; Vral, A.; Stamatiou, K.; Trinchieri, A. Green Tea Catechins for Chemoprevention of Prostate Cancer Patients with Histologically-Proven HG-PIN or ASAP. Concise Review and Meta-Analysis. *Arch. Ital. Urol. Androl.* **2019**, 153–156. [CrossRef] [PubMed]
74. Sun, C.L.; Yuan, J.M.; Koh, W.P.; Lee, H.P.; Yu, M.C. Green Tea and Black Tea Consumption in Relation to Colorectal Cancer Ri The Singapore Chinese Health Study. *Carcinogenesis* **2007**, *28*, 2143–2148. [CrossRef]
75. Yang, G.; Zheng, W.; Xiang, Y.B.; Gao, J.; Li, H.L.; Zhang, X.; Gao, Y.T.; Shu, X.O. Green Tea Consumption and Colorectal Can Risk: A Report from the Shanghai Men's Health Study. *Carcinogenesis* **2011**, *32*, 1684–1688. [CrossRef]
76. Chen, Y.; Wu, Y.; Du, M.; Chu, H.; Zhu, L.; Tong, N.; Zhang, Z.; Wang, M.; Gu, D.; Chen, J. An Inverse Association between T Consumption and Colorectal Cancer Risk. *Oncotarget* **2017**, *8*, 37367–37376. [CrossRef] [PubMed]
77. Ji, B.T.; Chow, W.H.; Hsing, A.W.; Mclaughlin, J.K.; Dai, Q.; Gao, Y.T.; Blot, W.J.; Fraumeni, J.F. Green Tea Consumption and t Risk of Pancreatic and Colorectal Cancers. *Int. J. Cancer* **1997**, *70*, 255–258. [CrossRef]
78. Luo, K.W.; Ye, W.; Li, N.; Cheng, B.H. Tea Polyphenol EGC Suppresses Colorectal Cancer Cell Proliferation Both in Vitro and Vivo via Downregulation of STAT3. *J. Funct. Foods* **2024**, *112*, 105977. [CrossRef]
79. Luo, K.W.; Xia, J.; Cheng, B.H.; Gao, H.C.; Fu, L.W.; Luo, X. Le Tea Polyphenol EGCG Inhibited Colorectal-Cancer-C Proliferation and Migration via Downregulation of STAT3. *Gastroenterol. Rep.* **2021**, *9*, 59–70. [CrossRef]
80. Zhang, Z.; Zhu, Q.; Wang, S.; Shi, C. Epigallocatechin-3-Gallate Inhibits the Formation of Neutrophil Extracellular Traps a Suppresses the Migration and Invasion of Colon Cancer Cells by Regulating STAT3/CXCL8 Pathway. *Mol. Cell. Biochem.* **20** *478*, 887–898. [CrossRef]
81. Jin, H.; Gong, W.; Zhang, C.; Wang, S. Epigallocatechin Gallate Inhibits the Proliferation of Colorectal Cancer Cells by Regulati Notch Signaling. *Onco. Targets. Ther.* **2013**, *6*, 145–153. [CrossRef] [PubMed]
82. Jung, Y.D.; Kim, M.S.; Shin, B.A.; Chay, K.O.; Ahn, B.W.; Liu, W.; Bucana, C.D.; Gallick, G.E.; Ellis, L.M. EGCG, a Major Compone of Green Tea, Inhibits Tumour Growth by Inhibiting VEGF Induction in Human Colon Carcinoma Cells. *Br. J. Cancer* **2001**, 844–850. [CrossRef] [PubMed]

3. Chen, Y.; Wang, X.Q.; Zhang, Q.; Zhu, J.Y.; Li, Y.; Xie, C.F.; Li, X.T.; Wu, J.S.; Geng, S.S.; Zhong, C.Y.; et al. (−)-Epigallocatechin-3-Gallate Inhibits Colorectal Cancer Stem Cells by Suppressing Wnt/β-Catenin Pathway. *Nutrients* **2017**, *9*, 572. [CrossRef] [PubMed]
4. Wubetu, G.Y.; Shimada, M.; Morine, Y.; Ikemoto, T.; Ishikawa, D.; Iwahashi, S.; Yamada, S.; Saito, Y.; Arakawa, Y.; Imura, S. Epigallocatechin Gallate Hinders Human Hepatoma and Colon Cancer Sphere Formation. *J. Gastroenterol. Hepatol.* **2016**, *31*, 256–264. [CrossRef] [PubMed]
5. Shimizu, M.; Deguchi, A.; Lim, J.T.E.; Moriwaki, H.; Kopelovich, L.; Weinstein, I.B. (−)-Epigallocatechin Gallate and Polyphenon E Inhibit Growth and Activation of the Epidermal Growth Factor Receptor and Human Epidermal Growth Factor Receptor-2 Signaling Pathways in Human Colon Cancer Cells. *Clin. Cancer Res.* **2005**, *11*, 2735–2746. [CrossRef] [PubMed]
6. Oh, S.; Gwak, J.; Park, S.; Yang, C.S. Green Tea Polyphenol EGCG Suppresses Wnt/β-Catenin Signaling by Promoting GSK-3β- and PP2A-Independent β-Catenin Phosphorylation/Degradation. *BioFactors* **2014**, *40*, 586–595. [CrossRef] [PubMed]
7. Morris, J.; Moseley, V.R.; Cabang, A.B.; Coleman, K.; Wei, W.; Garrett-Mayer, E.; Wargovich, M.J. Reduction in Promotor Methylation Utilizing EGCG (Epigallocatechin-3-Gallate) Restores RXRα Expression in Human Colon Cancer Cells. *Oncotarget* **2016**, *7*, 35313–35326. [CrossRef] [PubMed]
8. Umeda, D.; Yano, S.; Yamada, K.; Tachibana, H. Involvement of 67-KDa Laminin Receptor-Mediated Myosin Phosphatase Activation in Antiproliferative Effect of Epigallocatechin-3-O-Gallate at a Physiological Concentration on Caco-2 Colon Cancer Cells. *Biochem. Biophys. Res. Commun.* **2008**, *371*, 172–176. [CrossRef]
9. Lin, X.; Wang, G.; Liu, P.; Han, L.; Wang, T.; Chen, K.; Gao, Y. Gallic Acid Suppresses Colon Cancer Proliferation by Inhibiting SRC and EGFR Phosphorylation. *Exp. Ther. Med.* **2021**, *21*, 638. [CrossRef] [PubMed]
0. Shimizu, M.; Shirakami, Y.; Sakai, H.; Adachi, S.; Hata, K.; Hirose, Y.; Tsurumi, H.; Tanaka, T.; Moriwaki, H. (−)-Epigallocatechin Gallate Suppresses Azoxymethane-Induced Colonic Premalignant Lesions in Male C57BL/KsJ-Db/Db Mice. *Cancer Prev. Res.* **2008**, *1*, 298–304. [CrossRef] [PubMed]
1. Shimizu, M.; Shirakami, Y.; Sakai, H.; Yasuda, Y.; Kubota, M.; Adachi, S.; Tsurumi, H.; Hara, Y.; Moriwaki, H. (−)-Epigallocatechin Gallate Inhibits Growth and Activation of the VEGF/VEGFR Axis in Human Colorectal Cancer Cells. *Chem. Biol. Interact.* **2010**, *185*, 247–252. [CrossRef]
2. Shojaei-Zarghani, S.; Yari Khosroushahi, A.; Rafraf, M. Oncopreventive Effects of Theanine and Theobromine on Dimethylhydrazine-Induced Colon Cancer Model. *Biomed. Pharmacother.* **2021**, *134*, 111140. [CrossRef]
3. Adachi, S.; Nagao, T.; To, S.; Joe, A.K.; Shimizu, M.; Matsushima-Nishiwaki, R.; Kozawa, O.; Moriwaki, H.; Maxfield, F.R.; Weinstein, I.B. (−)-Epigallocatechin Gallate Causes Internalization of the Epidermal Growth Factor Receptor in Human Colon Cancer Cells. *Carcinogenesis* **2008**, *29*, 1986–1993. [CrossRef]
4. Huang, Y.; Chen, Q.; Liu, Y.; Tian, R.; Yin, X.; Hao, Y.; Yang, Y.; Yang, J.; Li, Z.; Yu, S.; et al. Association between Tea Consumption and Colorectal Cancer: A Systematic Review and Meta-Analysis of a Population-Based Study. *BMC Gastroenterol.* **2023**, *23*, 294. [CrossRef] [PubMed]
5. Zhu, M.-Z.; Lu, D.-M.; Ouyang, J.; Zhou, F.; Huang, P.-F.; Gu, B.-Z.; Tang, J.-W.; Shen, F.; Li, J.-F.; Li, Y.-L.; et al. Tea Consumption and Colorectal Cancer Risk: A Meta-Analysis of Prospective Cohort Studies. *Eur. J. Nutr.* **2020**, *59*, 3603–3615. [CrossRef] [PubMed]
6. Yu, F.; Jin, Z.; Jiang, H.; Xiang, C.; Tang, J.; Li, T.; He, J. Tea Consumption and the Risk of Five Major Cancers: A Dose-Response Meta-Analysis of Prospective Studies. *BMC Cancer* **2014**, *14*, 197. [CrossRef]
7. Wang, Z.H.; Gao, Q.Y.; Fang, J.Y. Green Tea and Incidence of Colorectal Cancer: Evidence from Prospective Cohort Studies. *Nutr. Cancer* **2012**, *64*, 1143–1152. [CrossRef] [PubMed]
8. Zhang, X.; Albanes, D.; Beeson, W.L.; Van Den Brandt, P.A.; Buring, J.E.; Flood, A.; Freudenheim, J.L.; Giovannucci, E.L.; Goldbohm, R.A.; Jaceldo-Siegl, K.; et al. Risk of Colon Cancer and Coffee, Tea, and Sugar-Sweetened Soft Drink Intake: Pooled Analysis of Prospective Cohort Studies. *J. Natl. Cancer Inst.* **2010**, *102*, 771–783. [CrossRef] [PubMed]
9. Wang, X.J.; Zeng, X.T.; Duan, X.L.; Zeng, H.C.; Shen, R.; Zhou, P. Association between Green Tea and Colorectal Cancer Risk: A Meta-Analysis of 13 Case-Control Studies. *Asian Pacific J. Cancer Prev.* **2012**, *13*, 3123–3127. [CrossRef] [PubMed]
00. Wang, Z.J.; Ohnaka, K.; Morita, M.; Toyomura, K.; Kono, S.; Ueki, T.; Tanaka, M.; Kakeji, Y.; Maehara, Y.; Okamura, T.; et al. Dietary Polyphenols and Colorectal Cancer Risk: The Fukuoka Colorectal Cancer Study. *World J. Gastroenterol.* **2013**, *19*, 2683–2690. [CrossRef]
01. Wada, K.; Oba, S.; Tsuji, M.; Goto, Y.; Mizuta, F.; Koda, S.; Uji, T.; Hori, A.; Tanabashi, S.; Matsushita, S.; et al. Green Tea Intake and Colorectal Cancer Risk in Japan: The Takayama Study. *Jpn. J. Clin. Oncol.* **2019**, *49*, 515–520. [CrossRef] [PubMed]
02. Shimizu, M.; Fukutomi, Y.; Ninomiya, M.; Nagura, K.; Kato, T.; Araki, H.; Suganuma, M.; Fujiki, H.; Moriwaki, H. Green Tea Extracts for the Prevention of Metachronous Colorectal Adenomas: A Pilot Study. *Cancer Epidemiol. Biomark. Prev.* **2008**, *17*, 3020–3025. [CrossRef] [PubMed]
03. Suzuki, Y.; Tsubono, Y.; Nakaya, N.; Koizumi, Y.; Suzuki, Y.; Shibuya, D.; Tsuji, I. Green Tea and the Risk of Colorectal Cancer: Pooled Analysis of Two Prospective Studies in Japan. *J. Epidemiol.* **2005**, *15*, 118–124. [CrossRef]

104. Kim, H.; Lee, J.; Oh, J.H.; Chang, H.J.; Sohn, D.K.; Shin, A.; Kim, J. Protective Effect of Green Tea Consumption on Colorectal Cancer Varies by Lifestyle Factors. *Nutrients* **2019**, *11*, 2612. [CrossRef]
105. Shin, C.M.; Lee, D.H.; Seo, A.Y.; Lee, H.J.; Kim, S.B.; Son, W.C.; Kim, Y.K.; Lee, S.J.; Park, S.H.; Kim, N.; et al. Green Tea Extract for the Prevention of Metachronous Colorectal Polyps among Patients Who Underwent Endoscopic Removal of Colorectal Adenomas: A Randomized Clinical Trial. *Clin. Nutr.* **2018**, *37*, 452–458. [CrossRef]
106. Chen, H.Y.; Sun, Z.J.; Li, C.H.; Chou, Y.T.; Chang, C.J.; Lu, F.H.; Yang, Y.C.; Wu, J.S. Cumulative Tea Consumption Is Inversely Associated with Colorectal Adenomas in Adults: A Cross-Sectional Study in a Taiwanese Population. *Cancer Epidemiol.* **2021**, *73*, 101945. [CrossRef]
107. Yang, G.; Shu, X.O.; Li, H.; Chow, W.H.; Ji, B.T.; Zhang, X.; Gao, Y.T.; Zheng, W. Prospective Cohort Study of Green Tea Consumption and Colorectal Cancer Risk in Women. *Cancer Epidemiol. Biomark. Prev.* **2007**, *16*, 1219–1223. [CrossRef] [PubMed]
108. Li, X.; Yu, C.; Guo, Y.; Bian, Z.; Shen, Z.; Yang, L.; Chen, Y.; Wei, Y.; Zhang, H.; Qiu, Z.; et al. Association between Tea Consumption and Risk of Cancer: A Prospective Cohort Study of 0.5 Million Chinese Adults. *Eur. J. Epidemiol.* **2019**, *34*, 753–763. [CrossRef] [PubMed]
109. Wu, Y.; Jin, M.; Liu, B.; Liang, X.; Yu, Y.; Li, Q.; Ma, X.; Yao, K.; Chen, K. The Association of XPC Polymorphisms and Tea Drinking with Colorectal Cancer Risk in a Chinese Population. *Mol. Carcinog.* **2011**, *50*, 189–198. [CrossRef] [PubMed]
110. Hua, R.X.; Zhuo, Z.J.; Zhu, J.; Zhang, S.D.; Xue, W.Q.; Zhang, J.B.; Xu, H.M.; Li, X.Z.; Zhang, P.F.; He, J.; et al. XPG Gene Polymorphisms Contribute to Colorectal Cancer Susceptibility: A Two-Stage Case-Control Study. *J. Cancer* **2016**, *7*, 1731–1739. [CrossRef] [PubMed]
111. Yu, Y.; Zhang, M.; Pan, Y.; Jin, M.; Jiang, X.; Zhang, S.; Wu, Y.; Ni, Q.; Li, Q.; Chen, K. PLA2G4A Mutants Modified Protective Effect of Tea Consumption against Colorectal Cancer. *Int. J. Color. Dis.* **2012**, *27*, 1005–1013. [CrossRef]
112. Zamora-Ros, R.; Cayssials, V.; Jenab, M.; Rothwell, J.A.; Fedirko, V.; Aleksandrova, K.; Tjønneland, A.; Kyrø, C.; Overvad, K.; Boutron-Ruault, M.C.; et al. Dietary Intake of Total Polyphenol and Polyphenol Classes and the Risk of Colorectal Cancer in the European Prospective Investigation into Cancer and Nutrition (EPIC) Cohort. *Eur. J. Epidemiol.* **2018**, *33*, 1063–1075. [CrossRef] [PubMed]
113. Dik, V.K.; Bueno-De-Mesquita, H.B.; Van Oijen, M.G.H.; Siersema, P.D.; Uiterwaal, C.S.P.M.; Van Gils, C.H.; Van Duijnhoven, F.J.B.; Cauchi, S.; Yengo, L.; Froguel, P.; et al. Coffee and Tea Consumption, Genotype-Based CYP1A2 and NAT2 Activity and Colorectal Cancer Risk—Results from the EPIC Cohort Study. *Int. J. Cancer* **2014**, *135*, 401–412. [CrossRef]
114. Seufferlein, T.; Ettrich, T.J.; Menzler, S.; Messmann, H.; Kleber, G.; Zipprich, A.; Frank-Gleich, S.; Algül, H.; Metter, K.; Odemar, F.; et al. Green Tea Extract to Prevent Colorectal Adenomas, Results of a Randomized, Placebo-Controlled Clinical Trial. *Am. J. Gastroenterol* **2022**, *117*, 884–894. [CrossRef]
115. Hartman, T.J.; Tangrea, J.A.; Pietinen, P.; Malila, N.; Virtanen, M.; Taylor, P.R.; Albanes, D. Tea and Coffee Consumption and Risk of Colon and Rectal Cancer in Middle-Aged Finnish Men. *Nutr. Cancer* **1998**, *31*, 41–48. [CrossRef]
116. Terry, P.; Wolk, A. Tea Consumption and the Risk of Colorectal Cancer in Sweden. *Nutr. Cancer* **2001**, *39*, 176–179. [CrossRef] [PubMed]
117. Shi, S.; Wang, K.; Zhong, R.; Cassidy, A.; Rimm, E.B.; Nimptsch, K.; Wu, K.; Chan, A.T.; Giovannucci, E.L.; Ogino, S.; et al. Flavonoid Intake and Survival after Diagnosis of Colorectal Cancer: A Prospective Study in 2 US Cohorts. *Am. J. Clin. Nutr.* **2023**, *117*, 1121–1129. [CrossRef]
118. Su, L.J.; Arab, L. Tea Consumption and the Reduced Risk of Colon Cancer—Results from a National Prospective Cohort Study. *Public Health Nutr.* **2002**, *5*, 419–425. [CrossRef] [PubMed]
119. Cerhan, J.R.; Putnam, S.D.; Bianchi, G.D.; Parker, A.S.; Lynch, C.F.; Cantor, K.P. Tea Consumption and Risk of Cancer of the Colon and Rectum. *Nutr. Cancer* **2001**, *41*, 33–40. [CrossRef]
120. Green, C.J.; de Dauwe, P.; Boyle, T.; Tabatabaei, S.M.; Fritschi, L.; Heyworth, J.S. Tea, Coffee, and Milk Consumption and Colorectal Cancer Risk. *J. Epidemiol.* **2014**, *24*, 146–153. [CrossRef]
121. Li, N.; Taylor, L.S.; Ferruzzi, M.G.; Mauer, L.J. Kinetic Study of Catechin Stability: Effects of Ph, Concentration, and Temperature. *J. Agric. Food Chem.* **2012**, *60*, 12531–12539. [CrossRef] [PubMed]
122. Wu, Q.Q.; Liang, Y.F.; Ma, S.B.; Li, H.; Gao, W.Y. Stability and Stabilization of (−)-Gallocatechin Gallate under Various Experimental Conditions and Analyses of Its Epimerization, Auto-Oxidation, and Degradation by LC-MS. *J. Sci. Food Agric.* **2019**, *99*, 5984–5993. [CrossRef] [PubMed]
123. Murakami, A. Dose-Dependent Functionality and Toxicity of Green Tea Polyphenols in Experimental Rodents. *Arch. Biochem. Biophys.* **2014**, *557*, 3–10. [CrossRef]
124. Vieira, I.R.S.; Tessaro, L.; Lima, A.K.O.; Velloso, I.P.S.; Conte-Junior, C.A. Recent Progress in Nanotechnology Improving the Therapeutic Potential of Polyphenols for Cancer. *Nutrients* **2023**, *15*, 3136. [CrossRef]
125. Granja, A.; Frias, I.; Neves, A.R.; Pinheiro, M.; Reis, S. Therapeutic Potential of Epigallocatechin Gallate Nanodelivery Systems. *Biomed Res. Int.* **2017**, *2017*, 5813793. [CrossRef] [PubMed]

6. Yang, Q.Q.; Wei, X.L.; Fang, Y.P.; Gan, R.Y.; Wang, M.; Ge, Y.Y.; Zhang, D.; Cheng, L.Z.; Corke, H. Nanochemoprevention with Therapeutic Benefits: An Updated Review Focused on Epigallocatechin Gallate Delivery. *Crit. Rev. Food Sci. Nutr.* **2020**, *60*, 1243–1264. [CrossRef] [PubMed]
7. Das, T.; Mondal, S.; Das, S.; Das, S.; Das Saha, K. Enhanced Anticancer Activity of (-)-Epigallocatechin-3-Gallate (EGCG) Encapsulated NPs toward Colon Cancer Cell Lines. *Free Radic. Biol. Med.* **2024**, *58*, 565–582. [CrossRef]
8. Borah, G.; Bharali, M.K. Green Tea Catechins in Combination with Irinotecan Attenuates Tumorigenesis and Treatment-Associated Toxicity in an Inflammation-Associated Colon Cancer Mice Model. *J. Egypt. Natl. Canc. Inst.* **2021**, *33*, 17. [CrossRef] [PubMed]
9. Xu, G.; Ren, G.; Xu, X.; Yuan, H.; Wang, Z.; Kang, L.; Yu, W.; Tian, K. Combination of Curcumin and Green Tea Catechins Prevents Dimethylhydrazine-Induced Colon Carcinogenesis. *Food Chem. Toxicol.* **2010**, *48*, 390–395. [CrossRef] [PubMed]

cancers

MDPI

Article

Evaluation of a Novel Lateral Emitting Laser Fiber for Near-Infrared Photoimmunotherapy

Motofumi Suzuki [1], Hisataka Kobayashi [2] and Hirofumi Hanaoka [1,*]

[1] Division of Fundamental Technology Development, Near InfraRed Photo-ImmunoTherapy Research Institute, Kansai Medical University, Hirakata 573-1010, Osaka, Japan; suzukimo@hirakata.kmu.ac.jp
[2] Molecular Imaging Branch, Center for Cancer Research, National Cancer Institute, National Institutes of Health, Bethesda, MD 20892-1088, USA; kobayash@mail.nih.gov
* Correspondence: hanaokah@hirakata.kmu.ac.jp; Tel.: +81-72-804-2454; Fax: +81-72-804-2671

Simple Summary: Near-infrared photoimmunotherapy (NIR-PIT) is a recently developed cancer therapy that uses NIR light and conjugates of a tumor-targeting monoclonal antibody and phthalocyanine dye. In clinical practice, frontal and cylindrical diffusers are used for NIR illumination. However, the light illumination of a narrow space is technically difficult with such diffusers. Therefore, we evaluated the usefulness of a lateral illumination system using lateral emitting laser (LEL) fiber. The LEL fiber was capable of illuminating only a certain area in a lateral direction. NIR-PIT with an LEL fiber showed cytotoxic effects in vitro and suppressed tumor proliferation in vivo. Moreover, the LEL fiber delivered NIR light to an entire tumor more uniformly than a frontal diffuser in a narrow-space model. These results suggest that the LEL fiber can be used for NIR-PIT and is suitable for NIR light illumination in a narrow space.

Abstract: Near-infrared photoimmunotherapy (NIR-PIT) is a new cancer therapy that uses NIR light and conjugates of a tumor-targeting monoclonal antibody and phthalocyanine dye. In clinical practice, frontal and cylindrical diffusers are the only options for NIR illumination. However, illumination in a narrow space is technically difficult with such diffusers. Therefore, we evaluated a lateral illumination system using a lateral emitting laser (LEL) fiber. The LEL fiber illuminated a certain area in a lateral direction. NIR-PIT with an LEL fiber reduced luciferase activity in a light-dose-dependent manner in A431-GFP-luc cells in vitro and significantly suppressed tumor proliferation in a xenograft mouse model. To evaluate the usefulness of the LEL fiber in the illumination of a narrow space, a tumor was illuminated from the inside of a cylinder, mimicking a narrow space, and the fluorescence intensity in the tumor was monitored. In the frontal diffuser, NIR light was unevenly delivered and little light reached a distal tumor area from the illuminated side. By contrast, the LEL fiber allowed a uniform illumination of the entire tumor, and a loss of fluorescence was observed even in distal areas. These findings suggested that the LEL fiber can be used for NIR-PIT and is suitable for NIR light illumination in a narrow space.

Keywords: cancer treatment; laser fiber; near-infrared photoimmunotherapy

1. Introduction

Near-infrared photoimmunotherapy (NIR-PIT) is a recently developed cancer therapy that uses NIR light at 690 nm and conjugates of a tumor-targeting monoclonal antibody and photo-absorbing phthalocyanine dye (IRDye700DX; IR700) [1]. IR700 is chemically decomposed by NIR light irradiation, which leads to aggregation of conjugates. When this change occurs on the membrane of target cells, it causes lethal membrane damage and leads to immunogenic cell death. Importantly, such conjugates and NIR light are minimally toxic; cell death is induced only when this conjugate binds to cancer cell antigens and is exposed to NIR light [2]. Therefore, NIR-PIT has the potential to provide a minimally invasive but highly selective treatment for many types of cancer. A global phase III randomized controlled trial of NIR-PIT was started in 2018 and is currently ongoing (NCT03769506). Moreover, NIR-PIT targeting the epidermal growth factor receptor (EGFR) using a cetuximab-IR700 conjugate was approved in Japan in September 2020 for the treatment of recurrent head and neck squamous cell carcinoma. In addition, NIR-PIT can target any surface membrane protein, regardless of carcinoma type. Therefore, preclinical experiments have shown that NIR-PIT can treat various types of cancers such as lung, pancreatic, ovarian, prostate, breast, gastric, colorectal, and bladder cancer, by targeting not only EGFR but also various molecules like human epidermal growth factor receptor 2, carcinoembryonic antigen, and prostate-specific membrane antigen [2]. Furthermore, the usefulness of NIR-PIT in targeting other than cancer cells, including cancer-associated fibroblasts and regulatory T cells, has been reported.

In NIR-PIT, since NIR light exposure is an essential part of the treatment, the use of an appropriate illumination device is important to maximize therapeutic efficacy. Light at a wavelength of approximately 700 nm penetrates approximately 1 cm into tissue [3]; thus, superficial tumors, such as those of the skin and mouth, can be easily treated by surface illumination with a frontal diffuser. For a more deeply located tumor, needle catheters are placed into the tumor and light is applied from inside the tumor with a cylindrical light diffuser [4,5]. Moreover, the use of an illuminating cylindrical light diffuser through an endoscope is effective in a preclinical model of gastrointestinal tract tumors and head and neck cancers [6,7]. Endoscopic illumination can be applied to various types of cancers, such as esophageal, colorectal, and urinary tumors [8,9]. Currently, forward-aiming laser fibers are used in endoscopic illumination in photodynamic therapy that uses laser illumination similarly to NIR-PIT [10,11]. Therefore, it is expected that the frontal diffuser currently used in clinical practice will be employed for illumination in NIR-PIT.

Although laser irradiation through an endoscope is a promising method, several challenges remain: A cylindrical light diffuser can irradiate in a 360-degree direction but it is difficult to limit the irradiation area. In addition, the NIR light is irradiated into normal areas where it is not required. Also, since EGFR is expressed, to some extent, in normal mucosal cells, excess irradiation to normal areas might cause side effects [12–14]. In the case of a frontal diffuser, when treating a prominent tumor, the laser energy is delivered only at the proximal side. Moreover, uniform illumination from an oblique angle is difficult in organs with narrow lumens, such as the esophagus [15]. Thus, a new type of diffuser is required to irradiate various areas, and a lateral emitting laser (LEL) fiber has potential in this regard. In the present study, we evaluated the usefulness of a novel LEL fiber for NIR-PIT by comparing forward-aiming laser fibers in vitro and in vivo.

2. Materials and Methods

2.1. Reagents and Antibodies

A water-soluble, silicon-phthalocyanine derivative IRDye700DX N-hydroxysuccinimide ester (IR700-NHS) was purchased from LI-COR Bioscience (Lincoln, NE, USA). Cetuximab, a chimeric monoclonal antibody directed against EGFR, was obtained from Merck Biopharma (Tokyo, Japan). All other chemicals were of reagent grade.

2.2. Beam Profile Determination

An LEL fiber was kindly supplied by Furukawa Electric Co., Ltd. (Tokyo, Japan). A microlens at the tip of the fiber allows lateral illumination. The beam profile of the laser with the LEL fiber was measured with a working distance of 4.5 mm using a BGP-USB-SP928-OSI beam profiling camera (Ophir; Tokyo, Japan). The laser was driven using an LDX-3230-685-HHL-400 semiconductor laser diode (LDX Optronics Inc; Maryville, TN, USA). The data were acquired in two separate sessions and a heatmap was drawn since the laser illumination field exceeded the active area of the beam profiler. Light intensity is shown as a relative value with background as zero.

2.3. Synthesis of IR700-Conjugated Cetuximab

Cetuximab (5 mg, 34.2 nmol) was incubated with IR700 NHS ester (669.2 µg, 342.5 nmol) in 0.1 M Na_2HPO_4 (pH 8.5) for 24 h at room temperature. The mixture was purified in a Sephadex G25 column (PD-10; GE Healthcare, Piscataway, NJ, USA) to obtain IR700-conjugated cetuximab (Cet-IR700). The protein concentration was determined with a protein assay BCA reagent (Fujifilm Wako Pure Chemical Industries; Osaka, Japan) by measuring the absorption at 562 nm with a spectrophotometer (V-730Bio; JASCO, Tokyo, Japan). The concentration of IR700 was measured by a spectrophotometer at an absorption wavelength of 689 nm to determine the number of fluorophore molecules conjugated to cetuximab. The synthesis was controlled so that an average of three IR700 molecules were bound to a single antibody.

2.4. Cell Culture

Human epidermoid carcinoma A431 cell lines expressing green fluorescence protein (GFP) and firefly luciferase (A431-GFP-luc) were established by Dr. Kobayashi's laboratory (National Institutes of Health, Bethesda, MD, USA). The cells were maintained in Dulbecco's modified Eagle's medium (DMEM; Fujifilm Wako Pure Chemical Industries) supplemented with 10% (v/v) fetal bovine serum (FBS; GE Healthcare, South Logan, UT, USA) and antibiotics (100 µg/mL penicillin and streptomycin) at 37 °C in a humidified atmosphere of 5% CO_2.

2.5. In Vitro NIR-PIT

Since the thickness of a normal culture dish was more than 10 mm, cells were seeded onto the inner lid of a culture dish coated with collagen (Cellmatrix Type I-C; Nitta Gelatin; Osaka, Japan) and then cultured overnight. After incubation with Cet-IR700 (10 µg/mL) for 1 h at 37 °C, NIR light was irradiated onto the cells using an LEL fiber from a height of 10 mm with an ML7710 laser system (Modulight; Tampere, Finland) at a dose of up to 32 J/cm^2. The actual power density was measured with an optical power meter (PM 100; Thorlabs, Newton, NJ, USA). For bioluminescence imaging (BLI), a medium containing D-luciferin (150 µg/mL) was administered to cells 1 h after NIR-PIT. Subsequently, BLI was performed using an IVIS Spectrum In Vivo Imaging System (PerkinElmer, Waltham, MA, USA).

2.6. In Vivo NIR-PIT

All animal experiments were performed according to the established guidelines of the "Law for The Care and Welfare of Animals in Japan" and approved by the Ethics Committee on Animal Experiments at Kansai Medical University (22-001 and 23-022). All animal experiments were conducted with a minimum of five mice. A431-GFP-luc cells were prepared by suspending 5×10^6 cells in 100 µL phosphate-buffered saline; the cell suspension was injected subcutaneously into 6-week-old female BALB/c nu/nu mice (The Jackson Laboratory Japan, Inc., Yokohama, Japan). When the tumor volume exceeded approximately 100 mm^3, the mice were randomized into three groups: (i) no treatment; (ii) NIR light (100 J/cm^2) without Cet-IR700; and (iii) NIR light (100 J/cm^2) with Cet-IR700 (NIR-PIT treatment). The NIR light exposure was performed using an LEL fiber with an ML7710 laser system. Imaging for IR700 fluorescence before and after NIR light exposure was conducted with a Pearl Trilogy small animal imaging system (LI-COR Biosciences) 24 h after the injection of Cet-IR700. Tumor volume was measured using calipers and calculated using the following formula: size (mm^3) = length (mm) × width (mm) × width (mm) × 0.5.

2.7. Evaluation of NIR Light Penetration into Tumors

To mimic the NIR light illumination of tumors in narrow lumens, a narrow acrylic cylinder (Φ 20 mm) with a square hole was prepared [16]. Mice were positioned so that each tumor was in the hole. Laser fibers were inserted in the cylinder and NIR light was illuminated at 100 J/cm^2. Comparisons were made for the following three different irradiation methods: LEL fiber illumination, frontal illumination with a frontal diffuser (Medlight; Ecublens, Switzerland), and angle illumination with a frontal diffuser bent 30 degrees to the cylinder, more closely resembling clinical conditions. For lateral illumination, the LEL fiber was placed directly above the tumor and irradiated from a height of 1 cm. For angle illumination, the light was illuminated using a frontal diffuser with the angle fixed at 30 degrees to deliver the light on the entire tumor. Five mice were used in each group.

Using a side view, IR700 fluorescence imaging was conducted with a LuminousQuester NI Near-Infrared Light Imaging System (Shimadzu Corporation, Kyoto, Japan). The camera of the system was placed approximately 50 cm from a tumor-bearing mouse [17]. For the comparison of pre- and post-treatment fluorescence imaging, fluorescence images of tumors were obtained before and after therapeutic NIR light illumination. To avoid the therapeutic effects of NIR light illumination for fluorescence imaging, illumination was performed for only 5 s from the same side as the imaging camera using a laser emitting light from a 690 nm continuous wave laser (MLL-III-690, CNI Technology, Changchun, China) with output timing controllable by a TTL control unit (Shimadzu Corporation). Three regions of interest (ROI) for each tumor area were set, and the pixel values were measured using ImageJ Fiji software (Version: 2.14.0/1.54f) [18]. The fluorescence intensity of pre-treatment was set to 100% and the loss of fluorescence was calculated.

2.8. Statistical Analysis

All results are expressed as mean ± standard deviation (SD) of the mean. Statistical analysis was carried out using GraphPad Prism 9.0 (GraphPad Software, Inc., San Diego, CA, USA). Comparisons of the two groups were performed using Student's *t*-tests. For multiple comparisons, a Tukey–Kramer test was used. The minimum level of significance was set at $p < 0.05$.

3. Results

3.1. Properties of LEL Fiber

As shown in Figure 1A, the frontal diffuser is illuminated in the forward direction only. By contrast, LEL fibers enabled illumination in a lateral direction only. The NIR light intensity using the LEL fiber on the illuminated field was measured with a working distance of 4.5 mm (Figure 1B). This beam profile showed that the LEL fiber enables uniform illumination in only a certain area to some extent.

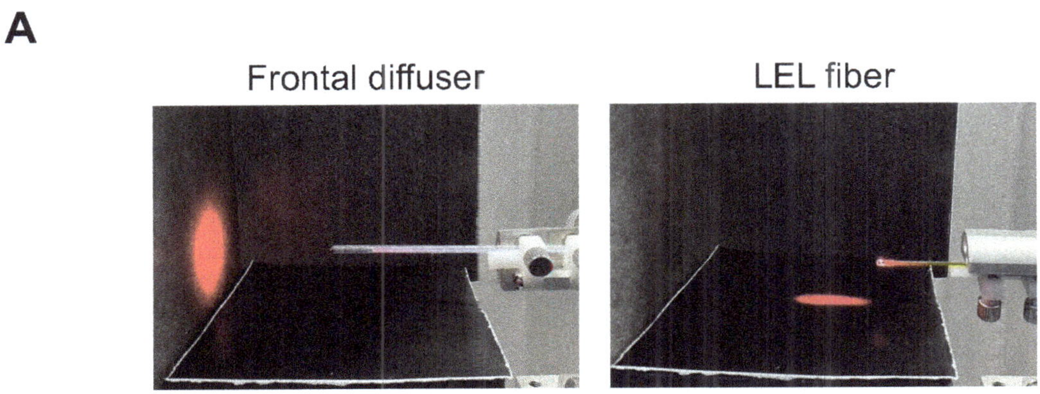

Figure 1. *Cont.*

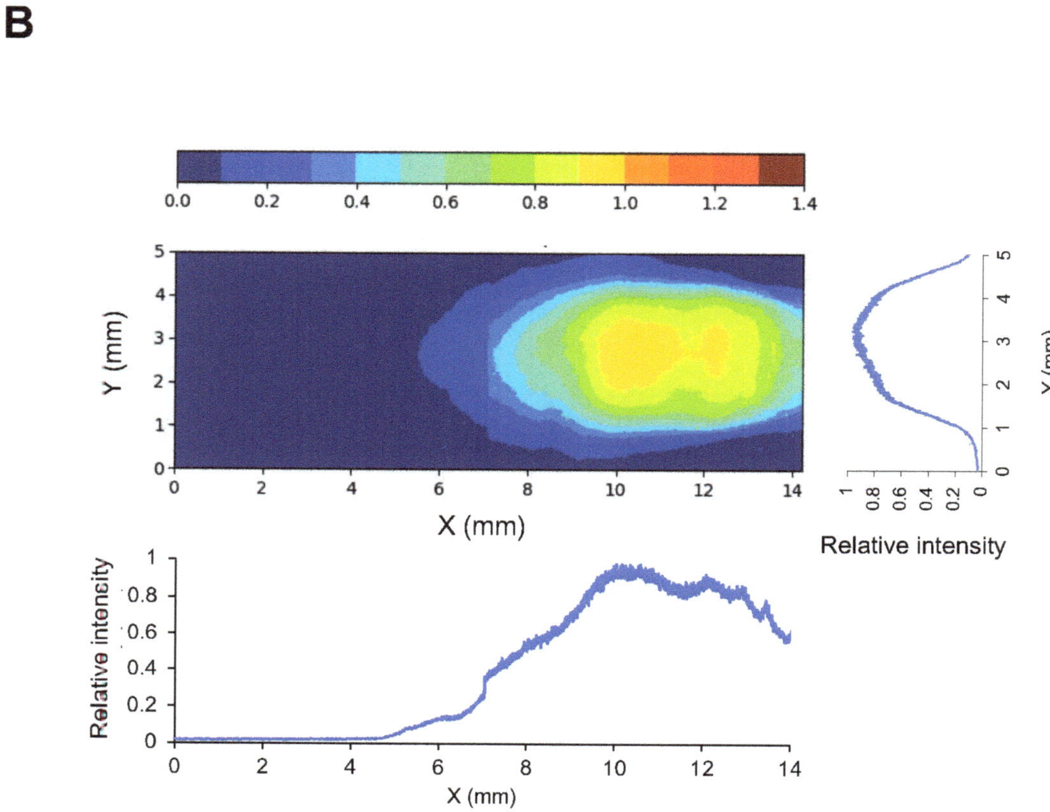

Figure 1. Properties of an LEL fiber. (**A**) Photographs of a laser spot using a frontal diffuser and LEL fiber. NIR light illumination with an LEL fiber was performed at a 10 mm height. (**B**) Uniformity of the illumination field provided by the LEL fiber. A heatmap image of the illumination intensity distribution viewed from above and a graph of the relative intensity profiles in the line of the fiber and perpendicular to the line of the fiber at a central illumination area, respectively.

3.2. *In Vitro NIR-PIT Assessed by BLI*

The cytotoxicity of NIR-PIT with an LEL fiber was assessed in Cet-IR700-treated cancer cells. Assuming illumination in a narrow space, such as the esophagus, the illumination was performed at a height of 10 mm. As shown in Figure 2A, the illumination area was ellipse-shaped, with a length of 21 mm and a width of 7 mm. After illumination, luciferase activity in A431-GFP-luc cells decreased within a certain area (Figure 2B). These results suggested that laser illumination with an LEL fiber induced a photochemical reaction and exhibited a cytotoxic effect on A431-GFP-luc cells. The area of the treatment effect was expanded in a dose-dependent manner up to 8 J/cm^2 (R^2 = 0.9877). However, above 16 J/cm^2, the effective area was not expanded beyond a certain size, which is consistent with the measured beam profile.

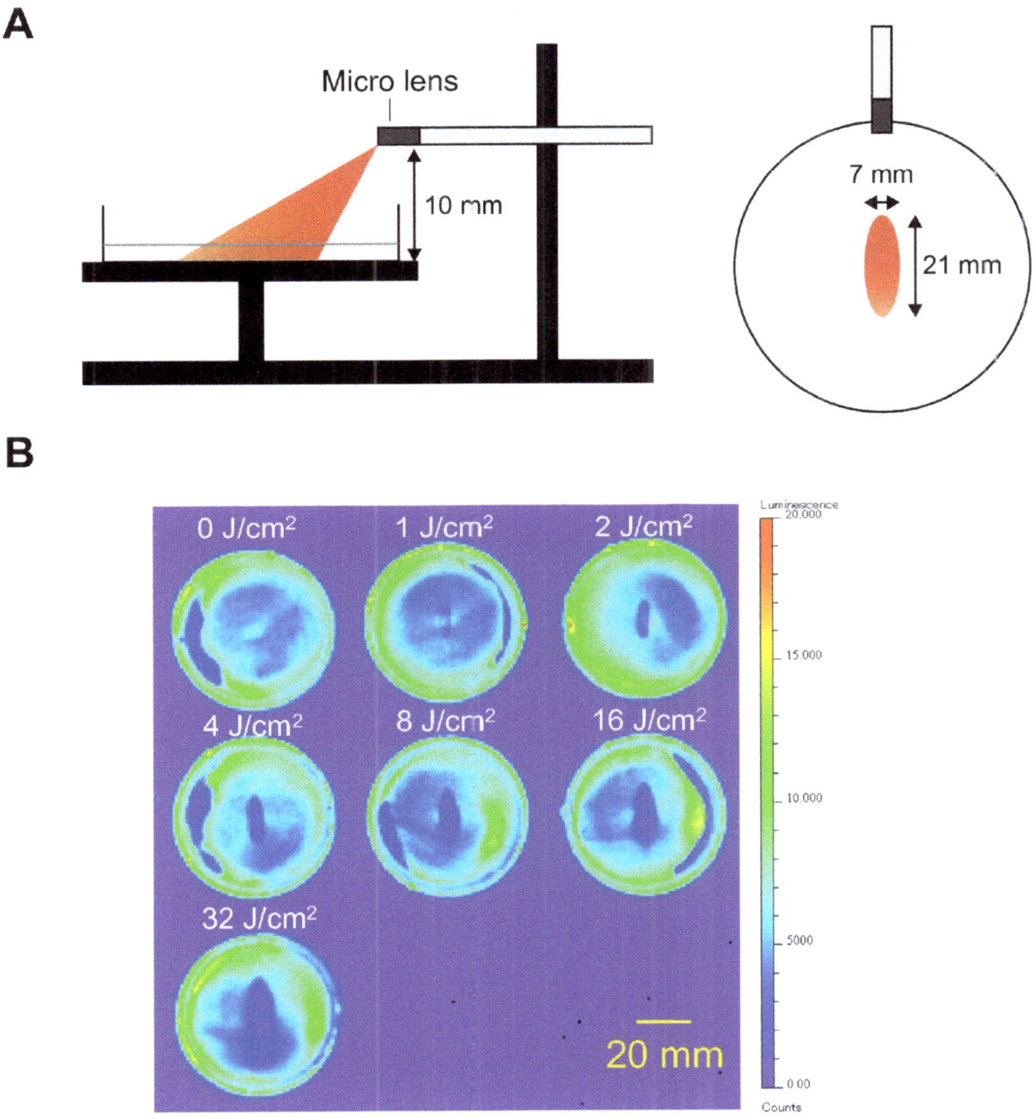

Figure 2. In vitro NIR-PIT in A431-GFP-luc. NIR light was illuminated using an LEL fiber in A431-GFP-luc cells treated with Cet-IR700. The cytotoxicity of NIR-PIT was assessed by bioluminescence imaging. (**A**) Schematic of in vitro NIR-PIT experimental settings. The illumination area from a 10 mm height was an ellipse, with a length of 21 mm and width of 7 mm. (**B**) Bioluminescence images of each dish after NIR-PIT at various doses in A431-GFP-luc cells. Scale bar, 20 mm.

3.3. *Therapeutic Efficacy of NIR-PIT In Vivo*

The therapeutic effect of lateral illumination with an LEL fiber was examined in a tumor-bearing mouse model. The treatment schema is shown in Figure 3A. The accumulation of Cet-IR700 in an A431-GFP-luc tumor was observed 24 h after intravenous injection (Figure 3B). In the NIR-PIT group, the IR700 fluorescence signal disappeared after NIR light exposure due to structural changes of the IR700 dye [19]. These results suggested that NIR light with an LEL fiber can induce NIR-PIT in vivo. After treatment, tumor growth was significantly suppressed in the NIR-PIT group by Days 6 and 9 (Figure 3C). No change in tumor volume was observed in the NIR-irradiated group only. These results suggested that the LEL fiber enabled the therapeutic efficacy of NIR-PIT in vivo as well as the use of a frontal light distributor used in previous experiments [20].

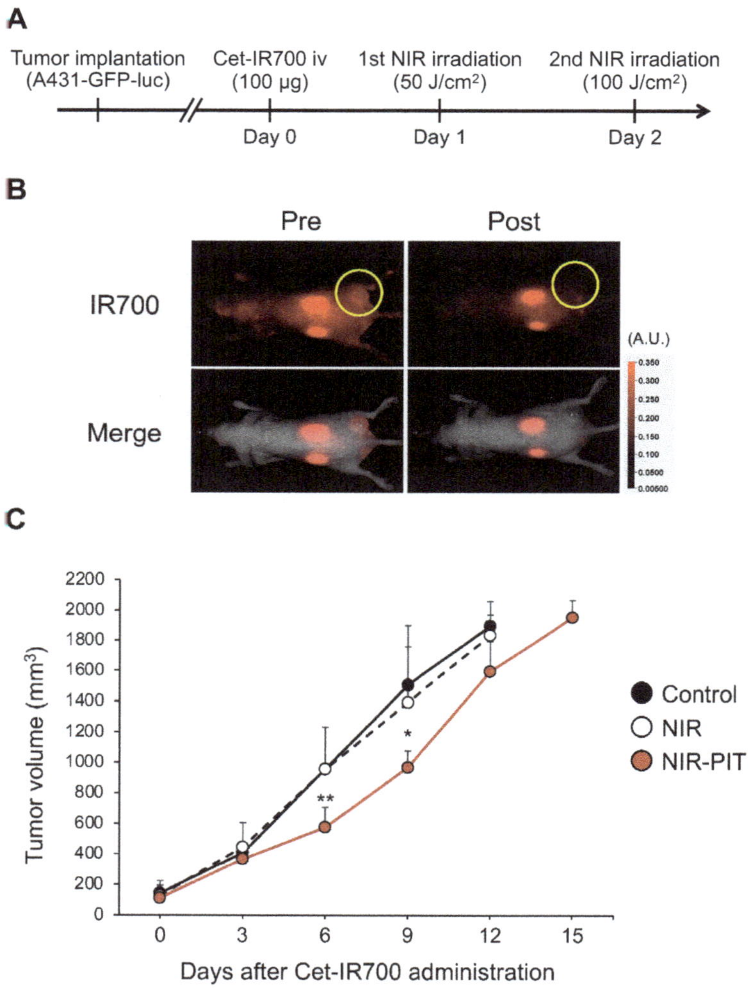

Figure 3. In vivo NIR-PIT using an LEL fiber in an A431-GFP-luc tumor-bearing mouse model. (**A**) Schematic diagram of Cet-IR700 injection and NIR light exposure. (**B**) Fluorescence images on Day 1 before and immediately after NIR-PIT. Yellow circles indicate tumor beds. (**C**) Tumor growth inhibition by NIR-PIT with LEL fiber. The graph represents tumor volumes after treatment. Data are expressed as the mean ± standard deviation (SD). * $p < 0.05$, ** $p < 0.01$ versus control (Tukey–Kramer test).

3.4. Comparison of Forward- and Lateral-Aiming Illumination

To undertake evaluations in a more clinically relevant study, an illumination model in a narrow cylinder was prepared for an in vivo study. A fiber was fixed in the center of the cylinder and the NIR illumination of a prominent tumor was undertaken (Figure 4A). In these experiments, the penetration of NIR light into the tumor using a lateral emitting fiber was compared with that of an existing frontal emitting diffuser. Moreover, oblique angle-aiming illumination was undertaken using a frontal emitting fiber, assuming endoscopic illumination using a frontal diffuser in clinical practice. Although NIR illumination was undertaken at a distance from the tumor to illuminate the entire tumor with a forward-aiming diffuser, it was obvious that the light was barely delivered to the distal regions of the tumor (Figure 4A). In angled illumination, increased delivery of the NIR light occurred compared to that in frontal illumination; however, the illumination was uneven and more was delivered to the proximal region. On the other hand, the NIR light was delivered to the entire tumor when using lateral illumination. To perform a more quantitative analysis, the fluorescence intensity of IR700 was compared before and after therapeutic NIR illumination (Figure 4B–D). After obtaining images, three ROIs (#1: shallow and proximal area, #2: middle area, and #3: deep and distal area) were set on the tumor and the loss of IR700 fluorescence by NIR-PIT was compared. In shallow (ROI #1) and intermediate (ROI #2) regions, the loss of fluorescence was significantly induced by angled and lateral illumination compared with frontal illumination (#1: frontal, 67.8 ± 9.3%; angled, 40.3 ± 7.5%; lateral, 36.6 ± 12.8%; and #2: frontal, 79.5 ± 8.5%; angled, 54.5 ± 17.1; lateral, 51.1 ± 12.1%). However, no significant differences were found between angle and lateral illuminations. In contrast, in deep (ROI #3) regions, lateral illumination occurred with more of a loss of fluorescence compared with angled illumination (#3: frontal, 86.5 ± 7.9%; angled, 67.0 ± 17.4%; lateral, 48.8 ± 10.9%). In a comparison within each tumor, the loss of fluorescence between ROIs #2 and #3 was almost the same as for the lateral illumination group. A decrease in the loss of fluorescence between ROI #2 and ROI #3 was found for the angled illumination group. These results suggested that NIR illumination with an LEL fiber allowed the uniform illumination of the entire tumor and is suitable for NIR light illumination in a narrow space compared to a frontal emitting laser.

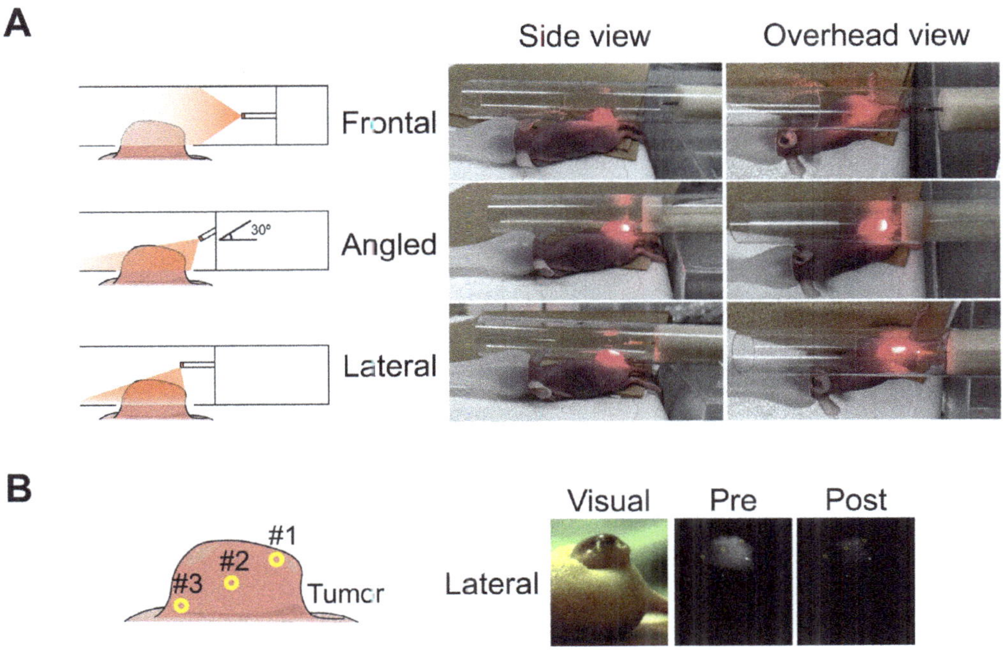

Figure 4. *Cont.*

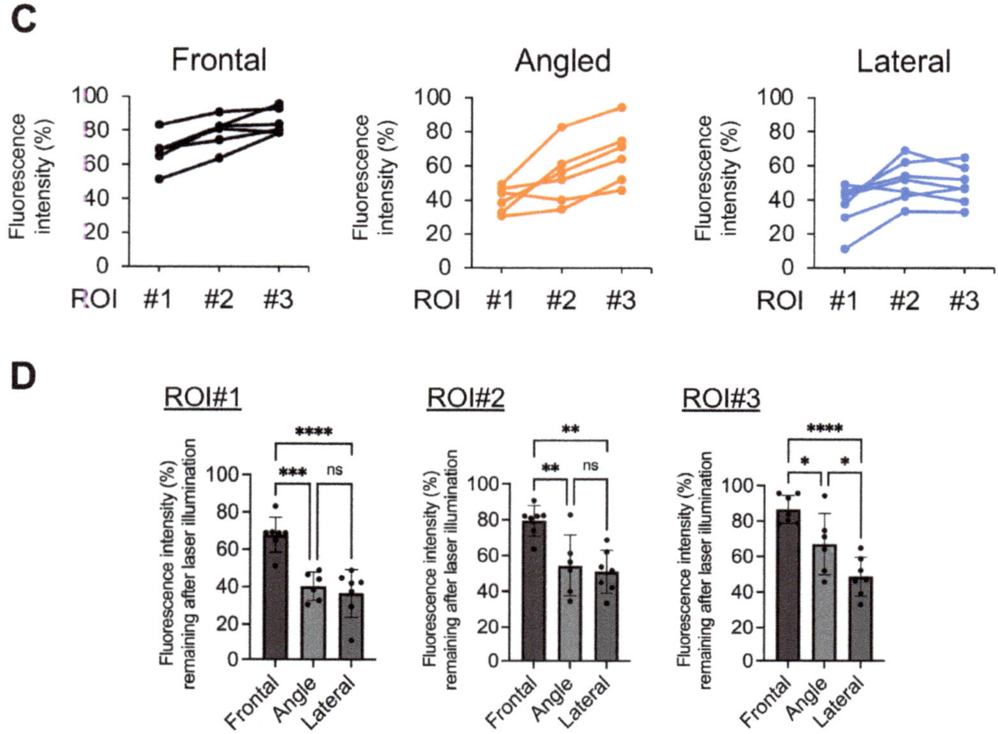

Figure 4. Evaluation of light penetration in a tumor model. Penetration of NIR light was compared using a frontal diffuser and LEL fiber. A fiber was fixed in the center of a 20 mm diameter cylinder, which mimicked a narrow lumen, with a hole to fit the tumor. The tumor was then illuminated by NIR light. In angled illumination, the frontal diffuser was fixed to the tube at an angle of 30 degrees. (**A**) Schematic and visual images of NIR illumination. (**B**) Schematic of three regions of interest (#1: shallow and proximal areas, #2: middle areas, and #3: deep and distal areas). Representative visual and fluorescence images of a tumor in the NIR-PIT with LEL fiber mouse treatment group. (**C**) Graphs represent the individual fluorescence intensity remaining after the illumination of each mouse at ROI #1–3. The fluorescence intensity of pre-treatment was set to 100% and the loss of fluorescence was calculated. (**D**) Graph represents fluorescence intensity remaining after illumination. Data are expressed as the mean ± SD. * $p < 0.05$, ** $p < 0.01$, *** $p < 0.001$, **** $p < 0.0001$ versus control (Tukey–Kramer test).

4. Discussion

As a therapy, NIR-PIT stands out as a promising cancer treatment strategy, with numerous drugs developed for photoimmunotherapy [2]. Despite this progress, the development of new diffusers for NIR light exposure, a crucial aspect of the treatment, has lagged. The LEL fiber assessed in this study illuminated a specific lateral area, inducing photochemical reactions and demonstrating cytotoxic effects on tumor cells. In vivo tumor-bearing models further demonstrated the therapeutic efficacy of NIR-PIT with the LEL fiber. The cytotoxic effects of NIR-PIT do not vary by illumination method as long as the same amount of light is irradiated. Therefore, the LEL fiber would enable the induction of therapeutic efficacy as same as the diffuser used in clinical practice. This indicates that the LEL fiber possesses sufficient properties that are effective for NIR-PIT use.

Gastrointestinal endoscopy has been widely used for diagnosis and treatment in clinical practice [21,22]. Therefore, illumination with endoscopic NIR light illumination makes it possible to expand the cancer indications of NIR-PIT [9]. However, endoscopic illumination using a forward-aiming diffuser is hard to apply to some tumors in narrow spaces. In the current study of a cylinder model, the forward-aiming illumination of NIR light required a longer focal length to illuminate the entire tumor. It is difficult to deliver adequate NIR light for treatment. Although we succeeded in shortening the focal length after illumination at an angle of 30 degrees, uniform light illumination was difficult to achieve, and most of the NIR light was delivered to a proximal area. Furthermore, although our cylinder model was 2 cm in diameter, actual anatomical lumens are often much thinner, and a 30-degree bend is assumed to be difficult. By contrast, NIR light illumination with an LEL fiber illuminated the entire tumor, showing that light could be delivered to distal depths. Even if the position of the fiber was closer to the tumor in a narrower lumen, it would be possible to irradiate the tumor without issue because of the lateral illumination. Accordingly, an LEL fiber is suitable for NIR light illumination in a narrow space and may maximize the therapeutic efficacy of NIR-PIT.

Head and neck cancers are relatively easy to illuminate with NIR light using existing light diffusers. A frontal diffuser is used for surface illumination, and a cylindrical diffuser is used for deep lesions, sometimes with the assistance of an imaging modality such as endoscopes, ultrasound, and computed tomography [23–25]. However, some body structures, such as the hypopharynx, are technically difficult to illuminate accurately with such diffusers [5]. The LEL fiber, enabling uniform tumor illumination while minimizing exposure to non-tumor tissue, is expected to reduce the risk of edema and side effects on normal organs.

Limiting the area of illumination of NIR light may reduce the adverse effects of NIR-PIT. Although NIR-PIT is a highly tumor cell-selective treatment and has few adverse effects on normal cells, edema around the tumor has been reported after treatment [26–28]. When NIR light reacts with IR700 under oxygen-rich conditions, reactive oxygen species are generated, leading to edema in normal tissue. Moreover, since EGFR is expressed to some extent in normal mucosal cells, excess irradiation to normal areas might cause side effects if using an EGFR-targeting drug. The LEL fiber allows uniform light illumination of the entire tumor and reduces light exposure to non-tumor tissue. Therefore, it is expected that NIR-PIT with an LEL fiber is less likely to cause edema and side effects in normal organs.

Several limitations exist in this study. We used only one cell line. However, since the aim of this study was to evaluate the usefulness of a novel LEL fiber, it is important to determine if it can properly illuminate the NIR light. Our results clearly showed that an LEL fiber can be used for photoimmunotherapy. Moreover, since the cytotoxic mechanism of NIR-PIT involves membrane damage after NIR light exposure, the therapeutic effects of NIR-PIT are thought to occur independently of cell type if the NIR light is properly exposed. Second, various novel illumination devices for NIR-PIT, such as a dedicated catheter with light-emitting diodes, a temperature sensor, and an endovascular therapy-based light illumination technology, have been developed [29,30]. It is difficult to compare these reported methods with an LEL fiber. However, it is important to increase the number of illumination options to accommodate various tumors. We are convinced that the LEL fiber is at least suitable for irradiating small lumens and, therefore, there are many indications that exist.

5. Conclusions

Side-firing illumination with an LEL fiber may illuminate NIR light to only a certain area of a lateral wall in a narrow space. The use of NIR-PIT with an LEL fiber showed cytotoxic effects in vitro and suppressed tumor proliferation in vivo. These results indicated that an LEL fiber may be a new option for NIR light illumination in NIR-PIT.

Author Contributions: M.S.: Conceptualization, Formal analysis, Funding acquisition, Investigation, Methodology, Visualization, Writing—original draft. H.H.: Conceptualization, Funding acquisition, Project administration, Resources, Supervision, Writing—review and editing. H.K.: Conceptualization, Supervision, Writing—review and editing. All authors have read and agreed to the published version of the manuscript.

Funding: This work was supported, in part, by Furukawa Electric Co., Ltd. H.K. was supported by the Intramural Research Program of the National Institutes of Health, National Cancer Institute, Center for Cancer Research (ZIABC011513, Funder Id: http://dx.doi.org/10.13039/100000054).

Institutional Review Board Statement: All animal experiments were conducted according to the established guidelines of the "Law for The Care and Welfare of Animals in Japan" and approved by the Ethics Committee on Animal Experiments at Kansai Medical University (protocol code 23-022, 1 April 2023).

Informed Consent Statement: Not applicable.

Data Availability Statement: The data presented in this study are available on request from the corresponding author.

Acknowledgments: We would like to thank Furukawa Electric Co., Ltd. personnel for providing the lateral emitting laser fiber: H. Hasegawa, J. Hasegawa, T. Hasegawa, and K. Take.

Conflicts of Interest: The authors declare no competing interests in relation to this study.

References

1. Mitsunaga, M.; Ogawa, M.; Kosaka, N.; Rosenblum, L.T.; Choyke, P.L.; Kobayashi, H. Cancer cell-selective in vivo near infrared photoimmunotherapy targeting specific membrane molecules. *Nat. Med.* **2011**, *17*, 1685–1691. [CrossRef] [PubMed]
2. Kobayashi, H.; Choyke, P.L. Near-Infrared Photoimmunotherapy of Cancer. *Acc. Chem. Res.* **2019**, *52*, 2332–2339. [CrossRef] [PubMed]
3. Gao, X.; Cui, Y.; Levenson, R.M.; Chung, L.W.; Nie, S. In vivo cancer targeting and imaging with semiconductor quantum dots. *Nat. Biotechnol.* **2004**, *22*, 969–976. [CrossRef]
4. Tahara, M.; Okano, S.; Enokida, T.; Ueda, Y.; Fujisawa, T.; Shinozaki, T.; Tomioka, T.; Okano, W.; Biel, M.A.; Ishida, K.; et al. A phase I, single-center, open-label study of RM-1929 photoimmunotherapy in Japanese patients with recurrent head and neck squamous cell carcinoma. *Int. J. Clin. Oncol.* **2021**, *26*, 1812–1821. [CrossRef]
5. Nishikawa, D.; Suzuki, H.; Beppu, S.; Terada, H.; Sawabe, M.; Hanai, N. Near-Infrared Photoimmunotherapy for Oropharyngeal Cancer. *Cancers* **2022**, *14*, 5662. [CrossRef] [PubMed]
6. Nagaya, T.; Okuyama, S.; Ogata, F.; Maruoka, Y.; Choyke, P.L.; Kobayashi, H. Endoscopic near infrared photoimmunotherapy using a fiber optic diffuser for peritoneal dissemination of gastric cancer. *Cancer Sci.* **2018**, *109*, 1902–1908. [CrossRef]
7. Okada, R.; Furusawa, A.; Inagaki, F.; Wakiyama, H.; Kato, T.; Okuyama, S.; Furumoto, H.; Fukushima, H.; Choyke, P.L.; Kobayashi, H. Endoscopic near-infrared photoimmunotherapy in an orthotopic head and neck cancer model. *Cancer Sci.* **2021**, *112*, 3041–3049. [CrossRef]
8. Fukushima, H.; Turkbey, B.; Pinto, P.A.; Furusawa, A.; Choyke, P.L.; Kobayashi, H. Near-Infrared Photoimmunotherapy (NIR-PIT) in Urologic Cancers. *Cancers* **2022**, *14*, 2996. [CrossRef]
9. Furumoto, H.; Kato, T.; Wakiyama, H.; Furusawa, A.; Choyke, P.L.; Kobayashi, H. Endoscopic Applications of Near-Infrared Photoimmunotherapy (NIR-PIT) in Cancers of the Digestive and Respiratory Tracts. *Biomedicines* **2022**, *10*, 846. [CrossRef]
10. Yano, T.; Minamide, T.; Takashima, K.; Nakajo, K.; Kadota, T.; Yoda, Y. Clinical Practice of Photodynamic Therapy Using Talaporfin Sodium for Esophageal Cancer. *J. Clin. Med.* **2021**, *10*, 2785. [CrossRef]
11. Overholt, B.F.; Lightdale, C.J.; Wang, K.K.; Canto, M.I.; Burdick, S.; Haggitt, R.C.; Bronner, M.P.; Taylor, S.L.; Grace, M.G.; Depot, M.; et al. Photodynamic therapy with porfimer sodium for ablation of high-grade dysplasia in Barrett's esophagus: International, partially blinded, randomized phase III trial. *Gastrointest. Endosc.* **2005**, *62*, 488–498. [CrossRef]
12. Zimmermann, M.; Zouhair, A.; Azria, D.; Ozsahin, M. The epidermal growth factor receptor (EGFR) in head and neck cancer: Its role and treatment implications. *Radiat. Oncol.* **2006**, *1*, 11. [CrossRef] [PubMed]
13. Grandis, J.R.; Tweardy, D.J. Elevated levels of transforming growth factor alpha and epidermal growth factor receptor messenger RNA are early markers of carcinogenesis in head and neck cancer. *Cancer Res.* **1993**, *53*, 3579–3584.
14. Carpenter, G.; Cohen, S. Epidermal growth factor. *Annu. Rev. Biochem.* **1979**, *48*, 193–216. [CrossRef] [PubMed]
15. Miyoshi, Y.; Nishimura, T.; Shimojo, Y.; Okayama, K.; Awazu, K. Endoscopic image-guided laser treatment system based on fiber bundle laser steering. *Sci. Rep.* **2023**, *13*, 2921. [CrossRef]
16. Kostic, S.V.; Rice, T.W.; Baker, M.E.; Decamp, M.M.; Murthy, S.C.; Rybicki, L.A.; Blackstone, E.H.; Richter, J.E. Timed barium esophagogram: A simple physiologic assessment for achalasia. *J. Thorac. Cardiovasc. Surg.* **2000**, *120*, 935–943. [CrossRef]

Takashima, K.; Koga, Y.; Anzai, T.; Migita, K.; Yamaguchi, T.; Ishikawa, A.; Sakashita, S.; Yasunaga, M.; Yano, T. Evaluation of Fluorescence Intensity and Antitumor Effect Using Real-Time Imaging in Photoimmunotherapy. *Pharmaceuticals* **2022**, *15*, 223. [CrossRef]
Schindelin, J.; Arganda-Carreras, I.; Frise, E.; Kaynig, V.; Longair, M.; Pietzsch, T.; Preibisch, S.; Rueden, C.; Saalfeld, S. Schmid, B.; et al. Fiji: An open-source platform for biological-image analysis. *Nat. Methods* **2012**, *9*, 676–682. [CrossRef]
Ogawa, M.; Takakura, H. Photoimmunotherapy: A new cancer treatment using photochemical reactions. *Bioorg. Med. Chem.* **2021**, *43*, 116274. [CrossRef] [PubMed]
Maruoka, Y.; Nagaya, T.; Nakamura, Y.; Sato, K.; Ogata, F.; Okuyama, S.; Choyke, P.L.; Kobayashi, H. Evaluation of Early Therapeutic Effects after Near-Infrared Photoimmunotherapy (NIR-PIT) Using Luciferase-Luciferin Photon-Counting and Fluorescence Imaging. *Mol. Pharm.* **2017**, *14*, 4628–4635. [CrossRef]
Dumoulin, F.L.; Hildenbrand, R.; Oyama, T.; Steinbruck, I. Current Trends in Endoscopic Diagnosis and Treatment of Early Esophageal Cancer. *Cancers* **2021**, *13*, 752. [CrossRef] [PubMed]
Tang, Y.; Anandasabapathy, S.; Richards-Kortum, R. Advances in optical gastrointestinal endoscopy: A technical review. *Mol. Oncol.* **2021**, *15*, 2580–2599. [CrossRef] [PubMed]
Omura, G.; Honma, Y.; Matsumoto, Y.; Shinozaki, T.; Itoyama, M.; Eguchi, K.; Sakai, T.; Yokoyama, K.; Watanabe, T.; Ohara, A.; et al. Transnasal photoimmunotherapy with cetuximab sarotalocan sodium: Outcomes on the local recurrence of nasopharyngeal squamous cell carcinoma. *Auris Nasus Larynx* **2023**, *50*, 641–645. [CrossRef] [PubMed]
Okamoto, I.; Okada, T.; Tokashiki, K.; Tsukahara, K. A Case Treated With Photoimmunotherapy Under a Navigation System for Recurrent Lesions of the Lateral Pterygoid Muscle. *In Vivo* **2022**, *36*, 1035–1040. [CrossRef] [PubMed]
Koyama, S.; Ehara, H.; Donishi, R.; Morisaki, T.; Ogura, T.; Taira, K.; Fukuhara, T.; Fujiwara, K. Photoimmunotherapy with surgical navigation and computed tomography guidance for recurrent maxillary sinus carcinoma. *Auris Nasus Larynx* **2023**, *50*, 646–651. [CrossRef] [PubMed]
Nakajima, K.; Sugikawa, A.; Yasui, H.; Higashikawa, K.; Suzuki, C.; Natsume, T.; Suzuki, M.; Takakura, H.; Tomita, M.; Takahashi, S.; et al. In vivo imaging of acute physiological responses after treatment of cancer with near-infrared photoimmunotherapy. *Mol. Imaging Biol.* **2023**, *25*, 648–658. [CrossRef] [PubMed]
Cognetti, D.M.; Johnson, J.M.; Curry, J.M.; Kochuparambil, S.T.; McDonald, D.; Mott, F.; Fidler, M.J.; Stenson, K.; Vasan, N.R.; Razaq, M.A.; et al. Phase 1/2a, open-label, multicenter study of RM-1929 photoimmunotherapy in patients with locoregional, recurrent head and neck squamous cell carcinoma. *Head Neck* **2021**, *43*, 3875–3887. [CrossRef] [PubMed]
Kato, T.; Okada, R.; Goto, Y.; Furusawa, A.; Inagaki, F.; Wakiyama, H.; Furumoto, H.; Daar, D.; Turkbey, B.; Choyke, P.L.; et al. Electron Donors Rather Than Reactive Oxygen Species Needed for Therapeutic Photochemical Reaction of Near-Infrared Photoimmunotherapy. *ACS Pharmacol. Transl. Sci.* **2021**, *4*, 1689–1701. [CrossRef] [PubMed]
Hirata, H.; Kuwatani, M.; Nakajima, K.; Kodama, Y.; Yoshikawa, Y.; Ogawa, M.; Sakamoto, N. Near-infrared photoimmunotherapy (NIR-PIT) on cholangiocarcinoma using a novel catheter device with light emitting diodes. *Cancer Sci.* **2021**, *112*, 828–838. [CrossRef]
Tsukamoto, T.; Fujita, Y.; Shimogami, M.; Kaneda, K.; Seto, T.; Mizukami, K.; Takei, M.; Isobe, Y.; Yasui, H.; Sato, K. Inside-the-body light delivery system using endovascular therapy-based light illumination technology. *eBioMedicine* **2022**, *85*, 104289. [CrossRef]

cancers

MDPI

Article

A Phthalimide-Functionalized Heptamethine Cyanine Dye for Tumor-Targeted Photothermal Therapy

Yoonbin Park [1,2], Juhui Yang [1,2] and Hoon Hyun [1,2,*]

1 Department of Biomedical Sciences, Chonnam National University Medical School, Hwasun 58128, Republic of Korea; unb1n0213@naver.com (Y.P.); joooh1008@gmail.com (J.Y.)
2 BioMedical Sciences Graduate Program (BMSGP), Chonnam National University, Hwasun 58128, Republic of Korea
* Correspondence: hhyun@chonnam.ac.kr; Tel.: +82-61-379-2652

Simple Summary: Most commercial heptamethine cyanine dyes have no tumor targetability owing to their limited intrinsic selectivity. The development of multifunctional heptamethine cyanine dyes that can enable both tumor-specific imaging and therapy is highly desirable for simple and effective cancer treatments. For this reason, we newly synthesized a tumor-targetable heptamethine cyanine dye which not only showed enhanced accumulation in the tumor site, but was also highly effective as an in vivo photothermal cancer therapy in a HT-29 human colon cancer xenograft mouse model. On the principle of the "structure-inherent targeting" approach, a single-molecule NIR dye could achieve tumor targeting, imaging, and therapy simultaneously. Hence, the present study suggests a simple but efficient strategy to design and synthesize the multifunctional small-molecule theranostic agents for tumor-targeted imaging and phototherapy.

Abstract: Background: A phthalimide-functionalized heptamethine cyanine dye, named Ph790H, is used for targeted photothermal cancer therapy in vivo. We highlight that the chemical structure of Ph790H is newly designed and synthesized for the first time in this study. Objectives: By possessing a rigid chloro-cyclohexenyl ring in the heptamethine cyanine backbone, the bifunctional near-infrared (NIR) fluorescent dye Ph790H can be preferentially accumulated in tumor without the need for additional targeting ligands, which is defined as the "structure-inherent tumor targeting" concept. Methods: The phototherapeutic effect of Ph790H is evaluated in HT-29 human colorectal cancer xenografts to be used as a cancer-targeting photothermal agent. Results: The results reveal that the Ph790H shows enhanced tumor accumulation in HT-29 xenografts 48 h post-injection with a high tumor-to-background ratio. After determination of the optimal timing for photothermal therapy (PTT), the HT-29 tumor-possessing nude mice pretreated with Ph790H are subsequently irradiated with an 808 nm NIR laser for 5 min. The tumor-targeted PTT treatment can efficiently inhibit the tumor development compared with that of control groups. Moreover, no tumor regrowth or Ph790H-induced mortality occurs after the treatment of Ph790H and laser irradiation during a period of monitoring. Conclusions: Therefore, this work demonstrates that the bifunctional phototheranostic agent Ph790H can be utilized for targeted cancer imaging and fluorescence-guided phototherapy simultaneously.

Keywords: heptamethine cyanine dyes; near-infrared fluorescence imaging; photothermal therapy; tumor-targeted imaging; structure-inherent targeting

1. Introduction

In recent years, photothermal therapy (PTT) combined with tumor-targeted near-infrared (NIR) fluorescent dyes has attracted much attention because it allows for fluorescence-guided real-time and in situ therapy [1–3]. PTT is a promising modality for effective cancer therapy owing to its intrinsic advantages in terms of cost, convenience, and safety with high tumor ablation efficiency [4–6]. Upon NIR light irradiation, the NIR fluorescent dyes, which are one type of a diverse range of PTT agents, convert the incident light into heat energy to physically damage the tumor cells. Most of the cyanine-based PTT agents are widely used in the NIR window (700–900 nm) because the multifunctional polymethine cyanine dyes have been continually developed for targeted cancer imaging and PTT/photodynamic therapy (PDT) by mostly using 671 and 808 nm laser sources [7–9]. The two most common laser-based phototherapies utilize light and photo-absorbing agents to produce cytotoxic reactive oxygen species or a local temperature increase, respectively. Since PDT application in cancer treatment has several limitations, mainly due to tumor hypoxia, PTT has attracted more attention owing to oxygen independence.

Importantly, among the various polymethine cyanine dyes developed so far, which mainly consist of heptamethine or pentamethine cyanine skeletons, the heptamethine cyanine structure having a chloro-cyclohexenyl ring is much more favorable than other types of cyanine dyes in terms of in vivo tumor targetability enhanced by the formation of albumin adducts [10–12]. On the other hand, the electron-withdrawing chloro-cyclohexenyl ring pulls electrons from the heptamethine chain to readily undergo nucleophilic substitution through a $S_{RN}1$ mechanism to replace the central chlorine atom with diverse functional groups, including alkyl-ether, aryl-ether, and aryl-amine linkages, for the conjugation of various biomolecules [8]. Additionally, it was previously reported that the commercial heptamethine cyanine dyes, including IR-780, IR-783, IR-786, and IR-820, possessing a rigid heptamethine backbone with the chloro-cyclohexenyl ring exhibited not only enhanced photostability and quantum yield, but also reduced photobleaching and aggregation behavior compared with that of indocyanine green (ICG) consisting of a heptamethine chain without the chloro-cyclohexenyl ring, which is the FDA-approved NIR fluorescent dye used for clinical diagnostic applications [8]. Thus, the heptamethine cyanine dyes containing the chloro-cyclohexenyl bridge are highly suitable for NIR fluorescence imaging and phototherapy applications.

Most heptamethine cyanine dyes available commercially have no tumor targetability. The development of multifunctional heptamethine cyanine dyes that can enable both tumor-specific imaging and therapy is highly desirable for simple and effective cancer treatments. Since the ICG has no chemical groups used for conjugation to other molecules, the various types of conjugatable heptamethine cyanine dyes, including Cy7, IRDye800CW, IR-808, and ZW800-1, have been commercially available for multipurpose NIR fluorescence imaging and other applications [13–15]. However, Choi et al. suggested that the targeting efficiency and therapeutic efficacy of the conjugates may be altered or reduced by the physicochemical properties (e.g., hydrophobicity, polarity, molecular conformation, and net surface charge) of conjugated NIR dyes [16].

An advanced strategy to overcome this limitation in target-specific NIR fluorescence imaging has been suggested to integrate the three functions (i.e., target specificity, fluorescence emission, and/or therapeutic potential) into a single-molecule NIR dye through the concept of "structure-inherent targeting" [17–19]. So far, various types of multifunctional heptamethine cyanine dyes, which showed tumor uptake natively without the conjugation to tumor-targeting ligands, have been designed and synthesized to be used for tumor-targeted fluorescence imaging [20–22]. Further, the tumor-targeted heptamethine cyanine dyes occasionally revealed potent therapeutic efficacy in various types of cancer. The key consideration for the design of multifunctional heptamethine cyanine dyes using this structure-inherent targeting strategy is the modulation of chemical structures (i.e., the heptamethine chain for PTT, indole side chains for the tumor-targeting domain, and *N*-alkyl side chains for chemotherapy).

In this study, we demonstrated the photothermal efficacy of the newly designed heptamethine cyanine dye, named Ph790H, combined with NIR laser irradiation in in vitro and in vivo studies using HT-29 cancer cells. The in vivo results revealed that Ph790H not only displayed enhanced and prolonged tumor accumulation in HT-29 xenografts but also exhibited superior photothermal conversion efficiency after laser irradiation for complete tumor ablation. Therefore, this work demonstrates that the bifunctional phototheranostic agent Ph790H can be utilized for targeted cancer imaging and fluorescence-guided phototherapy simultaneously.

2. Materials and Methods

2.1. Synthesis of the Heptamethine Cyanine Dye Ph790H

All chemicals and solvents were purchased from Sigma-Aldrich (St. Louis, MO, USA). 2,3,3-trimethylindolenine **3** (0.7 g, 4.3 mmol) and N-(3-bromopropyl)phthalimide **4** (1.34 g, 5 mmol) were mixed in toluene (50 mL) and heated at 100 °C for 72 h under a nitrogen atmosphere. After reaction, the mixture was washed with toluene several times to remove unreacted compounds. The solid was collected and used directly in the next step (0.76 g, 51.2%). Subsequently, compound **5** (0.76 g, 2.2 mmol), Vilsmeier-Haack reagent **6** (0.32 g, 1 mmol), and anhydrous sodium acetate (0.25 g, 3 mmol) were mixed in absolute ethanol (20 mL) and heated under reflux for 6 h under a nitrogen atmosphere. After the reaction, the residue was dissolved in dimethyl sulfoxide (DMSO) and purified by a preparative high-performance liquid chromatography (HPLC) system (Waters, Milford, MA, USA). The final compound Ph790H was collected as a dark green solid (0.68 g, 82% yield with 97% purity). The obtained Ph790H was confirmed with the Dionex UltiMate™ 3000 mass spectrometry system (Thermo Scientific, Waltham, MA, USA). The accurate mass HRMS (ESI) m/z $[M]^+$ was calculated for $[C_{52}H_{50}ClN_4O_4]^+$ 829.3515, and found $[M]^+$ 829.3500.

2.2. Optical and Physicochemical Property Analyses

Fetal bovine serum (FBS, pH 7.4) (Welgene, Gyeongsan, Republic of Korea) was used to dissolve Ph790H. A fiber optic FLAME spectrometer (Ocean Optics, Dunedin, FL, USA) was used to measure the absorbance of Ph790H. The Beer–Lambert equation was used to determine the molar extinction coefficient (ε) of Ph790H. A SPARK® 10M microplate reader (Tecan, Männedorf, Switzerland) was used to analyze the fluorescence emission spectrum excited at 740 nm. Indocyanine green (ICG) dissolved in DMSO (Φ = 13% as a calibration standard) was used to determine the fluorescence quantum yield (Φ) of Ph790H by matching the absorbance at 770 nm [16]. Marvin and JChem calculator plugins (under version 14.12.15.0, ChemAxon, Budapest, Hungary) were used to analyze in silico predictions of the distribution coefficient (log$D$ at pH 7.4) and topological polar surface area (TPSA).

2.3. In Vitro Cytotoxicity Assay

The human large-cell lung carcinoma cell line (NCI-H460), breast adenocarcinoma cell lines (MDA-MB-231 and MCF-7), colorectal adenocarcinoma cell line (HT-29), and mouse embryonic fibroblast cell line (NIH/3T3) were purchased from the American Type Culture Collection (ATCC; Manassas, VA, USA). Dulbecco's Modified Eagle's Medium (DMEM) or Roswell Park Memorial Institute (RPMI) 1640 media containing FBS, penicillin, streptomycin, and amphotericin B (Welgene) were used to maintain cultured cells. The cytotoxicity of Ph790H was examined using the 3-(4,5-dimethylthiazol-2-yl)-2,5-diphenyltetrazolium bromide (MTT) assay. Cells were seeded onto 96-well plates (1×10^4 cells per well). To evaluate the cytotoxicity depending on the concentrations of each sample, the cells were treated with Ph790H (2, 10, 20, and 50 μM) for 4 h and cultured for 24 h after treatment. At each time point, the incubation cell medium was replaced with 100 μL of fresh medium, and 10 μL of the MTT solution was directly added to each 100 μL well. Subsequently, the plates were then incubated for 4 h at 37 °C in a humidified 5% CO_2 incubator. Finally, the plates were placed in a microplate reader (SPARK® 10M, Tecan) to measure the absorption intensity at 570 nm. Cell viability was calculated using the following formula: cell viability (%) = $(A_{sample} - A_{blank})/(A_{control} - A_{blank}) \times 100$, where A is the average absorbance.

2.4. In Vitro Live-Cell Imaging

In total, 2 μM of Ph790H was added to HT-29 cells and stored at 37 °C in a humidified 5% CO_2 incubator for 1 h. After washing with phosphate-buffered saline (PBS, pH 7.4), the cells were observed by a 4-filter set of the Nikon Eclipse Ti-U inverted microscope system (Nikon, Seoul, Republic of Korea).

2.5. In Vitro Photothermal Conversion Efficiency

An NIR laser diode system (NaKu Technology Co., Ltd., Hangzhou, China) was used for irradiation of model substrates and mice tumors. The emission wavelength as provided by the supplier was 808 ± 5 nm. The emitted power and beam diameter were ~1000 mW and ~5 mm, respectively. In all cases, the distance between the laser output and the different irradiated targets was kept constant at 10 cm. Ph790H (50 μM) dissolved in FBS (100 μL, pH 7.4) was irradiated with an 808 nm laser (1.0 W/cm^2). A thermal imager (FLIR Systems, Wilsonville, OR, USA) was used to monitor temperature change in real time. The photothermal stability of Ph790H was tested by repeating the heating and cooling cycle three times. The photothermal conversion efficiency (η) of Ph790H was calculated based on the equations reported previously [23].

2.6. In Vitro Photothermal Cytotoxicity

Ph790H (10 μM) was added to HT-29 cancer cells and incubated for 4 h. After washing with PBS, the cells were irradiated with the 808 nm laser (1.0 W/cm^2) for 1 min. Subsequently, Calcein-AM and propidium iodide (PI; Thermo Fisher Scientific) were used to co-stain the cells for 30 min. After washing with PBS, the fluorescent microscope (Nikon) was used to confirm the stained cells.

2.7. HT-29 Xenograft Mouse Model

The animal experiment was approved by the Chonnam National University Animal Research Committee (CNU IACUC-H-2023-57). Male nude mice (6 weeks old, ≈25 g, 3 mice in a Ph790H-injected group used for time-dependent tumor imaging, and a total of 9 mice in three different treatment groups used for PTT treatment) were purchased from OrientBio (Gwangju, Republic of Korea). An HT-29 xenograft was prepared by the inoculation of HT-29 cancer cells (1×10^5 cells per mouse) subcutaneously into the right flank of each mouse. At 10 days post-inoculation, Ph790H was dissolved in a saline solution containing 5% wt./v bovine serum albumin (BSA) and injected intravenously.

2.8. In Vivo Time-Dependent Tumor Imaging

The FOBI fluorescence imaging system (NeoScience, Deajeon, Republic of Korea) was used to monitor the tumor-bearing mice in real time. The mice (3 mice in the PBS or Ph790H treatment group) were monitored for 72 h after intravenous injection of Ph790H. The tumor fluorescence intensity was measured using ImageJ software (National Institutes of Health, Bethesda, MD, USA, https://imagej.net/ij/, accessed on 1 October 2024). A thermal imager (FLIR Systems) was also used to monitor the tumor temperature for 5 min.

2.9. In Vivo Photothermal Therapeutic Efficacy

The HT-29 xenograft mouse model was used to inject the Ph790H or PBS (3 mice in the PBS with laser, Ph790H alone, or Ph790H with laser treatment group). At 48 h post-injection, the tumor area of each mouse was subjected to irradiate the 808 nm laser (1.0 W/cm^2) for 5 min. The temperature changes at the tumor site were observed using the thermal imager (FLIR Systems). Subsequently, the mice in each group were sacrificed to harvest the tumors for the histological confirmation using the hematoxylin and eosin (H&E) staining. In addition, mice were monitored to observe tumor development and body weight for 11 days. The calculation formula V = 0.5 × longest diameter × (shortest diameter)^2 was used to measure the tumor volume (V).

2.10. Histological Analysis

Tumor samples (n = 3) were collected from each mouse group and stored for H&E staining. The tumors were fixed in 4% paraformaldehyde and flash frozen in an optimal cutting temperature (OCT) compound using liquid nitrogen. Frozen samples were cryosectioned (10 μM thick), stained with H&E, and observed using a microscope. Histological confirmation of each tumor section was carried out using the Nikon Eclipse Ti-U inverted microscope system.

2.11. Statistical Analysis

Statistical analysis was determined by a one-way analysis of variance followed by Tukey's multiple comparison test using the Prism software version 5.01 (GraphPad, San Diego, CA, USA). The results were represented as mean ± standard deviation (S.D.), and a p-value less than 0.05 was statistically significant.

3. Results

3.1. Synthesis and Characterization of Ph790H

A phthalimide-functionalized heptamethine cyanine dye Ph790H was prepared according to the synthesis and purification process reported previously [24,25]. The heptamethine cyanine backbone was synthesized through the condensation reaction between phthalimide-functionalized heterocyclic salts **5** and the Vilsmeier–Haack reagent **6** (Figure 1a). After purification, the molecular weight of Ph790H was analyzed using mass spectrometry to confirm the successful synthesis of Ph790H (Figure 1b). As summarized in Figure 1c, the peak absorption at 790 nm and maximum fluorescence emission at 803 nm for the Ph790H dye dissolved in serum were detected with a typical Stokes shift of 13 nm. Ph790H exhibited a moderate molar extinction coefficient (ε = 185,000 $M^{-1}cm^{-1}$) and quantum yield (Φ = 8.6%) measured in serum, compared with that of ICG available clinically. Ph790H displayed a relatively higher hydrophobicity (log$D$ = 6.40) and lower polarity (TPSA = 81.01 Å^2), as compared with that of ICG.

a

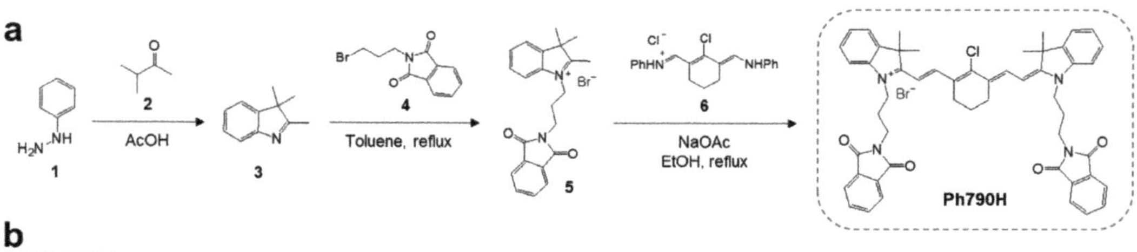

b

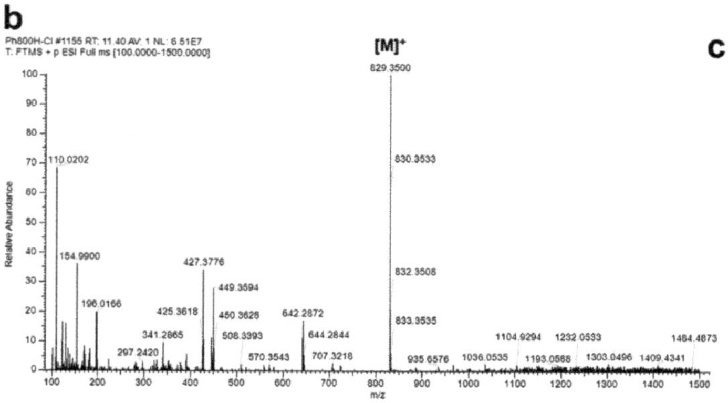

c

Property	ICG	Ph790H
MW (Da)	751.97	830.43
Log$D$, pH 7.4	4.91	6.40
TPSA (Å^2)	120.65	81.01
λ_{exc} (nm)	807	790
λ_{em} (nm)	822	803
Stokes shift (nm)	15	13
ε ($M^{-1}cm^{-1}$)	121,000	185,000
Φ (%)	9.3	8.6

Figure 1. (**a**) Synthetic process and (**b**) mass data of Ph790H. (**c**) Physicochemical and optical properties of ICG [24] and Ph790H.

3.2. In Vitro Cytotoxicity, Cell Binding, and Photothermal Effect

The absorption and fluorescence emission spectra of Ph790H were measured in serum due to its high hydrophobicity (Figure 2a). The cytotoxicity of Ph790H was found using the MTT assay in NIH/3T3, NCI-H460, MDA-MB-231, MCF-7, and HT-29 cells after the treatment of Ph790H (2–50 μM) (Figure 2b). To check the cytotoxic effect on normal cells, NIH/3T3 fibroblasts were treated with Ph790H under the same conditions. The NIH/3T3 cells showed no obvious cytotoxicity between 0 and 20 μM concentrations of Ph790H. Also, the NCI-H460, MDA-MB-231, and MCF-7 cancer cells showed no significant cytotoxicity in the concentration range of 0–20 μM of Ph790H, which is similar to that of NIH/3T3 cells. Interestingly, the cytotoxicity of HT-29 cancer cells treated with Ph790H increased depending on the concentrations. This result indicates that Ph790H has cell type-specific toxicity and can be used for the treatment of colon cancer. Moreover, we observed the cell binding of Ph790H after incubation in HT-29 cancer cells for 1 h (Figure 2c). As expected, Ph790H exhibited significant intracellular localization in the cytoplasm with high fluorescence intensity. This result indicates that Ph790H can deliver the photothermal energy into the HT-29 cancer cells during NIR laser irradiation for inducing cancer cell death.

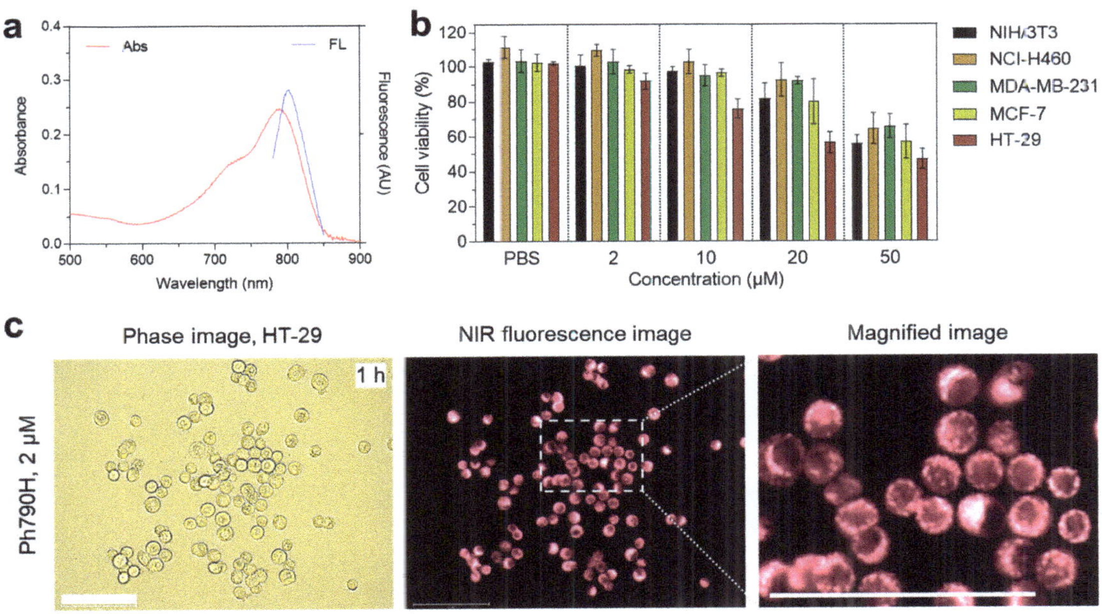

Figure 2. (**a**) Absorbance and fluorescence of Ph790H measured in serum. (**b**) Cytotoxicity analysis of Ph790H using NIH/3T3, NCI-H460, MDA-MB-231, MCF-7, and HT-29 cells. (**c**) Cell binding of Ph790H in HT-29 cells. Scale bars = 100 μM.

After confirming cytotoxicity and intracellular localization of Ph790H, the photothermal conversion efficiency of Ph790H was examined by analyzing the temperature changes in Ph790H (100 μM in serum) under 808 nm laser irradiation (1.0 W/cm^2) for 1 min. The thermal imager was used in real time to observe the temperature variation. The temperature of the Ph790H solution become elevated immediately, reaching up to 73.1 °C, after 1 min of irradiation, whereas the PBS solution exhibited little change in temperature after 1 min of irradiation (Figure 3a). Also, we examined the temperature changes in different concentrations of Ph790H upon 808 nm laser irradiation for 2 min. As expected, the Ph790H solutions showed a concentration-dependent increase in temperature after 2 min of irradiation (Figure 3b). Based on the temperature variation in the Ph790H solution displayed in Figure 3c, the photothermal conversion efficiency (η) of Ph790H was determined to be 31.1%, which is similar to that of the cyanine dyes reported previously [13,26,27]. This experiment demonstrates that Ph790H is suitable for photothermal tumor therapy in vivo. Additionally, we tested the photothermal stability of Ph790H by using the heating and cooling cycle (150 sec heating followed by 200 s cooling for each cycle). As shown in Figure 3d, the photostability of Ph790H in serum obviously declined during the second cycle. Moreover, the solution temperature of Ph790H was largely diminished during the third cycle of laser irradiation. This experiment result indicates that the heptamethine cyanine skeleton can irreversibly be damaged by irradiating the concentrated NIR light.

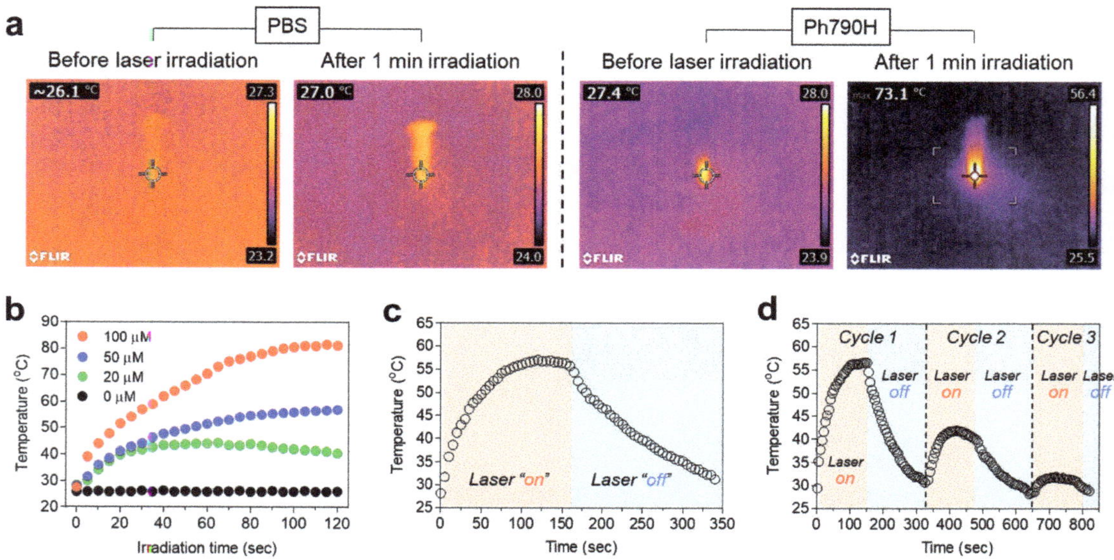

Figure 3. (**a**) Thermal images of PBS and Ph790H (100 μM) solutions before and after laser irradiation for 1 min. (**b**) Temperature changes in PBS and Ph790H (20, 50, and 100 μM) solutions were monitored for 120 s under laser irradiation. (**c**) Heating and cooling curve of Ph790H (50 μM) under laser irradiation. (**d**) Three on/off cycles of the Ph790H (50 μM) solution.

To observe the phototoxicity of HT-29 cells under laser irradiation, Calcein-AM (green for live cells) and propidium iodide (PI, red for dead cells) were used to stain HT-29 cancer cells before and after laser irradiation, respectively. The cancer cells were pretreated with Ph790H for 4 h and then exposed to 808 nm laser irradiation (1.0 W/cm^2) for 1 min. As shown in Figure 4, the HT-29 cancer cells treated with the combination of Ph790H and laser irradiation showed extensive cell death compared to other treatment groups (i.e., PBS without laser, PBS with laser, and Ph790H without laser irradiation), which are clearly confirmed by the PI and Calcein-AM staining. Thus, this result demonstrates that Ph790H (10 μM) alone or laser alone displayed no significant effect on cancer cell death. In this respect, high tumor accumulation of Ph790H is the most important requirement for maximizing the PTT effect in vivo.

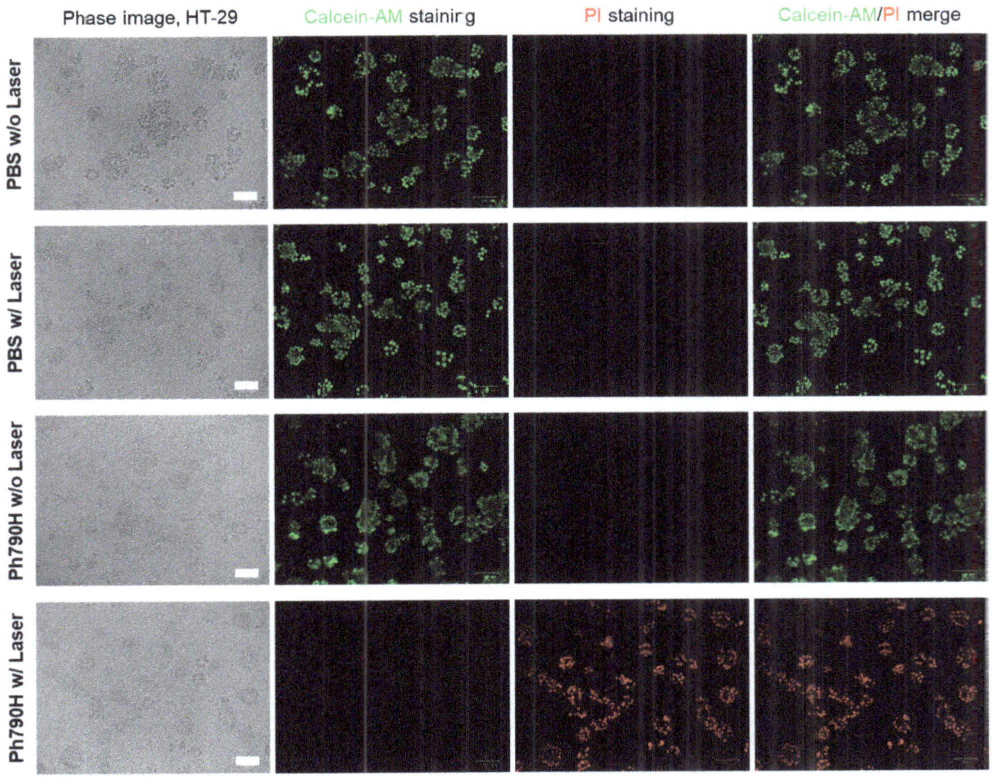

Figure 4. Fluorescence images of HT-29 cells treated with or without PTT treatment. Calcein-AM (green color) and propidium iodide (red color) were used to stain the HT-29 cells after each treatment. Scale bars = 100 μm.

3.3. Time-Dependent In Vivo Tumor Retention and Photothermal Effect

After confirming the effective cell phototoxicity of the combination of Ph790H and laser irradiation, we investigated the tumor-targeting efficiency of Ph790H using a HT-29 xenograft mouse model. The HT-29 xenografts were treated with Ph790H (0.6 mg/kg) and dissolved in a BSA solution through an intravenous administration. Remarkably, Ph790H showed high tumor accumulation over 72 h post-injection (Figure 5a). The tumor fluorescence intensity was continually elevated up to 48 h post-injection and gradually decreased after 72 h post-injection (Figure 5b). To find the optimal timing of laser irradiation, we determined the best time with high signals in the tumor but low signals in the surrounding tissues based on the tumor-to-background ratio over time (Figure 5c). The values of the tumor-to-background ratio were continuously elevated up to 72 h post-injection, owing to the maintained fluorescence signals in the tumor and decreased fluorescence signals in the surrounding tissues. Considering correlation between the tumor site and neighboring tissues, we carried out the laser irradiation on the tumor area at 48 h post-injection to avoid unnecessary damage to neighboring normal tissues.

Subsequently, we performed the 808 nm laser irradiation (1.0 W/cm^2) for 5 min using the HT-29 xenografts at 48 h post-injection of Ph790H. The power density of the NIR laser was optimized previously to prevent the normal tissue damage affected by the laser irradiation alone [28]. Interestingly, the temperature in the tumors treated with Ph790H and laser irradiation increased up to 56.2 °C during 5 min of laser irradiation, while the tumors injected with PBS exhibited no significant change in the temperature under the laser irradiation condition (Figure 5d). The temperature in tumors treated with Ph790H was gradually elevated by approximately 50 °C after 2 min of laser irradiation, and then the tumor temperature was maintained up to 55 °C for the next 3 min of laser irradiation (Figure 5e). These results demonstrate that the temperature (~55 °C) in tumors treated with the combination of Ph790H and laser irradiation is high enough to induce tumor necrosis for complete tumor ablation.

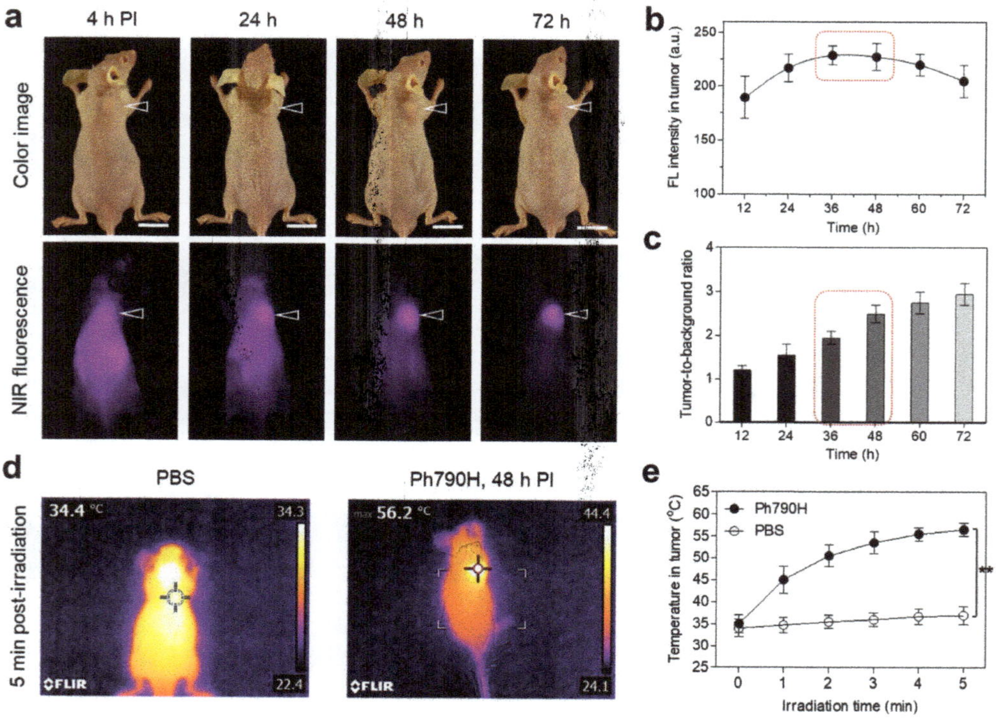

Figure 5. (**a**) In vivo NIR fluorescence imaging of HT-29 xenografts after injection of Ph790H. Arrowheads indicate the tumor area. Scale bars = 1 cm. (**b**) Fluorescence intensity and (**c**) tumor-to-background ratio at the tumor area for 72 h. (**d**) Thermal images and (**e**) temperature changes in each mouse group at the tumor area. Data are expressed as mean ± S.D. (n = 3). ** $p < 0.01$.

3.4. In Vivo Photothermal Therapeutic Efficacy

To evaluate the phototherapeutic effect of Ph790H in vivo, HT-29 xenografts were continuously observed for 11 days after treatment with Ph790H and laser irradiation (Figure 6a). As expected, the tumors treated with PBS and laser irradiation showed no therapeutic effect with a typical growth rate at the tumor site. Interestingly, the tumors treated with Ph790H alone exhibited slow development for the first 5 days after treatment, but the tumors began to grow over the next few days. This result suggests that the phthalimide moiety of Ph790H could inhibit the tumor growth for the early stage of treatment owing to its pharmacophore. More importantly, the tumors treated with the combination of Ph790H and laser irradiation showed a complete tumor ablation without tumor regrowth for 11 days (Figure 6b). Thus, these results demonstrate that Ph790H combined with laser irradiation can effectively ablate tumors based on the photothermal cell death. At day 11, a photograph of tumors collected from each mouse group displayed the obvious differences in tumor size (Figure 6c). Moreover, mice in each group showed no significant loss of body weight for 11 days, which indicates little systemic Ph790H toxicity (Figure 6d). As displayed in Figure 6e, the tumor section treated with PBS with laser irradiation showed normal shapes of cell proliferation. Although the tumor section treated with Ph790H alone exhibited little signs of apoptosis, the tumor section treated with Ph790H and laser irradiation showed apparent morphologic changes. This result suggests that the tumor-targetable Ph790H could be a safe and effective PTT agent for cancer phototherapy.

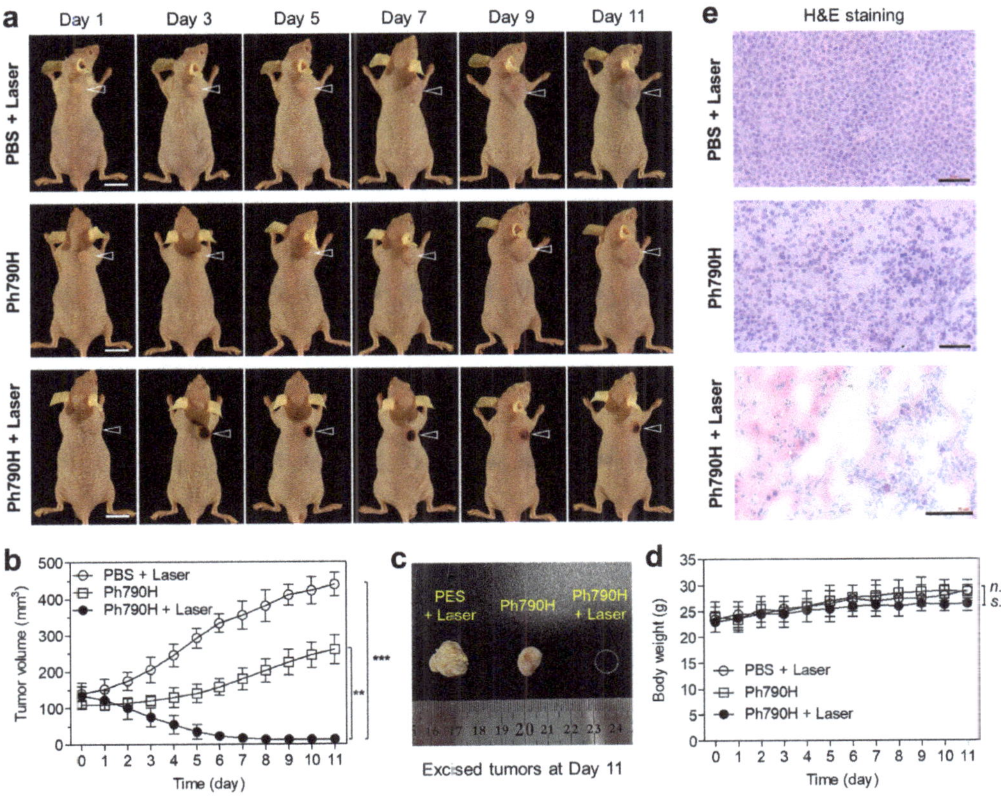

Figure 6. (**a**) Photographs of changes in tumor volume in HT-29 xenografts for 11 days. Arrowheads indicate the tumor area. Scale bars = 1 cm. (**b**) Tumor volumes, (**c**) a photograph of tumors collected from each mouse group at day 11, and (**d**) body weights were measured for 11 days. Data are expressed as mean ± S.D. (n = 3). ** $p < 0.01$; *** $p < 0.001$. (**e**) H&E staining of tumor sections collected from each mouse group (n = 3). Scale bars = 50 µm.

4. Discussion

Most studies have been focused mostly on nanomaterial-based systems for "all-in-one" theranostics by using relatively complicated methods. Although many different types of nanocarriers, including noble metal-, carbon-, protein-, and synthetic polymer-based nanomaterials, have been extensively developed for cancer-targeted chemo- and/or phototherapy, this approach is highly dependent on the tumor targetability of nanocarriers. That is, the size control and surface modification of nanomaterials are essentially needed to improve tumor accumulation and retention as well as clearance and excretion from the body for preventing systemic toxicity. However, there are still limitations to using the nanoparticles, such as complex synthetic processes for obtaining homogeneous products and unsolved biological safety [5] In this regard, the small-molecular cyanine dyes have recently attracted attention because of their excellent optical and physicochemical properties including noninvasive deep tissue imaging, tumor-specific uptake, and photothermal/photodynamic effect that can be used for simultaneous cancer-targeted imaging and therapy.

There are two major theories reported previously to explain how the heptamethine cyanine dyes can be accumulated in the tumor via the concept of "structure-inherent tumor targeting". Typically, a well-known mechanism is that the heptamethine cyanine dyes can be taken up by the organic anion transporting polypeptide (OATP) transporters into cancer cells [19]. On the other hand, Usama et al. recently suggested that the formation of covalent or noncovalent albumin adducts between them is a key trapping mechanism in tumors through albumin-mediated endocytosis [10–12]. Additionally, the albumin-mediated uptake in tumors is involved with the albumin binding proteins, including secreted protein acidic and rich in cysteine (SPARC) and membrane-associated glycoprotein, as well as the enhanced permeation and retention (EPR) effect [29]. Based on this theory, we understand that the rigid chloro-cyclohexenyl ring on the heptamethine backbone that forms the covalent albumin adducts plays an important role in tumor preferential accumulation of Ph790H.

The phthalimide group is known as a pharmacophore in medicinal chemistry owing to its antimicrobial, antioxidant, anticancer, and anti-inflammatory activities. The phthalimide group consisting of an isoindoline-1,3-dione core with an imide ring enables its movement across biological membranes in vivo. To date, many phthalimide derivatives have been developed and extensively investigated to confirm their therapeutic efficacy [30]. Interestingly, the phthalimide-functionalized heptamethine cyanine dye Ph790H showed delayed tumor development in the early stage of treatment for 5 days after a single-dose injection (0.6 mg/kg), compared to the control group. This suggests that the phthalimide moiety of Ph790H could contribute to its tumor suppressive activity in the early period. Based on this finding, the optimal dose of Ph790H could be made more effective in cancer chemotherapy. This can be further investigated in the next study, because PTT alone without a chemotherapeutic effect is still a limitation to carefully monitor and accurately treat primary tumor development and metastatic activity in orthotopic tumor models. Most importantly, the chemical structure of Ph790H may be a clue to provide a feasible approach for the development of multifunctional polymethine cyanine dyes enabling synergistic chemodynamic, photothermal, and photodynamic therapy.

5. Conclusions

In conclusion, we designed and synthesized a tumor-targetable heptamethine cyanine dye Ph790H which not only showed enhanced tumor accumulation 48 h post-injection but also excellent PTT efficacy in a HT-29 xenograft mouse model. On the basis of the "structure-inherent tumor targeting" concept, a single-molecule NIR dye Ph790H could achieve tumor targeting, imaging, and therapy simultaneously. Hence, the present study suggests a simple but efficient strategy to design and synthesize the multifunctional phototheranostic agents for cancer-targeted diagnosis and therapy. At present, most of the in vivo studies are still dependent on small animals, and a certain gap exists in the actual realization of extensive clinical application. With the progress and development of photo-based diagnosis and treatment methods, the multi-mode combined treatment strategy can provide a new opportunity and challenge for the next generation of cancer therapy.

Author Contributions: Conceptualization, H.H.; methodology, H.H.; validation, Y.P. and J.Y.; formal analysis, Y.P. and J.Y.; investigation, Y.P. and J.Y.; data curation, Y.P. and J.Y.; writing—original draft preparation, Y.P. and H.H.; writing—review and editing, Y.P. and H.H.; visualization, Y.P. and H.H.; supervision, H.H.; project administration, H.H.; funding acquisition, H.H. All authors have read and agreed to the published version of the manuscript.

Funding: This study was supported by the National Research Foundation of Korea (NRF) grant funded by the Korean government (MSIT) (No. RS-2024-00458419).

Institutional Review Board Statement: The study was conducted according to the guidelines of the Declaration of Helsinki, and approved by the Chonnam National University Animal Research Committee (protocol code CNU IACUC-H-2023-57, date of approval: 10 January 2024).

Informed Consent Statement: Not applicable.

Data Availability Statement: Data are contained within the article.

Conflicts of Interest: The authors declare no conflicts of interest.

ferences

Xie, Z.; Fan, T.; An, J.; Choi, W.; Duo, Y.; Ge, Y.; Zhang, B.; Nie, G.; Xie, N.; Zheng, T.; et al. Emerging combination strategies with phototherapy in cancer nanomedicine. *Chem. Soc. Rev.* **2020**, *49*, 8065–8087. [CrossRef] [PubMed]

Yin, X.; Cheng, Y.; Feng, Y.; Stiles, W.R.; Park, S.H.; Kang, H.; Choi, H.S. Phototheranostics for multifunctional treatment of cancer with fluorescence imaging. *Adv. Drug Deliv. Rev.* **2022**, *189*, 114483. [CrossRef] [PubMed]

Li, C.; Cheng, Y.; Li, D.; An, Q.; Zhang, W.; Zhang, Y.; Fu, Y. Antitumor Applications of Photothermal Agents and Photothermal Synergistic Therapies. *Int. J. Mol. Sci.* **2022**, *23*, 7909. [CrossRef]

Li, J.; Pu, K. Semiconducting Polymer Nanomaterials as Near-Infrared Photoactivatable Protherapeutics for Cancer. *Acc. Chem. Res.* **2020**, *53*, 752–762. [CrossRef]

Anilkumar, T.S.; Lu, Y.-J.; Chen, J.-P. Optimization of the Preparation of Magnetic Liposomes for the Combined Use of Magnetic Hyperthermia and Photothermia in Dual Magneto-Photothermal Cancer Therapy. *Int. J. Mol. Sci.* **2020**, *21*, 5187. [CrossRef] [PubMed]

Gallo, J.; Villasante, A. Recent Advances in Biomimetic Nanocarrier-Based Photothermal Therapy for Cancer Treatment. *Int. J. Mol. Sci.* **2023**, *24*, 15484. [CrossRef] [PubMed]

Li, Y.; Zhou, Y.; Yue, X.; Dai, Z. Cyanine conjugates in cancer theranostics. *Bioact. Mater.* **2021**, *6*, 794–809. [CrossRef]

Zhang, C.; Long, L.; Shi, C. Mitochondria-Targeting IR-780 Dye and Its Derivatives: Synthesis, Mechanisms of Action, and Theranostic Applications. *Adv. Therap.* **2018**, *1*, 1800069. [CrossRef]

Shi, C.; Wu, J.B.; Pan, D. Review on near-infrared heptamethine cyanine dyes as theranostic agents for tumor imaging, targeting, and photodynamic therapy. *J. Biomed. Opt.* **2016**, *21*, 050901. [CrossRef] [PubMed]

Usama, S.M.; Burgess, K. Hows and Whys of Tumor-Seeking Dyes. *Acc. Chem. Res.* **2021**, *54*, 2121–2131. [CrossRef] [PubMed]

Usama, S.M.; Lin, C.M.; Burgess, K. On the mechanisms of uptake of tumor-seeking cyanine dyes. *Bioconjugate Chem.* **2018**, *29*, 3886–3895. [CrossRef] [PubMed]

Usama, S.M.; Park, G.K.; Nomura, S.; Baek, Y.; Choi, H.S.; Burgess, K. Role of albumin in accumulation and persistence of tumor-seeking cyanine dyes. *Bioconjugate Chem.* **2020**, *31*, 248–259. [CrossRef] [PubMed]

Lu, Y.-J.; Anilkumar, T.S.; Chuang, C.C.; Chen, J.-P. Liposomal IR-780 as a Highly Stable Nanotheranostic Agent for Improved Photothermal/Photodynamic Therapy of Brain Tumors by Convection-Enhanced Delivery. *Cancers* **2021**, *13*, 3690. [CrossRef] [PubMed]

Thavornpradit, S.; Usama, S.M.; Park, G.K.; Shrestha, J.P.; Nomura, S.; Baek, Y.; Choi, H.S.; Burgess, K. QuatCy: A Heptamethine Cyanine Modification With Improved Characteristics. *Theranostics* **2019**, *9*, 2856–2867. [CrossRef]

Cooper, E.; Choi, P.J.; Denny, W.A.; Jose, J.; Dragunow, M.; Park, T.I.H. The use of heptamethine cyanine dyes as drug-conjugate systems in the treatment of primary and metastatic brain tumors. *Front. Oncol.* **2021**, *11*, 654921. [CrossRef] [PubMed]

Choi, H.S.; Gibbs, S.L.; Lee, J.H.; Kim, S.H.; Ashitate, Y.; Liu, F.; Hyun, H.; Park, G.; Xie, Y.; Bae, S.; et al. Targeted zwitterionic near-infrared fluorophores for improved optical imaging. *Nat. Biotechnol.* **2013**, *31*, 148–153. [CrossRef] [PubMed]

Li, M.; Long, S.; Kang, Y.; Guo, L.; Wang, J.; Fan, J.; Du, J.; Peng, X. De Novo Design of Phototheranostic Sensitizers Based on Structure-Inherent Targeting for Enhanced Cancer Ablation. *J. Am. Chem. Soc.* **2018**, *140*, 15820–15826. [CrossRef]

Liu, T.; Luo, S.; Wang, Y.; Tan, X.; Qi, Q.; Shi, C. Synthesis and characterization of a glycine-modified heptamethine indocyanine dye for in vivo cancer-targeted near-infrared imaging. *Drug Des. Dev. Ther.* **2014**, *8*, 1287–1297.

Fukuda, T.; Yokomizo, S.; Casa, S.; Monaco, H.; Manganiello, S.; Wang, H.; Lv, X.; Ulumben, A.D.; Yang, C.; Kang, M.W.; et al. Fast and Durable Intraoperative Near-infrared Imaging of Ovarian Cancer Using Ultrabright Squaraine Fluorophores. *Angew. Chem. Int. Ed.* **2022**, *61*, e202117330. [CrossRef]

Zhang, C.; Liu, T.; Luo, P.; Gao, L.; Liao, X.; Ma, L.; Jiang, Z.; Liu, D.; Yang, Z.; Jiang, Q.; et al. Near-infrared oxidative phosphorylation inhibitor integrates acute myeloid leukemia-targeted imaging and therapy. *Sci. Adv.* **2021**, *7*, eabb6104. [CrossRef]

Kang, H.; Shamim, M.; Yin, X.; Adluru, E.; Fukuda, T.; Yokomizo, S.; Chang, H.; Park, S.H.; Cui, Y.; Moy, A.J.; et al. Tumor-Associated Immune-Cell-Mediated Tumor-Targeting Mechanism with NIR-II Fluorescence Imaging. *Adv. Mater.* **2022**, *34*, e2106500. [CrossRef] [PubMed]

Yang, X.; Shi, C.; Tong, R.; Qian, W.; Zhau, H.E.; Wang, R.; Zhu, G.; Cheng, J.; Yang, V.W.; Cheng, T.; et al. Near IR heptamethine cyanine dye-mediated cancer imaging. *Clin. Cancer Res.* **2010**, *16*, 2833–2844. [CrossRef] [PubMed]

Roper, D.K.; Ahn, W.; Hoepfner, M. Microscale Heat Transfer Transduced by Surface Plasmon Resonant Gold Nanoparticles. *J. Phys. Chem. C Nanomater. Interfaces* **2007**, *111*, 3636–3641. [CrossRef] [PubMed]

24. Choi, H.S.; Nasr, K.; Alyabyev, S.; Feith, D.; Lee, J.H.; Kim, S.H.; Ashitate, Y.; Hyun, H.; Patonay, G.; Strekowski, L.; et al. Synthe and In Vivo Fate of Zwitterionic Near-Infrared Fluorophores. *Angew. Chem. Int. Ed.* **2011**, *50*, 6258–6263. [CrossRef] [PubMe
25. Hyun, H.; Bordo, M.W.; Nasr, K.; Feith, D.; Lee, J.H.; Kim, S.H.; Ashitate, Y.; Moffitt, L.A.; Rosenberg, M.; Henary, M.; et cGMP-compatible preparative scale synthesis of near-infrared fluorophores. *Contrast Media Mol. Imaging* **2012**, *7*, 516–5 [CrossRef] [PubMed]
26. Jiang, Z.; Yuan, B.; Wang, Y.; Wei, Z.; Sun, S.; Akakuru, O.U.; Li, Y.; Li, J.; Wu, A. Near-infrared heptamethine cyanine dye-bas nanoscale coordination polymers with intrinsic nucleus-targeting for low temperature photothermal therapy. *Nano Today* **20** *34*, 100910. [CrossRef]
27. Qian, H.; Cheng, Q.; Tian, Y.; Dang, H.; Teng, C.; Yan, L. An anti-aggregation NIR-II heptamethine-cyanine dye with a ster specific cyanine for imaging-guided photothermal therapy. *J. Mater. Chem. B* **2021**, *9*, 2688–2696. [CrossRef]
28. Lee, S.; Jo, G.; Jung, J.S.; Yang, D.H.; Hyun, H. Near-infra-red fluorescent chitosan oligosaccharide lactate for targeted can imaging and photothermal therapy. *Artif. Cells Nanomed. Biotechnol.* **2020**, *48*, 1144–1152. [CrossRef]
29. Spada, A.; Emami, J.; Tuszynski, J.A.; Lavasanifar, A. The Uniqueness of Albumin as a Carrier in Nanodrug Delivery. *Mol. Pha* **2021**, *18*, 1862–1894. [CrossRef] [PubMed]
30. Fernandes, G.F.S.; Lopes, J.R.; Dos Santos, J.L.; Scarim, C.B. Phthalimide as a versatile pharmacophore scaffold: Unlocking diverse biological activities. *Drug Dev. Res.* **2023**, *84*, 1346–1375. [CrossRef]

cancers

MDPI

Article

In Vitro Evaluation of the Safety and Efficacy of *Cibisatamab* Using Adult Stem Cell-Derived Organoids and Colorectal Cancer Spheroids

Victor Anstett [1,†], Elisa Heinzelmann [1,†], Francesco Piraino [1], Aline Roch [1], Antonius Chrisnandy [1], Maxim Norkin [2], Virginie Garnier [1], Krisztian Homicsko [2], Sylke Hoehnel-Ka [1,*] and Nathalie Brandenberg [1,*]

1 Doppl SA, 1015 Lausanne, Switzerland; victor.anstett@doppl.ch (V.A.); elisa.heinzelmann@doppl.ch (E.H.); francesco.piraino@doppl.ch (F.P.); aline.roch@doppl.ch (A.R.)
2 Department of Oncology, CHUV, 1066 Lausanne, Switzerland; maxim.norkin@chuv.ch (M.N.); krisztian.homicsko@chuv.ch (K.H.)
* Correspondence: sylke@doppl.ch (S.H.-K.); nathalie.brandenberg@doppl.ch (N.B.)
† These authors contributed equally to this work.

Simple Summary: This study introduces an advanced 3D assay using colorectal cancer spheroids and adult stem cell-derived, healthy human organoids to evaluate the efficacy and safety of *Cibisatamab*, a bispecific antibody targeting carcinoembryonic antigens (CEAs) on cancer cells and CD3 on T cells. Our approach integrates live imaging and cytotoxicity analysis with standardized workflows, providing precise and reproducible insights into therapeutic effects. The findings reveal that *Cibisatamab* effectively targets high-CEA-expressing cancer cells and is dose-dependent on target, off-tumor binding and killing on non-cancerous cells. By capturing complex tumor-immune interactions and variations in therapeutic responses, this model offers a valuable tool for advancing immunotherapy research and predicting potential safety liabilities in clinical trials.

Abstract: Objectives: Developing ex vivo models that replicate immune–tumor interactions with high fidelity is essential for advancing immunotherapy research, as traditional two-dimensional in vitro systems often lack the complexity required to fully represent these interactions. **Methods**: In this study, we establish a comprehensive 3D redirect lysis (3D-RDL) assay using colorectal cancer spheroids and adult stem cell-derived, healthy human organoids to evaluate the efficacy and safety profile of *Cibisatamab*, a bispecific antibody targeting carcinoembryonic antigens (CEAs) on cancer cells and CD3 on T cells. This model allows us to assess cytotoxic activity and immune responses, capturing variations in therapeutic response not observable in simpler systems. Our model integrates live imaging and cytotoxicity analyses to enable precise, real-time tracking of antibody effects on CEA-expressing tumor cells compared to healthy cells. Additionally, by standardizing effector-to-target cell ratios in each co-culture, we establish a reproducible workflow that enhances data accuracy and comparability across assays. Flow cytometry and Granzyme B release profiling further allow us to characterize immune cell activation, revealing distinct T cell activation markers and Granzyme B release patterns tied to *Cibisatamab* treatment. **Results**: Our results show that *Cibisatamab* effectively induces cell death in cancer spheroids with high CEA expression while being dose-dependent on target, off-tumor binding and killing on non-cancerous cells of healthy organoids with intermediate CEA levels. This highlights our model's potential to predict clinical immunotherapy outcomes, capturing complex responses like immune activation, therapeutic selectivity, and potential resistance mechanisms. **Conclusions**: These findings underscore the utility of our model as a reliable, physiologically relevant tool for screening new immunotherapies and advancing our understanding of tumor-immune dynamics.

 https://doi.org/10.3390/cancers17020291

Keywords: T cell bispecific (TCB) antibodies; adult stem cell-derived organoids; co-culture; preclinical safety and potency assessment

1. Introduction

Immunotherapies have emerged as powerful approaches in cancer treatment, aiming to harness and direct the body's own immune cells to identify and attack malignant cells. These therapies can be broadly categorized into two classes: active and passive.

Active immunotherapies stimulate the body's own immune system to enhance antitumor responses, either by counteracting immunosuppressive mechanisms, as seen with immune checkpoint inhibitors (ICIs), or by amplifying immune activity through immunostimulatory pathways, as seen with T cell bispecific (TCB) antibodies or bispecific T cell engagers (BiTEs) [1–3]. In contrast, passive immunotherapies directly deliver immune components, such as chimeric antigen receptor T cell (CAR-T) therapy and tumor-infiltrating lymphocyte (TIL) therapy. These approaches do not rely on the patient's immune system to initiate the response but instead deliver agents that target and attack cancer cells directly [4].

Among active immunotherapies, ICIs have gained importance by blocking key immunosuppressive molecules, such as programmed cell death 1 (PD-1) or its ligand PD-L1, enabling cytotoxic T cells to attach tumor cells more effectively [5]. The approval of *Ipilimumab* by the Food and Drug Administration (FDA) in 2011 marked a significant milestone in ICI-based therapies [6]. *Ipilimumab* targets cytotoxic T-lymphocyte-associated protein 4 (CTLA-4) to potentiate the T cell-mediated antitumor immune response. Despite its success in extending survival in metastatic melanoma patients, ICIs like *Ipilimumab* are associated with immune-related adverse events (irAEs) and limited efficacy in tumors with insufficient tumor-infiltrating lymphocytes (TILs) [7,8].

Passive immunotherapies, such as CAR-T cell therapy, have also demonstrated remarkable potential. CAR-T cell therapy involves genetically engineered patient T cells to express chimeric antigen receptors (CARs) that recognize tumor-associated antigens (TAAs). This modification enables T cells to selectively eradicate cancer cells [9]. While this approach is effective, the process of isolating, modifying, and expanding CAR-T cells is complex, costly, and time-consuming.

An emerging alternative within immunotherapies is the use of multi-specific immune engagers and TCB antibodies. Multi-specific immune engagers can target multiple TAAs simultaneously [10,11], while TCB antibodies typically bind a single TAA on the tumor cell while, in parallel, engaging the CD3 epsilon chain of the T cell receptor (TCR) complex, bridging T cells to cancer cells and triggering a potent immune response [3].

The first TCB, *Blinatumomab*, was approved by the FDA for the treatment of hematological cancers that originate from B cells [12]. Since then, nine TCBs have been approved, with seven for hematological cancers and two for solid tumors [13]. *Blinatumomab*, a CD19 BiTE, belongs to the class of Fc-free TCBs. Even though this class of TCBs shows a high killing potency, there is a disadvantage of its clinical application due to its short half-life and the need for continuous infusion. In contrast, newer IgG-based TCBs exhibit longer half-lives, improving clinical practicality [14]. The carcinoembryonic antigen (CEA) TCB is a member of the IgG-based bispecific antibody class and targets the CEA glycoprotein (also known as CEACAM5 or CD66e), which is anchored in the cell surface through glycosylphosphatidylinositol (GPI) [15]. CEA is overexpressed in various solid tumors, including colorectal, pancreatic, gastric, and non-small-cell lung cancers. In normal tissues, CEA expression is low and restricted to the apical surface of glandular epithelia in the gastrointestinal tract, limiting the accessibility to therapeutic antibodies [16,17]. However, even low expression levels of the target antigen in healthy tissues can lead to on-target, off-tumor toxicity [18].

Assessing the safety and efficacy of CEA TCBs in preclinical models poses unique challenges. Conventional animal models are unsuitable, as the anti-CEA arm of the antibody lacks cross-reactivity with species such as cynomolgus monkeys, and there are no CEA orthologs in rodents or other species for preclinical safety evaluations [19].

To address these limitations, advanced in vitro models, such as three-dimensional (3D) cell cultures and spheroids, offer a promising alternative. Spheroids, derived from cancer cell lines, replicate some aspects of tumor architecture and cell–cell interactions better than two-dimensional (2D) cultures, though they lack cellular diversity and complex tissue architecture [20]. In the last two decades, organoids have emerged as powerful in vitro models. Derived from patient samples, organoids can retain the genetic and phenotypic features of both healthy and tumor tissues, providing a more physiologically accurate model for drug testing [21,22]. Importantly, organoids enable the evaluation of on-target, off-tumor toxicity within the same donor-derived normal and cancer tissues. The use of 3D spheroids and organoids, thus, enhances the predictive power of preclinical studies, helping to bridge the gap between conventional 2D in vitro experiments and clinical applications, while reducing reliance on animal models that may lack relevant cross-species targets.

In this study, we present the 3D redirect lysis (3D-RDL) assay, a novel approach for the preclinical assessment of immune cell engagers, such as CEA TCBs. This assay evaluates the potency of CEA TCBs using 3D cancer spheroids and healthy organoids co-cultured with human peripheral blood mononuclear cells (hPBMCs) from healthy human blood samples. By capturing the complexity of tumor–immune interactions, the 3D-RDL assay overcomes the limitations of traditional 2D in vitro systems and provides critical insights into both efficacy and safety. Specifically, we assessed the CEA TCB *Cibisatamab*, a 2:1 bispecific antibody targeting CEA on tumor cells and CD3ε on T cells, using cancer spheroids and healthy organoids [23]. This approach enables a more precise characterization of the therapeutic window and dose determination for such antibodies, which is essential for minimizing on-target, off-tumor toxicity

2. Materials and Methods

2.1. Healthy Human Peripheral Blood Mononuclear Cells (hPBMCs) Isolation and Culture

hPBMCs were isolated from Buffy Coats (Don du sang—Epalinges, Switzerland) using Ficoll gradients and SepMate isolation tubes (85450, STEMCELL Technologies, Vancouver, BC, Canada). hPBMCs were cultured in complete RPMI (cRPMI) composed of RPMI + 10% Fetal Bovin Serum (FBS) + MEM Non-Essential Amino Acids (1X) + 1% Pen Strep + HEPES (1X) + Sodium pyruvate (1X).

2.2. Human Primary Organoids and Cell Lines

Tissue samples and annotated data were obtained and experimental procedures were performed within the framework of the non-profit foundation Human Tissue and Cell Research (HTCR) including the informed patient's consent [24]. The generation and storage of organoids from these biopsies were approved by the Cantonal Ethics Committee (CER-VD AO_2021-00076) and conducted by Doppl SA in Lausanne, Switzerland. Biopsies were collected from a total of 7 patients (age 61 ± 16, sex 71% females). Biopsies were collected from normal adjacent tissue (NAT) upon surgical tumor resection of patients suffering from malignancies of colorectal, ileal, or pancreatic origin. Human intestinal crypts were isolated from rectum, colon (transversal, ascending and appendix), and small intestine (jejunum and duodenum) biopsies following previously established protocols [25]. The stem cells present in the crypts allowed the establishment of rectal, colon, and small intestinal organoids. For organoid cultures, we used a revised cell culture growth medium inspired from the group of Sato, containing advanced DMEM/F12 supplemented with GlutaMAX™ (35050-038, Thermo Fisher Scientific, Waltham, MA, USA), HEPES (15630-056, Thermo Fisher Scientific), penicillin/streptomycin (15140-122, Thermo Fisher Scientific), and B-27 Supplement without Vitamin A (12587-001, Thermo Fisher Scientific) [26]. The organoids were maintained and expanded as follows: cultured in a humidified atmosphere of 5% CO_2 at 37 °C, the full medium was changed every 2 days, and the cells were passaged every 3–4 days.

MKN-45 (ACC 409, DSMZ) and DLD-1 (ACC 278, DSMZ) were cultured in the medium recommended by the supplier in a humidified atmosphere of 5% CO_2 at 37 °C.

2.3. Organoid Characterization

The rectum organoid line and small intestinal (Small Intestine_2) organoid line used in this study were closer characterized through Immunofluorescent staining (Supplementary Figure S1A,B) and bulk RNA sequencing (Supplementary Figure S1C–E).

Immunofluorescent staining: Prior to Immunofluorescent staining, organoids were collected, washed once with PBS, and fixed with 1 mL of 4% paraformaldehyde (PFA) for 30 min at RT. After three 10 min PBS washes, the fixed organoids were then stored at 4 °C for any immunofluorescence analysis. The respective organoids were blocked with blocking solution, containing 10% donkey serum and 0.2% Triton in PBS, for 3 h at RT and incubated with primary antibodies (MUC2: anti MUCIN2, 555926 BD Bioscience, San Jose, CA, USA; FABP1: anti-FABP1, AF1565 R&D Systems, Minneapolis, MN, USA; Ki67: ant-KI67, MA5-14520 Thermo Fisher Scientific; LYZ: anti-lysozyme, GTX72913 GeneTex, Irvine, CA, USA), diluted 1:200 in blocking solution, at 4 °C, overnight. The following day, the organoids were washed three times with PBS and subsequently incubated with secondary antibodies (donkey anti-mouse Alexafluor 568, A10037 Thermo Fisher Scientific; donkey anti-goat Alexafluor 488, A-11055 Thermo Fisher Scientific; donkey anti-rabbit Alexafluor 647, A31573 Thermo Fisher Scientific; and DAPI, 62248 Thermo Fisher Scientific) diluted 1:500 in blocking solution. Stained organoids were then washed three times with PBS and mounted on a glass slide with Fluoromount-G (Thermo Fisher Scientific). The fluorescence images of whole organoids were taken on an inverted confocal microscope (Olympus FV 4000) with a 20×/0.80 objective (2048 × 2048 pixels, 5.9535/µm). Image processing was performed using ImageJ (version 1.54K).

Bulk RNA sequencing: For bulk RNA sequencing, three biological replicates of the small intestinal organoid line and two biological replicates of the rectum organoid line were harvested, and the RNA was isolated. RNA was sent to a third-party contractor (Alithea Genomics, SA, Épalinges, Switzerland) for bulk RNA barcoding and sequencing (BRB-seq) and library preparation [27]. Raw reads were then demultiplexed and aligned to the human genome. The data were analyzed in R (4.4.1) with edgeR methods [28]. In brief, the lowly expressed genes were filtered out, and the counts were normalized and shown as log2 count per million (CPM). Data visualization was performed both in R and GraphPad Prism (version 10).

2.4. Allogeneic Co-Culture of hPBMCs

In the context of allogeneic co-culture experiments, target cells (MKN-45, DLD-1, small intestinal and rectal organoids) were incubated with varying amounts of hPBMCs at effector-to-target (E:T) ratios of 10:1, 5:1, or 2.5:1 for 72 h in the absence of treatment. During these preliminary experiments, one out of four healthy hPBMC donors exhibited mild spontaneous killing of target cells without treatment. To ensure the validity of the results, this donor was excluded from subsequent studies. These initial tests were conducted across all spheroid and organoid lines to establish baseline compatibility before assessing the killing potency mediated by treatment.

To validate the specificity of the readouts for evaluating target cell killing, hPBMCs (at the highest E:T ratio) and target cells were separately treated overnight with 0.8% Triton X-100. ATP content and lactate dehydrogenase (LDH) release were quantified to confirm the accuracy of the assay metrics.

2.5. Three-Dimensional Redirected Lysis (3D-RDL) Assay on Three-Dimensional Grown Target Cells

Organoids grown in solidified extracellular matrix were collected, enzymatically dissociated into single cell suspension, filtered, counted, and seeded in 96-well plates (Gri3D®96 Wellplate, 31 microwells per array, SunBioscience, Lausanne, Switzerland). Cells lines were collected from the flasks and processed the same way as organoids. Then, 2 days after target cell seeding, hPBMCs were seeded into cell culture flasks at 1×10^6 cells per mL in cRPMI medium for overnight resting in an incubator at 5% CO_2, 37 °C. At day 3, the hPBMCs were collected from the flasks, washed with FBS, counted, resuspended in target cell media, and added to the already-formed organoids/spheroids. *Cibisatamab* was added for 72 h to the wells at 6 different doses ranging from 0.001 to 100 μM (1:10 dilution factor). Each condition was tested in duplicates on 3 different hPBMC donors.

2.6. In Vitro Assessment of CEA-Mediated Killing Through Multiplexed Readouts

After 72 h, the supernatant of the 3D-RDL assay was homogenized and transferred to another 96-well flat bottom plate for LDH quantification, and the remaining volume was frozen at −20 °C for later cytokine analysis. LDH quantification was performed following the manufacturer guidelines (G1780, Promega, Madison, WI, USA). For live/dead quantification, cells were stained with 1 μM Calcein-AM and 2 μM Ethidium homodimer-1 (LIVE/DEAD Viability/Cytotoxicity Kit, L3224, Thermo Fisher, Waltham, MA, USA) for 1 h at 37 °C and imaged on an IN Cell Analyzer (GE Healthcare, Waukesha, WI, USA). The fluorescence signal was acquired over 20 Z-stacks (10 μm), deconvoluted, and analyzed in maximum projection. Through Doppl's fully automated image analysis pipeline, organoids were automatically detected and segmented and the median fluorescence intensity and organoid/spheroid size were measured. Post-imaging, the supernatant was removed for ATP quantification following the manufacturer protocol (G9681, Promega). Absorbance and luminescence were measured on the multimode plate reader Spark (Tecan, Männedorf, Switzerland). For all the read-outs, lysis buffer (Promega) was used a positive control for total cell lysis.

2.7. T Cell Activation Analysis and Relative Target Expression by Spectral Flow Cytometry

After 72 h, hPBMCs were retrieved from the 3D-RDL assay, filtered, and transferred to a V-bottom 96-well plate and resuspended in cold PBS. Cells were stained in PBS with live/dead staining dye Zombie-NIR (423106, Biolegend, San Diego, CA, USA) and FcR blocking solution (130-059-90, Miltenyi, Bergisch Gladbach, Germany) for 20 min at 4 °C. Cells were washed in FACS Buffer (PBS + 2 mM EDTA + 2% FBS) and resuspended in FACS Buffer and extra-cellular staining cocktail, CD69-APC (310910, Biolegend), CD4-BV421 (300532, Biolegend), CD8-SparkUV 387 (100798, Biolegend), and CD25-PE (385608, Biolegend), for 30 min at 4 °C. Cells were washed twice and resuspended in FACS Buffer for spectral flow cytometry analysis.

To evaluate the relative target expression on single cell dissociated organoids and cell lines (DLD-1 and MKN-45), 1×10^5 cells were seeded and stained as previously described with CD66d/e-AF647 (392806, Biolegend). All the samples were acquired using the EPFL

Flow Cytometry platform on a 5-Laser Cytek-Aurora instrument, and FACS data were analyzed using FlowJo Software (version 10).

2.8. Soluble Granzyme B, TNF-α and IFN-γ Quantification and Analysis

Granzyme B, TNF-α and IFN-γ released in the supernatant were quantified using the LEGENDplex Multi-Analyte Flow Assay kit (BioLegend) according to the manufacturer's instructions. Human CD8/NK Panel Detection Abs V02 was used to measure Granzyme B, TNF-α, and IFN-γ. Samples were acquired using a LSRFortessa (Becton, Dickinson, BD, Franklin Lakes, NJ, USA) flow-cytometer, and FCS files were analyzed with Biolegend's LEGENDplex data analysis software (https://legendplex.qognit.com, assessed on 16 October 2024). Standard curve was generated from 8 standard doses with duplicates of each standard dose (Supplementary Figure S4) When cytokine release was below the LLOD or above, the ULOD the value of the sample was set at LLOD or ULOD, respectively.

2.9. CEA Expression Levels

Data were extracted from the Protein Atlas website and displayed using the Prism 10 software. The collected data showed the normalized expression (nTPM) values from two different RNA-seq datasets: HPA and GTEx. They also showed an estimate of the highest protein expression score. Bulk RNA-seq data, generated at Doppl SA on different healthy organoids lines, were displayed as normalized log2 counts per minutes (Log2 CPM). Following enzymatic digestion described above, the single-cell suspensions of healthy rectal and intestinal organoids were FACS-stained and relative protein CEA expression levels were displayed by Prism 10. DLD-1 and MKN-45 cells were stained in parallel and acquired at the same time as the dissociated organoid lines.

2.10. Specific Lysis

The percentage of specific lysis was calculated by using the following equation:

$$\% \text{ of specific lysis (ATP)} = \frac{\text{X} - \text{maxkilling}}{\text{target cells only} - \text{maxkilling}} \times 100\%$$

(X = sample value; target cells only = target cells without treatment and effector cells; maximum killing = target cells treated overnight with 0.8% Triton-X).

2.11. Statistical Analysis

Statistical analyses and graph generation were performed using GraphPad Prism software (GraphPad Software, San Diego, CA, USA). Cytotoxicity upon *Cibisatamab* treatment was assessed by analyzing the effects of increasing doses on MKN-45, DLD-1, and small intestinal and rectal 3D cultures. An ordinary one-way ANOVA followed by Dunnett's multiple-comparisons test was used to compare the mean of each condition to that of the reference condition (Target only). For the analysis of Granzyme B levels released in the supernatant following *Cibisatamab* treatment, the same statistical approach was applied, comparing the mean of each group to that of the reference group (100 nM on MKN-45).

Additionally, a two-way ANOVA with multiple-comparisons test was performed to compare the treated MKN-45 cell line/rectum organoids (reference column: 10 μM) for each column to the mean of the DLD-1 cell line/small intestinal organoids (Supplementary Tables S1 and S2).

Only statistically significant differences are shown in the graphs, with significance levels denoted as follows: * $p < 0.05$, ** $p < 0.01$, *** $p < 0.001$, and **** $p < 0.0001$.

3. Results

3.1. Target Validation

First, we validated the expression of the target cell-surface glycoprotein CEA in healthy human tissue in two conventional adenocarcinoma cell lines (MKN-45 and DLD-1), as well as in patient-derived organoids from different organs (colon, rectum and small intestine) of the gastrointestinal (GI) tract. Protein expression data from the Human Protein Atlas indicated that CEA expression was highest in the rectum and colon, with lower levels observed in the small intestine (Figure 1B–D) [29]. Similarly, bulk RNA sequencing analysis of seven healthy GI organoid lines from Doppl SA's biobank showed a comparable pattern, with the highest CEA expression detected in rectal and colon organoids compared to small intestinal organoids (Figure 1C).

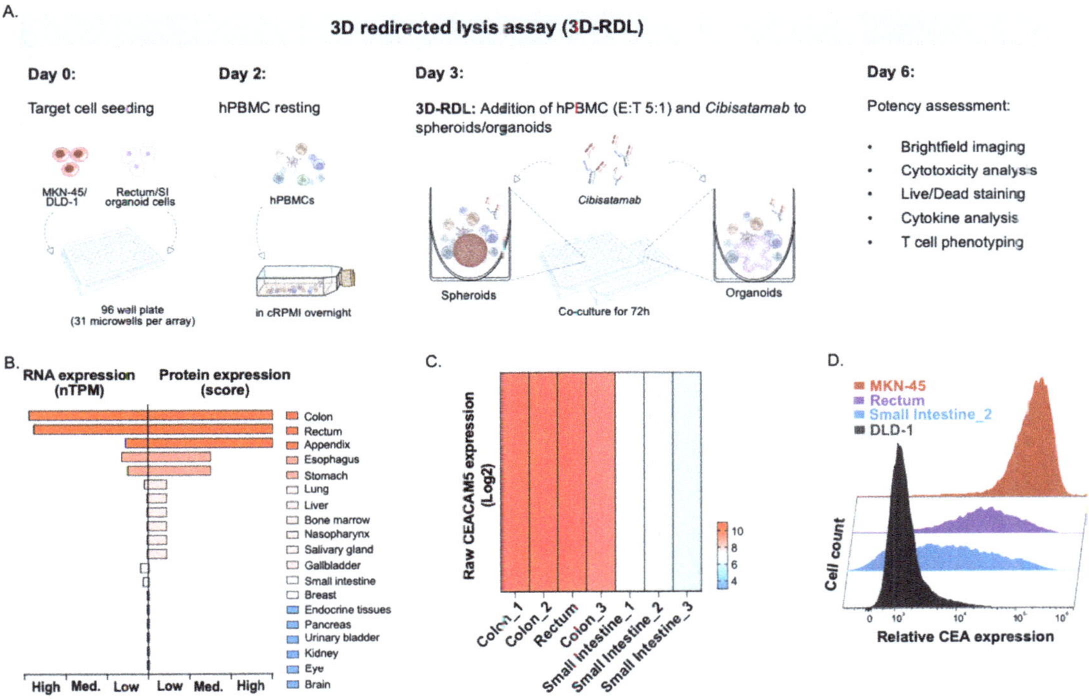

Figure 1. The experimental design of the 3D-RDL assay and CEA target expression validation: (**A**) In the preparation of the 3D-RDL assay, single cells from cell lines (MKN-45 and DLD-1) and organoids (Rectum and Small intestine_2) are seeded in 96-well microwell plates. At day 2, hPBMCs are rested overnight in cRPMI medium. At day 3, hPBMCs (E:T, 5:1) and *Cibisatamab* (six doses; 0.001–100 μM) are added to the aggregated spheroids and organoids and incubated for 72 h. (**B**) CEA protein and RNA expression levels in healthy human organs. CEA is expressed in a variety of epithelial tissues such as the urogenital, respiratory, and gastrointestinal tracts. In the healthy human colon, the CEA protein is restricted to the apical side of the differentiated epithelial cells forming the luminal surface. Source: Human Protein Atlas. (**C**) CEACAM5 bulk RNA expression levels in healthy colon, rectal, and intestinal organoids (Doppl SA Biobank). (**D**) Relative CEA protein expression levels of control cell lines MKN-45 (CEA_{high}) and DLD-1 (CEA_{low}) and 2 different healthy organoid lines (Rectum and Small intestine_2) analyzed by flow cytometry.

We selected one rectal organoid line (Rectum) and one small intestinal organoid line (Small Intestine_2) for detailed characterization within this study (Supplementary Figure S1A–E). The rectal and small intestinal organoid lines expressed markers for key intestinal cell types, including Goblet cells (MUC2+), enterocytes (FABP1+), and Paneth cells (LYZ+), confirming their differentiation status and suitability as models for healthy rectal and intestinal tissue (Supplementary Figure S1A,B). Additionally, bulk RNA sequencing analysis revealed the activation of distinct canonical pathways and hallmarks characteristic of small intestine and rectum tissues, further supporting the tissue-specific identity of the organoid lines used in this study (Supplementary Figure S1C–E).

Next, we directly compared and analyzed relative CEA protein expression levels between the organoid lines and the adenocarcinoma cell lines MKN-45 and DLD-1 using flow cytometry (Figure 1D). The adenocarcinoma MKN-45 cell line exhibited significant overexpression of CEA, consistent with published results characterizing MKN-45 as CEA-high (CEA_{high}) [30], making it a suitable positive control for this proof-of-concept study. It must be noted that alternative cell lines, including colorectal cancer (CRC)-specific cell lines, could also be tested in the proposed assay to evaluate the targeting efficacy of CEA TCBs in various tumor types.

The rectal organoid line showed the second-highest CEA expression, while the small intestinal organoid line demonstrated CEA levels lower than the rectal organoid but still higher than DLD-1, which had the lowest CEA expression overall (CEA_{low}) (Figure 1D). The two selected organoid lines were subsequently used for the co-culture and 3D-RDL assays to further evaluate the efficacy and safety of the CEA-targeting bispecific antibody *Cibisatamab*.

3.2. Probing Assay Specificity and Spontaneous Cytotoxicity of Allogeneic hPBMCs

To replicate the tumor-surrounding immune cell compartment, we utilized allogeneic hPBMCs obtained from four different healthy donors. For this study, the whole PBMCs were selected to replicate the intricate cellular environment of the immune system, enabling natural interactions between T cells and antigen-presenting cells (APCs), such as monocytes and dendritic cells. This approach offered a more physiologically relevant model for studying immune responses, as it closely mimicked the in vivo dynamics of immune cell interactions. It is important to note that isolated T cells or specific CD4/CD8 subpopulations could be employed to achieve more precise effector-to-target ratios and minimize potential interference from other cell types. Furthermore, autologous PBMCs matched to the organoid tissue could enhance the relevance of the experimental system; however, these approaches were beyond the scope of this proof-of-concept study.

Prior to establishing the 3D-RDL assay, we cultured hPBMCs alone to measure their intrinsic ATP production and lactate dehydrogenase (LDH) release, validating the specificity of our readout. All tested hPBMCs exhibited low levels of ATP and LDH compared to the target cells, confirming that this study primarily detected target cell death during the 3D-RDL assay (Supplementary Figure S2A,B).

To evaluate any potential spontaneous killing resulting from an allogeneic reaction, we co-cultured healthy human rectal or small intestinal organoids, as well as DLD-1 and MKN-45 spheroids, with the allogeneic hPBMCs. After 72 h, we found that the hPBMCs did not induce significant cell death in the healthy human rectal and small intestinal organoids, DLD-1, or MKN-45 spheroids. None of the hPBMCs caused significant depletion of ATP, LDH release, or cell death (Supplementary Figure S3A–C). Only hPBMCs derived from one of the healthy donors (Donor_1) showed a slight increase in LDH release, leading to the person's exclusion from further screenings in the presence of the compound.

3.3. Cibisatamab-Induced Selective Killing of CEA_{high} Expressing Cancer Spheroids

To evaluate the efficacy and safety of the CEA TCB *Cibisatamab*, we established the 3D-RDL assay (Figure 1A). In this assay, we first co-incubated *Cibisatamab* with 3D cancer spheroids derived from the adenocarcinoma cell lines MKN-45 and DLD-1. Our findings revealed that *Cibisatamab* showed high and selective potency against CEA-high-expressing MKN-45 (Figure 2A–D).

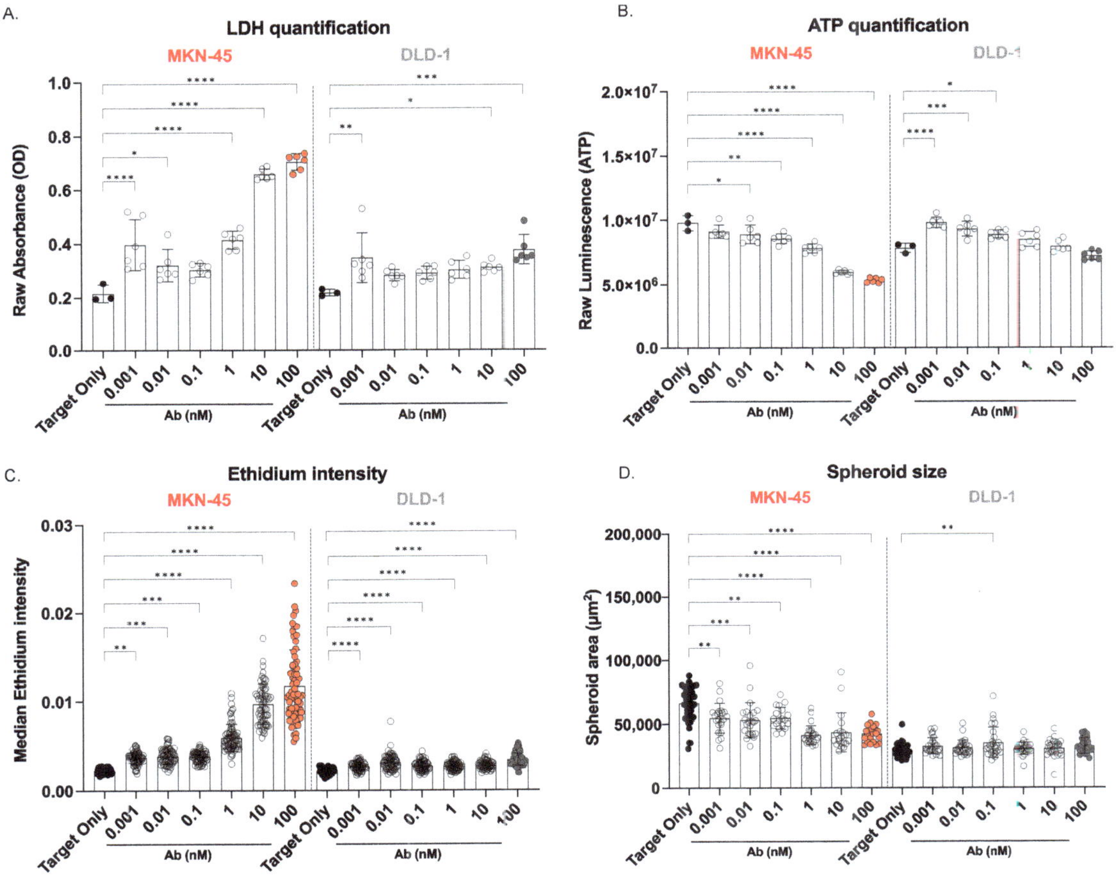

Figure 2. The 3D-RDL results of co-cultured 3D grown human adenocarcinoma cell lines MKN-45 and DLD-1 in the presence of hPBMCs treated at different doses of *Cibisatamab* for 72 h. (**A–D**) Cytotoxicity of MKN-45 and DLD-1 cells in the presence of *Cibisatamab* and healthy hPBMCs (E:T,5:1) for 72 h in a 3D-RDL assay. Each well was composed of 31 microwells containing one organoid/spheroid each. (**A**) Lactate dehydrogenase (LDH) released in the supernatant was quantified, and raw absorbance values was displayed. Increasing concentrations of LDH reflect an increase in target cell death. (**B**) ATP content was measured through luminescence release upon complete cell lysis. Decreasing ATP concentrations show increase in target cell death. (**C**) Ethidium median intensity was evaluated with our high-throughput automated analysis pipeline (imaged based). Increased ethidium signals correlated with increased cell death. (**D**) Our image-based analysis pipeline enabled the measurement of the size of each organoid. A decrease in spheroid size generally correlated with a decrease in cell viability. Data represent mean ± SD from 3 different hPBMCs donors tested in duplicate. Ordinary 1-way ANOVA with Dunnett's multiple-comparisons test was performed, and the mean of each column was compared to the mean of the reference column (Target only). * $p < 0.05$, ** $p < 0.01$, *** $p < 0.001$, and **** $p < 0.0001$. Only significant values are displayed ($p < 0.05$).

As a first indication for a selective cytotoxic effect of *Cibisatamab* on CEA_{high} cells, we observed a dose-dependent increase in LDH release from MKN-45 spheroids after 72 h treatment with escalating concentrations of *Cibisatamab*. LDH is a stable enzyme that is present in all cell types. It is rapidly released into the cell culture medium when the plasma membrane is damaged [31]. The rise in LDH levels in the supernatant is a direct indicator of increased target cell death, reflecting *Cibisatamab's* effectiveness in inducing cell lysis in CEA-high-expressing cells. In contrast, LDH release from DLD-1 spheroids did not show a significant increase, except at the highest concentration of CEA TCB (Figure 2A).

Additionally, ATP content was measured after 72 h of the 3D-RDL assay. Apoptotic cells stopped synthesizing ATP and ATP levels quickly degraded. Therefore, ATP was a reliable marker of cell viability [32]. We found a decrease in ATP levels in MKN-45 spheroids with higher doses of *Cibisatamab*, indicating reduced cellular viability and increased cell death. In comparison, there was no significant reduction in ATP content for DLD-1 spheroids, underscoring the selective impact of *Cibisatamab* on CEA_{high} cells (Figure 2B).

Cell death was further corroborated by live/dead staining analysis, where the ethidium intensity markedly increased in MKN-45 spheroids even at the lowest concentration of *Cibisatamab*. This increase in ethidium staining indicated a significant increase in cell death, which continued to escalate with higher antibody concentrations, reinforcing the potent killing effect of *Cibisatamab* on CEA-expressing tumor cells (Figure 2C).

The direct comparison between MKN-45 and DLD-1 revealed that the cytotoxic effect of *Cibisatamab* on MKN-45 was significantly higher than on DLD-1, as demonstrated by ATP production, LDH release, and live/dead staining. Specifically, MKN-45 cells treated with 10 µM of *Cibisatamab* showed significantly greater cytotoxicity compared to all treatment doses applied to the DLD-1 cell line (Supplementary Table S1).

To complement these findings, we analyzed the size of each spheroid using our image-based analysis pipeline. MKN-45 spheroids showed a noticeable decrease in size in response to increasing concentrations of *Cibisatamab*, which correlated with the observed reduction in viability. Conversely, the size of DLD-1 spheroids remained largely unaffected by *Cibisatamab* treatment, highlighting the minimal impact on cells with low CEA expression (Figure 2D).

Taken together, these results from the 3D-RDL assay provide a comprehensive demonstration of *Cibisatamab's* selective targeting and killing of CEA_{high} expressing cells, such as MKN-45, while sparing the CEA_{high} DLD-1 spheroids.

Next, we showed that the selective killing of CEA_{high}-expressing MKN-45 spheroids correlated with increased T cell activation upon *Cibisatamab* treatment. We investigated the T cell activation markers CD69, for early activation, and CD25, for late activation, by flow cytometry after 72 h co-culture of CEA_{high} MKN-45 spheroids with *Cibisatamab* and hPBMCs [32–34].

Due to the different kinetics of CD69 and CD25, we first analyzed their expression levels separately (Figure 3A,B) before evaluating the overall activation status of T cells by examining their co-expression (Figure 3C). To address potential overlap, we quantified T cells co-expressing CD69 and CD25, confirming that their co-expression increased with higher doses of *Cibisatamab* on both CD4+ and CD8+ T cell subpopulations (Figure 3C). This co-expression highlighted that T cells could simultaneously exhibit early and late activation states under appropriate stimulation conditions.

Overall, *Cibisatamab* induced a dose-dependent activation of both CD4+ and CD8+ T cell subpopulations, as evidenced by increased expression of CD25 and CD69 at the T cell surface (Figure 3A–C). Notably, CD8+ T cells showed higher upregulation of activation markers compared to CD4+ T cells.

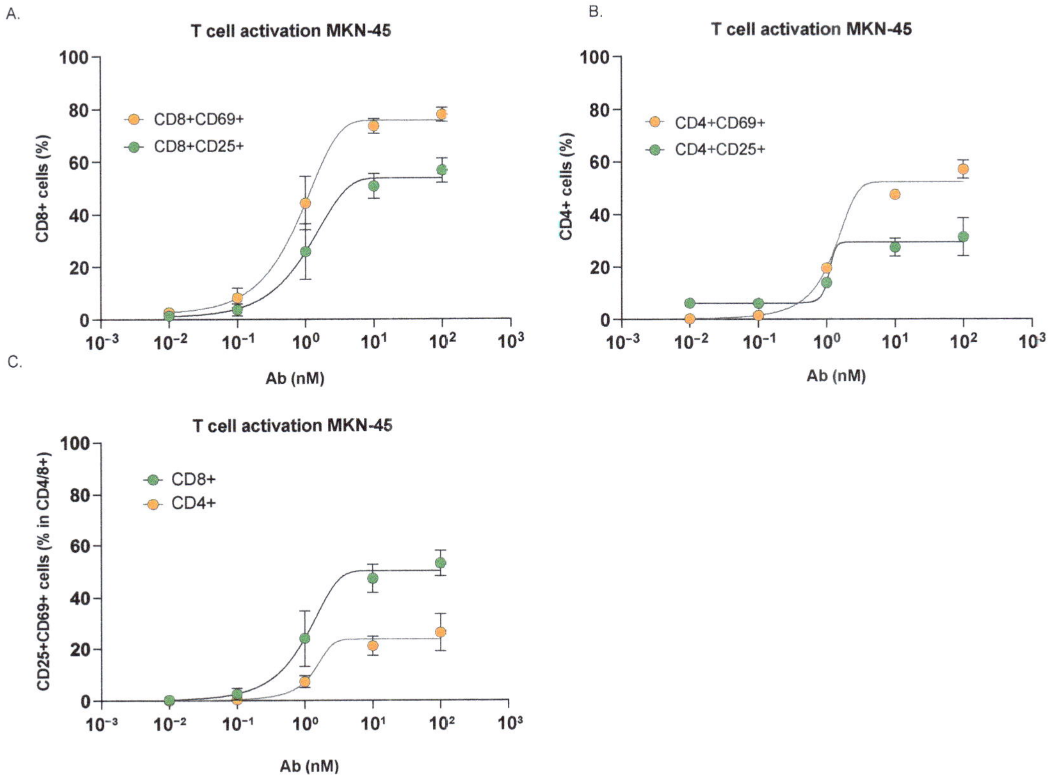

Figure 3. T cell activation after 72 h of co-culture of hPBMCs and MKN-45 cells in 3D-RDL assay format. (**A**) CD8+ and (**B**) CD4+ T cell response in the presence of healthy hPBMCs (E:T, 5:1), MKN-45 cells and increasing doses of CEA TCB antibodies. (**C**) CD8+ and CD4+ co-expression in the presence of healthy hPBMCs (E:T, 5:1), MKN-45 cells, and increasing doses of *Cibisatamab*. T cell activation upon *Cibisatamab* treatment was measured by flow cytometry while measuring the expression of CD25 (green) and CD69 (orange). Data represent non-linear regression analysis of the mean ± SD from 3 different hPBMCs donors tested in duplicate.

3.4. Cibisatamab-Induced Killing of Healthy Human Rectal Organoids

To evaluate the safety profile and potential on-target, off-tumor effects of the CEA TCB *Cibisatamab*, we expanded our 3D-RDL assay to include healthy organoid models, specifically rectal and small intestinal organoids. These models allowed us to assess the impact of *Cibisatamab* on non-cancerous tissues with varying levels of CEA expression, providing insight into the selectivity of the antibody.

Our results showed that high doses of *Cibisatamab* induced cell death in rectal organoids, which exhibited intermediate CEA expression levels (Figure 1C,D). The decrease in organoid viability was evident from the decrease in ATP levels of the organoids after 72 h of treatment in the 3D-RDL assay. ATP levels in the rectal organoids with intermediate CEA expression significantly decreased with the three highest doses of *Cibisatamab*, correlating with increased cell death. In contrast, no significant reduction in ATP content was observed in the small intestinal organoids, reinforcing the selectivity of *Cibisatamab* against cells with higher CEA expression (Figure 4A). Additionally, we observed a dose-dependent increase in ethidium intensity following live/dead staining in rectal organoids with increasing concentrations of *Cibisatamab* (Figure 4B). In comparison, the small intestinal organoids, which had lower CEA expression levels, did not show significant cell death even at the highest CEA TCB concentrations, indicating a higher resistance to off-tumor effects (Figure 4B). The direct comparison between rectal organoids and small intestinal organoids showed that the cytotoxic effect of *Cibisatamab* on rectal organoids was significantly higher than its effect on small intestinal organoids (Supplementary Table S2). In the ATP assay, the ATP production in rectal organoids treated with 10 μM of *Cibisatamab* was significantly lower compared to the three highest treatment doses in small intestinal organoids. Additionally, for the live/dead staining, the cytotoxic effect on rectal organoids treated with 10 μM of *Cibisatamab* was significantly higher compared to all doses tested for the small intestinal organoids.

The decrease in rectal organoid viability could also be detected in brightfield images of rectal organoids co-incubated with hPBMCs and *Cibisatamab* (Figure 4D); however, the reduced viability was not captured in the organoid size analysis using our image-based analysis pipeline. Neither the rectal nor the small intestinal organoids exhibited significant size changes upon *Cibisatamab* treatment, suggesting that the antibody's effect on organoid size was less pronounced in non-cancerous tissues (Figure 4C).

To directly compare the potency of the CEA TCB across the different cancer spheroids and healthy organoid lines, we analyzed the specific lysis after 72 h in the 3D-RDL assay, quantified by the ATP content, between the rectal and intestinal healthy organoids and the 3D cancer spheroids derived from MKN-45 and DLD-1. *Cibisatamab* selectively induced specific lysis in the CEA_{high}-expressing MKN-45 cancer spheroids, as well as milder specific lysis in rectal organoids with intermediate CEA expression, while sparing the small intestinal organoids and the DLD-1 spheroids, both of which had low CEA levels (Figure 4E).

These findings strongly support the specificity of CEA TCB *Cibisatamab* for targeting CEA-expressing cells, demonstrating its capacity to induce cell death in CEA_{high}- and intermediate CEA-expressing tissues, with limited off-tumor cytotoxicity in healthy tissues that express low levels of CEA.

3.5. Granzyme B Induction and Cytokine Release upon Cibisatamab Treatment

Next, we analyzed the release of Granzyme B in the supernatant collected from the 3D-RDL assay, which included the co-cultured hPBMCs together with MKN-45 and DLD-1 cancer spheroids, as well as healthy rectal and intestinal organoid lines, treated at two different doses (10 and 100 nM) of *Cibisatamab* for 72 h. Our results demonstrated a dose-dependent increase in Granzyme B levels across the tested models (Figure 5). At the higher dose (100 nM), hPBMCs co-cultured with MKN-45 spheroids and rectal organoids showed a pronounced increase in Granzyme B release compared to co-cultures with CEA_{low} DLD-1 spheroids and small intestinal organoids (Figure 5A). Granzyme B, a serine protease, played a pivotal role in inducing target cell apoptosis by facilitating cytotoxic T cell- and natural killer (NK) cell-mediated cytolytic activity [35]. The elevated Granzyme B release observed in the MKN-45 spheroids and rectal organoids aligned with the increased cell death induced by *Cibisatamab* in these models, highlighting a robust CEA TCB-mediated immune response. To further investigate this response, we examined Granzyme B release in the CEA_{high} MKN-45 spheroid co-culture across a range of *Cibisatamab* doses (0.001–100 nM) (Figure 5B). The dose-dependent increase in Granzyme B secretion corresponded with the cytotoxic effects of *Cibisatamab* observed at the three highest concentrations (compare Figure 2A–D).

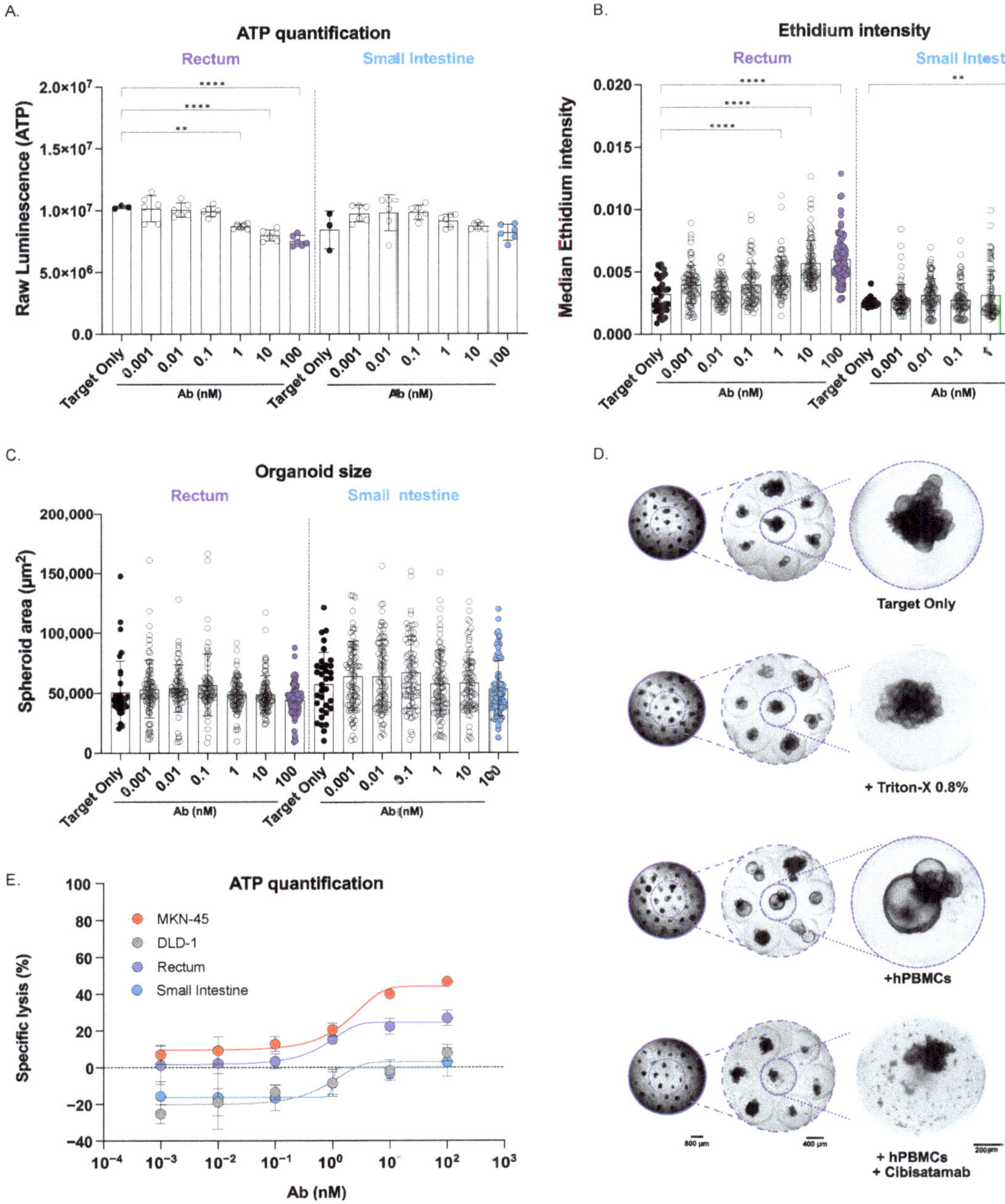

Figure 4. The 3D-RDL results of co-cultured rectal and intestinal healthy organoids in the presence of hPBMCs treated at different doses of *Cibisatamab* for 72 h. (**A–C**) Cytotoxicity of healthy rectal and intestinal organoids in the presence of *Cibisatamab* and healthy hPBMCs (E:T, 5:1) for 72 h in a RDL assay. Each well was composed of 31 microwells containing one organoid/spheroid each. (**A**) Ethidium median intensity was evaluated with our high-throughput automated analysis pipeline (imaged based). Increased ethidium signals correlated with increased cell death. (**B**) Our image-based analysis pipeline enabled the measurement of the size of each organoid. Decreases in spheroid size generally correlated with decreases in cell viability. (**C**) ATP content was measured through luminescence release upon complete cell lysis. Decreasing ATP concentrations showed increases in

target cell death. (**D**) Representative brightfield images of rectal organoids cultured in microwells in different conditions after 48 h: untreated rectal organoids (target cells alone); rectal organoids treated with Triton-X 0.8% (+Triton-X 0.8%); rectal organoids co-incubated with hPBMCs; rectal organoids co-incubated with hPBMCs and 100 nM of *Cibisatamab*. Scale bars: 800 µm, 400 µm, and 200 µm. (**E**) The cytotoxicity of cancer spheroids and healthy organoids in the presence of *Cibisatamab* and healthy hPBMCs (E:T, 5:1) for 72 h in a 3D-RDL assay. Data represent non-linear regression analysis of the mean ± SD from 3 different hPBMCs donors tested in duplicates. Specific lysis was calculated as follows: ((Sample—Target Only)/(Triton-X 0.8%—Target Only)). Data represent mean ± SD from 3 different hPBMCs donors tested in duplicate. Ordinary 1-way ANOVA with Dunnett's multiple-comparisons test was performed, and the mean of each column was compared to the mean of the reference column (Target only). ** $p < 0.01$, and **** $p < 0.0001$. Only significant values are displayed ($p < 0.05$).

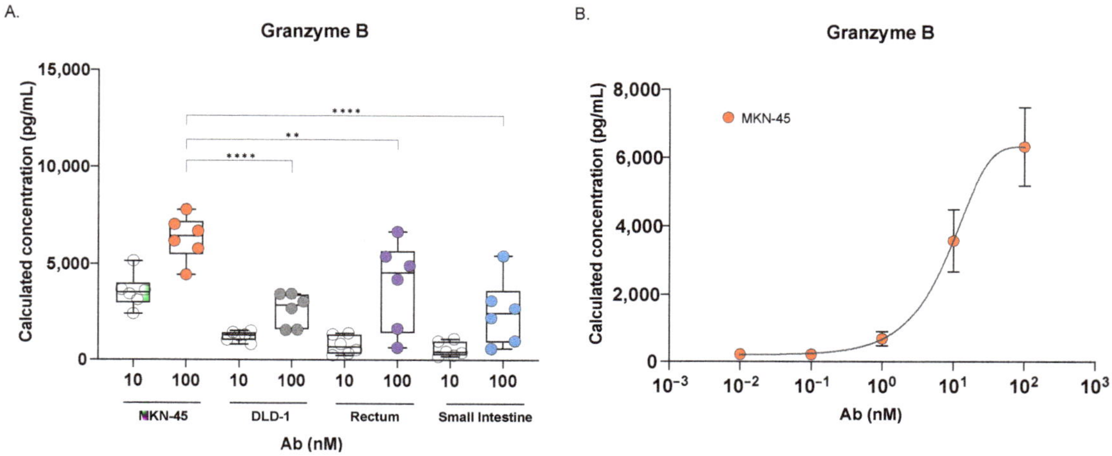

Figure 5. Granzyme B quantification after 72 h of co-culture of rectal/intestinal healthy organoids and 3D grown MKN-45/DLD-1 spheroids in the presence of hPBMCs treated at different doses of CEA TCB. (**A**,**B**) The quantification of Granzyme B released in the supernatant after 72 h RDL assay by flow cytometry. Data represent non-linear regression analysis of the mean ± SD from 3 different hPBMCs donors tested in duplicate. Ordinary 1-way ANOVA with Dunnett's multiple-comparisons test was performed, and the mean of each column compared to the mean of MKN-45 treated at 100 nM. ** $p < 0.01$ and **** $p < 0.0001$. Only significant values are displayed ($p < 0.05$).

Additionally, we quantified the soluble cytokines TNF-α and IFN-γ in the supernatants of MKN-45 co-cultures (Supplementary Figure S5). A dose-dependent increase was observed for both cytokines, with *Cibisatamab* inducing high levels at the two highest concentrations. This increase corresponded to the granzyme B secretion and apoptosis observed in MKN-45 spheroids. TNF-α and IFN-γ were key mediators of inflammatory responses, playing a central role in driving intestinal inflammation and epithelial damage, underscoring the immune activation induced by *Cibisatamab* [36].

Taken together, these results highlight the importance of Granzyme B, IFN-γ, and TNF-α as key players in the immune response triggered by *Cibisatamab*. Together, they provide insights into *Cibisatamab's* ability to activate the immune system and kill CEA_{high}-expressing tumor cells effectively.

4. Discussion

In this study, we developed and validated a 3D redirect lysis (3D-RDL) assay to comprehensively evaluate the potency and safety of CEA TCB antibodies. We used this assay to evaluate *Cibisatamab*, demonstrating its selective cytotoxic effects on CEA_{high}

expressing cancer cells while minimizing on-target, off-tumor activity on healthy tissues. The integration of this assay with various complementary analytical techniques, including bulk RNA sequencing, spectral flow cytometry, supernatant analysis, and high-throughput imaging, highlighted the versatility of our preclinical pipeline in dissecting the complex interactions between immune cells and tumor cells.

Our results revealed that *Cibisatamab* exhibited significantly higher potency against MKN-45 cancer spheroids, characterized by high CEA expression, compared to DLD-1 spheroids with lower CEA levels. This selective killing was confirmed through multiple readouts, including LDH release, ATP depletion, and image-based viability analysis, all of which consistently showed increased cell death in MKN-45 spheroids with escalating concentrations of *Cibisatamab*. Notably, MKN-45 spheroids also displayed a decrease in size, further supporting the correlation between antibody treatment and tumor cell viability reduction. In contrast, DLD-1 spheroids exhibited minimal changes across these metrics, underscoring the specificity of *Cibisatamab* for cells with CEA_{high} expression.

CEA TCB treatment led to the dose-dependent activation of both CD4+ and CD8+ T cells, measured through the upregulation of CD69 and CD25, with CD8+ T cells showing a more pronounced activation response. This finding aligns with the expected mechanism of action for TCBs, where T cell engagement is essential for the induction of targeted cytotoxicity.

The extension of our 3D-RDL assay to healthy organoid models allowed us to evaluate the on-target, off-tumor effects of *Cibisatamab*. We found that rectal organoids, which have intermediate CEA expression, showed a dose-dependent increase in cell death upon treatment, while small intestinal organoids, with low CEA levels, were not sensitive to the effects of CEA TCB. These observations were substantiated by ATP depletion and increased ethidium intensity in the rectal organoids but not in the small intestinal organoids. Importantly, the absence of significant size reduction in either organoid type suggests that *Cibisatamab's* impact on non-cancerous tissues might be limited to cell death without a higher impact on tissue viability.

When directly comparing the cytotoxic effects of *Cibisatamab* across cancer spheroids and healthy organoids, we observed a clear trend of specific lysis in CEA_{high} MKN-45 cancer spheroids and rectal organoids with intermediate CEA expression, while sparing DLD-1 spheroids and small intestinal organoids with lower CEA levels. This selective action supports the hypothesis that *Cibisatamab* predominantly targets cells with high CEA expression, thus reducing the risk of off-tumor toxicity in healthy tissues. Such specificity is crucial for the clinical application of TCBs, as it underlines the importance of targeting markers that are highly expressed on tumors but minimally present on normal tissues to minimize adverse effects.

Our analysis further revealed a dose-dependent increase in Granzyme B and cytokine release, particularly in CEA_{high} MKN-45 spheroids and rectal organoids. The elevated Granzyme B secretion strongly correlated with enhanced T cell-mediated cytotoxicity observed in these models, highlighting the robust immune activation induced by *Cibisatamab*. The pronounced Granzyme B induction in CEA_{high} models underscored its critical role as a key mechanism driving the therapeutic efficacy of *Cibisatamab*, distinguishing these models from CEA_{low} spheroids and organoids where the response was markedly less potent. Additionally, the dose-dependent increase in TNF-α and IFN-γ levels correlated with Granzyme B release and the observed apoptotic effects in MKN-45 spheroids, underscoring the strong immune activation induced by *Cibisatamab*.

These results validate our 3D-RDL assay as a tool for preclinical testing, enabling the evaluation of therapeutic candidates. This model offers advantages over traditional 2D co-culture systems, which often fail to capture the complex dynamics of immune–tumor interactions in a physiologically relevant manner. By replicating the 3D architecture of tissues and allowing for analysis of cellular responses, our platform provides a prediction of the efficacy and safety profiles of immunotherapies, thus aiding in the identification and optimization of therapeutic leads.

5. Conclusions

In conclusion, our study establishes the 3D-RDL assay as a preclinical tool for assessing the safety and potency of immune-modulatory agents, with a focus on CEA-targeting therapies like *Cibisatamab*. The integration of diverse analytical techniques within our pipeline—encompassing RNA sequencing, flow cytometry, viability assays, and high-throughput imaging—enables a multi-dimensional evaluation of both tumor and healthy tissue responses. This approach not only allows for precise characterization of on-target cytotoxic effects but also provides critical insights into potential off-tumor liabilities, enhancing the predictive accuracy of preclinical safety profiles.

Our findings with *Cibisatamab* illustrate its selective targeting capability, with potent killing of CEA-high-expressing cancer cells and minimal effects on healthy tissues with low CEA expression. This specificity is crucial for reducing on-target, off-tumor toxicity and improving the therapeutic window of CEA TCBs in clinical settings.

Beyond CEA TCBs, our adaptable platform is well suited for evaluating a wide array of immunotherapies, including immune checkpoint inhibitors, NK cell engagers, innate immune activators, and combination therapies, as well as cellular immunotherapies such as CAR-Ts and tumor-infiltrating lymphocytes (TILs).

This proof-of-concept study offers several avenues for extending the 3D-RDL assay to further enhance its translational relevance. For example, the use of tumoroids derived from colorectal cancer (CRC) patients together with autologous PBMCs or T cells will provide a more patient-specific model for studying immune–tumor interactions and treatment efficacy. Additional assay readouts can further enhance the study relevance. Live cell imaging and TIL mobility assays can be applied to directly analyze interactions between spheroids/organoids and T cells or TILs upon treatment. Moreover, readouts, such as Caspase-3 staining, could be incorporated to specifically monitor epithelial-cell or cancer-cell apoptosis.

The versatility of our assay provides an asset for preclinical research, offering robust datasets to support the development and optimization of next-generation cancer immunotherapies. Our approach not only facilitates the identification of promising therapeutic candidates but also aids in refining dosing strategies, ultimately contributing to the design of safer and more effective treatment options for patients with solid tumors.

6. Patents

The Ecole Polytechnique Fédérale de Lausanne has filed for patent protection on the technology described herein (WO2018/050862), and S.H. and N.B. are named as inventors on this patent; S.H. and N.B. are shareholders in Doppl SA, who is commercializing this patent.

Supplementary Materials: The following supporting information can be downloaded at: https://www.mdpi.com/article/10.3390/cancers17020291/s1, Figure S1: Characterization of rectal and small intestinal (Small intestine_2) organoid lines used in the study; Figure S2: Validation of read-out-specificity by single culture of human hPBMCs of 4 different donors; Figure S3: Validation of read-out-specificity by single culture of human hPBMCs of 4 different donors; Figure S4: Standard curves and 5PL fitting parameters of Granzyme B, TNF-α and INF-γ; Figure S5: TNF-α and INF-γ quantification after 72 h co-culture of 3D grown MKN-45 spheroids in the presence of hPBMCs treated at different doses of CEA TCB; Table S1: Two-way ANOVA with multiple comparisons test was performed and the mean of each column compared to the mean of the reference column (10 μM MKN-45); Table S2: Two-way ANOVA with multiple comparisons test was performed and the mean of each column compared to the mean of the reference column (10 μM Rectum).

Author Contributions: V.A., E.H., F.P. and N.B. conceived this study. V.A. conducted the experiments. All authors contributed to the writing of this manuscript. All authors have read and agreed to the published version of the manuscript.

Funding: This work was partially funded by Doppl SA.

Institutional Review Board Statement: Not applicable.

Informed Consent Statement: Tissue samples and annotated data were obtained and experimental procedures were performed within the framework of the non-profit foundation Human Tissue and Cell Research (HTCR) including the informed patient's consent [24]. The generation and storage of organoids from these biopsies have been approved by the cantonal ethics committee (CER-VD AO_2021-00076) and have been conducted by DoppL SA, in Lausanne, Switzerland.

Data Availability Statement: The data that support the findings of this study are available on reasonable request from the authors.

Acknowledgments: We acknowledge the support of the non-profit foundation HTCR, which holds human tissue collections on trust, making it broadly available for research on an ethical and legal basis.

Conflicts of Interest: V.A., E.H., F.P, A.R , A.C., V.G., N.B. and S.H.-K. report an employment relationship with Doppl SA. The Ecole Polytechnique Fédérale de Lausanne has filed for patent protection on the technology described herein (WO2018/050862), and S.H. and N.B. are named as inventors on this patent; S.H. and N.B. are shareholders in Doppl SA, who is commercializing this patent. If there are other authors, they declare that they have no known competing financial interests or personal relationships that could have appeared to influence the work reported in this paper.

References

1. Baumeister, S.H.; Freeman, G.J.; Dranoff, G.; Sharpe, A.H. Coinhibitory Pathways in Immunotherapy for Cancer. *Annu. Rev. Immunol.* **2016**, *34*, 539–573. [CrossRef] [PubMed]
2. June, C.H.; O'Connor, R.S.; Kawalekar, O.U.; Ghassemi, S.; Milone, M.C. CAR T cell immunotherapy for human cancer. *Science* **2018**, *359*, 1361–1365. [CrossRef] [PubMed]
3. Zhou, S.; Liu, M.; Ren, F.; Meng, X.; Yu, J. The landscape of bispecific T cell engager in cancer treatment. *Biomark. Res.* **2021**, *9*, 38. [CrossRef] [PubMed]
4. Sahu, M.; Suryawanshi, H. Immunotherapy: The future of cancer treatment. *J. Oral. Maxillofac. Pathol. JOMFP* **2021**, *25*, 371. [CrossRef]
5. Lao, Y.; Shen, D.; Zhang, W.; He, R.; Jiang, M. Immune Checkpoint Inhibitors in Cancer Therapy—How to Overcome Drug Resistance? *Cancers* **2022**, *14*, 3575. [CrossRef]
6. Ledford, H. Melanoma drug wins US approval. *Nature* **2011**, *471*, 561. [CrossRef]
7. Martins, F.; Sofiya, L.; Sykiotis, G.P.; Lamine, F.; Maillard, M.; Fraga, M.; Shabafrouz, K.; Ribi, C.; Cairoli, A.; Guex-Crosier, Y.; et al. Adverse effects of immune-checkpoint inhibitors: Epidemiology, management and surveillance. *Nat. Rev. Clin. Oncol.* **2019**, *16*, 563–580. [CrossRef]
8. Friedrich, M.; Jasinski-Bergner, S.; Lazaridou, M.-F.; Subbarayan, K.; Massa, C.; Tretbar, S.; Mueller, A.; Handke, D.; Biehl, K.; Bukur, J.; et al. Tumor-induced escape mechanisms and their association with resistance to checkpoint inhibitor therapy. *Cancer Immunol. Immunother. CII* **2019**, *68*, 1689–1700. [CrossRef]
9. Uscanga-Palomeque, A.C.; Chávez-Escamilla, A.K.; Alvizo-Báez, C.A.; Saavedra-Alonso, S.; Terrazas-Armendáriz, L.D.; Tamez-Guerra, R.S.; Rodríguez-Padilla, C.; Alcocer-González, J.M. CAR-T Cell Therapy: From the Shop to Cancer Therapy. *Int. J. Mol. Sci.* **2023**, *24*, 15688. [CrossRef]
10. Tapia-Galisteo, A.; Álvarez-Vallina, L.; Sanz, L. Bi- and trispecific immune cell engagers for immunotherapy of hematological malignancies. *J. Hematol. Oncol.* **2023**, *16*, 83. [CrossRef]
11. Staerz, U.D.; Kanagawa, O.; Bevan, M.J. Hybrid antibodies can target sites for attack by T cells. *Nature* **1985**, *314*, 628–631. [CrossRef] [PubMed]
12. Sanford, M. Blinatumomab: First global approval. *Drugs* **2015**, *75*, 321–327. [CrossRef] [PubMed]

13. Liu, J.; Zhu, J. Progresses of T-cell-engaging bispecific antibodies in treatment of solid tumors. *Int. Immunopharmacol.* **202** *138*, 112609. [CrossRef] [PubMed]
14. Bacac, M.; Klein, C.; Umana, P. CEA TCB: A novel head-to-tail 2:1 T cell bispecific antibody for treatment of CEA-positive soli tumors. *Oncoimmunology* **2016**, *5*, e1203498. [CrossRef]
15. Bacac, M.; Fauti, T.; Sam, J.; Colombetti, S.; Weinzierl, T.; Ouaret, D.; Bodmer, W.; Lehmann, S.; Hofer, T.; Hosse, R.J.; et al. A Nove Carcinoembryonic Antigen T-Cell Bispecific Antibody (CEA TCB) for the Treatment of Solid Tumors. *Clin. Cancer Res.* **2016**, *2* 3286–3297. [CrossRef]
16. Hammarström, S. The carcinoembryonic antigen (CEA) family: Structures, suggested functions and expression in normal an malignant tissues. *Semin. Cancer Biol.* **1999**, *9*, 67–81. [CrossRef]
17. Oikawa, S.; Inuzuka, C.; Kuroki, M.; Matsuoka, Y.; Kosaki, G.; Nakazato, H. Cell adhesion activity of non-specific cross-reactin antigen (NCA) and carcinoembryonic antigen (CEA) expressed on CHO cell surface: Homophilic and heterophilic adhesior *Biochem. Biophys. Res. Commun.* **1989**, *164*, 39–45. [CrossRef]
18. Wang, N.; Patel, H.; Schneider, I.C.; Kai, X.; Varshney, A.K.; Zhou, L. An optimal antitumor response by a novel CEA/CD bispecific antibody for colorectal cancers. *Antib. Ther.* **2021**, *4*, 90–100. [CrossRef]
19. Seckinger, A.; Majocchi, S.; Moine, V.; Nouveau, L.; Ngoc, H.; Daubeuf, B.; Ravn, U.; Pleche, N.; Calloud, S.; Broyer, L et al. Development and characterization of NILK-2301, a novel CEACAM5xCD3 κλ bispecific antibody for immunotherapy o CEACAM5-expressing cancers. *J. Hematol. OncolJ Hematol. Oncol.* **2023**, *16*, 117. [CrossRef]
20. Poornima, K.; Francis, A.P.; Hoda, M.; Eladl, M.A.; Subramanian, S.; Veeraraghavan, V.P.; El-Sherbiny, M.; Asseri, S.M Hussamuldin, A.B.A.; Surapaneni, K.M.; et al. Implications of Three-Dimensional Cell Culture in Cancer Therapeutic Researcl *Front. Oncol.* **2022**, *12*, 891673. [CrossRef]
21. Heinzelmann, E.; Piraino, F.; Costa, M.; Roch, A.; Norkin, M.; Garnier, V.; Homicsko, K.; Brandenberg, N. iPSC-derived anc Patient-Derived Organoids: Applications and challenges in scalability and reproducibility as pre-clinical models. *Curr. Re Toxicol.* **2024**, *7*, 100197. [CrossRef]
22. Piraino, F.; Costa, M.; Meyer, M.; Cornish, G.; Ceroni, C.; Garnier, V.; Hoehnel-Ka, S.; Brandenberg, N. Organoid models: Th future companions of personalized drug development. *Biofabrication* **2024**, *16*, 032009. [CrossRef] [PubMed]
23. Segal, N.H.; Melero, I.; Moreno, V.; Steeghs, N.; Marabelle, A.; Rohrberg, K.; Rodriguez-Ruiz, M.E.; Eder, J.P.; Eng, C.; Manj G.A.; et al. CEA-CD3 bispecific antibody cibisatamab with or without atezolizumab in patients with CEA-positive solid tumour Results of two multi-institutional Phase 1 trials. *Nat. Commun.* **2024**, *15*, 4091. [CrossRef]
24. Thasler, W.E.; Weiss, T.S.; Schillhorn, K.; Stoll, P.-T.; Irrgang, B.; Jauch, K.-W. Charitable State-Controlled Foundation Humar Tissue and Cell Research: Ethic and Legal Aspects in the Supply of Surgically Removed Human Tissue For Research in th Academic and Commercial Sector in Germany. *Cell Tissue Bank.* **2003**, *4*, 49–56. [CrossRef] [PubMed]
25. Brandenberg, N.; Hoehnel, S.; Kuttler, F.; Homicsko, K.; Ceroni, C.; Ringel, T.; Gjorevski, N.; Schwank, G.; Coukos, G.; Turcatti, G. et al. High-throughput automated organoid culture via stem-cell aggregation in microcavity arrays. *Nat. Biomed. Eng.* **2020**, *4* 863–874. [CrossRef] [PubMed]
26. Fujii, M.; Matano, M.; Toshimitsu, K.; Takano, A.; Mikami, Y.; Nishikori, S.; Sugimoto, S.; Sato, T. Human Intestinal Organoid Maintain Self-Renewal Capacity and Cellular Diversity in Niche-Inspired Culture Condition. *Cell Stem Cell* **2018**, *23*, 787–793.e
27. Alpern, D.; Gardeux, V.; Russeil, J.; Mangeat, B.; Meireles-Filho, A.C.; Breysse, R.; Hacker, D.; Deplancke, B. BRB-seq: Ultra affordable high-throughput transcriptomics enabled by bulk RNA barcoding and sequencing. *Genome Biol.* **2019**, *20*, 71. [CrossRef
28. Robinson, M.D.; McCarthy, D.J.; Smyth, G.K. edgeR: A Bioconductor package for differential expression analysis of digital gen expression data. *Bioinformatics* **2010**, *26*, 139–140. [CrossRef]
29. Uhlén, M.; Fagerberg, L.; Hallström, B.M.; Lindskog, C.; Oksvold, P.; Mardinoglu, A.; Sivertsson, Å.; Kampf, C.; Sjöstedt, E. Asplund, A.; et al. Proteomics. Tissue-based map of the human proteome. *Science* **2015**, *347*, 1260419. [CrossRef]
30. Gonzalez-Exposito, R.; Semiannikova, M.; Griffiths, B.; Khan, K.; Barber, L.J.; Woolston, A.; Spain, G.; von Loga, K.; Challoner, B. Patel, R.; et al. CEA expression heterogeneity and plasticity confer resistance to the CEA-targeting bispecific immunotherapy antibody cibisatamab (CEA-TCB) in patient-derived colorectal cancer organoids. *J. Immunother. Cancer* **2019**, *7*, 101. [CrossRef]
31. Kaja, S.; Payne, A.J.; Naumchuk, Y.; Koulen, P. Quantification of Lactate Dehydrogenase for Cell Viability Testing Using Cel Lines and Primary Cultured Astrocytes. *Curr. Protoc. Toxicol.* **2017**, *72*, 2.26.1–2.26.10. [CrossRef]
32. Maehara, Y.; Anai, H.; Tamada, R.; Sugimachi, K. The ATP assay is more sensitive than the succinate dehydrogenase inhibitior test for predicting cell viability. *Eur. J. Cancer Clin. Oncol.* **1987**, *23*, 273–276. [CrossRef] [PubMed]
33. Naghizadeh, M.; Hatamzade, N.; Larsen, F.T.; Kjaerup, R.B.; Wattrang, E.; Dalgaard, T.S. Kinetics of activation marker expression after in vitro polyclonal stimulation of chicken peripheral T cells. *Cytom. A* **2022**, *101*, 45–56. [CrossRef] [PubMed]
34. Antas, P.R.; Oliveira, E.B.; Milagres, A.S.; Franken, K.C.; Ottenhoff, T.H.; Klatser, P.; Sarno, E.N.; Sampaio, E.P. Kinetics of T cell-activation molecules in response to Mycobacterium tuberculosis antigens. *Mem. Inst. Oswaldo Cruz* **2002**, *97*, 1097–1099 [CrossRef]

Trapani, J.A. Granzymes: A family of lymphocyte granule serine proteases. *Genome Biol.* **2001**, *2*, reviews3014.1–reviews3014.7. [CrossRef] [PubMed]
Wang, F.; Graham, W.V.; Wang, Y.; Witkowski, E.D.; Schwarz, B.T.; Turner, J.R. Interferon-gamma and tumor necrosis factor-alpha synergize to induce intestinal epithelial barrier dysfunction by up-regulating myosin light chain kinase expression. *Am J. Pathol.* **2005**, *166*, 409–419. [CrossRef]

cancers

MDPI

Article

Targeting HSP90 with Ganetespib to Induce CDK1 Degradation and Promote Cell Death in Hepatoblastoma

Lea Amelie Jung [1], Salih Demir [1], Alina Hotes [1], Eiso Hiyama [2], Tomoro Hishiki [3], Emilie Indersie [4], Sophie Branchereau [5], Stefano Cairo [6,7] and Roland Kappler [1,*]

1. Department of Pediatric Surgery, Dr. von Hauner Children's Hospital, LMU University Hospital, 80337 Munich, Germany; jung.lea@campus.lmu.de (L.A.J.); salih.demir@med.uni-muenchen.de (S.D.); alina.hotes@med.uni-muenchen.de (A.H.)
2. Natural Science Center for Basic Research and Development, Hiroshima University, Hiroshima 739-0046, Japan; eiso@hiroshima-u.ac.jp
3. Department of Pediatric Surgery, Graduate School of Medicine, Chiba University, Chiba 260-8677, Japan; hishiki@chiba-u.jp
4. XenTech, 91000 Evry, France; emilie.indersie@xentech.eu
5. Bicêtre Hospital, AP-HP Paris Saclay University, 94270 Paris, France; sophie.branchereau@aphp.fr
6. Champions Oncology, Inc., Rockville, MD 20850, USA; scairo@championsoncology.com or s.cairo@irpcds.org
7. Pediatric Research Institute (IRP), 35127 Padova, Italy

* Correspondence: roland.kappler@med.uni-muenchen.de; Tel.: +49-89-4400-57810

Simple Summary: Hepatoblastoma is the most common pediatric liver cancer. Current treatments use strong drugs that can have serious side effects, so safer options are needed. In this study, we tested a group of drugs called HSP90 inhibitors, which work by blocking the proteins that cancer cells need to grow. Out of the five drugs we tested, one called ganetespib showed the best results. It was very effective at stopping the cancer cells from growing but had little impact on healthy cells. Even at low doses, ganetespib slowed down cancer cell growth, reduced their ability to survive over time, and caused the cancer cells to die. It works by disrupting a key protein named CDK1 that helps cancer cells divide. Our findings suggest that ganetespib could be a promising new treatment for hepatoblastoma, potentially offering a safer and less toxic alternative to current therapies.

Abstract: Background/Objectives: Hepatoblastoma, the most common malignant liver tumor in pediatric patients, is characterized by a remarkably low mutation rate, thereby impeding targeted therapies. Current treatment regimens rely on conventional cytotoxic agents that often cause severe adverse effects, necessitating the search for novel, less toxic therapeutic approaches. Methods: In this study, we explored the anti-tumor potential of heat shock protein 90 (HSP90) inhibitors using a unique collection of hepatoblastoma in vitro models. Results: Among the five tested inhibitors, we identified ganetespib as the most effective, significantly suppressing tumor cell growth while sparing healthy, non-tumor cells. Ganetespib treatment at low nanomolar concentrations markedly reduced cell proliferation, impaired long-term survival, and inhibited three-dimensional spheroid growth, ultimately leading to the induction of apoptosis. Mechanistically, ganetespib downregulated the expression of the HSP90 client protein cyclin-dependent kinase 1, a key cell cycle regulator controlling G2/M phase transition, which is heavily upregulated in hepatoblastoma. This disruption consequently resulted in cell cycle arrest, further contributing to its anti-tumor effects. Conclusions: HSP90 inhibition by ganetespib demonstrates significant potential as a novel therapeutic strategy for hepatoblastoma, offering a potential alternative to current cytotoxic treatments with fewer adverse effects.

Keywords: hepatoblastoma; HSP90; ganetespib; CDK1; apoptosis

1. Introduction

Hepatoblastoma (HB), though rare, is the most common malignant liver tumor in pediatric patients, with an incidence of 0.5–1.5 cases per million children, typically presenting within the first three years of life [1]. Generally, HB has a favorable long-term survival outcome of over 80% [2]. This has been achieved by the treatment of standard-risk patients with cisplatin monotherapy and using a combination regimen involving dose-dense cisplatin and doxorubicin for high-risk patients [3,4]. However, these chemotherapeutic agents show a variety of significant side effects, ranging from reversible symptoms such as nausea and vomiting to severe toxicities, including ototoxicity and nephrotoxicity [5,6]. Consequently, identifying alternative or adjunct therapies is critical to reducing the treatment burden and improving patient outcomes.

HB is characterized by a very low mutation rate [7], with comprehensive sequencing studies identifying only a very limited number of recurrent mutations, each occurring in subsets of patients [8,9]. Approximately two-thirds of HB cases harbor mutations in components of the Wnt signaling pathway, specifically in the β-catenin (CTNNB1) [10] or adenomatous polyposis coli (APC) [11] genes. Notably, the druggability of the Wnt signaling pathway has been challenging in the clinical setting [12]. Mutations of the nuclear factor erythroid-derived 2-like (NFE2L2) gene and in the telomerase reverse transcriptase (TERT) promoter have been detected only in 6% of HB cases [13]. As targeted therapies rely on the presence of specific, actionable mutations, identifying non-mutational dependencies, such as overexpressed proteins, might lead to the identification of molecular vulnerabilities in HB.

Heat shock protein 90 (HSP90) is a highly conserved and ubiquitous chaperone protein that functions as a homodimer, binding to so-called client proteins and facilitating their maturation into functionally active forms [14]. With several hundred proteins relying on HSP90, it serves as a key modulator of important processes that are integral to the hallmarks of cancer [15]. Consequently, HSP90 overexpression has been observed in a variety of cancers [16], including liver cancer [17]. Notably, HSP90 expression is increased along with the stepwise progression of hepatocarcinogenesis from low-grade dysplastic nodules to high-grade hepatocellular carcinomas and is associated with prognostic factors such as poorer histological grade of differentiation, microvascular invasion, and intrahepatic metastasis [18]. Evidence supporting the relevance of HSP90 in HB includes elevated HSP90 mRNA levels in genetically engineered HB mouse models [19] and preliminary in vitro studies using tanespimycin on patient-derived xenograft cell culture models [20]. The therapeutic potential of HSP90 inhibition has been demonstrated in several human cancers, including a phase II study of tanespimycin in breast cancer [21] and the approval of pimitespib for gastrointestinal stromal tumors that progress after chemotherapy [22]. Moreover, ganetespib showed a manageable safety profile in patients with advanced HCC [23]. Collectively, these data suggest that HSP90 inhibition could represent a potential therapeutic option for HB patients.

This study aimed to investigate the in vitro effects of five HSP90 inhibitors on HB cell lines and to evaluate the most effective candidate, ganetespib, in the preclinical setting.

2. Materials and Methods

2.1. Cell Lines

The HB cell lines HepG2 (American Type Culture Collection, Manassas, VA, USA), HepT1 (from Dr. T. Pietsch), HuH6 (Japanese Collection of Research Bioresources, Osaka,

Japan), the hepatocellular carcinoma cell lines Hep3B (American Type Culture Collection) and HUH7 (Japanese Collection of Research Bioresources), and the patient-derived xenograft cultures PDX214, PDX229, PDX233, PDX243, PDX282, PDX293B, PDX295, PDX303, PDX305, PDX310, PDX344, and PDX364 (XenTech, Evry, France) were used. As healthy controls, the human dermal fibroblasts HDFa (adult) and HDFn (neonatal), the embryonic kidney cell line HEK293 (all American Type Culture Collection), as well as the keratinocyte cell line HaCaT (CLS Cell Lines Service, Eppelheim, Germany) were used. The established tumor cell lines were cultured in RPMI 1640 containing 10% fetal bovine serum and 1% penicillin/streptomycin, the PDX cell lines in advanced DMEM/F12 with 10% fetal bovine serum, 1% penicillin/streptomycin, 1% L-glutamine (all from Thermo Fisher, Waltham, MA, USA), and 20 μM Y-27632 (Selleckchem, Chesterbrook, PA, USA), and the healthy controls in DMEM with 10% fetal bovine serum and 1% penicillin/streptomycin.

2.2. Western Blot Assay

Whole-cell lysates were collected in cell lysis buffer containing 30 mM TrisHCl/150 mM NaCl/1% Triton-X/10% glycerol/5 mM PMSF/0.5 M DTT/0.5 M Na_2VO_4/10 mM β-glycerophosphate/500 mM sodium fluoride/1:100 Halt™ protease and phosphatase inhibitor (Thermo Fischer). Denatured proteins were separated in Novex™ 4–20% gradient gel (Thermo Fischer) and transferred by Trans-Blot Turbo (Bio-Rad, San Francisco Bay Area, CA, USA) using a transfer cassette. Membranes were washed and blocked with PBS-T and 5% milk (Roth, Gießen, Germany) in PBS-T. The membranes were incubated at 4 °C overnight with the following primary antibodies: anti-HSP90 (Cell Signaling, Danvers, MA, USA) 1:1000, anti-CDC2 (Cell Signaling) 1:1000, anti-PLK1 (Cell Signaling) 1:1000, anti-AURKA (Cell Signaling) 1:1000, anti-alpha-tubulin (Sigma-Aldrich, St. Louis, MO, USA) 1:4000, anti-GAPDH (Thermo Fischer) 1:10,000, and anti-beta-actin (Cell Signaling) 1:2000 diluted in 5% milk in PBS-T. After washing, membranes were incubated with appropriate secondary antibodies: goat anti-mouse immunoglobulin HRP (Dako, Santa Clara, CA, USA) 1:1000 and anti-rabbit igG HRP-linked antibody (Cell Signaling) 1:10,000 in 5% milk in PBS-T. Protein detection was performed using enhanced chemiluminescence ECL reagent (Cytiva, Marlborough, MA, USA) with the ChemiDoc XRS + imaging system (Bio-Rad). The density of protein bands was measured with the ImageJ (version 1.54d) image processing tool.

2.3. Cell Viability Assay

MTT (3-(4,5-dimethylthiazol-2-yl)-2,5-diphenyltetrazoliumbromide) assays were performed on 5×10^5 cells per well in a 96-well-plate after treating cells for 48 h with ten increasing doses ranging from 5 nM to 100 μM of carboplatin, cisplatin, and doxorubicin as well as the HSP90 inhibitors geldanamycin, luminespib, pimitespib, tanespimicin, and ganetespib (all from Selleck Chemicals GmbH, Houston, TX, USA) as described earlier [24]. The viability of cells was measured after 48 h treatment by reading absorbance on a sunrise plate reader (Tecan, Männedorf, Switzerland). Area under the curve (AUC) was calculated using GraphPad Prism 8 (GraphPad Software, San Diego, CA, USA), and overall effectiveness was calculated as a Z-score using the formula: [(Individual AUC value)-(mean AUC)]/standard deviation. Therapeutic window scores (TWS) were calculated by using the formula: (mean AUC of healthy controls)/(mean AUC of tumor lines).

2.4. RNA Expression Analysis

RNA expression data (GSE131329) of 14 non-cancerous liver and 53 HB tissues were downloaded from the R2 genomics analysis and visualization platform on 16 November 2023 (http://r2.amc.nl) and used for gene set enrichment analysis (GSEA). Enriched pathways were detected by using the Enrichr web tool (https://maayanlab.cloud/Enrichr/), and enrichment plots were created by utilizing the Molecular Signature Database (MSigDB) (https://www.gsea-msigdb.org) both accessed on 18 December 2024. Candidate gene expression levels were compared between non-cancerous liver and tumor samples using an unpaired *t*-test, and $p < 0.05$ was a cutoff for significance.

Quantitative gene expression analysis of ganetespib (5 nM) and DMSO-treated cells was accomplished using total RNA according to procedures described earlier [25]. The following primer pairs (5′->3′orientation) were used: CDK1, GCATCCCATGT-CAAAAACTTGG, TTCAGTGCCATTTTGCCAGAA; PLK1, GGTTTTCGATTGCTCCCAGC, GGGGGTTCTCCAAGCCTTTA; AURKA, GAAGACTTGGGTCCTTGGGT, AAATATCCC-CGCACTCTGGC; and TBP, GCCCGAAACGCCGAATAT, CCGTGGTTCGTGGCTCTCT. Data were normalized to the expression level of the housekeeping gene TATA-Box-binding-Protein (TBP). Relative RNA expression was calculated using the ΔΔCT method and expressed as fold-change relative to the corresponding DMSO-treated sample [26].

2.5. Proliferation Assay

Cell proliferation was measured using the Click-iT EdU Imaging Kit (Thermo Fischer) according to the manufacturer's instructions. A total of 1×10^5 cells were seeded in a 24-well plate and incubated for 24 h. Cells were then labeled with the EdU working solution and treated with either 5 nM ganetespib or DMSO for 24 h. Cells were then fixed with 3.7% paraformaldehyde, permeabilized with 0.5% Triton-X in PBS, and finally stained with Hoechst 33342 (Thermo Fischer) and the Click-iT staining cocktail containing the Click-iT reaction buffer, $CuSO_4$, and Alexa Fluor acid (Thermo Fischer). Images were taken with the EVOS M7000 system (Thermo Fischer). Relative proliferation was calculated by referring the EdU-positive proliferating cells to the total Hoechst 33342-positive cells.

2.6. Colony Formation Assay

A total of 2×10^3 cells were seeded in a 6-well plate and after 72 h, they were treated with either 5 nM ganetespib or DMSO. The treatment was refreshed after another 72 h, and cells were incubated for a total of 9 days. Images were taken at a 4× magnification using the EVOS M7000 system (Thermo Fischer).

2.7. Spheroid Assay

A total of 1×10^3 cells were seeded in a 96 ultra-low attachment well plate (Corning Incorporated, Corning, NY, USA) and incubated for 72 h. Images were taken of the formed spheroids and then again after treatment with 5 nM of ganetespib or DMSO for 72 h. Both imaging and size measuring of spheroids on days 0 and 3 were performed by using the EVOS M7000 system (Thermo Fischer). The spheroid volume was calculated for 0 h and 72 h by the formula: $[(\text{length } (\mu m) \times \text{width } \mu m)^2)/2]$.

2.8. Live and Dead Staining

Treated spheroids were stained with 1 µM Calcein-AM (Biolegend, Hercules, CA, USA) and 1 mg/mL propidium iodide (Sigma-Aldrich) in serum-free medium (Thermo Fisher) for 20 min at 37 °C. Images were taken with the EVOS M7000 system (Thermo Fischer).

2.9. Cell Cycle Analysis

For synchronization, cells were cultured in a serum-free starvation medium for 48 h. The degree of synchronization was validated after 24 h using the Click-iT EdU Imaging Kit (Thermo Fischer) and after 48 h using standard immunofluorescent detection of the proliferation marker Ki-67 [27] and flow cytometry (Supplemental Figure S1). Then, serum-free medium was replaced with normal media and the cells were exposed to DMSO or 10 and 50 nM of ganetespib for 48 h. Cells were harvested and fixed overnight with 80%

ice-cold ethanol at −20 °C. The next day, 5×10^5 cells were incubated with 300 µL cell cycle staining buffer, consisting of 50 µg/mL propidium iodide (Sigma-Aldrich), 0.5% Triton-X-100 (Sigma-Aldrich) and 2 µg/mL RNAse (Qiagen, Hilden, Germany) in dH_2O, in the dark at 37 °C for 1 h. The distribution of cell cycle phases was measured by the LSRFortessa cell analyzer (BD Biosciences, East Rutherford, NJ, USA), and data analysis was conducted by FlowJo software v10.4 (BD Biosciences), applying the Watson Pragmatic model.

2.10. Apoptosis Assay

A total of 1×10^5 cells were seeded in a 24-well plate and incubated for 24 h. Cells were then treated with 5 nM ganetespib or DMSO for 24 h. Cells were then stained with Hoechst 33342 and the cell event caspase 3/7 detection reagent (both Thermo Fischer). Imaging was accomplished with the EVOS M7000 system (Thermo Fischer). Relative apoptosis was calculated by referring to the apoptotic caspase 3/7-positive cells to the total Hoechst 33342-positive cells.

2.11. Statistical Analyses

Data were visualized, showing the mean ± standard deviation (SD) or standard error of the mean (SEM), and analyzed by GraphPad Prism 8 (GraphPad Software). For the comparison of the two conditions, statistical significance was determined by Student's *t*-test. For Kaplan–Maier survival curves, Mantel–Cox log-rank analysis was applied. For all the statistical models, the significance was considered for values $p < 0.05$.

3. Results

3.1. Ganetespib Is the Most Effective HSP90 Inhibitor for Impeding HB Viability

In order to determine the therapeutic potential of HSP90 inhibition in HB patients, we tested a panel of HSP90 inhibitors—geldanamycin, luminespib, pimitespib, tanespimicin, and ganetespib—across 10 increasing concentrations on a comprehensive set of conventional liver tumor cell lines, patient-derived xenograft cultures, and healthy control cell lines using MTT-based cell viability assays. All HSP90 inhibitors demonstrated potent growth inhibitory effects on tumor cells while leaving normal cells almost unaffected (Figure 1A, left). Compared to conventional chemotherapy that is given to HB patients, such as carboplatin and cisplatin, all HSP90 inhibitors showed superior efficacy. Notably, most of the inhibitors even outperformed doxorubicin, a treatment reserved for high-risk HB patients [28]. Among the tested drugs, ganetespib emerged as the most effective, exhibiting remarkably high tumor-suppressive activity across all HB models. This was reflected in its smaller area under the curve (AUC) values and low Z-scores (Figure 1A, left and middle). Furthermore, ganetespib displayed the highest therapeutic window score, indicating its potential for flexible dose adjustments in the clinical setting (Figure 1A, right). The comparison of the drug inhibition curves between tumor models and healthy cells, as well as the corresponding half-maximal inhibitory concentrations (IC50), highlighted a >1000-fold difference in response, confirming the strong growth inhibitory effect of ganetespib with minimal toxicity to normal cells at comparable concentrations (Figure 1B,C).

We then integrated the response data of the HSP90 inhibitors into our recently published drug testing platform containing response data for 78 drugs [24]. By performing principal component analysis on the AUC values, we discovered HSP90 inhibitors as one of the most effective and tumor-suppressing drug categories, grouping them alongside HDAC and CDK9 inhibitors (Figure 1D). These categories demonstrated significantly greater efficacy compared to cisplatin, used as a reference drug.

To investigate the underlying mechanism of the HB-suppressive effects of HSP90 inhibitors, we examined the expression of HSP90 at the protein level. Western blot analysis of two normal liver tissues and four HB models revealed a pronounced overexpression of HSP90 in HB compared to mild expression observed in normal liver tissues (Figure 1E, Figure S2). Additionally, gene set enrichment analysis of publicly available RNA sequencing data from 53 HB patients and 14 non-cancerous liver samples (GSE131329) showed that the gene ontology term "heat shock protein binding" was significantly enriched in HB samples (Figure 1F). Moreover, HSP90 RNA levels were dramatically upregulated in HB samples (Figure 1G), and patients with high HSP90 expression showed a trend towards poor overall and event-free survival (Figure 1H).

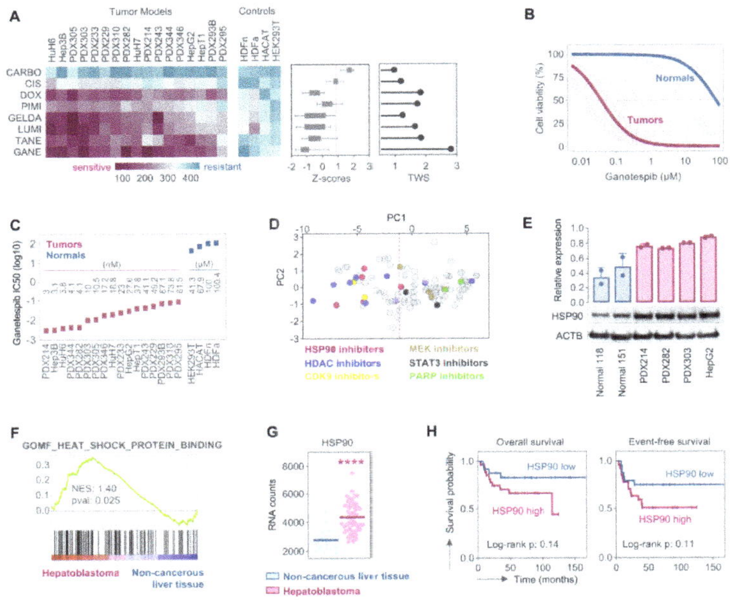

Figure 1. (**A**) A heat map showing the drug responses of 17 liver tumor models and 4 non-cancerous controls towards 5 HSP90 inhibitors (pimi)tespib, (gelda)namycin, (lumi)nespib, (tane)spimicin, and (gane)tespib as well as the standard-of-care drugs (carbo)platin, (cis)platin, and (dox)orubicin. Corresponding Z-scores and therapeutic window scores (TWS) are depicted as box-with-whiskers and lollipop plots, respectively. (**B**) Viability curves of liver tumor models towards 10 increasing concentrations of ganetespib (5 nm–100 μM). Dark pink and blue curves represent the mean response of the tumor models and healthy controls, respectively. (**C**) IC50 values of ganetespib-treated cell lines. (**D**) Principal component analysis of 83 compounds in our in vitro drug-testing platform. The drugs positioned on the right-hand side of the dashed line, which represents the PC1 of cisplatin, were considered ineffective. (**E**) Western blot analysis showing the HSP90 protein expression levels in two normal liver tissues and four hepatoblastoma models. Beta actin (ACTB) was used as a loading control. Bar graph shows the mean relative band density of two independent experiments ± SD. (**F**) Enrichment plot of gene ontology obtained from gene set enrichment analysis (GSEA) of 53 HB patients versus 14 non-cancerous livers from the publicly available GSE131329 data set. (**G**) Comparative RNA expression analysis of HSP90 in 53 HB patients and 14 non-cancerous liver samples from the GSE131329 data set, utilizing R2 Genomics Analysis and Visualization Platform. (**H**) Kaplan-Meier curves displaying overall survival (OS) and event-free survival (EFS) probabilities for patients with HB of the GSE131329 data set with either high (n = 27) or low (n = 25) HSP90 expression. Median expression was considered as cutout, and the log-rank Mantel–Cox test was used to calculate significance. **** $p < 0.0001$.

Together, these findings suggest that elevated HSP90 levels and increased heat shock protein binding render HB highly susceptible to HSP90 inhibition, resulting in suppressed cell growth.

3.2. Ganetespib Significantly Suppresses HB Growth Potential

To evaluate the impact of ganetespib, the most effective HSP90 inhibitor, on the growth properties of HB, we treated four high-risk HB models with the drug at IC50 concentrations and assessed growth-related features using multiple approaches. Proliferative activity of tumor cells, detected by the incorporation of nucleoside analogs into newly synthesized DNA of actively dividing cells, was markedly inhibited after 24 h of ganetespib treatment (Figure 2A).

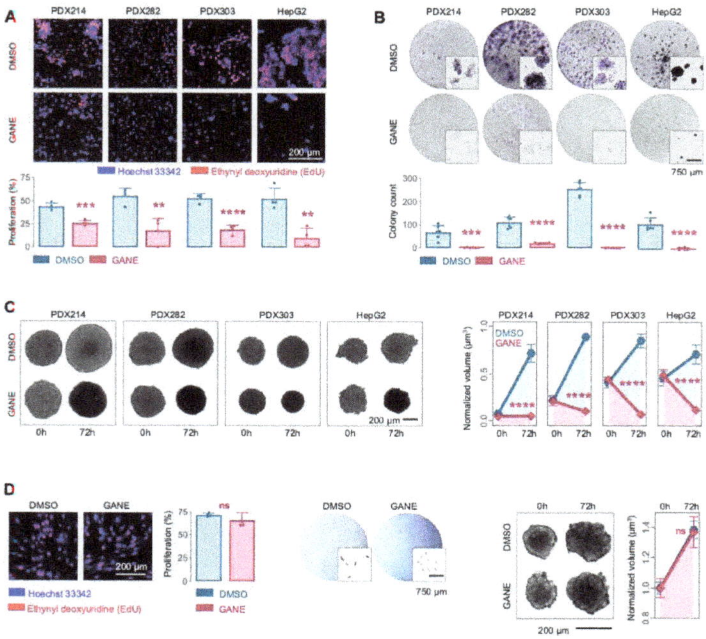

Figure 2. (**A**) Immunofluorescence images (top) and corresponding calculation of the proliferating cells (red) in relation to total cells (blue) upon DMSO or 5 nM (gane)tespib exposure for 24 h. (**B**) Crystal violet stainings of formed colonies (top) and absolute colony counts (bottom) are demonstrated upon DMSO or 5 nM (gane)tespib treatment for 10 days. (**C**) HB spheroids images and respective normalized spheroid volumes were displayed following DMSO or 5 nM (gane)tespib exposure for 72 h. (**D**) Proliferation, colony formation, and spheroid assays as in A-C on neonatal fibroblasts. Bar and line graphs represent the mean ± SEM of two independent experiments. Statistics were calculated using a two-tailed unpaired Student's *t*-test, with ** $p < 0.01$, *** $p < 0.001$, **** $p < 0.0001$, ns = not significant.

Furthermore, the long-term survival potential of HB cells was significantly reduced, as evidenced by fewer and smaller colonies formed over a 10-day period in ganetespib-treated wells (Figure 2B). Notably, in a three-dimensional spheroid model, ganetespib treatment resulted in substantially smaller spheroid volumes, indicating compromised cell viability and growth (Figure 2C). In contrast, normal fibroblasts maintained their proliferative capacity in response to ganetespib, although they did not form colonies (Figure 2D). These results underscore the potent ability of ganetespib to hinder the growth and survival of HB cells across different experimental models while sparing normal cells.

3.3. Ganetespib Triggers Tumor Cell Death in HB Models

Treatment of HB spheroids with ganetespib not only inhibited growth but also led to volume shrinkage, a hallmark of cell death in three-dimensional models. To confirm this observation, we performed live and death staining on ganetespib-treated spheroids. As anticipated, a 72 h exposure to ganetespib resulted in a significant increase in propidium iodide-positive cells, particularly in the central and proliferative zones of the spheroids (Figure 3A). Additionally, all HB models treated with ganetespib for 24 h showed significantly higher levels of active caspase 3 and 7 substrates, highlighting the induction of apoptosis (Figure 3B). Of note, normal fibroblasts were unaffected by ganetespib treatment (Figure 3C). These findings collectively demonstrate ganetespib's ability to induce selective tumor cell death through activation of caspase-dependent apoptotic pathways.

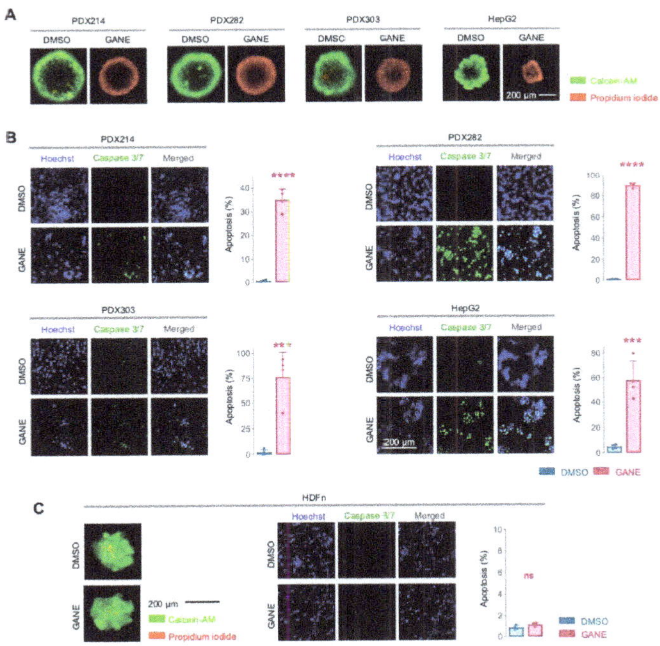

Figure 3. (**A**) Microscopy images of three-dimensional hepatoblastoma spheroids upon 72 h exposure to DMSO or 5 nM (gane)tespib. Cell death was detected by live (Calcein-AM, green) and dead (propidium iodide, red) staining. (**B**) Portions of apoptotic cells were determined by active caspase 3/7 staining after 24 h treatment with DMSO or 5 nM (gane)tespib. (**C**) Cell death and caspase activation assays as in (**A**,**B**) on neonatal fibroblasts. Bar graphs represent the mean ± SEM of two independent experiments with duplicates. Statistics were calculated using a two-tailed unpaired Student's *t*-test, with *** $p < 0.001$, **** $p < 0.0001$, ns = not significant.

3.4. Elevated Expression of the HSP90 Client CDK1 Drives Hyperactivation of Cell Cycle Mechanisms in HB

HSP90 inhibition is known to disrupt interactions between HSP90 and its client proteins [14]. To investigate this mechanism in HB, we first examined the impact of ganetespib exposure on HSP90 protein expression. Western blot analysis revealed no significant changes in HSP90 expression levels when HB cells were treated with 10 or 50 nM ganetespib (Figure 4A, Figure S2). This suggests that ganetespib does not induce degradation of HSP90 itself but likely exerts its effect by preventing ATP-dependent binding of HSP90 to its client proteins (Figure 4B). To identify HSP90 client proteins implicated in HB, we conducted KEGG pathway analysis on 767 upregulated genes (p-value < 0.05, fold change > 2) retrieved from publicly available RNA sequencing data (GSE131329) comparing 53 HB patients to 14 non-cancerous liver samples. This analysis highlighted "DNA replication" and "cell cycle" as the most significantly enriched terms (Figure 4C,D), consistent with the established role of numerous HSP90 clients in cell cycle regulation [29]. Utilizing the same RNA sequencing dataset, we further explored transcriptional changes in known HSP90 client proteins associated with the cell cycle [30]. Among these, cyclin-dependent kinase 1 (CDK1), polo-like kinase 1 (PLK1), and aurora kinase A (AURKA) exhibited dramatically increased transcriptional expression in HB patients compared to non-cancerous liver samples, suggesting a pivotal role as HSP90 clients in maintaining cell cycle activity in HB (Figure 4E). Protein-level analysis corroborated these findings, as all three candidate proteins were undetectable in non-cancerous liver tissues but highly expressed in HB models, as shown by Western blot analysis (Figure 4F, Figure S2). Notably, CDK1, PLK1, and AURKA showed strong co-expression with HSP90 in HB patients (Figure 4G). Collectively, these data confirm that cell cycle is profoundly activated in HB tumors, with elevated client protein levels as key drivers of dysregulated cell cycle regulation, potentially through their interaction with HSP90.

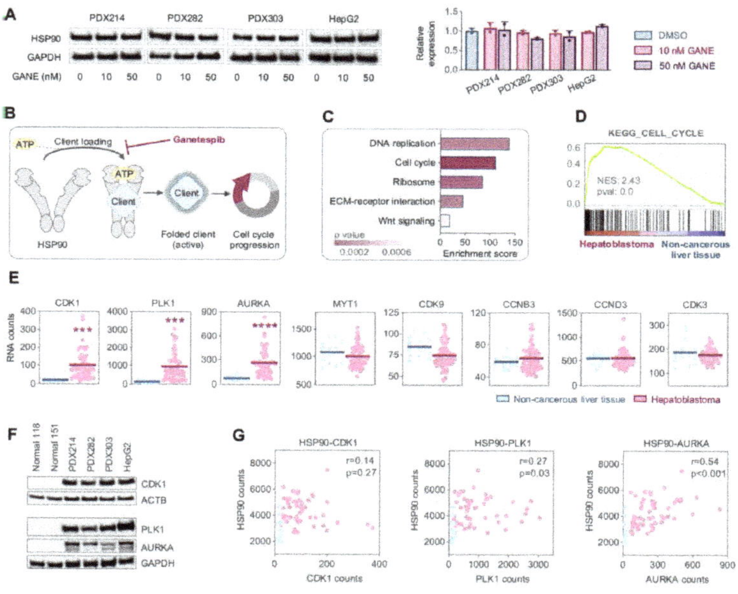

Figure 4. (**A**) Western blot analysis (left) and the bar graphs with corresponding density calculation of the bands (right) showing alterations in heat shock protein 90 (HSP90) protein levels upon increasing concentrations of (gane)tespib for 24 h. Glycerinaldehyd-3-phosphat-dehydrogenase (GAPDH) was used as loading control. (**B**) Schematic illustration of recruitment of a client by HSP90. (**C**) KEGG pathway enrichment scores of 767 significantly upregulated ($p < 0.05$, fold > 2) genes from RNA sequencing analysis of GSE131329 data set. (**D**) Enrichment plot showing the KEGG category cell cycle, obtained from gene set enrichment analysis (GSEA) of the GSE131329 data set. (**E**) Comparative RNA expression analysis of HSP90 clients from the GSE131329 data set, utilizing R2 Genomics Analysis and Visualization Platform. Statistics were calculated using a two-tailed unpaired Student's t-test, with *** $p < 0.001$, **** $p < 0.0001$. (**F**) Western blot showing the expression of cyclin-dependent kinase 1 (CDK1), polo-like kinase 1 (PLK1), and aurora kinase A (AURKA) in non-cancerous liver tissue and hepatoblastoma models. Beta actin (ACTB) and GAPDH served as loading controls. (**G**) Correlation of gene expression between CDK1, PLK1, and AURKA with HSP90 of the GSE131329 data. Two-tailed Pearson test was performed, and correlation coefficient was calculated.

3.5. Ganetespib-Induced Inhibition CDK1 Leads to G2/M Arrest

Given the observed upregulation of several cell cycle-associated client proteins in HB, we wanted to investigate the functional consequences of their inhibition by ganetespib. When HB models were treated with increasing doses of ganetespib, CDK1 protein expression levels were drastically decreased (Figure 5A, Figure S2) while leaving its transcriptional levels almost unaffected (Figure 5B), suggesting that ganetespib prevents HSP90-mediated recruitment and stabilization of CDK1. In addition, we examined the two other HSP90 client proteins involved in the cell cycle and hepatoblastoma, namely the polo-like kinase 1 (PLK1) and the aurora kinase A (AURKA) [31,32]. However, ganetespib treatment reduced their protein levels in only two of the four HB models (Figure 5A, Figure S2), again with no consequences in RNA expression (Figure 5B). This indicates that while HSP90 inhibition principally affects both clients, they may not be the primary drivers of the dramatic tumor growth suppression and cell death observed across all HB models.

To investigate the impact of HSP90 inhibition on the cell cycle, we next synchronized HB models in the G0/G1 phase through serum starvation, as evidenced by the dramatic reduction in proliferating cells after 24 h, which became even more pronounced at 48 h (Supplemental Figure S1). Most importantly, subsequent ganetespib treatment caused significantly increased proportions of cells arrested in the G2/M phase of the cell cycle in all HB models after 48 h (Figure 5C,D), an effect already evident in fast-growing HepG2 and PDX303 cells after 24 h (Supplementary Figure S1D). This suggests that HSP90 inhibition by ganetespib leads to impairment of cell cycle progression by preventing the stabilization of CDK1, which is critical for G2/M transition.

Altogether, our data demonstrate that HSP90-mediated stabilization of CDK1 is an essential step for cell cycle progression and is hyperactivated in HB tumors. Moreover, we showed that ganetespib primarily targets CDK1 function, identifying CDK1 as a putative marker for response to ganetespib and the outcome of HB patients.

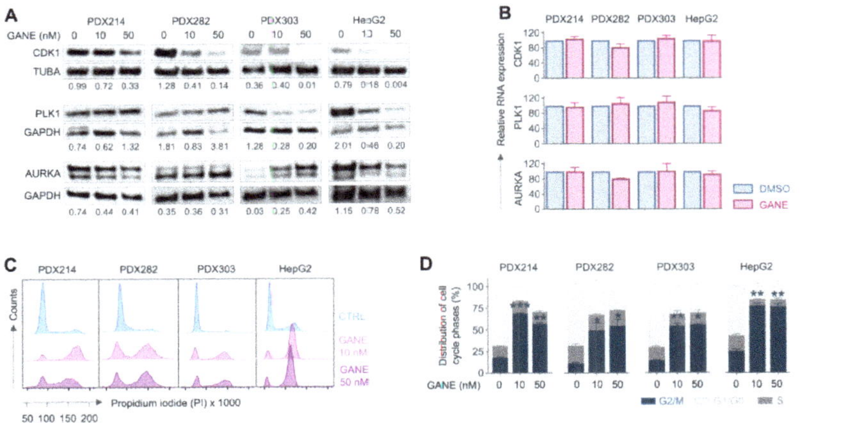

Figure 5. (**A**) Western blot and calculation of the normalized band densities demonstrate the expression of CDK1, PLK1, and AURKA proteins upon increasing concentrations of (gane)tespib for 24 h. Tubulin alpha 1a (TUBA) and glycerinaldehyd-3-phosphat-dehydrogenase (GAPDH) were used as loading controls. (**B**) PCR-based relative RNA expression analysis of gantespib treated tumor cells normalized to the housekeeping gene TBP. (**C**) Histograms showing the propidium iodide-labeled DNA distribution in synchronized hepatoblastoma models after 48 h treatment with DMSO, 10 nM or 50 nM (gane)tespib. (**D**) Stacked bar graphs displaying the distributions of cell cycle phases calculated according to the histograms. Significance was calculated in comparison to the DMSO control, using Student's *t*-test with * $p < 0.05$, ** $p < 0.01$, *** $p < 0.001$.

4. Discussion

The combination of cisplatin and doxorubicin has led to significant success in the treatment of HB [3,4]. However, these therapies are associated with severe adverse effects [33], underscoring the urgent need for novel, less toxic treatment options. In this study, we evaluated the efficacy of five HSP90 inhibitors in suppressing tumor growth using a comprehensive pediatric liver cancer testing platform. Our results showed that HSP90 is overexpressed in HB, positioning its inhibition as a promising target for cancer therapy. Ganetespib proved to be the most effective HSP90 inhibitor in HB models, exhibiting minimal impact on normal cells. We identified CDK1 as the key HSP90 client protein targeted by ganetespib in HB, leading to pronounced G2-M phase arrest and, ultimately, cell death.

Since the discovery of the first naturally occurring HSP90 inhibitor geldanamycin, HSP90 inhibition has been explored as an effective strategy for treating various malignancies, with 22 inhibitors tested in 186 cancer clinical trials [34]. Similarly, our in vitro screening with first-generation (geldanamycin), second-generation (tanespimycin and ganetespib), and third-generation (luminespib and pimitespib) HSP90 inhibitors demonstrated remarkable anti-tumor activity on HB models. Among the five tested inhibitors, ganetespib was the most promising candidate for the treatment of HB, showing potent growth inhibitory activity at very low nanomolar concentrations in all tested models while leaving normal cells nearly unaffected. Previous studies have shown the strong activity of ganetespib against various cancer types in the preclinical setting, including hepatocellular carcinoma [35]. More importantly, initial phase clinical trials in solid tumors showed that ganetespib is well tolerated [36]. Altogether, ganetespib demonstrates exceptional promise as a new therapy for HB, combining potent anti-tumor efficacy with minimal effects on normal cells. Coupled with its favorable safety profile in early-phase clinical trials, these findings position ganetespib as a leading candidate in the quest for safer and more precise therapies for HB.

HSP90 plays a key role in protein folding, facilitating the maturation of client proteins into their molecularly active conformations [14]. Proteomic studies have revealed more than 6000 HSP90 client proteins, including receptor tyrosine kinases, signaling proteins, transcription factors, and cell cycle regulatory proteins [37]. Malignant cells exploit the highly active HSP90 chaperone system to prevent misfolding and degradation of oncogenes and growth-promoting factors, leading to their constitutive activation [38]. Accordingly, elevated HSP90 expression is observed in the majority of cancer types, including liver cancer [17]. This dysregulation supports uncontrolled growth and disruption of cell cycle processes, a phenomenon underscored by the involvement of multiple HSP90 clients in cell signaling [29]. In line with these observations, our transcriptomic analysis of HB patients revealed gene signatures enriched for heat-shock protein binding and cell cycle pathways. One of the most important HSP90 clients as a key mediator of cell cycle progression is CDK1 [39]. Our investigation for relevant HSP90 clients revealed that CDK1 transcript levels are dramatically elevated in HB patients, while non-cancerous liver samples showed no detectable CDK1 protein expression. Overexpression of CDK1 has been detected in multiple cancer types, identifying it as a universal pan-cancer biomarker [40]. More importantly, survival analysis of 53 HB patients from the GSE131329 data set showed a clear association between high CDK1 expression and shorter survival periods. This is in line with previous work linking increased levels of CDK1 to inferior survival rates and unfavorable clinical outcomes, solidifying CDK1 as an important prognostic marker for many cancers [41]. Our data also indicated a massive decrease in CDK1 protein levels upon ganetespib treatment without altering its transcription, suggesting that CDK1 is post-translationally degraded due to blocking of HSP90 client recruiting activity. We further showed that ganetespib exposure led to G2/M arrest in all HB models, supporting the repression of CDK1 function upon ganetespib-mediated HSP90 inhibition, since CDK1 regulates transition between G2 phase and mitosis [42]. The downregulation of CDK1 as a consequence of exposure to the HSP90 inhibitors has been reported in different studies, such as by ganetespib in mantle cell lymphoma [43], tanespimycin in colon cancer [44], or by 17DMAG, a semi-synthetic derivative of geldanamycin in hepatocellular carcinoma [45].

One limitation of our study is that we do not provide direct evidence of a physical interaction between HSP90 and CDK1, nor confirm that HSP90 inhibition causes CDK1 proteasomal degradation. While our findings provide strong preclinical evidence supporting ganetespib as a potential therapy, additional studies are necessary to fully elucidate its mechanisms of action, particularly its effects on CDK1 and cell cycle progression.

Overall, these findings demonstrate the critical role of HSP90 in supporting cancer cell survival through its cell cycle-associated client proteins, particularly CDK1. Targeting HSP90 with inhibitors such as ganetespib offers a promising strategy for HB treatment, disrupting cancer's dependency on HSP90 while inducing cell cycle arrest and apoptosis. Well-designed in vivo studies that closely mimic the clinical situation are essential to demonstrate the potential of HSP90 inhibition as a therapeutic option for HB.

5. Conclusions

This study highlights the potential of HSP90 inhibitors, particularly ganetespib, in suppressing HB tumor growth while sparing normal cells. HSP90 is overexpressed in HB and aids cancer cell survival by stabilizing proteins like CDK1, a critical regulator of cell cycle progression. Ganetespib effectively reduces CDK1 levels, induces G2-M arrest, and promotes cell death. With potent anti-tumor activity and tolerability in early clinical trials, ganetespib emerges as a promising candidate for HB therapy, targeting a crucial cancer dependency.

Supplementary Materials: The following supporting information can be downloaded at https://www.mdpi.com/article/10.3390/cancers17081341/s1, Figure S1: Cell synchronization and cell cycle analysis, Figure S2: Original Western blot images.

Author Contributions: Conceptualization, S.D. and R.K.; methodology, L.A.J., S.D., A.H. and E.I.; validation, S.D. and R.K.; formal analysis, L.A.J., E.H., T.H. and S.D.; investigation, L.A.J., S.D. and A.H.; resources, E.H., T.H., S.B. and S.C.; data curation, S.D. and R.K.; writing—original draft preparation, L.A.J., S.D. and R.K.; writing—review and editing, S.D. and R.K.; visualization, L.A.J. and S.D.; supervision, R.K.; project administration, R.K.; funding acquisition, R.K. All authors have read and agreed to the published version of the manuscript.

Funding: This research was partially supported by the Fight Kids Cancer Funding Programme (BT4ChildLC) and received funding from the Bettina Bräu Foundation Munich and the Gänseblümchen Association Voerde.

Institutional Review Board Statement: Not applicable.

Informed Consent Statement: Not applicable.

Data Availability Statement: No new data were created.

Acknowledgments: We acknowledge Torsten Pietsch for providing the HepT1 cell line and Xentech for providing the patient-derived xenograft models.

Conflicts of Interest: The authors declare no conflicts of interest.

Abbreviations

The following abbreviations are used in this manuscript:

HB	Hepatoblastoma
HSP90	Heat shock protein 90
CDK1	Cyclin-dependent kinase 1
AUC	Area under the curve
TWS	Therapeutic window scores
GSEA	Gene set enrichment analysis

References

1. Spector, L.G.; Birch, J. The epidemiology of hepatoblastoma. *Pediatr. Blood Cancer* **2012**, *59*, 776–779. [CrossRef] [PubMed]
2. Feng, T.C.; Zai, H.Y.; Jiang, W.; Zhu, Q.; Jiang, B.; Yao, L.; Li, X.Y.; Wang, Z.M. Survival and analysis of prognostic factors hepatoblastoma: Based on SEER database. *Ann. Transl. Med.* **2019**, *7*, 555. [CrossRef] [PubMed]
3. Perilongo, G.; Maibach, R.; Shafford, E.; Brugieres, L.; Brock, P.; Morland, B.; de Camargo, B.; Zsiros, J.; Roebuck, D.; Zimmerma A.; et al. Cisplatin versus cisplatin plus doxorubicin for standard-risk hepatoblastoma. *N. Engl. J. Med.* **2009**, *361*, 1662–16 [CrossRef]
4. Zsiros, J.; Brugieres, L.; Brock, P.; Roebuck, D.; Maibach, R.; Zimmermann, A.; Childs, M.; Pariente, D.; Laithier, V.; Otte, J.B.; et Dose-dense cisplatin-based chemotherapy and surgery for children with high-risk hepatoblastoma (SIOPEL-4): A prospecti single-arm, feasibility study. *Lancet Oncol.* **2013**, *14*, 834–842. [CrossRef]
5. Carvalho, C.; Santos, R.X.; Cardoso, S.; Correia, S.; Oliveira, P.J.; Santos, M.S.; Moreira, P.I. Doxorubicin: The good, the bad a the ugly effect. *Curr. Med. Chem.* **2009**, *16*, 3267–3285. [CrossRef]
6. Tsang, R.Y.; Al-Fayea, T.; Au, H.J. Cisplatin overdose: Toxicities and management. *Drug Saf.* **2009**, *32*, 1109–1122. [CrossR [PubMed]
7. Eichenmüller, M.; Trippel, F.; Kreuder, M.; Beck, A.; Schwarzmayr, T.; Häberle, B.; Cairo, S.; Leuschner, I.; von Schweinitz, Strom, T.M.; et al. The genomic landscape of hepatoblastoma and their progenies with HCC-like features. *J. Hepatol.* **2014**, 1312–1320. [CrossRef]
8. Hirsch, T.Z.; Pilet, J.; Morcrette, G.; Roehrig, A.; Monteiro, B.J.E.; Molina, L.; Bayard, Q.; Trepo, E.; Meunier, L.; Caruso, S.; et Integrated Genomic Analysis Identifies Driver Genes and Cisplatin-Resistant Progenitor Phenotype in Pediatric Liver Canc *Cancer Discov.* **2021**, *11*, 2524–2543. [CrossRef]
9. Nagae, G.; Yamamoto, S.; Fujita, M.; Fujita, T.; Nonaka, A.; Umeda, T.; Fukuda, S.; Tatsuno, K.; Maejima, K.; Hayashi, A.; et Genetic and epigenetic basis of hepatoblastoma diversity. *Nat. Commun.* **2021**, *12*, 5423. [CrossRef]
10. Koch, A.; Denkhaus, D.; Albrecht, S.; Leuschner, I.; von Schweinitz, D.; Pietsch, T. Childhood hepatoblastomas frequently carry mutated degradation targeting box of the beta-catenin gene. *Cancer Res.* **1999**, *59*, 269–273.
11. Giardiello, F.M.; Petersen, G.M.; Brensinger, J.D.; Luce, M.C.; Cayouette, M.C.; Bacon, J.; Booker, S.V.; Hamilton, S.R. Hepatobl toma and APC gene mutation in familial adenomatous polyposis. *Gut* **1996**, *39*, 867–869. [CrossRef] [PubMed]
12. Shaw, H.V.; Koval, A.; Katanaev, V.L. Targeting the Wnt signalling pathway in cancer: Prospects and perils. *Swiss Med. Wk* **2019**, *149*, w20129. [CrossRef] [PubMed]
13. Cairo, S.; Armengol, C.; Maibach, R.; Haberle, B.; Becker, K.; Carrillo-Reixach, J.; Guettier, C.; Vokuhl, C.; Schmid, I.; Buend M.A.; et al. A combined clinical and biological risk classification improves prediction of outcome in hepatoblastoma patients. *E J. Cancer* **2020**, *141*, 30–39. [CrossRef] [PubMed]
14. Schopf, F.H.; Biebl, M.M.; Buchner, J. The HSP90 chaperone machinery. *Nat. Rev. Mol. Cell Biol.* **2017**, *18*, 345–360. [CrossRef]
15. Hanahan, D. Hallmarks of Cancer: New Dimensions. *Cancer Discov.* **2022**, *12*, 31–46. [CrossRef]
16. Whitesell, L.; Lindquist, S.L. HSP90 and the chaperoning of cancer. *Nat. Rev. Cancer* **2005**, *5*, 761–772. [CrossRef]
17. Pascale, R.M.; Simile, M.M.; Calvisi, D.F.; Frau, M.; Muroni, M.R.; Seddaiu, M.A.; Daino, L.; Muntoni, M.D.; De Miglio, M. Thorgeirsson, S.S.; et al. Role of HSP90, CDC37, and CRM1 as modulators of P16(INK4A) activity in rat liver carcinogenesis a human liver cancer. *Hepatology* **2005**, *42*, 1310–1319. [CrossRef]
18. Lim, S.O.; Park, S.G.; Yoo, J.H.; Park, Y.M.; Kim, H.J.; Jang, K.T.; Cho, J.W.; Yoo, B.C.; Jung, G.H.; Park, C.K. Expression heat shock proteins (HSP27, HSP60, HSP70, HSP90, GRP78, GRP94) in hepatitis B virus-related hepatocellular carcinomas a dysplastic nodules. *World J. Gastroenterol.* **2005**, *11*, 2072–2079. [CrossRef]
19. Hurley, E.H.; Tao, J.; Liu, S.; Krutsenko, Y.; Singh, S.; Monga, S.P. Inhibition of Heat Shock Factor 1 Signaling Decreas Hepatoblastoma Growth via Induction of Apoptosis. *Am. J. Pathol.* **2023**, *193*, 148–160. [CrossRef]
20. Kissoondoyal, A.; DeSousa, E.; Quaglietta, P.R.; Indersie, E.; Cairo, S.; Malkin, D.; Luques, L.; Baertschiger, R.M. Reduction in c viability of hepatoblastoma cells following inhibtion of Hsp90. *Cancer Res.* **2023**, *83* (Suppl. S7), 3540. [CrossRef]

21. Modi, S.; Stopeck, A.; Linden, H.; Solit, D.; Chandarlapaty, S.; Rosen, N.; D'Andrea, G.; Dickler, M.; Moynahan, M.E.; Sugarman, S.; et al. HSP90 inhibition is effective in breast cancer: A phase II trial of tanespimycin (17-AAG) plus trastuzumab in patients with HER2-positive metastatic breast cancer progressing on trastuzumab. *Clin. Cancer Res.* **2011**, *17*, 5132–5139. [CrossRef] [PubMed]
22. Hoy, S.M. Pimitespib: First Approval. *Drugs* **2022**, *82*, 1413–1418. [CrossRef]
23. Goyal, L.; Wadlow, R.C.; Blaszkowsky, L.S.; Wolpin, B.M.; Abrams, T.A.; McCleary, N.J.; Sheehan, S.; Sundaram, E.; Karol, M.D.; Chen, J.; et al. A phase I and pharmacokinetic study of ganetespib (STA-9090) in advanced hepatocellular carcinoma. *Investig. New Drugs* **2015**, *33*, 128–137. [CrossRef]
24. Demir, S.; Hotes, A.; Schmid, T.; Cairo, S.; Indersie, E.; Pisano, C.; Hiyama, E.; Hishiki, T.; Vokuhl, C.; Branchereau, S.; et al. Drug prioritization identifies panobinostat as a tailored treatment element for patients with metastatic hepatoblastoma. *J. Exp. Clin. Cancer Res.* **2024**, *43*, 299. [CrossRef] [PubMed]
25. Eichenmüller, M.; Gruner, I.; Hagl, B.; Häberle, B.; Müller-Höcker, J.; von Schweinitz, D.; Kappler, R. Blocking the hedgehog pathway inhibits hepatoblastoma growth. *Hepatology* **2009**, *49*, 482–490. [CrossRef] [PubMed]
26. Pfaffl, M.W. A new mathematical model for relative quantification in real-time RT-PCR. *Nucleic Acids Res.* **2001**, *29*, e45. [CrossRef]
27. Demir, S.; Razizadeh, N.; Indersie, E.; Branchereau, S.; Cairo, S.; Kappler, R. Targeting G9a/DNMT1 methyltransferase activity impedes IGF2-mediated survival in hepatoblastoma. *Hepatol. Commun.* **2024**, *8*, e0378. [CrossRef]
28. Zsiros, J.; Maibach, R.; Shafford, E.; Brugieres, L.; Brock, P.; Czauderna, P.; Roebuck, D.; Childs, M.; Zimmermann, A.; Laithier, V.; et al. Successful treatment of childhood high-risk hepatoblastoma with dose-intensive multiagent chemotherapy and surgery: Final results of the SIOPEL-3HR study. *J. Clin. Oncol.* **2010**, *28*, 2584–2590. [CrossRef]
29. Burrows, F.; Zhang, H.; Kamal, A. Hsp90 activation and cell cycle regulation. *Cell Cycle* **2004**, *3*, 1530–1536. [CrossRef]
30. Echeverria, P.C.; Bernthaler, A.; Dupuis, P.; Mayer, B.; Picard, D. An interaction network predicted from public data as a discovery tool: Application to the Hsp90 molecular chaperone machine. *PLoS ONE* **2011**, *6*, e26044. [CrossRef]
31. Kats, D.; Ricker, C.A.; Berlow, N.E.; Noblet, B.; Nicolle, D.; Mevel, K.; Branchereau, S.; Judde, J.G.; Stiverson, C.D.; Stiverson, C.L.; et al. Volasertib preclinical activity in high-risk hepatoblastoma. *Oncotarget* **2019**, *10*, 6403–6417. [CrossRef] [PubMed]
32. Wang, F.; Zhang, H.; He, B.; Liu, Z.; Wu, X.; Liu, Y.; Xu, X.; Gou, X.; Wang, H.; Yang, Q. Heat shock protein 90 inhibitors induce cell differentiation via the ubiquitin-dependent aurora kinase A degradation in a MPLW515L mouse model of primary myelofibrosis. *Hematol. Oncol.* **2023**, *41*, 474–486. [CrossRef] [PubMed]
33. van Kalsbeek, R.J.; Hudson, M.M.; Mulder, R.L.; Ehrhardt, M.; Green, D.M.; Mulrooney, D.A.; Hakkert, J.; den Hartogh, J.; Nijenhuis, A.; van Santen, H.M.; et al. A joint international consensus statement for measuring quality of survival for patients with childhood cancer. *Nat. Med.* **2023**, *29*, 1340–1348. [CrossRef]
34. Rastogi, S.; Joshi, A.; Sato, N.; Lee, S.; Lee, M.J.; Trepel, J.B.; Neckers, L. An update on the status of HSP90 inhibitors in cancer clinical trials. *Cell Stress Chaperones* **2024**, *29*, 519–539. [CrossRef]
35. Saber, S.; Hasan, A.M.; Mohammed, O.A.; Saleh, L.A.; Hashish, A.A.; Alamri, M.M.S.; Al-Ameer, A.Y.; Alfaifi, J.; Senbel, A.; Aboregela, A.M.; et al. Ganetespib (STA-9090) augments sorafenib efficacy via necroptosis induction in hepatocellular carcinoma: Implications from preclinical data for a novel therapeutic approach. *Biomed. Pharmacother.* **2023**, *164*, 114918. [CrossRef]
36. Goldman, J.W.; Raju, R.N.; Gordon, G.A.; El-Hariry, I.; Teofilivici, F.; Vukovic, V.M.; Bradley, R.; Karol, M.D.; Chen, Y.; Guo, W.; et al. A first in human, safety, pharmacokinetics, and clinical activity phase I study of once weekly administration of the Hsp90 inhibitor ganetespib (STA-9090) in patients with solid malignancies. *BMC Cancer* **2013**, *13*, 152. [CrossRef] [PubMed]
37. Wu, Z.; Gholami, A.M.; Kuster, B. Systematic identification of the HSP90 candidate regulated proteome. *Mol. Cell Proteom.* **2012**, *11*, M111.016675. [CrossRef]
38. Workman, P.; Burrows, F.; Neckers, L.; Rosen, N. Drugging the cancer chaperone HSP90: Combinatorial therapeutic exploitation of oncogene addiction and tumor stress. *Ann. N. Y. Acad. Sci.* **2007**, *1113*, 202–216. [CrossRef]
39. Bhattacharyya, N.; Gupta, S.; Sharma, S.; Soni, A.; Bagabir, S.A.; Bhattacharyya, M.; Mukherjee, A.; Almalki, A.H.; Alkhanani, M.F.; Haque, S.; et al. CDK1 and HSP90AA1 Appear as the Novel Regulatory Genes in Non-Small Cell Lung Cancer: A Bioinformatics Approach. *J. Pers. Med.* **2022**, *12*, 393. [CrossRef]
40. Yang, Y.; Liu, Q.; Guo, X.; Yuan, Q.; Nian, S.; Kang, P.; Xu, Z.; Li, L.; Ye, Y. Systematic Pan-Cancer Analysis Identifies CDK1 as an Immunological and Prognostic Biomarker. *J. Oncol.* **2022**, *2022*, 8115474. [CrossRef]
41. Li, M.; He, F.; Zhang, Z.; Xiang, Z.; Hu, D. CDK1 serves as a potential prognostic biomarker and target for lung cancer. *J. Int. Med. Res.* **2020**, *48*, 300060519897508. [CrossRef]
42. Sunada, S.; Saito, H.; Zhang, D.; Xu, Z.; Miki, Y. CDK1 inhibitor controls G2/M phase transition and reverses DNA damage sensitivity. *Biochem. Biophys. Res. Commun.* **2021**, *550*, 56–61. [CrossRef]
43. Liu, H.; Lu, Z.; Shi, X.; Liu, L.; Zhang, P.; Golemis, E.A.; Tu, Z. HSP90 inhibition downregulates DNA replication and repair genes via E2F1 repression. *J. Biol. Chem.* **2021**, *297*, 100996. [CrossRef] [PubMed]

44. Chai, Z.; Wang, L.; Zheng, Y.; Liang, N.; Wang, X.; Zheng, Y.; Zhang, Z.; Zhao, C.; Zhu, T.; Liu, C. PADI3 plays an antitumor rol via the Hsp90/CKS1 pathway in colon cancer. *Cancer Cell Int.* **2019**, *19*, 277. [CrossRef]
45. Huang, Z.; Zhou, X.; He, Y.; Ke, X.; Wen, Y.; Zou, F.; Chen, X. Hyperthermia enhances 17-DMAG efficacy in hepatocellula carcinoma cells with aggravated DNA damage and impaired G2/M transition. *Sci. Rep.* **2016**, *6*, 38072. [CrossRef]

cancers

MDPI

ticle

argeting Hsp70 Immunosuppressive Signaling Axis with Lipid Janovesicles: A Novel Approach to Treat Pancreatic Cancer

hmet Kaynak [1], Subrahmanya D. Vallabhapurapu [1], Eric P. Smith [1], Harold W. Davis [1], Clayton S. Lewis [1], seph Ahn [1], Petr Muller [2], Borek Vojtesek [2], Keith F. Stringer [3], Robert S. Franco [1], Vladimir Y. Bogdanov [4], en-Hai Shao [5] and Xiaoyang Qi [1,*]

1 Division of Hematology & Oncology, Department of Internal Medicine, College of Medicine, University of Cincinnati, Cincinnati, OH 45267, USA; kaynakat@ucmail.uc.edu (A.K.); subrahmaya@gmail.com (S.D.V.); smithep4554@gmail.com (E.P.S.); harold.davis19@gmail.com (H.W.D.); lewis2c3@ucmail.uc.edu (C.S.L.); jha9504@gmail.com (J.A.); francors@ucmail.uc.edu (R.S.F.)

2 Masaryk Memorial Cancer Institute, Research Centre for Applied Molecular Oncology, Zluty Kopec 7, 656 53 Brno, Czech Republic; pmuller@post.cz (P.M.); vojtesek@mou.cz (B.V.)

3 Division of Pathology and Laboratory Medicine, Cincinnati Children's Hospital Medical Center, Cincinnati, OH 45229, USA; keith.stringer@cchmc.org (K.F.S.)

4 Center for Scientific Review, National Institutes of Health, Bethesda, MD 20892, USA; (formerly at University of Cincinnati) vladimir.bogdanov@nih.gov (V.Y.B.)

5 Division of Rheumatology, Allergy & Immunology, Department of Internal Medicine, College of Medicine, University of Cincinnati, Cincinnati, OH 45267, USA; shaowi@ucmail.uc.edu (W.-H.S.)

* Correspondence: xiaoyang.qi@uc.edu (X.Q.); Tel.: +1-5135584025

Simple Summary: Pancreatic ductal adenocarcinoma (PDAC) is the twelfth most frequent cancer worldwide and the fourth leading cause of cancer-related deaths in the USA, and is projected to be ranked second by 2030. FOLFIRINOX or gemcitabine-abraxane, the current standard treatments for advanced PDAC, prolong survival by only several months in chemo-sensitive patients. Thus, there remains an urgent medical need to develop novel approaches to treat PDAC. Since the PDAC tumor microenvironment (TME) is highly immunosuppressive with M2 polarized macrophages (MΦ), inhibition of immunosuppressive M2 polarization offers a potentially robust approach to reactivate the immune system against PDAC. In this study, we developed a Saposin C (SapC)-coupled dioleoylphosphatidylglycerol (SapC-DOPG) that binds secreted heat shock protein 70 (Hsp70), which was recently shown to be secreted by PDAC cells and to be a potent mediator of TME immunosuppression. Our preclinical studies using mouse models of pancreatic cancer show that SapC-DOPG blocks Hsp70 actions and inhibits tumor growth. These preclinical studies suggest a promising immunotherapeutic approach to treat PDAC.

Abstract: Background: Despite many efforts to effectively treat PDAC, PDAC carries one of the highest mortality rates of all major cancers. Thus, there is a critical unmet need to develop novel approaches to improve the clinical outcome of PDAC. It is well known that many cancers, including PDAC, generate a local TME that allows cancer to escape normal immune surveillance. Phosphatidylserine (PS), a negatively charged phospholipid that is abundant on the cancer cell membrane and with known actions to promote the secretion of immunomodulatory proteins, may provide a mechanism to regulate the TME. This study explored that possibility. Methods: MΦ differentiation and polarization were assessed by Western blotting and flow cytometric approaches. PS exposure and surface markers were analyzed by flow cytometry. Protein–protein and protein–lipid interactions were analyzed by immunofluorescence and enzyme-linked immunosorbent assay (ELISA). Phospholipid and SapC-DOPG treatment were employed to assess target protein functions in MΦ polarization, tumor growth, and survival in subcutaneous and orthotopic tumor models. The PK-PD and safety of SapC-DOPG were tested on orthotopic mouse models.

Results: Our studies show that PDAC secretes Hsp70 that stimulates the MΦ polarization to the immunosuppressive M2 phenotype. We found that high surface PS on cancer cells correlates with increased secretion of Hsp70 and is associated with higher MΦ differentiation activity in vitro and in vivo. Furthermore, blocking cancer cell-secreted Hsp70 with SapC-DOPG reverses the immune suppression and reduces tumor growth. Conclusions: These preclinical results reveal a novel immunotherapeutic approach to potentially improve the outcome of PDAC treatment in humans.

Keywords: cancer cell-secreted Hsp70; phosphatidylserine; M2 macrophage polarization; dioleoylphosphatidylglycerol (DOPG); Saposin C-coupled dioleoylphosphatidylglycerol (SapC-DOPG); immunotherapy; targeting therapy; pancreatic cancer

1. Introduction

Cancer cells have evolved mechanisms to establish an immunosuppressive tumor microenvironment (TME) that promotes escape from host immune surveillance, resulting in sustained tumor growth [1–4]. Reversal of this immunosuppression is an increasingly employed modality for cancer treatment [1,3,5,6]. The TME is composed of a variety of cells, soluble factors, and tumor-associated macrophages (TAMs). TAMs are mostly immunosuppressive M2 polarized macrophages (MΦs) derived from resident MΦs.

Recent studies have demonstrated that the inhibition of M2 MΦ polarization through interleukin (IL)-27 leads to a reduction in the proliferation, migration, and metastatic capacity of pancreatic cancer cells in vitro [7]. Furthermore, colony stimulating factor 1 receptor (CSF-1R) inhibitors, which selectively target CSF-1R, and are predominantly expressed by monocytes and macrophages, result in reduced M2 MΦ polarization and enhanced survival in a pancreatic ductal adenocarcinoma (PDAC) mouse model [8]. Therefore, strategies that block the development of immunosuppressive M2 TAMs and facilitate immune activation against the tumor provide a promising therapeutic approach for pancreatic cancer. However, increasing the effectiveness of this approach will require (1) the identification of cancer cell-derived factors that regulate M2 MΦ polarization and (2) approaches to therapeutically block identified mediators.

Despite decades of efforts to develop effective therapies against PDAC, it continues to manifest high morbidity and mortality; ~90% of patients diagnosed with PDAC succumb within five years of diagnosis [9]. The resistance to treatment with the resulting high mortality rate is, in part, explained by a highly immune-suppressive TME [10]. The specific mechanisms underlying the unusually robust immunosuppressive nature of PDAC tumors and the secreted factors that comprise this uniquely inhibitory TME are poorly understood. Recent evidence from multiple investigators suggests phosphatidylserine (PS), a negatively charged phospholipid that is highly expressed in PDAC and other cancer cells compared to healthy cells [11], may have a role in regulating the TME [12]. High surface PS leads to immune suppression and the promotion of tumor growth [12], and the level of PS in cancer cells is highly associated with malignancy and metastatic ability [13]. Several groups, including ours, have demonstrated that cancer surface PS can be targeted by PS-directed drugs [14–16]. Saposin C (SapC) is a small, stable lysosomal protein whose physiological role is to activate glucocerebrosidase [17] and, in the context of cancer biology, has a higher binding affinity for PS on cell membranes in the acidic environment typical of tumors [18]. Predictably, SapC specifically targets cancer cells that have higher PS exposure on the outer cell membrane compared to healthy cells. Preclinical and early-phase human studies have shown that the lipid nanovesicle-based PS-targeting drug, SapC coupled with dioleoylphosphatidylserine (SapC-DOPS), localizes to PS-expressing tumor cells and tumor blood vessel endothelial cells [19]. SapC-DOPS is internalized into the cell through endocytosis. After fusion with lysosomes, SapC leads to the activation of acid sphingomyelinase and induces ceramide accumulation in the lysosomes. The elevated ceramide levels then initiate a cascade of events, including caspase activation, which ultimately leads to apoptosis [19,20]. SapC-DOPS has demonstrated a robust anti-tumor effect in solid tumors, including pancreatic cancer [14].

Another component of the TME recently discovered to influence the immune environment within the TME is heat shock protein 70 (Hsp70). Although Hsp70 was initially identified as an intracellular chaperone involved in the cellular stress response [21], Hsp70 is increasingly appreciated to have many other functions [22,23], including in cancer promotion [24,25]. Hsp70 secretion is thought to occur via a non-conventional mode involving lysosomal endosomes or by association with membrane rafts and other secretory proteins [26–28]. Although post-translational modifications of Hsp70 such as phosphorylation play critical roles in chaperone function [29,30], a role for Hsp70 phosphorylation in its secretion and or extracellular function is largely unknown. Hsp70 has been shown to be overexpressed in various malignancies [31,32] associated with increased cancer cell proliferation [29,31]. Of specific relevance to this study, elevated levels of both membrane-bound and circulating Hsp70 have been documented in patients diagnosed with pancreatic cancer [32]. In relation to a potential role in the TME, we recently published a study revealing that Hsp70, secreted by PDAC cancer cells, promotes M2 MΦ polarization through the induction of Mer receptor tyrosine kinase (MerTK) upregulation [33].

In this study, we sought to explore the molecular underpinnings of cancer cell innate immune cell crosstalk relevant to pancreatic cancers in which, not only a high surface PS cancer cell phenotype, but also increased expression of Hsp70, is observed. We identified a potentially significant mechanism through which PS operates in the context of PDAC. Elevated levels of surface PS on cancer cells appear to facilitate a non-conventional mode of secretion for Hsp70, thereby enhancing previously reported effects [33] of Hsp70 to promote immunosuppressive M2 MΦs monocyte differentiation. Importantly, we demonstrated a therapeutic strategy to prevent this M2 MΦs induction.

2. Materials and Methods

2.1. Cell Lines

The cell lines studied were primary human astrocytes (ScienCell Research Laboratories, Carlsbad, CA, USA), primary human pancreatic ductal epithelial (HPDE) (CVCL_0P38) cells, human pancreatic cancer cell lines (AsPC-1 (ATCC-CRL-1682), MiaPaCa-2 (ATCC-CRL-1420), cfPac1-Luc3, Rink-1 and KPC-1624), melanoma cell line (A2058) (ATCC-CRL-3601), lung cancer cell line (H1299 (ATCC-CRL-5803), LLC-GFP), glioblastoma cell lines (Gli36, U373 (ATCC HTB-17), U87ΔEGFR, LN229 (ATCC-CRL-2611)), and human leukemia monocytic cell line THP-1(ATCC-TIB202). Gli36 and U87ΔEGFR cells were a gift from Dr. Balveen Kaur to Dr. Qi; the human pancreatic cfPac1-Luc3 cells were a kind gift from Dr. O. Wildner to Dr. Qi; LLC-GFP (CSC-RRO525, Creative Biogene Inc., Shirley, NY, USA) cells were a kind gift from Dr. John C. Morris to Dr Qi. Rink-1 cells were a kind gift from Drs. Murray Korc and Fazlul Sarkar to Dr. Bogdanov; KPC-1624 cells were a kind gift from Dr. Jen Jen Yeh to Dr. Bogdanov.

We used a range of cell lines to investigate the correlation between PS, Hsp70 expression, and M2 polarization. We aimed to assess whether this relationship holds true across different types of cancer cells. Our findings showed that the correlation was indeed consistent across various high- and low-PS cell lines, including those from glioblastoma, melanoma, and PDAC. While these broader observations were important for our initial analysis, we chose to focus on PDAC in subsequent experiments because of the availability of appropriate mouse models for this cancer type, allowing us to explore the underlying mechanisms in vivo.

2.2. Cell Culture

J774 MΦ, KPC-1624, AsPC-1, cfPac-1 Luc, MiaPaCa-2, Rink-1, and THP-1 cells were cultured in RPMI with 25 mM HEPES. Human astrocytes were cultured in an astrocyte cell medium (ScienCell Research Laboratories, Carlsbad, CA, USA) supplemented with the provided growth factor supplements, FBS, and antibiotics. HPDE cells were cultured in a keratinocyte cell medium (Sigma-Aldrich, St. Louis, MO, USA) supplemented with the provided growth factor supplements, FBS and antibiotics. All other cell lines were cultured in DMEM, and all media were supplemented with 10% FBS and 1% penicillin/streptomycin. All cells were cultured in a 5% CO_2 incubator at 37 °C. Cells were routinely tested for mycoplasma contamination. No cross-contamination was observed in the cell lines, as determined by cellular morphology and growth parameters.

2.3. Generation of Exosome/Microparticle-Free Conditioned Media (EMD-CM) from Cancer Cell Lines

Cancer cell lines were grown in their respective media until 70% confluency in 10 cm Corning tissue culture plates (ThermoFisher, Waltham, MA, USA). Then, the media were removed, and the cells were washed twice with serum-free media to remove remnants of serum and dead cells and replenished with serum-free media. After 24 h, conditioned media were collected, centrifuged at 10,000× g to remove cellular debris, and followed by ultracentrifugation at 100,000× g to remove extracellular exosomes and microparticles (EMD-CM).

2.4. THP-1 Differentiation Assay

THP-1 cells (2×10^5) were cultured in 1 mL of EMD-CM adjusted to 100 μg of total cellular protein derived from the specified cancer cell lines. Control THP-1 cells were maintained in DMEM. Following a 24 h incubation period, both control and EMD-CM-treated cells were centrifuged and subsequently incubated with CD14-PE conjugated antibody (eBioscience, San Diego, CA, USA) and PI (ThermoFisher, Waltham, MA, USA) in 100 μL of FACS buffer for 30 min on ice. The cells were then washed with flow cytometry buffer (phosphate-buffered saline (PBS) with 2% FBS), and CD14 expression was evaluated using a BD Fortessa X-20 flow cytometer (BD Biosciences, Cambridge, UK). In experiments designed to assess the role of exosomes and microparticles in THP-1 differentiation, THP-1 cells were cultured for 24 h in unfractionated CM, EMD-CM, or with the exosome/microparticle fraction of the CM. Differentiation was assessed via flow cytometric quantification of CD14 expression as previously described.

2.5. Quantification of Hsp70 in Cancer Cells Conditioned Media by ELISA

The concentration of Hsp70 in the EMD-CM was measured using an Hsp70 ELISA Kit (ThermoFisher, Waltham, MA, USA) in accordance with the manufacturer's protocol.

2.6. Preparation of Phospholipid Lipid Vesicles and Treatment of THP-1 Cells

Phospholipid vesicle (PL LV) stock solutions were prepared at a concentration of 1 mM. To prepare the stock, 32 μL of dioleylphosphatidylglycerol (DOPG, Avanti Polar Lipids, Alabaster, AL, USA) (25 mg/mL), 28.8 μL of dioleoylphosphatidic acid (DOPA, Avanti Polar Lipids, Alabaster, AL, USA) (25 mg/mL), 32.5 μL of dioleoylphosphatidylserine (DOPS, Avanti Polar Lipids, Alabaster, AL, USA) (25 mg/mL), 31.5 μL of distearoylphosphatidylcholine (DSPC, Avanti Polar Lipids, Alabaster, AL, USA) (25 mg/mL), or 32 μL of distearoylphosphoglycerol (DSPG, Avanti Polar Lipids, Alabaster, AL, USA) (25 mg/mL) were added into separate glass tubes. The final volume was brought up to 1 mL with chloroform. For the preparation of PL LVs at 50 μM and 100 μM final concentrations, 50 μL and 100 μL of the 1 mM stock solution, respectively, were transferred into glass vials. The organic solvent (chloroform) was removed using nitrogen gas, and a dry lipid film was formed. Following film formation, 1 mL of PBS was added to each vial, and the mixture was bath sonicated in an ice-water bath for 30 min to generate PL LVs. THP-1 cells were treated for 24 h with the indicated cancer cell EMD-conditioned medium (EMD-CM) in the presence of 50 μM to 100 μM of the designated PL LVs.

2.7. Preparation of Saposin C-Coupled 1,2-Dioleoyl-sn-glycero-3-phospho-rac-(1-glycerol) (SapC-DOPG) Lipid Vesicles (LVs)

A quantity of 6.4 μL DOPG (25 mg/mL) was dried under a stream of N_2 gas and mixed with 0.4 mg recombinant SapC protein in 20 μL of citrate-phosphate buffer (pH 5.0). A quantity of 1 mL of PBS was added to the mixture. The Saposin C-coupled DOPG (SapC-DOPG) mixture was then gently sonicated in ice water for 30 min. To fluorescently label the SapC-DOPG lipid vesicles (LVs), we added 30 μL CellVue Maroon (CVM; 1 mM stock in ethanol; Molecular Targeting Technologies, West Chester, PA, USA) to DOPG and dried them together before the addition of SapC and 1 mL PBS. After bath sonication, unbound CVM was removed from the nanovesicle suspension by filtration through a Sephadex G25 column (PD-10; Amersham–Pharmacia Biotech, Piscataway, NJ, USA).

2.8. Characterization of Lipid Vesicles (LVs)

Particle sizes and zeta potential were measured by a Nano partica SZ-100V2 (Horiba, Irvine, CA, USA) nanoparticle analyzer.

2.9. Cell Sorting

The indicated cancer cell lines were stained with annexin V-FITC (Invitrogen, Waltham, MA, USA) and propidium iodide (PI, BD PharMingen, San Jose, CA, USA), according to the manufacturer's protocol. Briefly, 1×10^6 cells were incubated with annexin V binding buffer (Invitrogen, Waltham, MA, USA) together with PI for 30 min at room temperature. Subsequently, the cells were washed with annexin V buffer and resuspended in this buffer. Cells with low and high annexin V signals were then gated and sorted with a BD FACS Aria II (BD Bioscience, San Jose, CA, USA).

2.10. Immunofluorescence Staining

Cells were seeded on gelatin (0.01%)-coated coverslips. After a 5 h incubation, cells were washed twice with PBS. Then, the cells were incubated overnight at 4 °C with anti-Hsp70 antibody (Abcam, Cambridge, UK) followed by anti-rabbit IgG (H+L), F(ab')2 fragment AlexaFluor 555 (Cell Signaling Technologies, Danvers, MA, USA) for 1 h at room temperature. After washing with PBS, the cells were stained with FITC-conjugated Annexin V (ABP Biosciences, Rockville, MD, USA), and then mounted using Fluoro-gel II with DAPI and analyzed using a BX51 fluorescence microscope (Olympus, Tokyo, Japan).

2.11. Western Blotting Analyses

For Western blot analysis, 50 μg of whole cell lysates in RIPA buffer (Sigma-Aldrich, St. Louis, MO, USA) were denatured using sodium dodecyl sulfate (SDS)-loading dye (Bio-Rad Laboratories, Hercules, CA, USA) and subsequently loaded onto 4–15% denaturing polyacrylamide gradient gels (Bio-Rad Laboratories, Hercules, CA, USA). Following electrophoresis, proteins were transferred to nitrocellulose membranes. The membranes were blocked with 5% non-fat dry milk in PBS with 0.1% Tween-20, after which protein-specific antibodies were added, and the membranes were incubated overnight at 4 °C. The blots were washed three times with PBS-Tween-20 and then incubated with HRP-coupled secondary antibodies. After washing with PBS-Tween-20 three times, the blots were developed using SuperSignal West Dura (ThermoFisher, Waltham, MA, USA). The images were developed at Biorad GelDoc XR+ (Bio-Rad Laboratories, Hercules, CA, USA).

2.12. Cell Viability Assay

The MTT (3-[4,5-dimethylthiazol-2yl]-2,5-diphenyl-tetrazolium bromide) assay was performed using the manufacturer's protocol (Roche Applied Science, Mannheim, Germany) to evaluate cell viability. Briefly, cancer cells (5×10^3) were seeded onto 96-well plates and incubated overnight in a 5% CO_2 incubator at 37 °C. The cells were treated with DOPG or SapC-DOPG LVs for 48 h, and a viability assay was performed.

2.13. Subcutaneous Cancer Cell Implantation and DOPG Lipid Vesicle (LVs) Treatment

Lewis lung carcinoma cells expressing green fluorescent protein (LLC-GFP) (1×10^5) were subcutaneously implanted in 6–8-week-old C57BL/6J mice, with five mice per group, comprising both male and female. The treatment group received DOPG nanoparticles in 200 μL of PBS at a dosage of 1.6 mg/kg per mouse, while the control group received 200 μL of PBS per mouse. Treatment commenced for two days post-implantation and was administered three times a week until the conclusion of the study. Tumor growth was monitored daily by measuring tumor volume with Vernier calipers. Tumor volumes were calculated using the formula $V = (\pi/6)LW^2$ (where V is volume, L is length, and W is width). Tumors were harvested once the volume reached 500 mm^3 for subsequent histological and tumor MΦ analyses.

2.14. Generation of Orthotopic PDAC Tumors in Mice

Rink-1 or KPC-1624 cells (2×10^6) were injected into the pancreata of 6–8-week-old C57BL/6J mice (10 mice per group, comprising both male and female). The treatment regimen involved the administration of DOPG liposomes (in 200 μL PBS, DOPG dosage of 1.6 mg/kg per mouse), SapC-DOPG liposomes (in 200 μL PBS, containing SapC at 4 mg/kg and DOPG at 1.6 mg/kg per mouse), or control saline (200 μL saline per mouse) via tail vein injection. Treatments started one day after the implantation and were administered three times a week until the treatment was completed. For the KPC-1624 and Rink-1 models, 5 and 10 mice were used per treatment group, respectively. All animal procedures followed the guidelines set forth by the National Research Council's "Guide for the Care and Use of Laboratory Animals" and were approved by the University of Cincinnati Institutional Animal Care and Use Committee (IACUC) (Protocol No. 22-10-25-01).

2.15. Phospholipids and SapC-DOPG LVs Binding Assays

PL and SapC-DOPG LVs were prepared as described above, except that HEPES buffered saline was used instead of PBS. Cancer cell EMD-CMs were incubated with indicated PL and SapC-DOPG LVs with the indicated concentration for 3 h rotating at RT, then ultracentrifuged at 100,000× *g* for 1 h. Both pellet and supernatant fractions were used for ELISA. Prior to the analysis of the pellets, the pellets were incubated with ice-cold lysis buffer to release the protein.

2.16. PK/PD and Tissue Distribution of SapC-DOPG LVs

SapC-DOPG-CVM LVs was generated using DOPG labeled with a CVM. SapC-DOPG-CVM LVs were assembled using 0.13 mg SapC, 0.08 mg DOPG, and 30 μg CVM per mouse and were used to study organ distribution in mice after intravenous injection into orthotopic PDAC tumor-bearing C57BL/6 mice at different time points. A PDAC tumor model was established as mentioned above. After the tumor was established, tumor-bearing C57BL/6 mice were organized into three mice/group for seven different time points: 10 min, 30 min, 1, 2, 4, 24, and 48 h. SapC-DOPG-CVM LVs (0.13 mg SapC, 0.08 mg DOPG, and 30 μg CVM per mouse) were administered through the tail vein. Blood and organs (brain, ovary, prostate, pancreas, liver, lung, intestine, and heart) were collected at the indicated time points. The Spectrum In Vivo Imaging System (PerkinElmer, Hopkinton, MA, USA) was used for imaging with the Cy5 filter ranging from 640 to 680 nm. Non-compartmental analysis was performed using PKanalix software (version 2019R2, Lixoft SAS, Antony, France). Maximum observed plasma concentration (i.e., C_{max}), the time taken to reach C_{max} (i.e., T_{max}), the time required to reduce the concentration of drug by half (i.e., $T_{1/2}$), and the average time a molecule stays in the body (i.e., MRT) were determined directly from the individual data. The cumulative area under the plasma concentration–time curve (AUC) from time zero to infinity (i.e., $AUC_{0\text{-}inf}$) was derived from the clearance and the dose. AUC was evaluated using the trapezoidal rule.

2.17. Toxicity Studies of SapC-DOPG LVs

SapC-DOPG LVs were prepared as mentioned above. C57BL/6 mice at six weeks of age were randomly grouped into two groups of six mice each (three males and three females/group). One group was injected via the tail vein with SapC-DOPG LVs (SapC = 34.4 mg/kg and DOPG = 21.4 mg/kg). Human equivalent dose (HED) was calculated based on body surface area using the equations (Equations (1) and (2)) [34] shown below:

HED: Human equivalent dose;

Km: Correction factor;

NOAEL: No adverse effect levels of the drug from preclinical toxicological studies were observed.

$$HED\left(\frac{\text{mg}}{\text{kg}}\right) = Animal\ NOAEL\left(\frac{\text{mg}}{\text{kg}}\right) \times (Weight_{animal}[\text{kg}]/Weight_{human}[\text{kg}])^{(1-0.67)} \quad (1)$$

$$HED\left(\frac{\text{mg}}{\text{kg}}\right) = Animal\ does\left(\frac{\text{mg}}{\text{kg}}\right) \times (Animal\ K_m/Human\ K_m) \quad (2)$$

The NOAEL of SapC-DOPS, which is in Phase II clinical trials, was used as a reference. The injections were administered daily for the first week and then every two days on the following week. Mouse weights were measured daily, and blood was collected from all mice after the second week. The mice were euthanized after 21 days, and their organs were collected for toxicity analyses.

2.18. Quantitative Analysis of Fluorescence Microscopy Images

Image J software (version 1.38e, NIH, Bethesda, MD, USA) was used to calculate the area of total cells and colocalization of PS and Hsp70. Ten randomly selected cells were used for calculations. Statistical analysis was performed using GraphPad Prism 6 Software (GraphPad Software Inc., San Diego, CA, USA). The data were analyzed using a Student's paired *t*-test or ANOVA followed by Bonferroni's post hoc test.

2.19. Hematological Analyses

Blood samples were collected from six mice, comprising three males and three females, utilizing the retro-orbital method. Briefly, a capillary tube was carefully inserted behind each eye into the retrobulbar venous sinus, facilitating blood collection through capillary action into a collection tube. A total of 20 μL of blood was obtained from each mouse, and hematological analysis was conducted using the Hemavet 950 (Drew Scientific, Plantation, FL, USA).

2.20. Statistical Analysis

All statistical analyses were conducted using GraphPad Prism 6 software. The data were analyzed using a Student's paired t-test or ANOVA followed by Bonferroni's post hoc test. In vitro experiments were performed two to three times. In vivo experiments were conducted at least three times. * $p < 0.05$, ** $p < 0.01$, *** $p < 0.001$, and **** $p < 0.0001$.

3. Results

3.1. PS^{high} Cancer Cells Promote Monocyte Differentiation and M2 Polarization via Secreted-Hsp70

In our previous study, we identified a non-chaperone function of cancer cell-secreted Hsp70. Our results revealed that cancer cell-secreted Hsp70 induces MerTK upregulation and promotes polarization of immunosuppressive M2 macrophages [33]. The promotion of an immunosuppressive TME by Hsp70 released from cancer cells suggests a highly regulated process. Hsp70 is thought to be secreted through a non-classical mechanism involving membrane rafts [35]. Previous studies also showed that PS derived from apoptotic tumor cells correlated with the M2-like phenotype of peritoneal MΦs in vitro [36]. Given that PS is highly expressed on the surface of cancer cells [37,38], correlated with MΦ polarization, and documented to capture Hsp70 [39], we examined the correlation between cancer cell surface PS, the secretion of Hsp70, and MΦ differentiation. Cancer cells were grouped into PS^{low} and PS^{high} lines by annexin V staining (Figure 1A). As expected, EMD-CMs from PS^{high} cells induced robust THP-1 differentiation based on CD14 expression compared to PS^{low} cells such as primary human astrocytes, primary human pancreatic duct epithelial (HPDE) cells, and AsPC-1 cancer cells (Figure 1B). A positive correlation was evident between cancer cell surface PS and CD14 expression induced on the MΦs ($r^2 = 0.8396$, $p = 0.0006$) (Figure 1C). Using a Hsp70 ELISA, we assessed whether PS^{low} and PS^{high} cancer cells differed in the amount of secreted Hsp70 and showed that released Hsp70 levels strongly correlated with cell surface PS (Figure 1D; $r^2 = 0.6747$, $p = 0.0066$). Moreover, Hsp70 correlated with THP-1 differentiation (Figure 1E; $r^2 = 0.8044$, $p = 0.001$).

Since PS^{high} cells secreted more Hsp70, we assessed whether Hsp70 and membrane PS colocalize on the cell surface. We compared the extent of colocalization of PS and Hsp70 in HPDE, PS^{low} AsPC-1 cells, PS^{high} MiaPaCa-2, Gli36, and U87ΔEGFR cells. Colocalization of Hsp70 and PS on the cell membrane was greater in PS^{high} cells (Gli36, U87ΔEGFR, and MiaPaCa-2) (Figure 1F and Figure S1) compared to normal HPDE cells and PS^{low} cells (AsPC-1) (Figure 1F and Figure S1). Interestingly, the secreted Hsp70 was phosphorylated at T636 and tyrosine residues and more abundant in PS^{high} cell culture media (Figure 1G). Notably, J774 MΦ cells treated with the EMD-CMs from PS^{high} cells (Gli36 and LLC-GFP) showed marked increases in the M2 polarization markers, arginase 1, and transglutaminase 2 (TGM2), with corresponding decreases in the M1 markers NOS2(iNOS) and SOCS3, compared to PS^{low} cell lines AsPC-1 and H1299 (Figure 1H). These results support a positive correlation between cancer cell surface PS and secreted Hsp70.

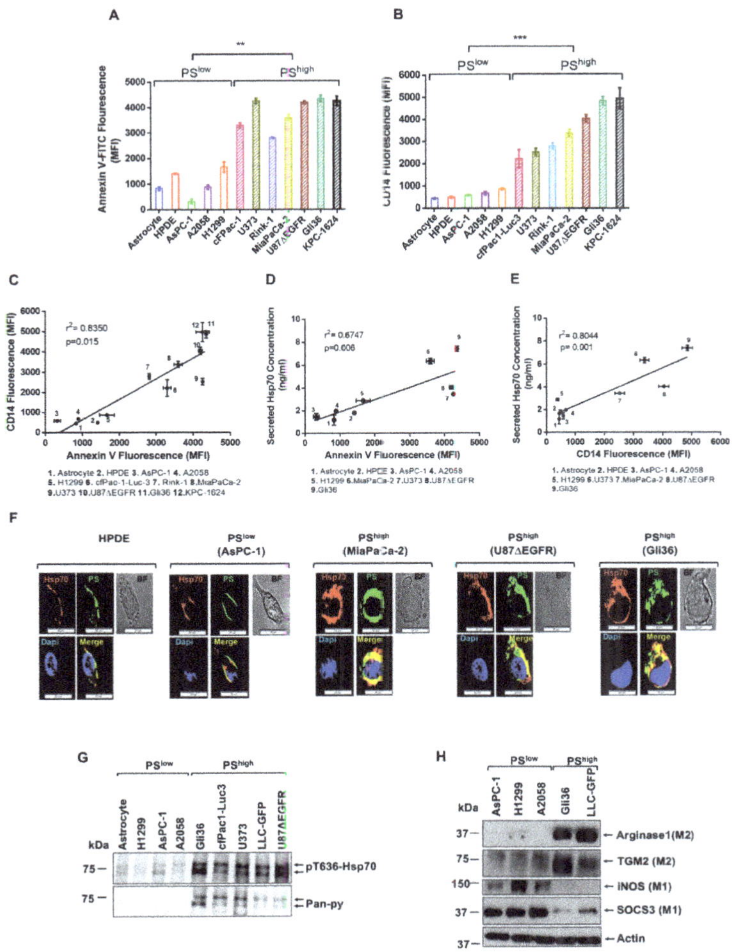

Figure 1. High-PS (PS^{high}) cancer cell EMD-CM induces THP-1 differentiation and MΦ M2 polarization. (**A**) Flow cytometric measurement of annexin V on indicated primary cells and cancer cell lines. (**B**) Flow cytometric measurement of CD14 on THP-1 cells cultured for 24 h with EMD-CM from PS^{low} and PS^{high} cancer cells. (**C**) Correlation between surface PS and THP-1 differentiation ability from indicated primary and cancer cell lines, as measured by flow cytometric measurement of annexin V on cancer cell lines and CD14 on EMD-CM-treated THP-1 cells. (**D**) Correlation between surface PS and secreted Hsp70 from indicated primary and cancer cell lines, as measured by flow cytometric measurement of annexin V and ELISA-quantified Hsp70 from the EMD-CM of indicated cells. (**E**) Correlation between THP-1 differentiation ability and secreted Hsp70 from indicated primary and cancer cell lines, as measured by flow cytometric measurement of CD14 on EMD-CM-treated THP-1 cells and ELISA-quantified Hsp70 from the EMD-CM of indicated cells. (**F**) Immunofluorescence microscopic analyses of surface expression of Hsp70 (red), surface PS (green), and their colocalization in HPDE, AsPC-1, MiaPaCa-2, U87ΔEGFR, and Gli36 cells. Yellow and blue colors represent Hsp70-PS colocalized area and nuclei, respectively. The bright field image is shown to the right of each panel. (**G**) Western blot analyses of the EMD-CM from indicated PS^{low} and PS^{high} cell lines probed with indicated mAbs. (**H**) Western blot analyses of J774 cell lysates made from cells cultured for 24 h with the EMD-CM from indicated cancer cell lines probed with indicated mAbs. The experiments were repeated at least twice. ** $p < 0.01$, and *** $p < 0.001$. Original western blots are presented in Supplementary Materials File Blots (File S1).

Recent studies have revealed that tumor cells are highly heterogeneous with distinct cellular subsets and distinct functions [40,41]. We hypothesized the presence of subsets of cells that differ in surface PS and secretion of Hsp70. To test this possibility, PS^{high} cell lines were sorted by flow cytometry after annexin V staining into two subgroups, $PS^{high-low}$ and $PS^{high-high}$ (Figure 2A). Controls were DMEM, unsorted, and a mixture of the two sorted subgroups from individual cancer cell lines (cfPac1-Luc3, Gli36, and LLC-GFP). The $PS^{high-high}$ cells exhibited higher THP-1 differentiation capacity compared to $PS^{high-low}$ cells from cfPac1-Luc3, Gli36, and LLC-GFP (Figure 2B–D). Furthermore, we observed that $PS^{high-high}$ LLC-GFP cells induced stronger M2 polarization when injected into the peritonea of C57BL/6J mice compared to $PS^{high-low}$ LLC-GFP cells (Figure 2E,F). The sorted $PS^{high-high}$ LLC-GFP cells initiated tumors faster than $PS^{high-low}$ cells when injected subcutaneously into C57BL/6J mice (Figure 2G). These findings reveal the correlation between surface PS and CD14 expression and tumor initiation.

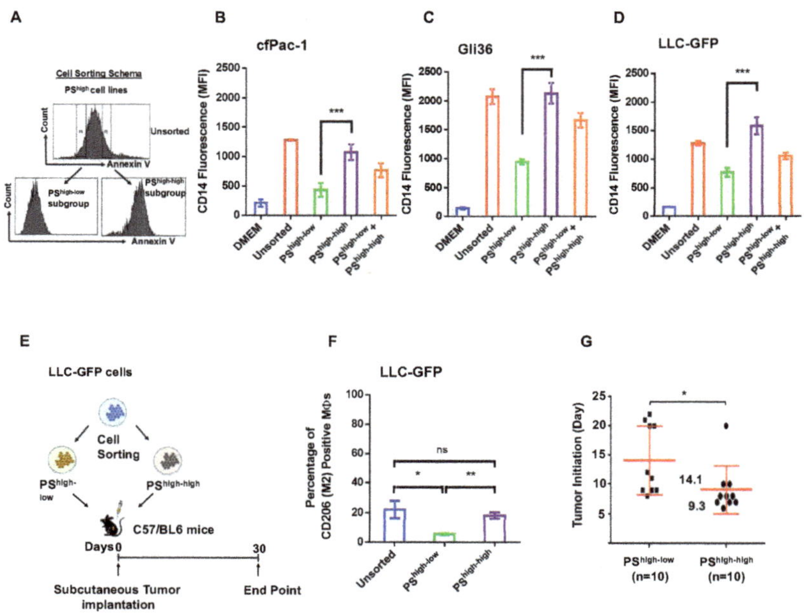

Figure 2. Sorted $PS^{high-high}$ cells from individual cancer cell lines induce strong THP-1 differentiation and M2 MΦ polarization, and initiate tumor faster. (**A**) Scheme showing flow cytometric sorting of cells. Flow cytometric measurement of CD14 on THP-1 cells cultured with DMEM medium, unsorted, sorted $PS^{high-low}$, $PS^{high-high}$ cells, and a mix of both $PS^{high-low}$ and $PS^{high-high}$ cells from cfPac-Luc3 (**B**), Gli36 (**C**), and LLC-GFP (**D**) cell lines. (**E**) Scheme showing subcutaneous implantation in mice with sorted $PS^{high-low}$ or $PS^{high-high}$ LLC-GFP cells. (**F**) CD206 expression on mouse peritoneal MΦs polarization in response to unsorted, $PS^{high-low}$, and $PS^{high-high}$ LLC-GFP cells, measured by flow cytometry. (**G**) Tumor initiation in mice by sorted $PS^{high-low}$ and $PS^{high-high}$ LLC-GFP cells. The experiments were repeated at least twice. ns: not significant, * $p < 0.05$, ** $p < 0.01$, and *** $p < 0.001$.

3.2. DOPG Lipid Vesicles (LVs) Block EMD-CM-Induced Monocyte Differentiation and Increase Survival in Mice

Our previous study demonstrated that Hsp70, secreted by cancer cells, interacts with toll-like receptor 2 (TLR2) and triggers the upregulation of MerTK, thereby stimulating M2 MΦ polarization and contributing to tumor growth [33]. The identification of this novel immunosuppressive pathway suggests innovative therapeutic potential. Hsp70 is known to bind phospholipids (PLs) [39,42]. Thus, we hypothesized that this might serve to inhibit the extracellular bioactivity of phosphorylated Hsp70. Consequently, we analyzed LVs composed of a panel of PLs differing in saturation and head groups for their ability to block THP-1 differentiation induced by cancer cell EMD-CM. MiaPaCa-2 EMD-CM cultured THP-1 cells treated with DOPG LVs showed substantial dose-dependent inhibition of CD14 expression (Figure 3A,B). In contrast, DOPA, DOPS, DSPC, and DSPG LVs elicited only mild reductions in CD14 expression (Figure 3A). Since PI staining of THP-1 cells treated with 100 μM DOPG LVs was comparable to PI staining of THP-1 cells without LV treatment (Figure 3C), the observed effect of DOPG LVs on THP-1 differentiation is unlikely to be attributable to toxicity.

Next, we measured the particle size and zeta potential of LVs to evaluate their physical properties. The results revealed that the average size of DOPG LVs was 258 nm (standard deviation (S.D): 42.1 nm) (Table S1). In contrast, SapC-DOPG LVs were significantly smaller, with an average size of 116 nm (S.D: 40 nm). The smaller size of the SapC-DOPG LVs suggests that SapC may be influencing lipid organization, likely by reducing aggregation or enhancing vesicle formation, leading to smaller, more stable particles. Smaller vesicles are characterized by enhanced cellular uptake, easier penetration of biological barriers, and improved stability. We also evaluated the zeta potential of LVs. The zeta potential of DOPG was −10.1 mV. After coupling with SapC, the zeta potential became less negative, at −7.4 mV. This indicates that SapC leads to reduced electrostatic repulsion between vesicles, suggesting that the coupling of SapC has altered the surface charge of DOPG.

We then tested whether DOPG LVs bind to Hsp70 present in EMD-CM. After incubation of MiaPaCa-2 EMD-CM with PL LVs, we applied ultracentrifugation to separate LVs-bound-Hsp70 in the pellet from its free form in the supernatant. ELISA results demonstrated that Hsp70 bound specifically to DOPG LVs compared to the other PL LVs, as shown by a significant increase of Hsp70 in the DOPG LVs pellet (Figure 3D, left panel) and its reduction in the supernatant (Figure 3D, right panel). These results suggested that DOPG LVs may sequester Hsp70 secreted by cancer cells.

Next, the efficacy of the DOPG LVs was tested in a subcutaneous mouse model (Figure 3E). Intravenously administered DOPG LVs did not alter mouse body weight but substantially reduced tumor growth compared to sham mice (Figures S2 and 3F). Importantly, the reduction in tumor growth by DOPG LVs was associated with a significant reduction in the intra-tumor M2 polarized MΦs (Figure 3G). To support the possibility that the immunosuppressive effect of DOPG is at least in part due to binding to cancer cell-secreted Hsp70 (Figure 3D, left and right panels), we used a syngeneic PDAC orthotopic tumor model to test the efficacy of DOPG LVs. The KPC-1624 cell line was used for in vivo study because of its high surface PS and robust CD14-inducing activity (Figure 1A,B). Briefly, 2×10^6 KPC-1624 cells were implanted into C57BL/6 mice pancreata. The mice were treated with DOPG LVs or saline every other day (Figure 3H). A Kaplan–Meier survival analysis demonstrated that DOPG LVs (26 days of median survival) significantly increased the survival of mice compared to the saline control (23 days of median survival) (Figure 3I). Cell viability analysis showed that 100 μM DOPG LVs showed no killing effect in KPC-1624 cells in vitro (Figure 3J).

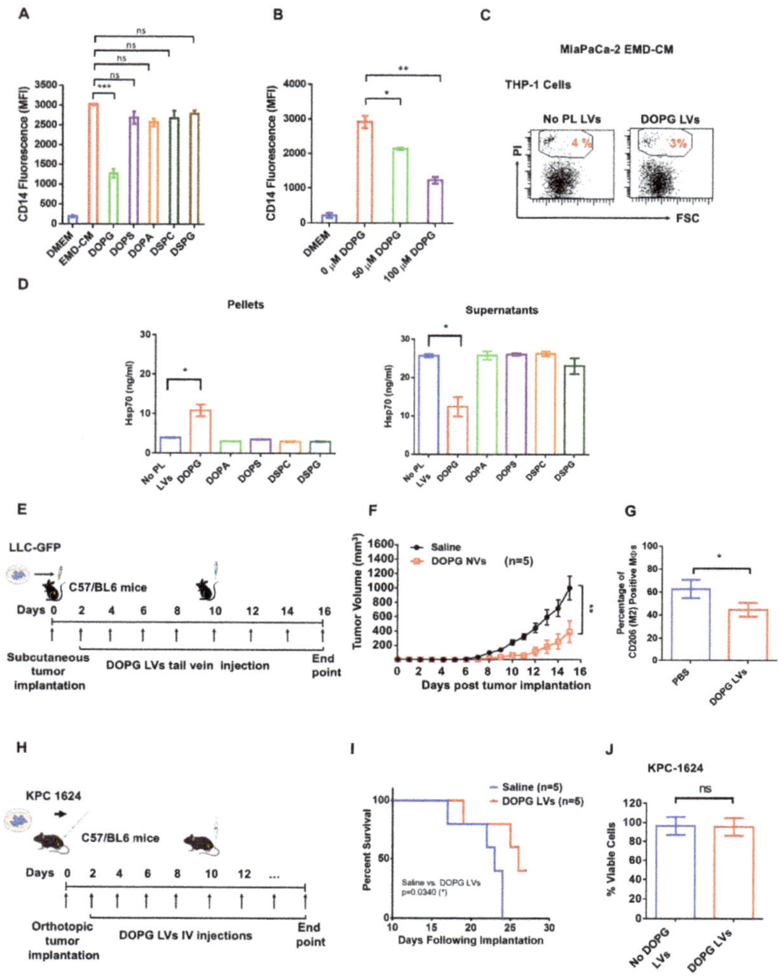

Figure 3. DOPG LVs reduce M2 MΦ, inhibit tumor growth, and increase overall survival in subcutaneous and PDAC orthotopic mouse models. CD14 expression of THP-1 cells incubated with MiaPaCa-2 EMD-CM and indicated LVs at 100 µM (**A**) and DOPG LVs with indicated concentrations (**B**). (**C**) Flow cytometric dot plots showing the percentage of PI positive cells upon incubation of THP-1 cells with MiaPaCa-2 EMD-CM and MiaPaCa-2 EMD-CM together with 100 µM of the DOPG LVs. (**D**) Quantification of Hsp70 by ELISA in pellets (left panel) and supernatant (right panel) obtained after 3 h incubation of MiaPaCa-2 EMD-CM with 100 µM of the indicated PL LVs and ultracentrifugation at 100,000× *g* for 1 h. (**E**) Schematic showing implantation of LLC-GFP cells into mouse flank and DOPG LV treatment regimen and (**F**) tumor volume in mice in response to PBS injection or DOPG LV injection. DOPG concentration was 1.6 mg/kg per mouse. (**G**) Percentage of intra-tumoral MΦs that are CD206-positive (gated on F4/80-positive cells) in mice injected with PBS/DOPG LVs. (**H**) The schematic showing orthotopic implantation of KPC-1624 cells into the mouse pancreas and the DOPG LVs treatment regimen. DOPG concentration was 1.6 mg/kg per mouse. (**I**) The Kaplan–Meier survival curve of the KPC-1624 orthotopic mice treated with saline or DOPG LVs. (**J**) MTT viability analysis of 0 and 100 µM DOPG LVs on KPC-1624 cells. The experiments were repeated at least twice. ns: not significant, * $p < 0.05$, ** $p < 0.01$, and *** $p < 0.001$.

3.3. SapC-DOPG LVs Block EMD-CM-Induced Monocyte Differentiation and Inhibit Tumor Growth

Although our toxicity and viability results suggested that DOPG had minimal toxicity to either THP-1 cells or KPC-1624 PDAC cells (Figure 3C,J), studies by others showed that DOPG could induce apoptosis in RAW 264.7 murine macrophage-like cells by air oxidation in vitro [43]. Therefore, we prepared SapC-DOPG LVs. Importantly, the safety of SapC-DOPS (BXQ-350) has already been established in a Phase I clinical trial. First, we tested whether SapC-DOPG LVs block THP-1 differentiation induced by cancer cell EMD-CM. Incubation of MiaPaCa-2 EMD-CM-treated THP-1 cells with 100 μM SapC-DOPG LVs led to substantial inhibition of CD14 expression on THP-1 cells (Figure 4A). Toxicity and cell viability results showed that SapC-DOPG had no toxicity to either THP-1 or Rink-1 PDAC cells (Figure 4B,C). Next, we assessed whether SapC-DOPG LVs bind to cancer cell-secreted Hsp70. ELISA assays demonstrated that SapC-DOPG LVs bound to cancer cell-secreted Hsp70 as indicated by a significant increase of Hsp70 in the ultracentrifuged pellet (Figure 4D, left panel) and its reduction in the supernatant (Figure 4D, right panel). Further, the efficacy of SapC-DOPG LVs was evaluated in a Rink-1 PDAC orthotopic mouse model. Briefly, 2×10^6 Rink-1 cells were orthotopically implanted into the pancreate of C57BL/6 mice. The mice were treated with SapC-DOPG LVs, saline, or SapC every other day for 16 days (Figure 4E) and euthanized, and the tumors were dissected and weighed. Results showed that orthotopic PDAC tumor growth was reduced by about 60% in SapC-DOPG LV-treated mice compared to saline or SapC-treated mice (Figure 4F).

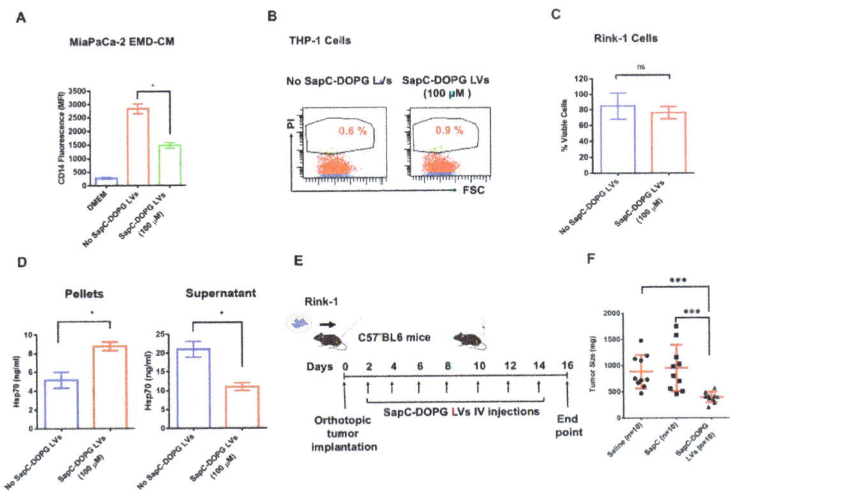

Figure 4. SapC-DOPG LVs block EMD-CM-induced monocyte differentiation and inhibit tumor growth in a PDAC orthotopic mouse model. (**A**) CD14 expression of THP-1 cells incubated MiaPaCa-2 EMD-CM with SapC-DOPG LVs at 0 and 100 μM concentrations. (**B**) Flow cytometric dot plots showing the percent of PI positive cells upon incubation of THP-1 cells with MiaPaCa-2 EMD-CM and MiaPaCa-2 EMD-CM together with 0 and 100 μM of SapC-DOPG LVs. FCS stands for forward scatter. (**C**) MTT viability analysis of 0 and 100 μM SapC-DOPG LVs on Rink-1 PDAC cells. (**D**) Quantification of Hsp70 by ELISA in pellets (left panel) and supernatant (right panel) obtained after 3 h incubation of MiaPaCa-2 EMD-CM with 0 and 100 μM SapC-DOPG LVs and ultracentrifugation at 100,000× g for 1 h. (**E**) The schematic showing orthotopic implantation of Rink-1 cells into the mouse pancreata and the SapC-DOPG LVs treatment regimen. SapC-DOPG concentration was SapC = 4 mg/kg and DOPG = 1.6 mg/kg per mouse. (**F**) Orthotopic tumor weights from mice. The experiments were repeated at least twice. ns not significant, * $p < 0.05$, and *** $p < 0.001$.

3.4. SapC-DOPG LVs Target Pancreatic Tumors Without Body Weight Loss or Organ Toxicity in Mice

SapC demonstrates a greater binding affinity for phosphatidylserine (PS) on cell membranes in the acidic environment typical of tumors [18]. It has been shown that SapC-DOPS localizes to PS-expressing tumor cells and tumor blood vessel endothelial cells [18,19]. To test whether Sap-DOPG LVs specifically target tumors in PDAC orthotopic mice, 2 $\times$ 10^6 Rink-1 cells were implanted orthotopically into C57BL/6 mice. After the tumor was established, fluorescently labeled SapC-DOPG-CVM LVs were administered through the tail vein. The blood and body organs were collected at the indicated time points (Figure 5A). Fluorescent intensity of blood and organs was measured and quantified using the Spectrum in vivo Imaging System. Pancreatic tumors showed a gradual increase in CVM signal from SapC-DOPG-CVM LVs, which reached a maximum by 24–48 h, suggesting specific targeting to the tumor (Figure 5B, top and bottom panels). The pharmacokinetic profile of SapC-DOPG-CVM-LVs indicated that the peak concentration of SapC-DOPG-CVM-LVs in the tumor was 3.08 $\times$ 10^9 (C_{max}) after 40 h (T_{max}) (Table 1). An initial peak of CVM signal intensity in the liver 1 h after SapC-DOPG-CVM LV injection was observed (Figure 5C top and bottom panels). This was diminished and reached the background level by 48 h (Figure 5C, top and bottom panels). A sharp peak of CVM signals in blood was detected within 1 h after injection, followed by a rapid decline within 4 h, and reaching non-detectable levels by 48 h (Figure 5D).

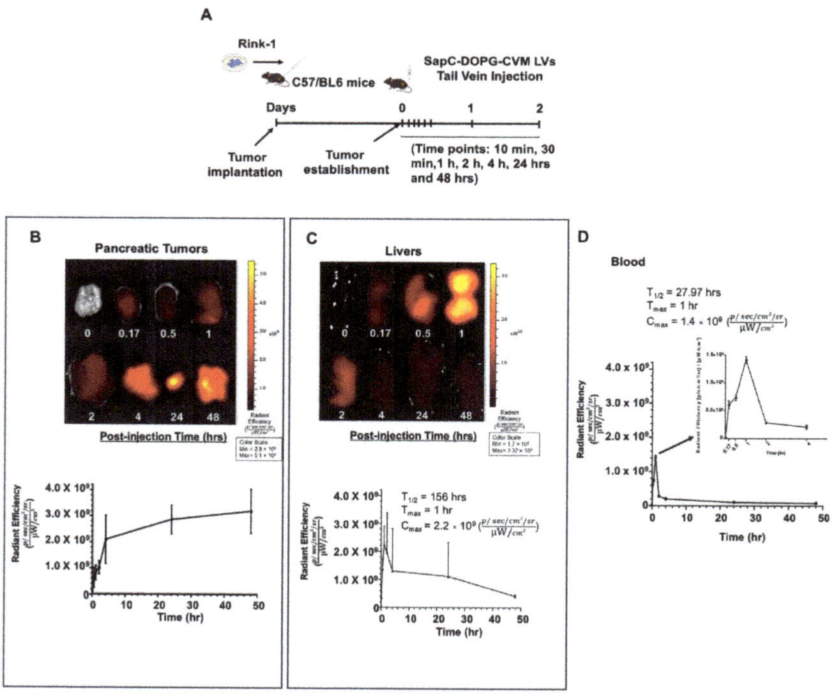

Figure 5. SapC-DOPG LVs target pancreatic tumors. (**A**) Schematic showing orthotopic implantation of Rink-1 cells into mouse pancreas and SapC-DOPG-CVM LV treatment regimen (SapC 6.4 = mg/kg, DOPG = 4 mg/kg, CVM = 0.3 mg/kg per mouse.) (**B**) Fluorescent images of SapC-DOPG-CVM from

PDAC tumors at the indicated time points between 0 and 48 h (top panel). The signal of SapC-DOPG-CVM from PDAC tumors at the indicated time points between 0 and 48 h (bottom panel). The signal was normalized based on the area. (**C**) Fluorescent images of SapC-DOPG-CVM from mouse livers at the indicated time points between 0 and 48 h (top panel). The signal of SapC-DOPG-CVM from mouse livers at the indicated time points between 0 and 48 h (bottom panel). The signal was normalized based on the area. (**D**) The signal of SapC-DOPG-CVM from the blood of mice at the indicated time points between 0 and 48 h. The signal was normalized based on the area. The experiments were repeated at least twice.

Table 1. Pharmacokinetic/pharmacodynamic (PK/PD) analysis of pancreatic tumors and organs of mice treated with SapC-DOPG.

Group	AUC_{INF}	AUC_{last}	C_{max}	T_{max}	$T_{1/2}$	MRT
	Radiant Efficiency [p/s/cm^2/sr]*h/[μW/cm^2]	**Radiant Efficiency [p/s/cm^2/sr]*h/[μW/cm^2]**	**Average Radiant Efficiency [p/s/cm^2/sr]/[μW/cm^2]**	**h**	**h**	**h**
Blood (SEM)	8.622×10^9 (7.947×10^8)	6.170×10^9 (3.326×10^8)	1.184×10^9 (2.129×10^8)	0.83 (0.17)	29.2 (6.08)	37.05 (9.01)
Tumor (SEM)	N/A	1.209×10^{11} (1.757×10^{10})	3.085×10^9 (4.749×10^8)	40 (8)	N/A	N/A
Liver (SEM)	N/A	4.640×10^{10} (2.693×10^{10})	2.596×10^9 (6.091×10^8)	1.33 (0.33)	N/A	N/A
Kidney (SEM)	N/A	9.312×10^9 (3.097×10^9)	5.821×10^8 (1.301×10^8)	8.83 (7.6)	N/A	N/A
Spleen (SEM)	N/A	1.739×10^{10} (2.409×10^9)	5.295×10^8 (8.229×10^7)	32 (8)	N/A	N/A
Brain (SEM)	8.431×10^9 (1.712×10^8)	5.360×10^9 (8.997×10^8)	3.086×10^8 (4.907×10^7)	8.83 (7.6)	38.38 (14.95)	51.7 (23.08)
Heart (SEM)	5.577×10^9 (3.748×10^8)	4.320×10^9 (5.650×10^8)	2.480×10^8 (5.119×10^7)	9.33 (7.33)	29.37 (4.73)	42.09 (3.32)
Lung (SEM)	N/A	2.119×10^{10} (8.078×10^9)	1.051×10^9 (1.821×10^8)	1.17 (0.44)	N/A	N/A
Ovary (SEM)	N/A	1.464×10^{10} (4.112×10^9)	4.912×10^8 (1.469×10^8)	9.33 (7.33)	N/A	N/A
Intestine (SEM)	2.351×10^{10} (8.036×10^9)	1.302×10^{10} (1.034×10^9)	7.441×10^8 (2.281×10^8)	2.33 (0.88)	30.78 (8.61)	47.24 (16.14)
Bladder (SEM)	N/A	1.433×10^{10} (7.469×10^9)	6.028×10^8 (1.992×10^8)	40 (8)	N/A	N/A

* AUC, C_{max}, T_{max}, T1/2 and Mean residence time (MRT) were calculated using the NCA module of PKanalix (PKanalix version 2020R1. Antony, France: Lixoff SAS, 2020). N/A: not available.

We also examined other organs like the lung, brain, bladder, ovary, kidneys, and heart. Background levels of the CVM signal were recorded from SapC-DOPG-CVM LVs (Figure S3A–F).

We further tested whether SapC-DOPG LVs have adverse effects on organs and body weight. Briefly, 6-week-old C57BL/6 mice were randomly divided into two groups of six (three males and three females per group). One group was injected with SapC-DOPG LVs via the tail vein (SapC = 34.4 mg/kg and DOPG = 21.4 mg/kg per mouse) (Figure 6A). Mean body weights of both sham and SapC-DOPG LV-treated mice were comparable (Figure 6B). These results indicate that SapC-DOPG LVs do not have an adverse effect on overall body weight. Next, hematological analyses were conducted using a Hemavet 950. SapC-DOPG LVs did not induce significant changes in the levels of WBC, RBC, HGB, HCT, MCV, MCH, MCHC, or PLT (Table S2). Next, we determined whether SapC-DOPG LVs cause functional damage to the liver and/or kidneys of mice. For liver function, we compared the levels of aspartate aminotransferase (AST) and alanine transaminase (ALT) activity in the plasma of mice injected with saline or SapC-DOPG LVs. Even though there was a slight change

(p = 0.045) in AST activity between SapC-DOPG female and Sham female, the enzyme activity was still within the normal range (Figure 6C). Further, ALT activity in the plasma of mice revealed that SapC-DOPG LVs showed no significant changes (Figure 6D). Kidney function assessed by creatinine and urea levels showed no significant changes in either the mice treated with SapC-DOPG LVs or the saline group, regardless of sex, as presented in Figure 6E,F. Moreover, histopathology results revealed no significant tissue damage in the organs of mice treated with SapC-DOPG LVs or saline, as assessed by a pathologist (Figure 6G). The livers from both groups showed no sign of fatty change, sinus congestion, hemorrhage, necrotic foci, neoplastic change, microgranulomas, or Kupffer cell swelling (Figure 6G, top panel). Similarly, the kidneys of the control groups displayed no tubular necrosis, widened sinusoids, or neoplastic changes (Figure 6G, middle panel). No evidence of hemorrhage, granuloma, or neoplastic changes was observed in the lungs (Figure 6G bottom panel).

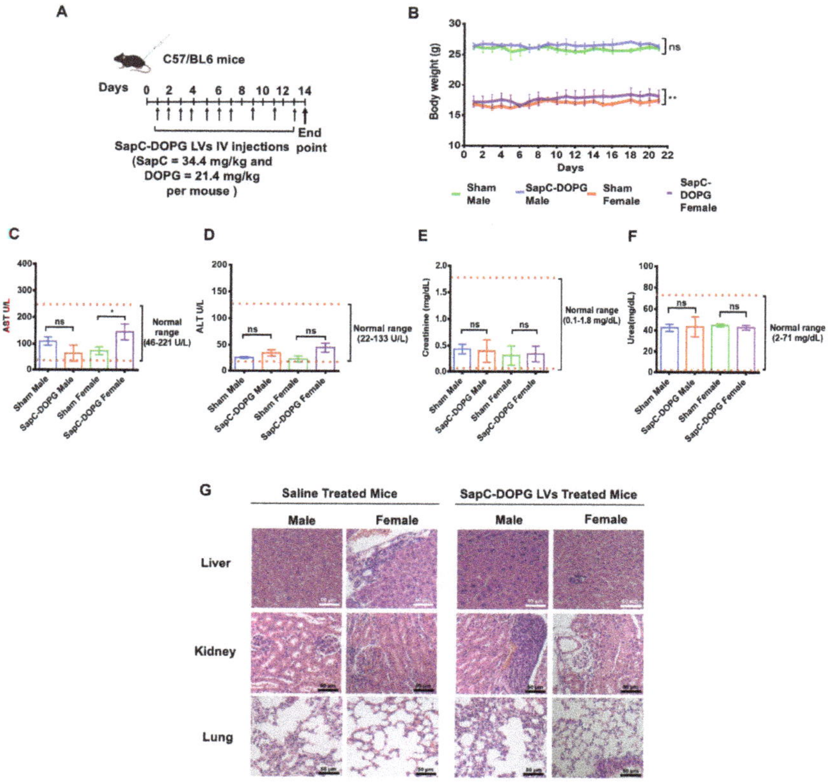

Figure 6. SapC-DOPG LVs have no significant adverse effect on the overall body weight and organs of the mice. (**A**) Schematic showing SapC-DOPG LV treatment regimen. (**B**) Body weights of sham and SapC-DOPG LV-treated male and female mice. ALT (**C**) and AST (**D**) enzymatic analysis from the blood plasma of sham and SapC-DOPG LV-treated mice. Creatinine (**E**) and urea (**F**) analysis from the blood plasma of sham and SapC-DOPG LV-treated mice. (**G**) H&E staining of the liver (top panel), kidney (middle panel), and lung (bottom panel) tissues from sham and SapC-DOPG LV-treated mice. The experiments were repeated at least twice. ns: not significant, * $p < 0.05$, and ** $p < 0.01$.

Together, these data indicate that in a mouse model, SapC-DOPG LVs specifically target pancreatic tumors without significant toxicity.

4. Discussion

In this study, we show that cancer cells with elevated PS on the cell surface secrete a phosphorylated Hsp70 (pT636-Hsp70) that is soluble and not associated with exosomes or microparticles. Moreover, reinforcing our recent study [33], EMD-CM harvested from cancer cells with elevated surface PS induced robust monocyte differentiation and M2 MΦ polarization. Consistent with a positive correlation between cancer cell surface PS, secreted Hsp70, and CD14 upregulation in THP-1 cells, subpopulations of cancer cells with higher cell surface PS sorted from their original more heterogeneous original lines possessed higher MΦ differentiation and earlier tumor initiation ability. Further, we showed that DOPG LVs are non-toxic to normal cells, bind to cancer-secreted Hsp70, increase the survival of KPC 1624 orthotopic mice model, and reduce the tumor volume and M2 MΦs in LLC-GFP mouse subcutaneous tumors.

Herein, we found larger areas of surface colocalization of Hsp70 and PS in PS^{high} cancer cell lines compared to PS^{low} cancer cells, indicating a PS-associated externalization of Hsp70 in cancer cells. Furthermore, we found a striking correlation between cancer cell surface PS, secreted Hsp70, and MΦ differentiation. Interestingly, cancer cells are known to express higher levels of surface PS compared to normal cells; the higher intracellular calcium in cancer cells reduces flippase activity, leading to higher PS on the cell surface [11]. We have previously found that different types of cancers and even cells within individual cancer types vary in surface PS exposure [11]. As Hsp70 lacks a conventional secretory signal, the mechanism of secretion involves a non-classical pathway via membrane rafts by binding to other proteins. The current study strongly suggests that PS exerts an important role in facilitating non-classical Hsp70 secretion. Consistent with these findings, the endolysosomal route has been described as a release mechanism for membrane-bound, exosomal, and soluble forms of Hsp70 [35].

Our results may reveal an important regulatory mechanism that underlies the cancer-promoting actions of Hsp70. Specifically, we found a positive correlation between cancer cell PS exposure and the quantity of secreted M2 markers, such as arginase 1 and TGM2. Arginase 1 is a key marker for M2 polarization in cancers, significantly contributing to the immunosuppressive TME by depleting L-arginine [44], while TGM2 plays a supporting role in M2 polarization by facilitating the extracellular matrix remodeling and promoting cell adhesion, thereby further enhancing tumor progression [45]. In particular, we demonstrated that cancer subcellular populations with the higher PS cellular fraction possess higher MΦ differentiation ability and faster tumor initiation ability. This suggests that higher surface PS cells in individual cancer types could be targeted to inhibit cancer immunosuppression.

SapC-DOPS is a PS-targeting drug. In preclinic models, it has been shown that SapC-DOPS targets a variety of primary and metastatic tumors including pancreatic cancer [19], glioblastoma multiforme (GBM) [46], and lung cancer [47]. We have shown that SapC-DOPS has an additive response when used in combination with gemcitabine or gemcitabine-abraxane in PDAC models [14]. Also, SapC-DOPS has been shown to cross the blood–brain tumor barrier (BBTB) and has a strong synergistic effect in combination with temozolomide in the glioblastoma model [43]. SapC-DOPS is non-toxic to normal cells or tumor-free organs. BXQ-350 (SapC-DOPS) has completed a Phase I clinical trial. It was well tolerated in humans without any dose-limiting toxicity in the Phase I safety trials in brain tumors in both children and adults [48,49] and has now progressed to a Phase Ib/II clinical trial for newly diagnosed metastatic colorectal carcinoma [50].

It is known that Hsp70 interacts with negatively charged phospholipids in the cell membrane, and this serves to fold nascent Hsp70 proteins or prevent aggregation of membrane-bound proteins. In this study, in vitro binding assay results revealed that negatively charged DOPG has an affinity for positively charged Hsp70. The electrostatic attraction between the positive charges on Hsp70 and the negative charges on DOPG helps to mediate the binding of Hsp70 to DOPG. DOPG LVs are also superior to DOPS LVs in reducing M2 MΦ polarization in THP-1 cells (Figure 3A). Therefore, DOPG LVs were used in vivo efficacy studies. DOPG has demonstrated therapeutic potential. However, concerns regarding its toxicity in certain contexts have been raised in the literature [43]. SapC exhibits a higher binding affinity for phosphatidylserine on cell membranes within the acidic environment of tumors [18]. To enhance the targeted delivery of DOPG to the tumor microenvironment, ensuring that the treatment is more selective to the PDAC cells while minimizing off-target effects and potential toxicity, we coupled SapC to DOPG (SapC-DOPG). Indeed, our results showed that SapC-DOPG specifically targeted the pancreatic tumor and significantly reduced the size of tumor in a Rink-1 orthotopic mouse model. Moreover, SapC-DOPG demonstrated safety in a mouse model, with no observed adverse effects, reinforcing its potential as a therapeutic agent. Additionally, one of the analogs of SapC-DOPG, SapC-DOPS, is already undergoing Phase II clinical trials, and its safety profile has been approved in patients, further supporting the viability and safety of using SapC in combination with phospholipids for targeted therapeutic delivery.

Unlike for SapC-DOPS, we do not have data indicating the potential synergistic or additive effects of SapC-DOPG when combined with existing chemotherapeutic or immunotherapeutic regimens. As SapC-DOPG affects immune surveillance functions, therapies with mechanisms distinct from those primarily affecting immune pathways, such as gemcitabine that directly impair DNA synthesis, are likely to be at least additive and potentially synergistic. Further investigation is required in future studies to clarify this aspect.

Although our study has some limitations, none significantly weaken the overall credibility of the primary findings. First, the role of PS for the secretion of Hsp70 is only partially defined. The inhibition of M2 MΦ polarization in PDAC cancer cells by sequestering Hsp70 via DOPG LVs was largely correlative. While our results demonstrated a positive correlation between cancer cell surface PS and secreted Hsp70, as well as a physical interaction between PS and Hsp70 observed through fluorescence imaging, the specific binding must be further elucidated using the microscale thermophoresis technique (MST). Second, the precise mechanism through which Hsp70 induces the M2 MΦ polarization and the role of different phosphorylated forms are not completely dissected. Third, we do not suggest that Hsp70 is the only and/or the most potent secreted factor in the PDAC TME. In our recent study [11], we show that Hsp70, with its identity confirmed by LC-MS, is preferentially secreted by PDAC cancer cells. However, immuno-neutralizing monoclonal antibodies to Hsp70 resulted in at most a 50% reduction in CD14 expression in THP-1 monocytes. Consistent with these prior results, several studies have indicated that soluble programmed death ligand 1 (sPD-L1), which is a protein secreted by cancer cells and is detectable in the TME, may adversely modulate T-cell functionality. In patients diagnosed with gastric cancer [51] and non-small cell lung cancer [52], elevated serum levels of sPD-L1 have been associated with enhanced metastasis and an increased risk of mortality. Also, recent studies revealed the effect of cancer cell-secreted microRNAs (miRNAs) on immune cell function [53,54]. Fourth, although we observed a positive correlation between PS exposure and secreted Hsp70 in PDAC cells, the specific contribution of PS exposure to the secretion of Hsp70 was not directly investigated. Moreover, how these two phenomena regulate the immunosuppressive landscape of the PDAC tumor microenvironment remains speculative, and future research efforts should aim to clarify this pathway. Fifth, the high-high PS population appears similar in CD14 expression to that of the non-sorted population; we speculate this may be due to the threshold effect of PS exposure. Once this threshold is reached, Hsp70 release is triggered with subsequent macrophage activation. While the low-low cells do not reach this threshold, the non-sorted cells do. Finally, most experiments were performed using EMD-CM from cancer cells rather than a purified or semi-purified form of Hsp70; a more definitive approach and one intended for future studies would be to use purified bioactive Hsp70 that could be quantified by a specific ELISA.

5. Conclusions

Our studies reveal a new and important function for Hsp70 protein released from cancer cells. Our previous study revealed that cancer-secreted Hsp70 exerts a marked immunosuppressive effect in TME [33] mediated through MerTK and TLR2 [33]. Our current study extends this previously unknown function of Hsp70 by demonstrating (1) the role of PS in Hsp70 secretion, (2) the inhibition of M2 MΦ polarization in PDAC cancer cells by sequestering Hsp70 via DOPG LVs, and (3) the inhibition of the growth of tumors and the sequestration of Hsp70 by these lipid-binding nanovesicles. Indeed, our findings that cancer cell surface PS is associated with induction of M2 MΦ polarization through cancer cell-secreted Hsp70 opens a novel functional role for tumor cell PS in communication with immune cells. Because elevated cell-surface PS is almost a unique feature of tumor cells, PS, together with cancer cell-secreted Hsp70, provides critical targets to block tumor-induced immune suppression.

Supplementary Materials: The following supporting information can be downloaded at: https://www.mdpi.com/article/10.3390/cancers17071224/s1, Figure S1: Quantification of exposed PS area, HSP70 area, PS and Hsp70 co-localized areas on cancer cells; Figure S2: Body weights of LLC-GFP subcutaneous mouse model following injection of PBS or DOPG LVs; Figure S3: The signal of SapC-DOPG-CVM measured at mice organs; Table S1. Particle size distribution of the synthesized nanoparticles measured by DLS. Table S2. Hematological analysis of mice injected with SapC-DOPG LVs; File S1: Original western blots.

Author Contributions: Conceptualization, A.K., S.D.V. and X.Q.; methodology, A.K., S.D.V., H.W.D. and X.Q.; software, A.K.; validation, A.K., S.D.V., H.W.D. and X.Q.; formal analysis, A.K., S.D.V., H.W.D. and K.F.S., and X.Q.; investigation, A.K., S.D.V. and H.W.D.; resources, X.Q.; data curation, A.K. and S.D.V.; writing—original draft preparation, A.K.; writing—review and editing, S.D.V., H.W.D., E.P.S., C.S.L., J.A., P.M., B.V., R.S.F., W.-H.S., V.Y.B. and X.Q.; visualization, A.K.; supervision, X.Q.; project administration, X.Q.; funding acquisition, P.M., W.-H.S. and X.Q. All authors have read and agreed to the published version of the manuscript.

Funding: This research was funded by 2020 Pancreatic Cancer Action Network Translational Research, 20-65-QIXI (X.Q.); 2022 Pancreatic Cancer Action Network Translational Research Grant Extension funded by Kenneth D. Custance and Gladys C. Custance, honoring the memory of Martha 'Molly' Reed Woodroofe (X.Q.); NIH, R01-CA158372 (X.Q.), 2UL1TR001425-05A1 (X.Q.), and R01-DK116789 (W.S.); Give Hope Foundation Award (X.Q.); and Czech Science Foundation grant 22-17102S (P.M.).

Institutional Review Board Statement: The animal study protocol was approved by the Institutional Review Board (or Ethics Committee) of University of Cincinnati (22-10-25-01 and 5 December 2022).

Informed Consent Statement: Not applicable.

Data Availability Statement: No new datasets were created that are posted.

Acknowledgments: The authors thank the Proteomics and Mass Spectrometry Laboratory core facility at the University of Cincinnati for LC-MS analyses.

Conflicts of Interest: The authors declare no conflicts of interest.

Disclaimer: The authors declare no conflicts of interest. This article was prepared while Dr. Vladimir Y. Bogdanov was employed at University of Cincinnati. The opinions expressed in this article are the author's own and do not reflect the view of the National Institutes of Health, the Department of Health and Human Services, or the United States Government.

References

1. Binnewies, M.; Roberts, E.W.; Kersten, K.; Chan, V.; Fearon, D.F.; Merad, M.; Coussens, L.M.; Gabrilovich, D.I.; Ostrand-Rosenberg, S.; Hedrick, C.C.; et al. Understanding the tumor immune microenvironment (TIME) for effective therapy. *Nat. Med.* **2018**, *24*, 541–550. [CrossRef] [PubMed]
2. Rabinovich, G.A.; Gabrilovich, D.; Sotomayor, E.M. Immunosuppressive strategies that are mediated by tumor cells. *Annu. Rev. Immunol.* **2007**, *25*, 267–296. [CrossRef] [PubMed]
3. Bejarano, L.; Jordão, M.J.C.; Joyce, J.A. Therapeutic Targeting of the Tumor Microenvironment. *Cancer Discov.* **2021**, *11*, 933–959. [CrossRef]
4. Yap, T.A.; Parkes, E.E.; Peng, W.; Moyers, J.T.; Curran, M.A.; Tawbi, H.A. Development of Immunotherapy Combination Strategies in Cancer. *Cancer Discov.* **2021**, *11*, 1368–1397. [CrossRef]
5. Mantovani, A.; Marchesi, F.; Malesci, A.; Laghi, L.; Allavena, P. Tumour-associated macrophages as treatment targets in oncology. *Nat. Rev. Clin. Oncol.* **2017**, *14*, 399–416.
6. Germano, G.; Frapolli, R.; Belgiovine, C.; Anselmo, A.; Pesce, S.; Liguori, M.; Erba, E.; Uboldi, S.; Zucchetti, M.; Pasqualini, F.; et al. Role of macrophage targeting in the antitumor activity of trabectedin. *Cancer Cell* **2013**, *23*, 249–262. [CrossRef]
7. Yao, L.; Wang, M.; Niu, Z.; Liu, Q.; Gao, X.; Zhou, L.; Liao, Q.; Zhao, Y. Interleukin-27 inhibits malignant behaviors of pancreatic cancer cells by targeting M2 polarized tumor associated macrophages. *Cytokine* **2017**, *89*, 194–200. [CrossRef]
8. Candido, J.B.; Morton, J.P.; Bailey, P.; Campbell, A.D.; Karim, S.A.; Jamieson, T.; Lapienyte, L.; Gopinathan, A.; Clark, W.; McGhee, E.J.; et al. CSF1R(+) Macrophages Sustain Pancreatic Tumor Growth through T Cell Suppression and Maintenance of Key Gene Programs that Define the Squamous Subtype. *Cell Rep.* **2018**, *23*, 1448–1460. [CrossRef]
9. Siegel, R.L.; Miller, K.D.; Fuchs, H.E.; Jemal, A. Cancer statistics, 2022. *CA Cancer J. Clin.* **2022**, *72*, 7–33. [CrossRef]
10. Capurso, G.; Sette, C. Drug resistance in pancreatic cancer: New player caught in act. *EBioMedicine* **2019**, *40*, 39–40. [CrossRef]
11. Vallabhapurapu, S.D.; Blanco, V.M.; Sulaiman, M.K.; Vallabhapurapu, S.L.; Chu, Z.; Franco, R.S.; Qi, X. Variation in human cancer cell external phosphatidylserine is regulated by flippase activity and intracellular calcium. *Oncotarget* **2015**, *6*, 34375–34388. [CrossRef]
12. Birge, R.B.; Boeltz, S.; Kumar, S.; Carlson, J.; Wanderley, J.; Calianese, D.; Barcinski, M.; Brekken, R.A.; Huang, X.; Hutchins, J.T.; et al. Phosphatidylserine is a global immunosuppressive signal in efferocytosis, infectious disease, and cancer. *Cell Death Differ.* **2016**, *23*, 962–978. [CrossRef]
13. Riedl, S.; Rinner, B.; Asslaber, M.; Schaider, H.; Walzer, S.; Novak, A.; Lohner, K.; Zweytick, D. In search of a novel target—Phosphatidylserine exposed by non-apoptotic tumor cells and metastases of malignancies with poor treatment efficacy. *Biochim. Biophys. Acta (BBA) Biomembr.* **2011**, *1808*, 2638–2645. [CrossRef]
14. N'Guessan, K.F.; Davis, H.W.; Chu, Z.; Vallabhapurapu, S.D.; Lewis, C.S.; Franco, R.S.; Olowokure, O.; Ahmad, S.A.; Yeh, J.J.; Bogdanov, V.Y.; et al. Enhanced Efficacy of Combination of Gemcitabine and Phosphatidylserine-Targeted Nanovesicles against Pancreatic Cancer. *Mol. Ther. J. Am. Soc. Gene Ther.* **2020**, *28*, 1876–1886. [CrossRef]
15. Phinney, N.Z.; Huang, X.; Toombs, J.E.; Brekken, R.A. Development of betabodies: The next generation of phosphatidylserine targeting agents. *J. Biol. Chem.* **2024**, *300*, 107681. [CrossRef]
16. Hsiehchen, D.; Beg, M.S.; Kainthla, R.; Lohrey, J.; Kazmi, S.M.; Khosama, L.; Maxwell, M.C.; Kline, H.; Katz, C.; Hassan, A.; et al. The phosphatidylserine targeting antibody bavituximab plus pembrolizumab in unresectable hepatocellular carcinoma: A phase 2 trial. *Nat. Commun.* **2024**, *15*, 2178. [CrossRef]
17. Qi, X.; Leonova, T.; Grabowski, G.A. Functional human saposins expressed in Escherichia coli. Evidence for binding and activation properties of saposins C with acid beta-glucosidase. *J. Biol. Chem.* **1994**, *269*, 16746–16753. [CrossRef]
18. Qi, X.; Chu, Z.; Mahller, Y.Y.; Stringer, K.F.; Witte, D.P.; Cripe, T.P. Cancer-selective targeting and cytotoxicity by liposomal-coupled lysosomal saposin C protein. *Clin. Cancer Res.* **2009**, *15*, 5840–5851. [CrossRef]
19. Chu, Z.; Abu-Baker, S.; Palascak, M.B.; Ahmad, S.A.; Franco, R.S.; Qi, X. Targeting and cytotoxicity of SapC-DOPS nanovesicles in pancreatic cancer. *PLoS ONE* **2013**, *8*, e75507. [CrossRef]
20. Sulaiman, M.K.; Chu, Z.; Blanco, V.M.; Vallabhapurapu, S.D.; Franco, R.S.; Qi, X. SapC-DOPS nanovesicles induce Smac- and Bax-dependent apoptosis through mitochondrial activation in neuroblastomas. *Mol. Cancer* **2015**, *14*, 78. [CrossRef]
21. Ellis, J. Proteins as molecular chaperones. *Nature* **1987**, *328*, 378–379. [CrossRef]
22. Rosenzweig, R.; Nillegoda, N.B.; Mayer, M.P.; Bukau, B. The Hsp70 chaperone network. *Nat. Rev. Mol. Cell Biol.* **2019**, *20*, 665–680. [CrossRef] [PubMed]
23. Mayer, M.P.; Bukau, B. Hsp70 chaperones: Cellular functions and molecular mechanism. *Cell. Mol. Life Sci.* **2005**, *62*, 670–684. [CrossRef] [PubMed]
24. Albakova, Z.; Armeev, G.A.; Kanevskiy, L.M.; Kovalenko, E.I.; Sapozhnikov, A.M. HSP70 Multi-Functionality in Cancer. *Cells* **2020**, *9*, 587. [CrossRef] [PubMed]
25. Sherman, M.Y.; Gabai, V.L. Hsp70 in cancer: Back to the future. *Oncogene* **2015**, *34*, 4153–4161. [CrossRef]

Mambula, S.S.; Calderwood, S.K. Heat shock protein 70 is secreted from tumor cells by a nonclassical pathway involving lysosomal endosomes. *J. Immunol.* **2006**, *177*, 7849–7857. [CrossRef]

Blott, E.J.; Griffiths, G.M. Secretory lysosomes. *Nat. Rev. Mol. Cell Biol.* **2002**, *3*, 122–131. [CrossRef]

Broquet, A.H.; Thomas, G.; Masliah, J.; Trugnan, G.; Bachelet, M. Expression of the molecular chaperone Hsp70 in detergent-resistant microdomains correlates with its membrane delivery and release. *J. Biol. Chem.* **2003**, *278*, 21601–21606. [CrossRef]

Nitika; Truman, A.W. Cracking the Chaperone Code: Cellular Roles for Hsp70 Phosphorylation. *Trends Biochem. Sci.* **2017**, *42*, 932–935.

Zemanovic, S.; Ivanov, M.V.; Ivanova, L.V.; Bhatnagar, A.; Michalkiewicz, T.; Teng, R.J.; Kumar, S.; Rathore, R.; Pritchard, K.A., Jr.; Konduri, G.G.; et al. Dynamic Phosphorylation of the C Terminus of Hsp70 Regulates the Mitochondrial Import of SOD2 and Redox Balance. *Cell Rep.* **2018**, *25*, 2605–2616. [CrossRef]

Rohde, M.; Daugaard, M.; Jensen, M.H.; Helin, K.; Nylandsted, J.; Jaattela, M. Members of the heat-shock protein 70 family promote cancer cell growth by distinct mechanisms. *Genes Dev.* **2005**, *19*, 570–582. [CrossRef] [PubMed]

Juhasz, K.; Lipp, A.M.; Nimmervoll, B.; Sonnleitner, A.; Hesse, J.; Haselgruebler, T.; Balogi, Z. The complex function of hsp70 in metastatic cancer. *Cancers* **2013**, *6*, 42–66. [CrossRef] [PubMed]

Kaynak, A.; Vallabhapurapu, S.D.; Davis, H.W.; Smith, E.P.; Muller, P.; Vojtesek, B.; Franco, R.S.; Shao, W.-H.; Qi, X. TLR2-Bound Cancer-Secreted Hsp70 Induces MerTK-Mediated Immunosuppression and Tumorigenesis in Solid Tumors. *Cancers* **2025**, *17*, 450. [CrossRef]

Nair, A.B.; Jacob, S. A simple practice guide for dose conversion between animals and human. *J. Basic Clin. Pharm.* **2016**, *7*, 27–31. [CrossRef] [PubMed]

Mambula, S.S.; Stevenson, M.A.; Ogawa, K.; Calderwood, S.K. Mechanisms for Hsp70 secretion: Crossing membranes without a leader. *Methods* **2007**, *43*, 168–175. [CrossRef]

Liang, X.; Luo, M.; Shao, B.; Yang, J.Y.; Tong, A.; Wang, R.B.; Liu, Y.T.; Jun, R.; Liu, T.; Yi, T.; et al. Phosphatidylserine released from apoptotic cells in tumor induces M2-like macrophage polarization through the PSR-STAT3-JMJD3 axis. *Cancer Commun.* **2022**, *42*, 205–222. [CrossRef]

Schilling, D.; Gehrmann, M.; Steinem, C.; De Maio, A.; Pockley, A.G.; Abend, M.; Molls, M.; Multhoff, G. Binding of heat shock protein 70 to extracellular phosphatidylserine promotes killing of normoxic and hypoxic tumor cells. *FASEB J. Off. Publ. Fed. Am. Soc. Exp. Biol.* **2009**, *23*, 2467–2477. [CrossRef]

Utsugi, T.; Schroit, A.J.; Connor, J.; Bucana, C.D.; Fidler, I.J. Elevated expression of phosphatidylserine in the outer membrane leaflet of human tumor cells and recognition by activated human blood monocytes. *Cancer Res.* **1991**, *51*, 3062–3066.

McCallister, C.; Kdeiss, B.; Nikolaidis, N. Biochemical characterization of the interaction between HspA1A and phospholipids. *Cell Stress Chaperones* **2016**, *21*, 41–53. [CrossRef]

Dagogo-Jack, I.; Shaw, A.T. Tumour heterogeneity and resistance to cancer therapies. *Nat. Rev. Clin. Oncol.* **2018**, *15*, 81–94. [CrossRef]

Jacquemin, V.; Antoine, M.; Dom, G.; Detours, V.; Maenhaut, C.; Dumont, J.E. Dynamic Cancer Cell Heterogeneity: Diagnostic and Therapeutic Implications. *Cancers* **2022**, *14*, 280. [CrossRef] [PubMed]

Smulders, L.; Daniels, A.J.; Plescia, C.B.; Berger, D.; Stahelin, R.V.; Nikolaidis, N. Characterization of the Relationship between the Chaperone and Lipid-Binding Functions of the 70-kDa Heat-Shock Protein, HspA1A. *Int. J. Mol. Sci.* **2020**, *21*, 5995. [CrossRef]

Kuo, J.H.; Jan, M.S.; Jeng, J.; Chiu, H.W. Induction of apoptosis in macrophages by air oxidation of dioleoylphosphatidylglycerol. *J. Control. Release* **2005**, *108*, 442–452. [CrossRef] [PubMed]

Menjivar, R.E.; Nwosu, Z.C.; Du, W.; Donahue, K.L.; Hong, H.S.; Espinoza, C.; Brown, K.; Velez-Delgado, A.; Yan, W.; Lima, F.; et al. Arginase 1 is a key driver of immune suppression in pancreatic cancer. *eLife* **2023**, *12*, e80721. [CrossRef] [PubMed]

Tempest, R.; Guarnerio, S.; Maani, R.; Cooper, J.; Peake, N. The Biological and Biomechanical Role of Transglutaminase-2 in the Tumour Microenvironment. *Cancers* **2021**, *13*, 2788. [CrossRef]

Wojton, J.; Meisen, W.H.; Jacob, N.K.; Thorne, A.H.; Hardcastle, J.; Denton, N.; Chu, Z.; Dmitrieva, N.; Marsh, R.; Van Meir, E.G.; et al. SapC-DOPS-induced lysosomal cell death synergizes with TMZ in glioblastoma. *Oncotarget* **2014**, *5*, 9703–9709. [CrossRef]

Zhao, S.; Chu, Z.; Blanco, V.M.; Nie, Y.; Hou, Y.; Qi, X. SapC-DOPS nanovesicles as targeted therapy for lung cancer. *Mol. Cancer Ther.* **2015**, *14*, 491–498. [CrossRef]

Abdelbaki, M.S.; DeWire Schottmiller, M.D.; Cripe, T.P.; Curry, R.C.; Cruze, C.A.; Her, L.; Demko, S.; Casey, D.; Setty, B. An open-label multi-center phase 1 safety study of BXQ-350 in children and young adults with relapsed solid tumors, including recurrent malignant brain tumors. *Heliyon* **2022**, *8*, e12450. [CrossRef]

Rixe, O.; Villano, J.L.; Wesolowski, R.; Noonan, A.M.; Puduvalli, V.K.; Wise-Draper, T.M.; Curry, R., 3rd; Yilmaz, E.; Cruze, C.; Ogretmen, B.; et al. A First-in-Human Phase I Study of BXQ-350, a First-in-Class Sphingolipid Metabolism Regulator, in Patients with Advanced/Recurrent Solid Tumors or High-Grade Gliomas. *Clin. Cancer Res.* **2024**, *30*, 5053–5060. [CrossRef]

50. Patel, R.A.; Baron, A.D.; Boulmay, B.C.; Flora, D.B.; Gemmill, J.A.L.; Lee, F.C.; Sohal, D.; Williams, G.R.; Curry, R.C.; Gazda, ; et al. BXQ-350: A phase 1b/2 placebo controlled, double blinded study on the efficacy and safety of BXQ-350 in combinati with mFOLFOX7 and bevacizumab in newly diagnosed metastatic colorectal carcinoma (mCRC). *J. Clin. Oncol.* **2024**, *42*, TPS2 [CrossRef]
51. Shin, K.; Kim, J.; Park, S.J.; Lee, M.A.; Park, J.M.; Choi, M.-G.; Kang, D.; Song, K.Y.; Lee, H.H.; Seo, H.S.; et al. Prognostic value soluble PD-L1 and exosomal PD-L1 in advanced gastric cancer patients receiving systemic chemotherapy. *Sci. Rep.* **2023**, *13*, 69 [CrossRef]
52. Okuma, Y.; Hosomi, Y.; Nakahara, Y.; Watanabe, K.; Sagawa, Y.; Homma, S. High plasma levels of soluble programmed cell de ligand 1 are prognostic for reduced survival in advanced lung cancer. *Lung Cancer* **2017**, *104*, 1–6. [CrossRef] [PubMed]
53. Paladini, L.; Fabris, L.; Bottai, G.; Raschioni, C.; Calin, G.A.; Santarpia, L. Targeting microRNAs as key modulators of tum immune response. *J. Exp. Clin. Cancer Res.* **2016**, *35*, 103. [CrossRef] [PubMed]
54. Khayati, S.; Dehnavi, S.; Sadeghi, M.; Tavakol Afshari, J.; Esmaeili, S.-A.; Mohammadi, M. The potential role of miRNA regulating macrophage polarization. *Heliyon* **2023**, *9*, e21615. [CrossRef]

cancers

MDPI

Article

Chondrosarcoma: Multi-Targeting Therapeutic Effects of Doxorubicin, BEZ235, and the Small Molecule Aspartyl-Asparaginyl-β-hydroxylase Inhibitor SMI1182

Megan Fife [1], Ming Tong [2], Bhaskar Das [3], Rene Rodriguez [4,5], Parthiban Chokkalingam [3], Rolf I. Carlson [2] and Suzanne M. de la Monte [6,*]

1 Molecular Pharmacology, Physiology, and Biotechnology Graduate Program, Brown University, Providence, RI 02912, USA
2 Department of Medicine, Rhode Island Hospital, Brown University Health, Providence, RI 02912, USA; rolf_carlson@brown.edu (R.I.C.)
3 Department of Drug and Biotherapeutic Discovery, School of Pharmacy and Pharmaceutical Sciences, University at Buffalo, State University of New York (SUNY), Buffalo, NY 14201, USA; bhaskard@buffalo.edu (B.D.)
4 Health Research Institute of Asturias (ISPA), University Institute of Oncology of Asturias (IUOPA), 33001 Oviedo, Spain; rene.rodriguez@ispasturias.es
5 CIBER Enoncologia (CIBERONC), 28054 Madrid, Spain
6 Departments of Pathology and Laboratory Medicine, Neurology, and Neurosurgery, Rhode Island Hospital, Women & Infants Hospital, Brown University Health, Alpert Medical School of Brown University, Providence, RI 02912, USA
* Correspondence: suzanne_delamonte_md@brown.edu

Simple Summary: This study compares molecular features in intermediate and high-grade conventional chondrosarcomas and their responses to individual, combination, and novel chemotherapeutic agents. The findings may account for differences in aggressive tumor behavior in relation to histopathological grade and suggest that chondrosarcoma tumor grade should be factored in for predicting the efficacy of emerging chemotherapeutic approaches.

Abstract: Background/Objectives: Chondrosarcoma (CS), the most common malignant bone tumor in adults, exhibits a poor prognosis due to high rates of post-surgical recurrence and metastasis, and resistance to chemotherapy. CS's abundant expression of aspartyl-asparaginyl-β-hydroxylase (ASPH), which drives invasive tumor growth via Notch and PI3K/mTOR activation, opens opportunities for treatment in combination with standard Doxorubicin (DOX) chemotherapy. We hypothesized that the small molecule inhibitor SMI1182, which targets the catalytic domain of ASPH, and BEZ235, which targets PI3K/mTOR, could enhance the chemotherapeutic effects of DOX. Human CS1 (Grade 3) and CDS11 (Grade 2) conventional CS cell lines were treated with broad dose ranges of DOX, BEZ235, or SMI1182 as mono- or combination therapy to assess their anti-tumor effects on cell viability, toxicity, and motility. Methods: Mechanistic studies included the analysis of ASPH expression, Notch signaling, and insulin/IGF/IRS pathway activation through mTOR. DOX, BEZ235, or SMI1182 treatments caused dose-dependent cell loss and cytotoxicity. Results: SMI1182 and BEZ235, with or without DOX, significantly reduced directional motility. Combined treatments had additive cytotoxic effects linked to the reduced expression of ASPH, Notch transcription factors, and insulin receptor substrate type I, which positively regulates both ASPH and Notch. Conclusions: Triple-drug treatment with DOX, SMI1182, and BEZ235 could potentially improve disease-free survival with CS by the simultaneous targeting of multiple upstream mediators of aggressive malignant tumor cell behavior.

Keywords: aspartyl-asparaginyl-β-hydroxylase; BEZ235; cancer treatment; chondrosarcoma; doxorubicin; Notch; small molecule inhibitor

1. Introduction

Chondrosarcoma (CS) is the most common bone sarcoma in adults [1] and accounts for 20–25% of osseous malignancies. The initial diagnosis of CS is typically made in people 30 to 60 years old, but 80% are older than 40 (https://www.bcrt.org.uk/information/information-by-type/chondrosarcoma/, accessed on 20 April 2025). Among the four main histopathological subtypes, conventional CS accounts for 75%, and de-differentiated, mesenchymal, and clear cell subtypes together account for approximately 25% of CS. In the U.S.A., the crude annual incidence of CS increased over time through 2013 [2], but subsequently, the rates leveled off. In a later national study, Thorkildsen and Myklebust showed that the prevalence rate of CS progressively increased between 2000 and 2020 (Figure 1), although age-adjusted incidence and mortality did not fluctuate significantly over that same interval [3]. Higher prevalence vis-à-vis stable incidence rates may suggest improvements in survival or the detection of treatable recurrent disease.

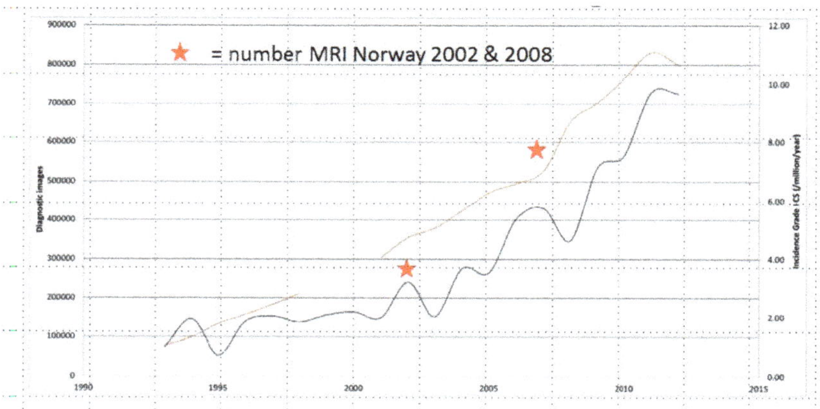

Figure 1. Incidence of ACT/CS grade 1 in the Netherlands between 1989 and 2013, number of MRI examinations over time. Adapted from Thorkildesen and Myklebust (2023) [3]. The gold line corresponds to the number of diagnostic imaging studies performed. The charcoal line reflects the crude incident rates of Grade I CS over time.

The standard treatment for CS is wide surgical resection, including the affected bone and surrounding tissue, with wide negative margins [4]. In patients with unresectable chondrosarcoma or metastatic disease, the clinical prognosis is poor [4]. The first-line chemotherapeutic agent for high-grade and advanced sarcomas, including chondrosarcoma, consists of Doxorubicin (DOX) [5,6]. However, high-grade conventional chondrosarcomas are known to be resistant to radiotherapy and chemotherapy [4], and DOX mono-chemotherapy is suboptimum in terms of efficacy, as evidenced by the universally poor outcomes due to pulmonary metastases and mortality [7]. Current standard treatment regimens mainly consist of multimodal combination treatment with 2 or 3 drugs, including DOX plus DNA-damaging agents such as cisplatin or isofamide [4]. Unresectable recurrences and metastasis linked to post-treatment resistance to anticancer agents render chondrosarcoma clinical outcomes discouragingly poor [4,8].

The five-year survival rate with CS is between 26% and 32%, with metastatic disease further reducing mean survival to 23% [9]. Importantly, prognosis is linked to tumor grade such that high-grade CSs exhibit the highest recurrence rates and shortest survivals [9]. Moreover, there is no effective systemic treatment for chondrosarcoma. These statistics, combined with the minimal progress in CS therapeutics, led the American Academy of Orthopaedic Surgeons to prioritize the development of CS treatment strategies. Most likely, newer, more effective therapeutics will require molecular targeting to inhibit CS growth and metastasis. Potential strategies to improve survival with CS include the use of chemotherapeutic agents that target signaling pathways that have roles in chondrosarcoma malignant behavior, such as the PI3K-Akt-mTOR pathway and angiogenesis [10] [11], in addition to genomic mutations of isocitrate dehydrogenase (IDH) IDH1/2, exostosin-1/2 (EXT1/2), COL2A1, TP53, telomerase reverse transcriptase (TERT), and cyclin-dependent kinase inhibitor 2A/B (CDKN2A/B) [4,10,12]. To this end, there are many ongoing but early-stage clinical trials designed to assess the therapeutic efficacy in targeting angiogenesis (PD-L1), mTOR, PI3K, CDK4/6, or IDH1 [4]. In a retrospective study, Molho et al. demonstrated efficacy in treating CS with the mTOR inhibitor sirolimus in combination with cyclophosphamide [8,13].

Another likely important therapeutic target for chondrosarcoma is Notch, which regulates cell proliferation, differentiation, migration, infiltrative growth, and cell fate, and has potential therapeutic efficacy for a broad range of cancers [14–18]. Aberrantly increased Notch activity has been implicated in many malignancies, including chondrosarcoma [19]. The Notch signaling inhibitor LY3039478 was initially reported to be well tolerated as monotherapy for various carcinomas [20], but in a later Phase 1b/II randomized study, although intra-tumoral Notch and downstream signaling through phosphorylated Akt were reduced, there were no significant effects on progression-free or overall survival in a heterogeneous group of sarcomas, including CS [21]. Furthermore, enthusiasm for pursuing Notch-inhibitor chemotherapeutic drugs for advanced disease and forwarding their use in clinical trials has been dampened by the significant off-target effects [17,22–24].

Emerging data point toward aspartyl-asparaginyl-β-hydroxylase (ASPH) as an excellent candidate for strategically targeting Notch signaling networks in malignancies. ASPH, a type II transmembrane protein of the α-ketoglutarate-dependent dioxygenase family [25,26], has functional roles in cell motility utilized for infiltrative and metastatic tumor growth [27]. Mechanistically, the C-terminal region of ASPH contains a catalytic site that hydroxylates epidermal growth factor (EGF)-like domains expressed in molecules such as Notch and Jagged [28–31]. EGF-like domains have important roles in regulating localized signaling of growth and differentiation [32]. ASPH's interactions with Notch lead to the nuclear translocation of the Notch intracellular domain [33], followed by the activation of transcription factors such as HES and HEY [29,33]. The abolishment of ASPH's catalytic activity by either site-directed mutagenesis of its critical His-675 residue or deletion of the molecule's C-terminus prevents Notch activation and reduces cell motility and tumor growth [28].

Previous studies demonstrated high levels of ASPH expression in clinically aggressive malignancies, such as those originating from the gastrointestinal tract, lung, liver, pancreas, breast, or central nervous system [29,33–37]. Aberrantly increased Notch activity has been implicated in many malignancies, including chondrosarcoma [19]. Furthermore, high levels of ASPH expression were shown to correlate with worse clinical outcomes [29,38], including abbreviated survival [36,39]. In contrast, most normal cells and tissues express very low or non-detectable levels of ASPH [34]. Similarly, preliminary studies also suggest that high-grade conventional chondrosarcomas express higher levels of ASPH than low-grade chondrosarcomas [40]. In light of the well-documented roles of Notch in many malignancies, including chondrosarcoma, ASPH could potentially serve as an excellent target for chondrosarcoma diagnostics and therapeutics [41–44].

One evolving approach for cancer therapeutics is the use of small molecule inhibitors that target critical signaling pathways utilized in the growth and survival of malignant cells, including inhibitors of tyrosine and serine/threonine kinases, proteosomes, apoptosis, or matrix metalloproteinases [45]. Of note, a small molecule inhibitor of ADAM (INCB7839), designed to release the extracellular domain of Notch, was shown to be an effective anti-cancer agent in epithelial neoplasms [46,47]. Regarding ASPH, previous studies led to the design and characterization of unique small molecule inhibitors that fit in the pocket of ASPH's catalytic domain and diminish or abolish its enzymatic (hydroxylase) activity [27,35,36,38,48,49]. The ASPH small molecule inhibitor 1182 (SMI1182) was shown to inhibit ASPH's functions linked to Notch activation [36,48,49] and produce anti-tumor effects in experimental models of cholangiocarcinoma [48], hepatocellular carcinoma [38], pancreatic carcinoma [35], breast cancer [27], glioblastoma [36], and more recently, chondrosarcoma [40]. However, despite encouraging results, none of the studies demonstrated complete cancer cell killing with SMI1182 used as a monotherapy. The preliminary finding that the combined use of SMI1182 with DOX in an experimental breast cancer model was more effective than SMI1182 monotherapy [50] inspired further investigation of this concept in relation to chondrosarcoma.

We hypothesized that SMI1182 could potentially be used as an adjuvant chemotherapeutic agent to enhance DOX-mediated CS killing. In addition, we postulated that the simultaneous targeting of Notch and mTOR would further enhance the killing of CS cells. These concepts have clinical relevance because if SMI1182 and BEZ235, a potent inhibitor of PI3K/mTOR, could additively increase the chemotherapeutic effects of DOX, then more effective CS treatment could be provided at lower, less toxic doses of DOX. Herein, we report results from in vitro experimental models to assess the potential for future in vivo therapeutics to improve clinical outcomes in patients with CS.

2. Materials and Methods

The studies utilized three mycoplasma-free human CS cell lines: CS1 (RRID: CVCL_T022), CDS11 (RRID: CVCL_WJ30), and CDS17 (T-CDS17: CVCL_WJ32; CDS17: CVCL_WJ31). The CS1 cell line originated from a high-grade (Grade 3) conventional CS tumor in an untreated patient with metastatic disease [51]. CDS11 cells correspond to slow-growing, less invasive secondary conventional CS [12,52]. CDS17 cells are de-differentiated chondrosarcoma cells with a more invasive phenotype compared with CDS11 [52]. All three cell lines were authenticated using short tandem repeat profiling, and all studies were performed with cells passaged 10 or fewer times. CS1 cells were maintained in Roswell Park Memorial Institute (RPMI) 1640 medium (Cytiva/Hyclone, Logan, UT, USA). CDS11 and CDS17 cells were grown in Dulbecco's Modified Eagle's Medium (DMEM) with high glucose (Cytiva/Hyclone, Logan, UT, USA). Culture medium was supplemented with 5% fetal bovine serum (FBS) and 2 mM L-Glutamine (Corning, Glendale, AZ, USA), and all in vitro experiments were conducted at 37 °C in a standard humidified 5% CO_2 incubator.

The investigations were focused on CS1 and CDS11 cells to compare responses between intermediate and high-grade conventional chondrosarcomas. To study the effects of DOX, SMI1182, BEZ235 (see chemical structures in Figure 2), and combination treatments, 24 h old sub-confluent CS cultures were treated with vehicle (DMSO; negative control), 0.0001–10 μM DOX (Pfizer Inc., New York, NY, USA), 0–100 μM SMI1182, or 0.002–2.0 μM BEZ235 by direct addition to the medium. We compared the dose-related effects of monotherapy with the effects of dose-range monotherapy plus a fixed-dose second or third compound (combination therapy). After 48 h of treatment, the cultures were analyzed for cell viability, cytotoxicity, morphology, ASPH immunoreactivity, directional motility, mRNA transcripts related to Notch and insulin, insulin-like growth factor (IGF)/insulin receptor substrate (IRS) networks, and Akt-mTOR pathway signaling. The rationale for these approaches was that previous studies linked ASPH regulation and function to the activation of insulin/IGF-Akt and Notch in various cancers [29,53–55].

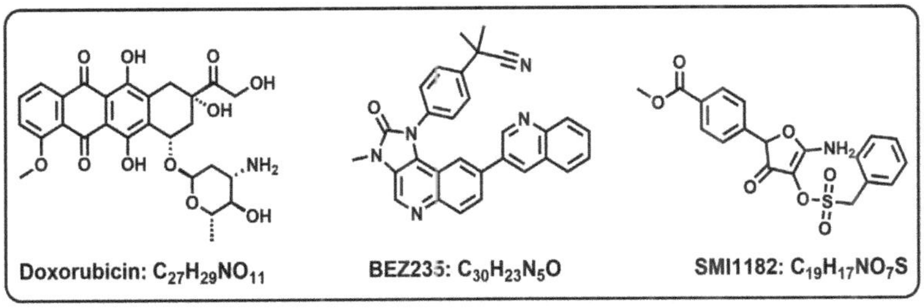

Figure 2. Chemical structures and formulas of Doxorubicin, BEZ235, and SMI1182.

2.1. Cell Viability and Cytotoxicity Assays

Cell viability and metabolic function were measured with the 3-(4,5-dimethylthiazol-2-yl)-2,5-diphenyltetrazolium bromide (MTT) (#M5655, Sigma-Aldrich, St. Louis, MO, USA) assay [56–58]. Cell density was assessed by Hoechst 33342 (H3570, Invitrogen, Carlsbad, CA, USA) fluorescence. Cytotoxicity was measured with the CyQUANT Cytotoxicity Assay Kit (G6PD Release) (Thermo Fisher Scientific, Bedford, MA, USA), which is based on glucose-6-phosphate dehydrogenase release into the culture supernatant. All three assays were performed in 96-well cultures with 3000 viable (Trypan blue-excluded) CS cells seeded in 100 µL of culture medium. The MTT assay was performed by adding 10 µL/well of freshly prepared MTT solution (5 mg/mL in MEM without phenol red) and incubating the plate for 20 min at 37 °C in a 5% CO_2 atmosphere. After aspirating the culture medium, the substrate was eluted by adding 100 µL of acidic isopropanol elution buffer (0.04 mol/L HCl/isopropanol) and 5 min of gentle platform agitation at room temperature. Absorbances were measured at 540 nm in a Spectra-Max M5 Multimode Plate Reader (Molecular Devices, Sunnyvale, CA, USA) [56–58]. To measure cell number in the same wells as MTT activity, the eluate buffer was replaced with 50 µL of 10 µg/mL Hoechst 33342 dye in phosphate-buffered saline (PBS). Fluorescence intensity (Ex360nm/Em460nm) was measured in a Spectra-Max M5 after 5 min of room temperature incubation with light-shielding. Cytotoxicity was quantified in 50 µL of culture supernatant added to CyQUANT assay solution and incubated for 30 min. Fluorescence intensity (Ex530nm/Em590nm) was measured in a Spectra-Max M5.

2.2. Sample Preparation for Immunoassays

CS cells seeded into 6-well plates (1×10^5/mL) were treated for 48 h with vehicle (DMSO), DOX, SMI1182, BEZ235, or combinations. For protein assays, the cells were harvested in weak lysis buffer (50 mM Tris (pH 7.5), 150 mM NaCl, 5 mM EDTA (pH 8.0), 50 mM NaF, 0.1% Triton X-100) containing protease inhibitor cocktail (1mM PMSF, 0.1 mM TPCK, 2 µg/mL aprotinin, 2 µg/mL pepstatin A, 1 µg/mL leupeptin). After centrifuging the homogenates at 14,000× *g* rpm for 10 min at 4 °C, the supernatant fractions were divided and stored at −80 °C. Prior to freezing, small aliquots were set aside to measure protein concentration with the bicinchoninic acid (BCA) assay. Proteins harvested from 6-well cultures and treated as described above were used for Western blot analysis and enzyme-linked immunosorbent assays (ELISAs) to measure immunoreactivity to ASPH (two antibodies: A85G6 and FB50), vimentin, GAPDH, β-Actin, and large acidic ribonuclear protein (RPLPO), as previously described [28,59–61]. (See Table 1 for antibody sources, characteristics, validations, and conditions of use. RPLPO served as a loading control due to its linear correlation with protein content [62].)

Table 1. Antibodies used: characteristics, sources, and concentrations.

Antibody Name *	Antibody Type	Source	Catalog #	Concentration
A85G6-ASPH)	Mouse Monoclonal	Liver Research Center, Brown University Health, RI USA	A85G6	1.3 µg/mL
FB50-ASPH)	Mouse Monoclonal	Liver Research Center, Brown University Health, RI USA	FB50	0.845 µg/mL
GAPDH	Mouse Monoclonal	Santa Cruz Biotechnology	sc-365062	0.067 µg/ml
β-Actin	Mouse Monoclonal	Santa Cruz Biotechnology	sc-47778	0.2 µg/mL
Vimentin	Mouse Monoclonal	RV-202; Abcam	ab8978	2.5 µg/mL
RPLPO	Mouse Monoclonal	Santa Cruz Biotechnology	sc-293260	0.1 µg/mL
HRP-conjugated secondary	Goat Polyclonal	Thermo Fisher Scientific;	Anti-mouse-31430Anti-rabbit-31461	0.04 µg/mL

* Abbreviations: ASPH = aspartyl-asparaginyl-β-hydroxylase; GAPDH = glyceraldehyde-3-phosphate dehydrogenase; HRP = horseradish peroxidase; RPLPO = large acidic ribonuclear protein. # Has been defined to reflect abbreviations in the table.

2.3. *Western Blot Analysis*

Western blot analysis was performed by fractionating protein homogenates (30 µg samples) in 10% SDS-PAGE gels along with pre-stained molecular weight standards (Precision Plus Protein Dual Color Standards, Bio-Rad Laboratories, Hercules, CA, USA; Catalog #161-0374). The samples were electroblot transferred to polyvinylidene difluoride (PVDF) membranes (Bio-Rad Laboratories, Hercules, CA, USA) and incubated with SuperBlock-TBS (Thermo Fisher, Bedford, MA, USA) to mask non-specific binding sites. The membranes were then incubated with primary antibodies overnight at 4 °C with gentle platform agitation. Immunoreactivity was detected with horseradish peroxidase (HRP)-conjugated secondary antibody and Femto-chemiluminescence reagent (Thermo Fisher Scientific; Prod #34096). Immunoreactivity was imaged using the Intelligent Dark Box II (FUJIFILM, Tokyo, Japan). The original, uncropped Western blot membrane can be found in Supplementary Figure S1.

2.4. *Enzyme-Linked Immunosorbent Assay (ELISA)*

Triplicate 50 ng aliquots of protein were adsorbed to the bottom surfaces of enzyme immunoassay (EIA) MaxiSorp flat-bottom 96-well plates (Thermo Fisher Scientific; Catalog #436110) by overnight incubation at 4 °C. Non-specific sites were masked with SuperBlock-TBS (Thermo Fisher Scientific; Catalog #37535). Primary antibodies were incubated overnight at 4 °C. After thorough rinsing of the wells, the samples were sequentially incubated with horseradish peroxidase (HRP)-conjugated secondary antibody and Amplex UltraRed fluorophore with TBS rinses between steps. Fluorescence intensity (Ex530nm/Em590nm) was measured in a Spectra-Max M5 instrument. Negative controls included the omission of the primary antibody, secondary antibody, or Amplex UltraRed. The results were normalized to RPLPO as a sample loading control because previous studies showed that the levels of RPLPO immunoreactivity correlate with protein content and do not vary with experimental conditions [61]. In contrast, GAPDH, which is regulated by insulin/IGF stimulation [63] and oxidative stress [64], was found to vary with cytotoxic agent treatment and, therefore, could not be used as a loading control.

2.5. Immunocytochemistry

Cytospin preparations of cultured cells were generated using a Shandon Cytospin 3 Cytology Centrifuge with Rotor (Marshall Scientific, Hampton, NH, USA). Cells grown in 6-well plates were trypsinized and re-suspended to 0.5×10^6 cells/mL in culture medium containing 0.5% FBS to prevent non-specific lysis. Approximately 0.5×10^5 cells were centrifuged onto Plus-charged glass microscope slides at 600× *g* rpm for 5 min at RT, and then immediately fixed in 10% neutral buffered formalin. Replicate slides were stained with crystal violet (Thermo Scientific Chemicals, Haverhill, MA, USA) to examine cell morphology or immunostained with the FB50-ASPH monoclonal antibody (Laboratory-made). Immunoreactivity was detected with the ImmPRESS peroxidase polymer detection reagents (Vector Laboratories, Newark, CA, USA) and diaminobenzidine (DAB) as the chromogen according to the manufacturer's protocol. The immunostained cells were lightly counterstained with Gill's Hematoxylin 2, dehydrated through graded ethanol solutions, and preserved under coverglass with Cytoseal 60 mounting media (Epredia, Kalamazoo, MI, USA; REF# 8310-4).

2.6. Motility Assay

Directional motility was measured using the ATP Luminescence-Based Motility/Invasion (ALMI) assay [65]. This assay was constructed with blind well chambers (Neuro Probe, Gaithersburg, MD, USA) divided into upper and lower segments using 13 mm diameter, 8 μm pore diameter polycarbonate membranes. Motility was measured using uncoated membranes. The motility/invasion assay was assembled by placing culture medium (200 μL) containing 1% FBS as the trophic factor in the blind well (lower part of the chamber below the membrane) and 1.0×10^5 viable cells in 100 μL of 0.1% FBS medium in the upper chamber. The assays were incubated for 30 min at 37 °C in a standard CO_2 cell culture incubator. Results were analyzed by measuring ATP luminescence in the cells remaining in the upper chamber (non-migrated), distributed on the undersurface of the membrane (migrated adherent), and distributed in the lower chamber (migrated non-adherent). ATP content with measured with ATP-Lite reagents (Perkin-Elmer, Waltham, MA, USA), as previously described [65]. Luminescence was quantified in a TopCount NXT Microplate Scintillation and Luminescence Counter (GMI, Ramsey, MN USA). The percentages of non-motile, motile adherent, and motile non-adherent cells in replicate assays were calculated and used for statistical analysis.

2.7. Quantigene 2.0 RNA Multiplex Assay

Total RNA was extracted from 6-well CS cultures using QIAzol Lysis Reagent (Germantown, MD USA). Messenger RNA transcripts corresponding to the human insulin/Notch pathway signaling molecules were quantified with a custom Quantigene 2.0 Multiplex (QGP) Assay (Affymetrix Inc., Santa Clara, CA, USA). RPL13a served as the internal control gene for normalizing results. Cooperative hybridization and quantification were performed following the manufacturer's protocol. Briefly, the assay allows for direct mRNA quantification using xMAP Luminex beads and reporter signal amplification. A working bead mix containing lysis mixture, blocking reagent, capture beads, and a 2.0 probe set was prepared and distributed in a 96-well format plate. Total RNA (1 μg) was added to the sample wells and incubated overnight with the xMAP fluorescent beads. Sterile nuclease-free water was used as the negative control. The samples were first incubated with a set of oligonucleotide probes (pre-amplifier, amplifier, and biotin-label), followed by streptavidin-conjugated R-Phycoerythrin (SAPE). The resulting fluorescent signals were detected with a Luminex MAGPIX instrument (Diasorin, Austin, TX, USA). MAGPIX calibration and verification standards were used throughout, ensuring the levels of SAPE fluorescence were proportional to RNA transcript abundance captured by the beads. After subtracting the probe-related background from the target median fluorescence intensity (MFI), the results were normalized to RPL13a.

2.8. Multiplex ELISAs (Akt-mTOR Pathway)

We used 11-Plex MILLIPLEX Akt/mTOR Total and Phosphoprotein Magnetic Bead Kits (MilliporeSigma, Burlington, MA, USA) to evaluate the effects of treatment on mTOR signaling mechanisms in CS cells (Table 2). Multiplex ELISAs were performed according to the manufacturer's protocol except that the reactions were optimized by using 12.5 µg protein/sample. Following incubation with antibody-bound beads, immunoreactivity in the captured antigens was detected with biotinylated secondary antibodies and phycoerythrin-conjugated streptavidin and measured in a MAGPIX with xPONENT software (https://int.diasorin.com/en/luminex-ltg/reagents-accessories/software). Standard curves were generated for each analyte.

Table 2. The 11-Plex Akt-mTOR pathway (proteins and phosphoproteins).

Akt/mTOR Pathway Molecules	Protein Abbreviation	Phosphoprotein
Insulin Receptor	Insulin-R	$pYpY^{1162/1163}$-Insulin R
Insulin-Like Growth Factor Receptor Type 1	IGF-1R	$pYpY^{1135/1136}$-IGF-1R
Insulin Receptor Substrate, Type 1	IRS1	pS^{636}-IRS-1
Akt (Protein Kinase B)	Akt	pS^{473}-Akt
Phosphatase and tensin homolog	PTEN	pS^{380}-PTEN
Glycogen Synthase Kinase 3α	GSK-3α	pS^{21}-GSK3α
Glycogen Synthase Kinase 3β	GSK-3β	pS^{9}-GSK3β
Tuberous Sclerosis Complex 2	TSC2	pS^{939}-TSC2
Mechanistic Target of Rapamycin	mTOR	pS^{2448}-mTOR
p70 Ribosomal S6 kinase	P70S6K	pT^{412}-p70S6K
Ribosomal Protein S6	RPS6	$pS^{235/S236}$-RPS6

2.9. Data Analysis

All assays were repeated at least 3 times, with 3 or 4 technical and experimental replicates per sample. Violin plots depict the distribution of results with the median (horizontal bars) and 95% confidence interval limits. The data were analyzed by *t*-test or Analysis of Variance (ANOVA) with corrections for multiple comparisons. Statistical analyses and graphical presentation of results were performed using GraphPad Prism 10.4 (GraphPad Software Inc., San Diego, CA, USA).

2.10. Reagent Sources

The ELISA MaxiSorp 96-well plates, bicinchoninic acid (BCA) reagents, horseradish peroxidase (HRP)-conjugated secondary antibodies, and Superblock (TBS) were purchased from Thermo-Fisher Scientific (Bedford, MA, USA). The soluble fluorophores, Amplex UltraRed and 4-Methylumbelliferyl phosphate (4-MUP), were from Life Technologies (Carlsbad, CA, USA). All other fine reagents were purchased from CalBiochem/Millipore Sigma (Burlington, MA, USA), Pierce Chemical (Dallas, TX, USA), or Sigma-Aldrich Co. (St. Louis, MO, USA). SMI1182 was synthesized, crystallized, and validated using NMR, mass spectrometry, and thin-layer chromatography [66] (see Figure S1).

3. Results

3.1. ASPH Immunoreactivity Detected in CS Cells

Western blot analysis was performed with two ASPH monoclonal antibodies, FB50 and A85G6. FB50 binds to the N-terminal region of ASPH, which overlaps with the sequence for Humbug [67,68], a truncated form of ASPH lacking the C-terminal catalytic domain [69–71]. A85G6 binds to the C-terminus of ASPH [68], which contains a unique catalytic domain for Notch and Jagged hydroxylation [67,69]. Western blot analysis detected ASPH immunoreactivity in CS1 (Grade 3 conventional chondrosarcoma), CDS11 (Grade 2 conventional chondrosarcoma), CDS17 (de-differentiated chondrosarcoma), and CYZ/PNET2 neuroblastoma cells (positive control) (Figure 3A). A85G6 detected the expected ~86kD full-length ASPH protein, as well as super-shifted, likely phosphorylated forms, and cleavage products in all four cell lines (Figure 3A, left panel). FB50 detected abundant ASPH immunoreactivity in the three CS cell lines but very low levels in CYZ/PNET2 cells (Figure 3A, middle panel). The dominant FB50-positive band corresponds to Humbug. Re-probing the blot with antibodies to the p85-subunit of PI3 Kinase detected p85 and its p50 spliced variant and demonstrated comparable sample protein loading in each lane [72] (Figure 3A, right panel). ELISA studies confirmed the presence of A85G6 ASPH immunoreactivity in CS1 and CDS11 cells and more clearly demonstrated higher levels of A85G6-ASPH immunoreactivity in CS1 compared with CDS11 cells (Figure 3B).

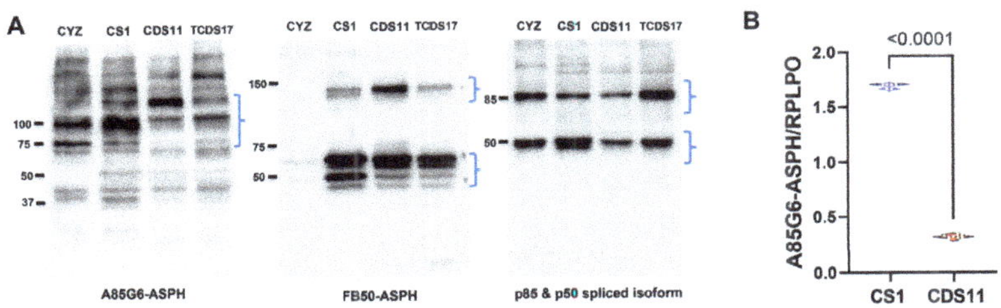

Figure 3. (**A**) The A85G6 or FB50 monoclonal antibody, which binds to the C-terminal or N-terminal region of ASPH, was used for Western blot analysis of CYZ (neuroblastoma; positive control), CS1, CDS11, and TCDS17 cells. The same blot was stripped and re-probed. The third probing with polyclonal antibodies to p85 served as a loading control. The positions of molecular weight standards are indicated. Brackets mark the expected positions of A85G6-ASPH, FB50-ASPH/Humbug, and p85 together with the p50 subunit. (**B**) ELISA results comparing levels of A85G6-ASPH immunoreactivity in CS1 and CDS11 cells. The levels of immunoreactivity were normalized to RPLPO measured in the same samples. Intergroup comparisons were made by t-test.

3.2. Mechanistic and Functional Differences Between CS1 and CDS11 Cells

Since higher levels of ASPH have been associated with higher-grade tumors [31,35,36,42,73–76] and abbreviated survival and post-treatment, disease-free intervals [31,36], we compared the CS1 and CDS11 cells with respect to directional motility (Figure S2), mRNA expression of insulin/IGF/IRS pathway genes that regulate ASPH (Figure S3), and mRNA expression of Notch pathway molecules (Figure 4). Using the ATP-Lite Motility assay, we detected similar percentages of motile adherent, motile non-adherent, non-motile, and total motile populations in CS1 versus CDS11 cultures (Figure S2). Multiplex RNA hybridization studies demonstrated that *INSULIN*, *IGF1*, and *IGF2* were expressed at low levels in CS cells, *IGF2R* was more abundantly expressed than *INSULIN-R* and *IGF1R*, and *IRS1* was 10-fold more abundant than *IRS2* and *IRS4* (Figure S3). Intergroup statistical comparisons revealed that CDS11 cells had significantly ($p \leq 0.05$) or statistical trendwise ($0.10 < p < 0.05$) lower expression of *INSULIN-R* (Figure S2D), *IGF2R* (Figure S2F), and IRS1 (Figure S2G) but higher levels of *INSULIN* (Figure S1A) compared with CS1 cells. The other five *INSULIN/IGF/IRS* pathway genes were similarly expressed in CS1 and CDS11 cells (Figures S2B,C,E,H,I). Concerning Notch pathway mRNA transcripts, CDS11 cells expressed significantly lower levels of *NOTCH1* (Figure 4B), *JAGGED1* (Figure 4C), *HES1* (Figure 4D), and *HIF1a* (Figure 4F) compared to CS1 cells. In contrast, there were no significant inter-group differences in mRNA expression of *ASPH* (Figure 4A) or *HEY1* (Figure 4E).

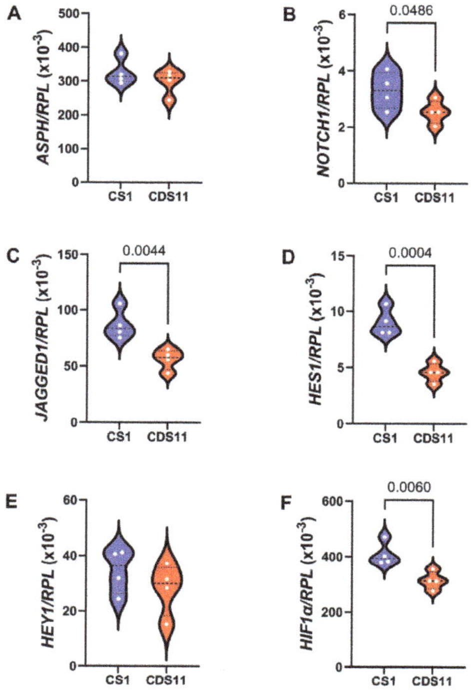

Figure 4. ASPH and Notch network gene expression in Grade 3 versus Grade 2 conventional chondrosarcoma. CS1 (Grade 3) and CDS11 (Grade 2) cells were analyzed for (**A**) *ASPH*, (**B**) *NOTCH1*, (**C**) *JAGGED1*, (**D**) *HES1*, (**E**) *HEY1*, and (**F**) *HIF1α* mRNA expression using a multiplex bead-based RNA hybridization protocol with results normalized to Ribosomal Protein L13a (RPL). Inter-group comparisons were made by *t*-test analysis. Significant *p*-values ($p \leq 0.05$) are shown in the panels.

3.3. DOX, BEZ235, and SMI1182 Monotherapy and Combination Therapy Effects on Metabolic Activity, Cell Viability, and Cytotoxicity

This segment of the study examined CS1 and CDS11 dose-responses to DOX, BEZ, and SMI1182 monotherapy and combination therapy using the MTT activity, H33342 fluorescence (index cell number), and CYQUANT assays. All assays were performed in 96-well cultures that were treated for 48 h with dose ranges of DOX (0.0001–1 μM), BEZ235 (BEZ) (0.002–2.0 μM), or SMI1182 (0 to 100 μM), or with vehicle (DMSO control). The effects of combination therapy were assessed by combining the dose range of one compound with a fixed dose of one or two of the other compounds. In addition to dose–response curves, treatment effects were compared using area-under-curve (AUC) analysis and one-way ANOVA tests. The MTT assay provides a measure of mitochondrial function, proliferation, and cell viability. However, under conditions of mild oxidative stress, MTT activity can increase, potentially leading to erroneous interpretations about cellular responses, e.g., proliferation [58]. To complement the MTT studies, the cells were also stained with H33342 dye, which labels DNA; H33342 fluorescence is linearly correlated with cell number.

Dead or dying cells detach from the plate and thus are not counted/measured in the assay. Cellular cytotoxic responses were evaluated by measuring G6PD release using the CYQUANT assay.

DOX (Figure 5): In CS1 cells, DOX monotherapy had minimal effects on MTT activity or H33342 fluorescence over most of the dose range tested (Figure 5A,B). Co-treatment with SMI1182, BEZ, or SMI1182+BEZ slightly increased MTT activity but profoundly reduced H33342 relative to vehicle, as demonstrated by ANOVA tests of the AUC analyses (Figure 5E,F) (MTT: F = 4.465; $p = 0.008$; H33342: F = 231.6; $p < 0.0001$) with post hoc Tukey tests. The H33342 results showed that cell loss was greatest in cultures co-treated with DOX+SMI+BEZ, followed by DOX+BEZ. In contrast, CDS11 cells exhibited similar dose-dependent reductions in MTT and H33342 following treatment with DOX monotherapy, DOX+BEZ, or DOX+BEZ+SMI1182 (Figure 5C,D,G,H). However, co-treatment with DOX plus SMI1182 was protective in maintaining relatively high levels of MTT and H33342, except at the highest doses of DOX. ANOVA tests of the AUC analyses demonstrated significant inter-group differences for both MTT (F = 37.35, $p < 0.0001$) and H33342 (F = 37.35, $p < 0.0001$).

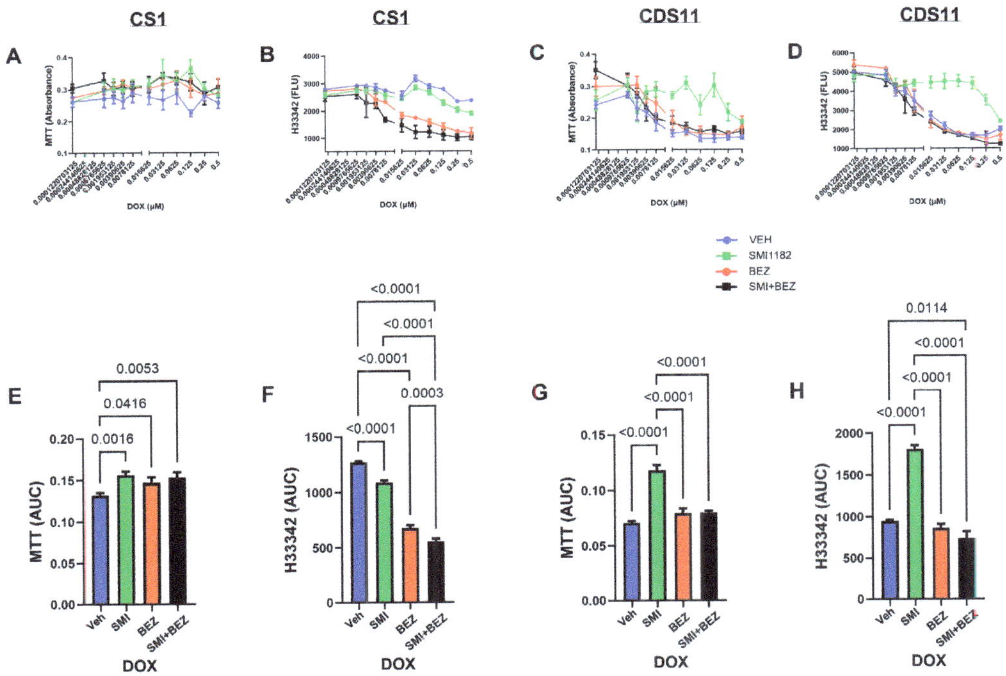

Figure 5. Dose effects of DOX+Vehicle, DOX+SMI1182, DOX+BEZ, or DOX+SMI+BEZ on MTT activity (**A**,**C**) and H33342 fluorescence (viability; **B**,**D**) in CS1 (**A**,**B**) and CDS11 (**C**,**D**) cells. The assays were performed with cells seeded in 96-well micro-cultures and treated for 48 h with dose ranges of DOX plus fixed doses of SMI1182 (0.5 µM), BEZ (2 µM), or SMI+BEZ. Panels (**E**–**H**) depict area-under-curve analysis results corresponding to the curves in Panels (**A**–**D**), respectively. Data points and bar graphs show mean ± S.D of results of 4 replicate cultures. Inter-group differences were compared by ANOVA with post hoc tests. Significant *p*-values ($p \leq 0.05$) are shown in the panels.

BEZ (Figure 6): CS1 cells treated with BEZ or BEZ+SMI1182 exhibited similar dose-dependent reductions in MTT activity, whereas co-treatment with BEZ+DOX or BEZ+DOX+SMI resulted in sustained high levels of MTT over the full BEZ dose range, as confirmed by AUC analysis (Figure 6A,E). Correspondingly, ANOVA tests were significant for the inter-group differences in MTT (F = 132.7, $p < 0.0001$) and H33342 (F = 26.12, $p < 00001$). In contrast, in CS1 cells treated with BEZ alone or in combination with SMI1182, DOX, or DOX+SMI1182, H33342 fluorescence (cell viability) progressively declined (Figure 6B), with significant additive effects of the co-treatments, as demonstrated by AUC analysis (Figure 6F). CDS11 cells treated with BEZ alone exhibited reduced MTT activity at the lowest doses but minimal further reduction except at the highest doses (1–2 µM) (Figure 6C). Co-treatment with SMI1182 reduced MTT such that the mean AUC was significantly below the other three treatments (F = 30.17, $p < 0.0001$) (Figure 6C,G). BEZ+DOX or BEZ+DOX+SMI1182 caused similar dose-dependent and overall reductions in MTT relative to BEZ-only treatment (Figure 6C,G). Regarding H33342, BEZ-only had a minimal effect on CDS11 cell viability, but the addition of SMI1182, DOX, or DOX+SMI1182 sharply reduced cell viability by 50% or more (F = 221.6, $p < 0.0001$) (Figure 6D,H).

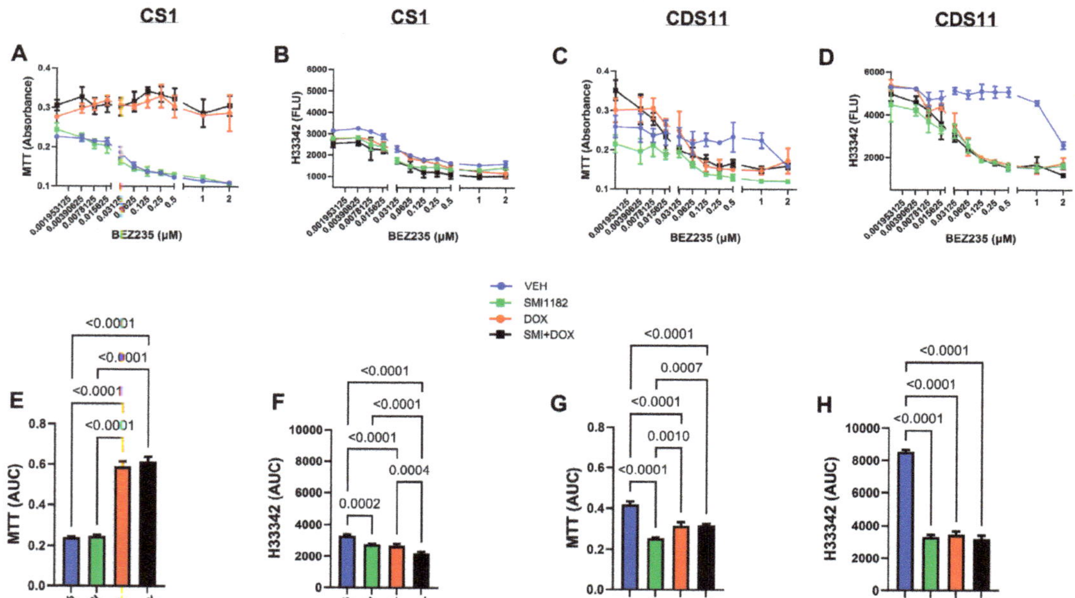

Figure 6. Dose effects of BEZ+Vehicle (Veh), BEZ+SMI1182, BEZ+DOX, or BEZ+SMI+DOX on MTT activity (**A**,**C**) and H33342 fluorescence (viability; **B**,**D**) in CS1 (**A**,**B**) and CDS11 (**C**,**D**) cells. The assays were performed with cells seeded in 96-well micro-cultures and treated for 48 h with dose ranges of BEZ plus fixed doses of SMI1182 (0.5 µM), DOX (0.05 µM), or SMI+DOX. Panels (**E–H**) depict area-under-curve analysis results corresponding to the curves in Panels (**A–D**), respectively. Data points and bar graphs show the mean ± S.D. of results from 4 replicate cultures. Inter-group differences were compared by ANOVA with post hoc Tukey tests. Significant p-values ($p \leq 0.05$) are shown in the panels.

SMI1182 (Figure 7): Previous studies showed that SMI1182 monotherapy functions by reducing invasive growth and cell motility [77,78]. In the present study, the treatment of CS1 cells with 10 µM or lower concentrations of SMI1182 caused only modest reductions in MTT activity (Figure 7A) and H33342 fluorescence (Figure 7B) and minimal increases in G6PD release (cytotoxicity) (Figure 7C). However, SMI1182 treatments above 10 µM sharply reduced viability and increased cytotoxicity. Dual SMI1182+BEZ or SMI1182+DOX significantly reduced MTT activity relative to SMI1182 monotherapy, as shown in both the dose–response curves and AUC analyses (F = 34.62, $p < 0.0001$) (Figure 7A,D). In addition, SMI1182 + DOX reduced H33342 (F = 23.30, $p < 0.0001$) (Figure 7B,E) and increased cytotoxicity (F = 10.69, $p = 0.0003$) (Figure 7C,F) compared with SMI1182 monotherapy and SMI1182+BEZ. Although at lower doses BEZ co-treatment reduced CS1 cell viability, at higher doses there was no significant difference relative to SMI1182 alone, accounting for the similar AUC calculated results. Correspondingly, there were no significant effects of SMI1182+BEZ on G6PD release relative to SMI1182 treatment alone (Figure 7C,F).

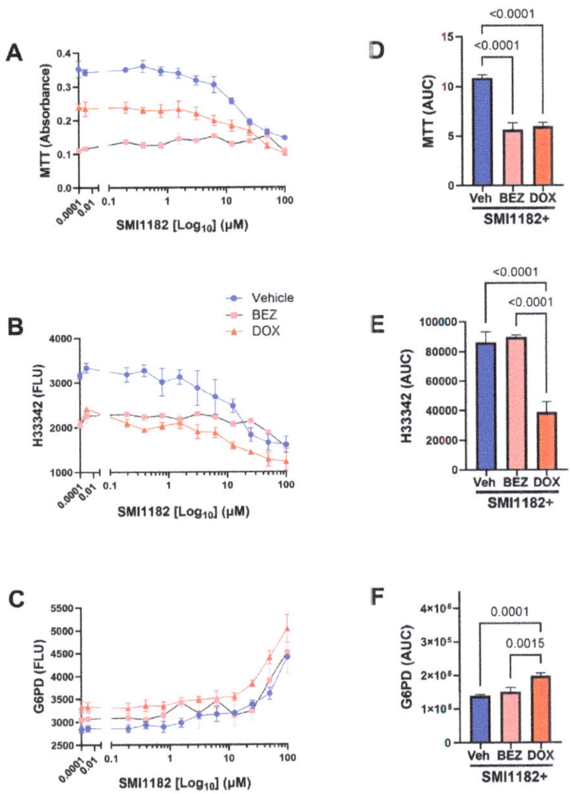

Figure 7. Dose effects of SMI1182 plus vehicle, BEZ, or DOX on MTT activity, H33342 fluorescence (viability), and G6PD release (cytotoxicity). CS1 96-well micro-cultures were treated with a dose range of SMI1182 and co-treated with vehicle or fixed doses of BEZ or DOX for 48 h. (**A**) MTT, (**B**) H33342, and (**C**) G6PD assay results were measured in 4 replicate cultures per data point. Panels (**D–F**) depict area-under-curve analyses. Data points and bar graphs show mean ± S.D. of results. by ANOVA with post hoc Tukey tests. Significant p-values ($p \leq 0.05$) are shown in the panels.

3.4. Effects of SMI1182, DOX, and BEZ on CS Morphology

Crystal violet-stained chamber slide cultures (Figure 8) and cytospin preparations (Figure 9) were used to evaluate cellular morphological responses to 48 h of treatment with vehicle, SMI1182, BEZ, or DOX. The chamber slides showed that the vehicle- and SMI1182-treated cultures grew to confluence with high cellular densities, and they exhibited similar nuclear and cytoplasmic morphologies (Figure 8). In contrast, parallel cultures treated with BEZ or DOX exhibited reduced cell densities, increased nuclear hyperchromasia (dark, densely stained), numerous cells with narrow, tapered morphologies, and scattered cellular debris (Figure 8). In addition, DOX increased the variability in cell size manifested by irregular nuclear and cytoplasmic enlargements, whereas BEZ-treated cells often exhibited small shrunken nuclei (pre-apoptotic).

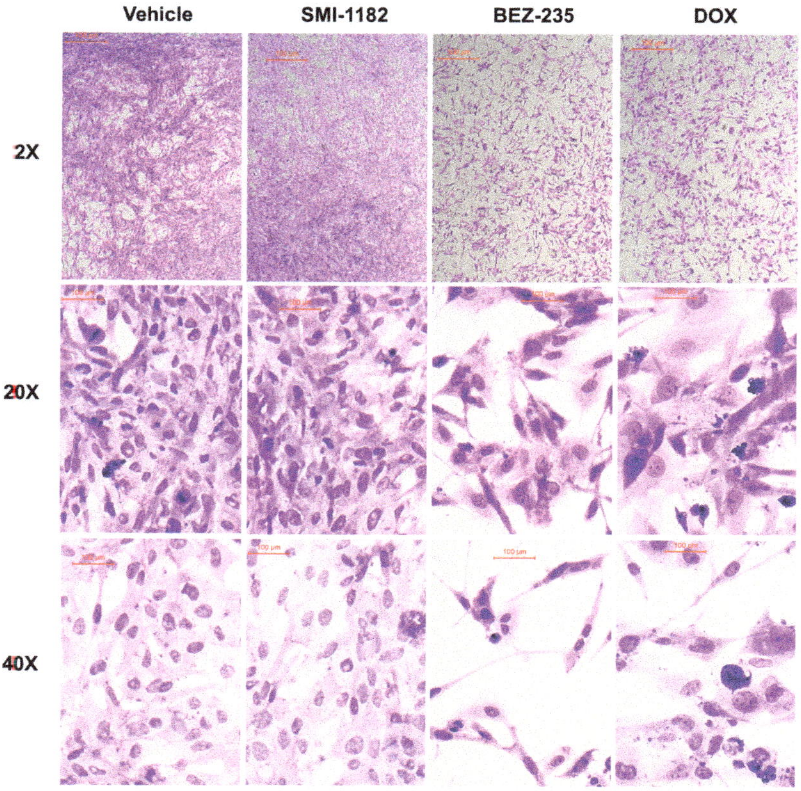

Figure 8. Cytomorphological effects of vehicle, SMI1182, BEZ235, or DOX treatment of CS1 chamber slide cultures. CS1 cells were treated for 48 h, then fixed in formalin, stained with crystal violet, and examined by light microscopy. Representative areas of the cultures were photographed with the 2×, 20×, or 40× objective. Scale bars (100 µm) are included in each panel. Note the similarly high cell densities and relatively uniform cellular and nuclear morphologies in cultures treated with vehicle or SMI1182, versus lowered cell densities and greater variation in nuclear and cellular sizes and the tapered, spindled appearance of cells in BEZ235- and DOX-treated cultures.

Cytospin preparations were generated by trypsin-dissociating the cultures and centrifuging the cells onto glass slides in a 5 mm focused area. This approach allows the visualization of individual cells apart from the monolayers. Control (vehicle-treated) cells had pleomorphic polygonal morphologies with high nuclear-to-cytoplasmic ratios, variability in cell size, scattered cells with nuclear condensation (pyknosis), and readily detected nuclear mitoses (Figure 9, Upper Left). SMI1182 treatment resulted in conspicuous cellular enlargement and surface blebbing, nucleolar prominence, and cytoplasmic vacuolation (Figure 9, Upper Right). DOX increased cell size, shape variability, nuclear condensation, apoptosis, cellular fragmentation, and atypical nuclear mitoses (Figure 9, Lower Left). BEZ treatment resulted in pronounced cellular/nuclear condensation (pre-apoptotic stage), modest surface blebbing, and cells with abundant fine hair-like surface processes (Figure 9, Lower Right). Combined SMI1182+DOX, SMI1182+BEZ, or BEZ+DOX treatments produced overlapping cytomorphological changes corresponding to the effects of each drug.

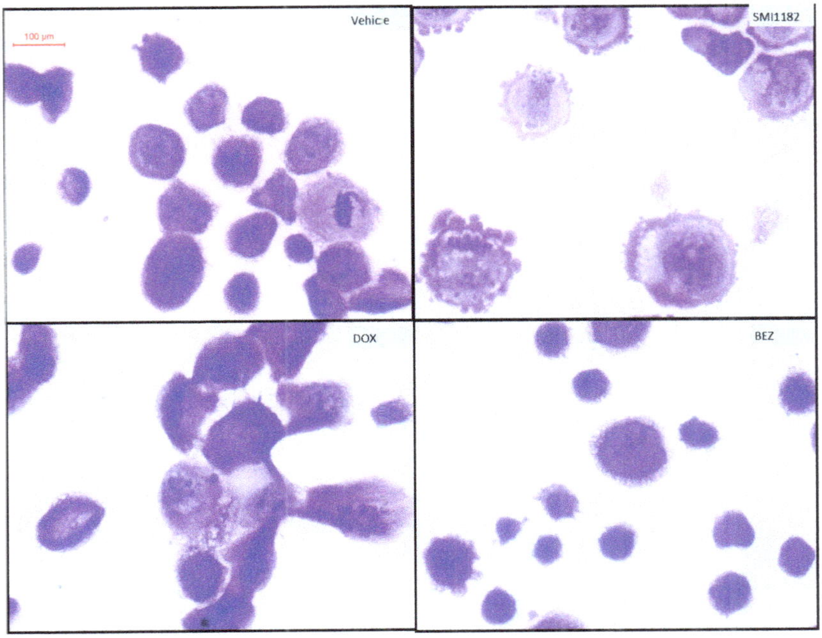

Figure 9. Cytomorphological effects of vehicle, SMI1182, DOX, or BEZ235 treatment of CS1 cells. After 48 h of treatment in 6-well plates, the cells were harvested and used to generate cytospin preparations, which were fixed in formalin, stained with crystal violet, and examined by light microscopy. (**Upper Left**) Vehicle/control-treated cells exhibited variation in cell size, prominent nucleoli, and nuclear condensation. (**Upper Right**) SMI1182 treatment resulted in prominent cytoplasmic surface blebbing, cellular enlargement, and both nuclear and cytoplasmic vacuolation. (**Lower Left**) DOX treatment increased nucleolar prominence and cytoplasmic vacuolation. (**Lower Right**) BEZ caused abundant nuclear condensation and necrosis, as well as conspicuous fine, hair-like surface processes.

3.5. *Effects of SMI1182, DOX, and BEZ on ASPH Immunoreactivity*

ELISA and immunocytochemical staining studies were used to compare the effects of SMI1182, DOX, and BEZ on ASPH immunoreactivity. Parallel studies examined immunoreactivity to cytoskeletal proteins for correlation with the cytomorphologic treatment effects. In CS1 cells, BEZ and SMI1182 significantly reduced A85G6-ASPH immunoreactivity (Figure 10A), whereas SMI1182 singularly increased FB50-ASPH (Humbug) (Figure 10B), GAPDH (Figure 10C), and β-Actin (Figure 10D) immunoreactivities relative to vehicle. In addition, vimentin immunoreactivity was elevated by each of the treatments relative to the control (Figure 10E). Immunocytochemical staining with the FB50-ASPH antibody revealed cell surface, cytoplasmic, perinuclear, and beaded fibrilla process labeling in vehicle-treated cultures (Figure 11-Left). SMI1182 treatment resulted in ASPH immunoreactivity tightly localized along the inner subjacent region of the plasma membrane or in the nucleus rather than on the outer membrane surfaces (Figure 11-Center Left). DOX treatment resulted in prominently fibrillar and beaded cell surfaces rather than cytoplasmic or perinuclear FB50-ASPH immunoreactivity (Figure 11-Center Right). BEZ treatment resulted in similar patterns of FB50-ASPH immunoreactivity as observed in the control CS1 cells, but with notably increased dense nuclear, short surface fibrillar, and surface bleb staining patterns (Figure 11-Right).

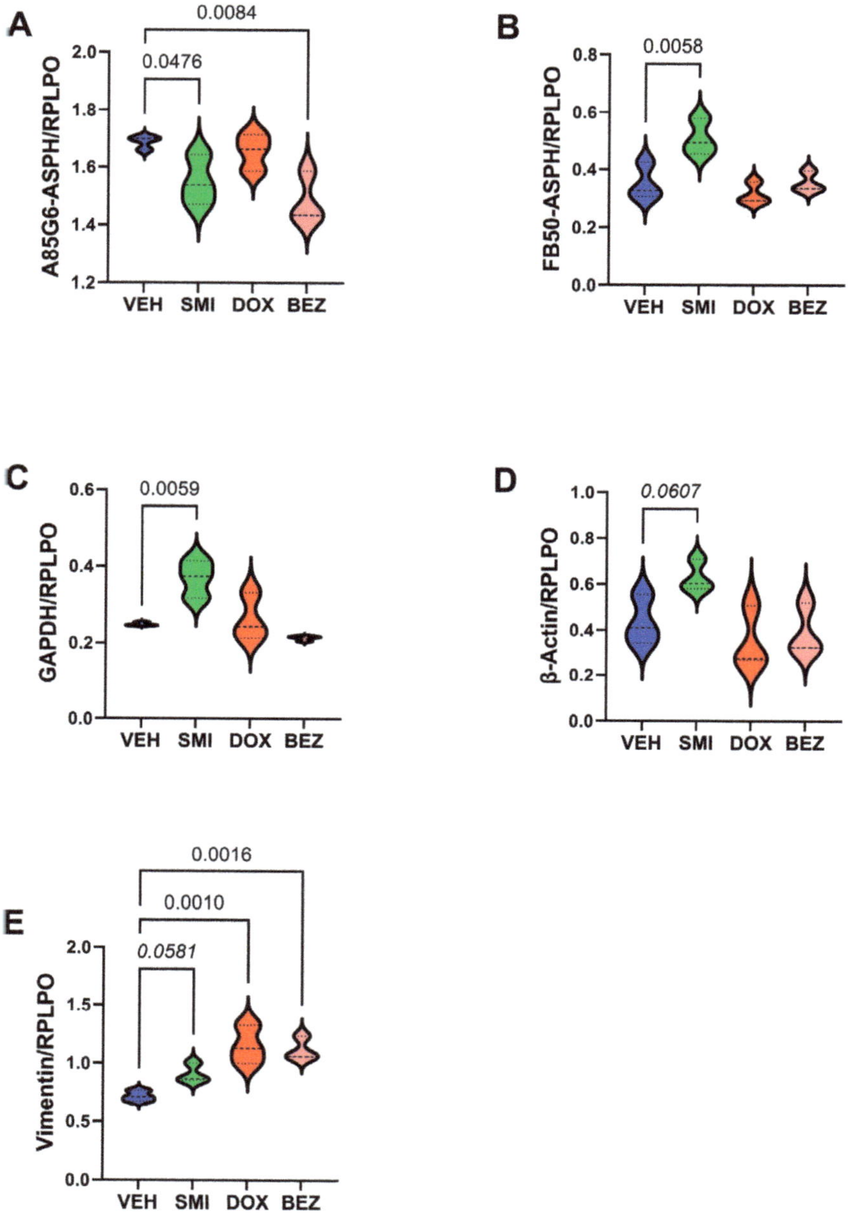

Figure 10. ELISA results following treatment of CS1 cells with SMI1182, DOX, or BEZ. Immunoreactivity to (**A**) A85G6-ASPH, (**B**) FB50-ASPH, (**C**) GAPDH, (**D**) β-Actin, and (**E**) vimentin was measured in 4 replicate 50 ng protein samples by ELISA, with results normalized to RPLPO. Violin plots depicting the median (mid-level line) and 95% confidence interval limits (upper and lower lines). Inter-group statistical comparisons were made using one-way ANOVA with post hoc Tukey tests. Significant *p*-values ($p \leq 0.05$) are shown in the panels. Statistical trend effects ($0.10 < p < 0.05$) are indicated with italic font.

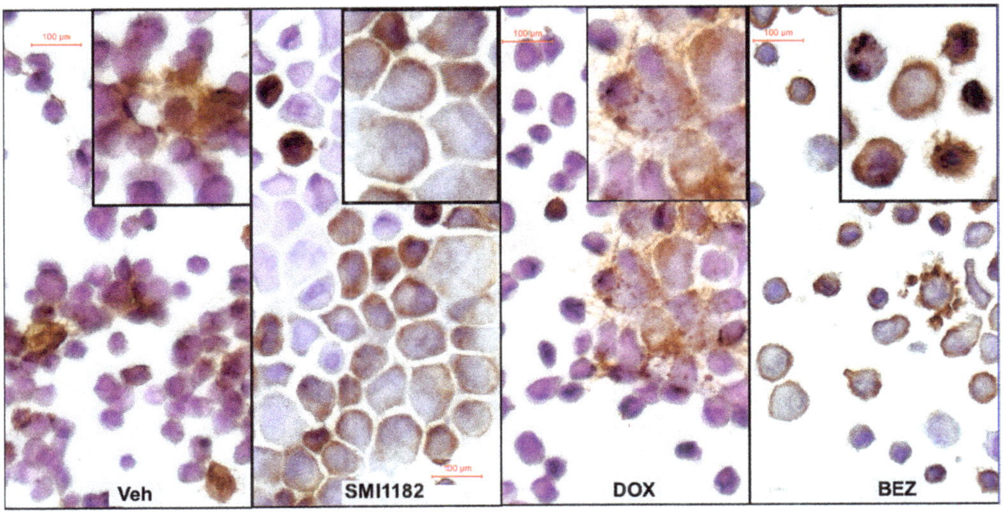

Figure 11. Altered cellular distributions of ASPH immunoreactivity in CS1 cells treated with vehicle (Veh), 3.125 µM SMI1182 (SMI1182), 0.05 µM DOX, or 2 µM BEZ for 48 h. Cytospin preparations were immunostained with FB50-ASPH monoclonal antibody, and immunoreactivity was revealed with biotinylated secondary antibody, horseradish peroxidase-conjugated streptavidin, and DAB (brown precipitate). All images were obtained at 400× magnification. Insets show higher magnification images of FB50-ASPH-positive labeling. Vehicle-treated cells show surface, cytoplasmic, and perinuclear FB50-ASPH immunoreactivity. SMI1182 treatment resulted in FB50-ASPH immunoreactivity localized along the inner surface of the plasma membrane or perinuclear rather than the cell surface. DOX treatment resulted in conspicuous bead-like fibrillar labeling of cell surface processes, particularly in larger tumor cells (see inset). BEZ treatment resulted in high percentages of pyknotic/shrunken cells with condensed immunostained nuclei and increased surface process and bleb FB50-ASPH immunoreactivity.

3.6. Cell Motility Studies

The effects of 3.125 µM SMI1182, 0.05 µM DOX, and 2 µM BEZ on directional motility were evaluated in CS1 cells using the ATP-Lyte Motility assay with uncoated polycarbonate membranes. Two-way ANOVA revealed significant effects of treatment (F = 175.3; DFn, DFd: 3,48; $p < 0.0001$) and treatment × motility index interactions (F = 5.561; DFn, DFd: 9,48; $p < 0.0001$). The post hoc multiple comparisons tests demonstrated that SMI1182 reduced the percentages of migrated adherent and increased the migrated non-adherent percentages relative to vehicle, DOX, and BEZ (Figure 12A,B). DOX significantly increased the percentage of migrated adherent cells and the overall total percentage of migrated cells relative to SMI1182 and BEZ (Figure 12A,D), and it increased the percentage of migrated adherent cells relative to vehicle (Figure 12A). In contrast, BEZ did not significantly alter motility indices relative to any of the treatments, including vehicle (Figure 12A–D). In essence, the effects of SMI1182 and DOX were opposite from one another; SMI1182 reduced cell motility and adhesion, whereas DOX increased adherent cell motility and overall cell motility.

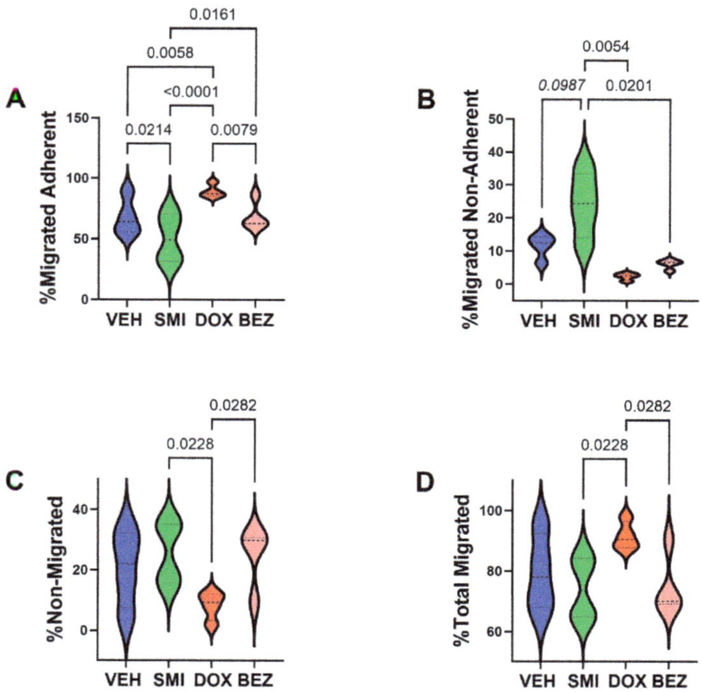

Figure 12. Effects of SMI1182, DOX, and BEZ on directional cell motility and adhesion. CS1 cells were treated for 48 h with vehicle (VEH), SMI1182 (SMI), DOX, or BEZ, then transferred to a Boyden chamber-type apparatus containing uncoated 8 µM pore diameter polycarbonate filters. Cell migration was allowed to proceed for 2 h in a standard cell culture incubator. The ATPLite luminescence assay was used to measure the motility index, i.e., the percentages of (**A**) motile adherent (migrated through the pores but still adherent to the undersurface of the membrane), (**B**) motile non-adherent (migrated to the bottom well of the chamber), (**C**) non-migrated (remaining on the upper membrane surface), and (**D**) total motile (motile adherent + motile non-adherent). Results were analyzed by ANOVA with post hoc multiple comparisons. Significant ($p \leq 0.05$) and statistical trendwise ($0.05 < p < 0.10$) p-values are displayed. Trendwise effects have italic font.

3.7. Molecular Correlates of SMI1182-, DOX-, and BEZ-Mediated Effects in CS Cells

The goal of these studies was to gain an understanding of the molecular distinctions between the high-grade CS1 and intermediate-grade CDS11 cells for correlation with malignant behavior and responsiveness to treatment. Such analyses could impact future strategies to improve the effectiveness of CS chemotherapy. Quantigene multiplex RNA hybridization assays were used to examine the effects of DOX, BEZ, and DOX+BEZ on the expression of the ASPH-Notch pathway (Figures 13 and 14) and insulin/IGF signaling networks (Figures S3 and S4) that mediate cell growth, survival, migration, and invasion. In addition, we used multiplex bead-based ELISAs to measure total and phosphorylated molecules in the insulin/IGF-Akt-mTOR pathway in cells treated with vehicle, 10 µM SMI1182, 0.05 µM DOX, or 2 µM BEZ (Figures 15–20). The treatment doses were based on the 50% inhibitory concentrations of the compounds in CS1/CDS11 cells.

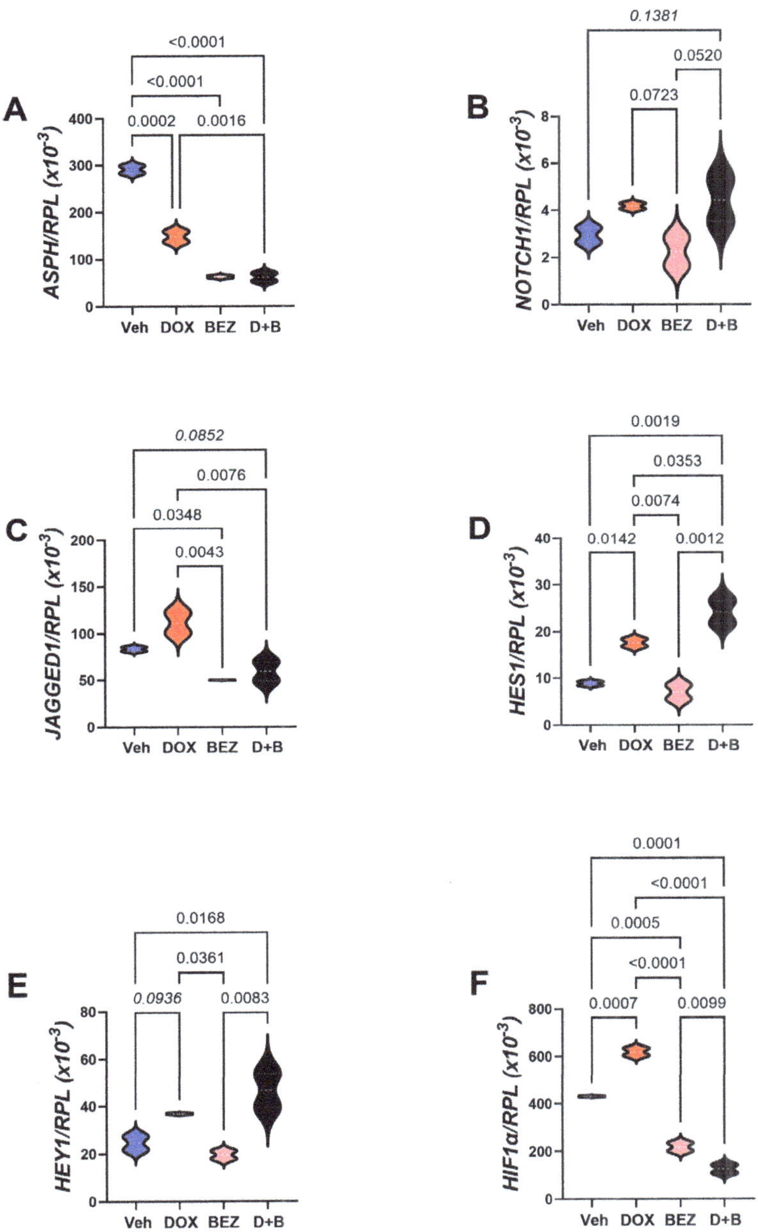

Figure 13. Molecular pathways of DOX-, BEZ-, and DOX+BEZ (D+B)-mediated CS1 cytotoxicity via ASPH and Notch networks. Cells (n = 4 cultures/group) were treated for 48h with Vehicle (Veh), DOX (0.05 µM), BEZ (2 µM), or D+B. (**A**) ASPH, (**B**) NOTCH1, (**C**) JAGGED1, (**D**) HES1, (**E**) HEY1, and (**F**) HIF-1α mRNA levels were measured using a multiplex bead-based RNA hybridization protocol with results normalized to Ribosomal Protein L13a (RPL) (see Table 3). Inter-group comparisons were made by one-way ANOVA with a post hoc repeated measures test. Significant ($p \leq 0.05$) and statistical trendwise ($0.05 < p < 0.10$) p-values are displayed. Trendwise effects have italic font.

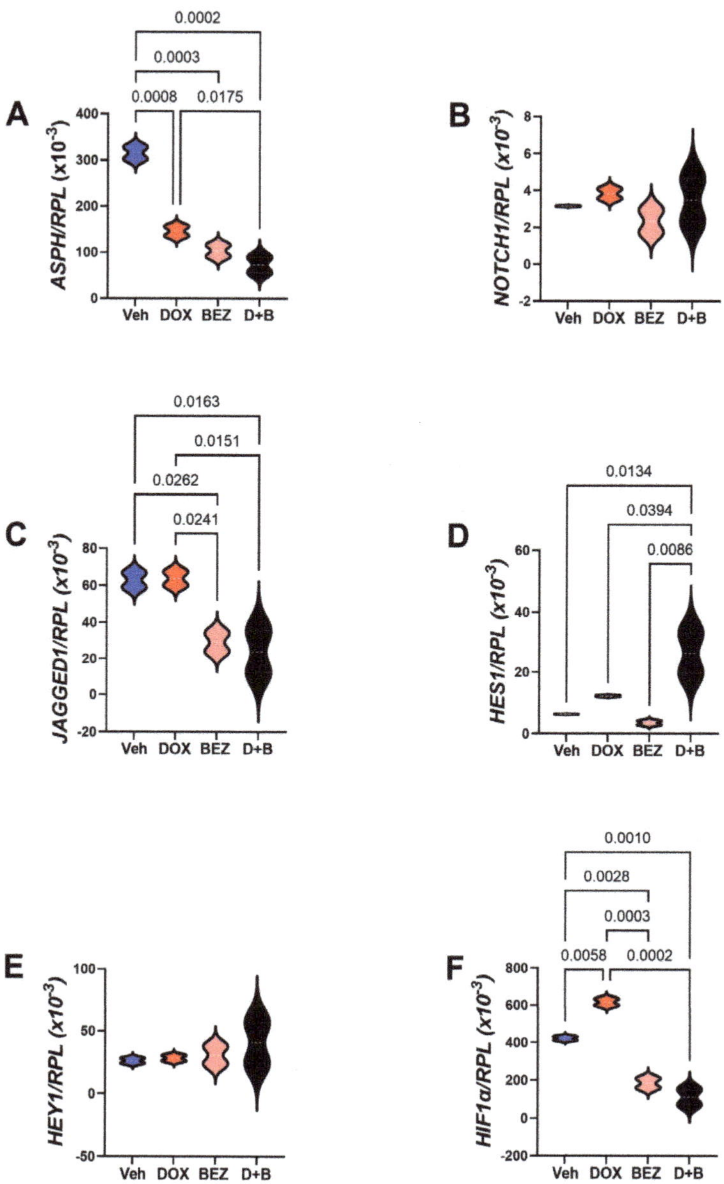

Figure 14. Molecular pathways of DOX-, BEZ-, and DOX+BEZ (D+B)-mediated CDS11 cytotoxicity via ASPH and Notch networks. Cells (n = 4 cultures/group) were treated for 48h with Vehicle (Veh), DOX (0.05 μM), BEZ (2 μM), or D+B. (**A**) ASPH, (**B**) NOTCH1, (**C**) JAGGED1, (**D**) HES1, (**E**) HEY1, and (**F**) HIF-1α mRNA levels were measured using a multiplex bead-based RNA hybridization protocol with results normalized to Ribosomal Protein L13a (RPL) (see Table 3). Inter-group comparisons were made by one-way ANOVA with a post hoc repeated measures test. Significant ($p \leq 0.05$) and statistical trendwise ($0.05 < p < 0.10$) p-values are displayed. Trendwise effects have italic font.

Table 3. CS1 multiplex RNA hybridization assay ANOVA test results.

Gene Code	Gene Name	F-Ratio	*p*-Value
ASPH	Aspartyl-Asparaginyl-β-Hydroxylase	178.8	0.0001
NOTCH1	Notch 1	6.260	0.05
JAGGED1	Jagged 1	13.66	0.014
HES1	Hairy and enhancer of split-1	43.44	0.0016
HEY1	HES-related family bHLH transcription factor	9.259	0.0268
HIF1α	Hypoxia-Inducible Factor 1-alpha	242.2	<0.0001
INS	Insulin	1.234	N.S.
IGF1	Insulin-Like Growth Factor 1	1.336	N.S.
IGF2	Insulin-Like Growth Factor 2	2.577	N.S.
INSR	Insulin Receptor	2.637	N.S.
IGF1R	Insulin-Like Growth Factor 1 Receptor	3.164	N.S.
IGF2R	Insulin-Like Growth Factor 2 Receptor	9.891	0.0254
IRS1	Insulin Receptor Substrate, Type 1	16.01	0.0108
IRS2	Insulin Receptor Substrate, Type 2	0.667	N.S.
IRS4	Insulin Receptor Substrate, Type 4	0.487	N.S.

Molecular pathways of DOX-, BEZ-, and DOX+BEZ (D+B)-mediated CS1 cytotoxicity via insulin, IGF, and IRS pathways. The mRNA levels were measured with a multiplex bead-based RNA hybridization protocol, with results normalized to Ribosomal Protein L13a (RPL). Inter-group comparisons were made by one-way ANOVA. See Figure S4 for the results of the Tukey post hoc repeated measures tests of all inter-group comparisons. n = 4 cultures/group. Significant ($p \leq 0.05$) *p*-values are displayed. N.S. = not significant.

ASPH-NOTCH Panel: Comparisons were made with cells treated with DOX, BEZ, or DOX+BEZ (D+B). SMI1182 treatment effects were not studied because preliminary studies failed to demonstrate any changes in mRNA expression. One-way ANOVA tests detected significant inter-group differences with respect to *ASPH, NOTCH1, JAGGED1,* HES1, HEY1, and HIF-1α in CS1 cells (Table 3), and ASPH, *JAGGED1, HES1,* and *HIF-1α* for CDS11 cells (Table 4). Post hoc tests demonstrated that DOX treatment of CS1 cells significantly reduced ASPH (Figure 13A) but either significantly or trendwise increased *NOTCH 1* (Figure 13B), *HES1* (Figure 13D), *HEY1* (Figure 13E), and *HIF1α* (Figure 13F) relative to vehicle. BEZ significantly reduced *ASPH* (Figure 13A), *JAGGED1* (Figure 13C), and *HIF-1a* (Figure 13F). Dual DOX+BEZ mimicked the inhibitory effects of BEZ monotherapy on *ASPH, JAGGED1,* and *HIF1a,* as well as DOX's upregulated expression of *NOTCH1, HES1,* and *HEY1.* In CDS11 cells, DOX significantly inhibited *ASPH* (Figure 14A) but had no significant effects on *NOTCH1* (Figure 14B), *JAGGED1* (Figure 14C), *HES1* (Figure 14D), or *HEY1* (Figure 14E), and increased *HIF1α* (Figure 14F) relative to vehicle. BEZ significantly reduced *ASPH, JAGGED1,* and *HIF1α* but not *NOTCH1, HES1, or HEY1.* Dual DOX+BEZ treatments significantly reduced *ASPH, JAGGED1,* and *HIF1α* but increased *HES1* relative to the other treatments in CDS11 cells.

INSULIN/IGF/IRS pathway mRNA transcripts were measured with a custom QuantiPlex assay, and the results were normalized to *RPL13a*. One-way ANOVA tests detected significant treatment effects on *IGF2R* and *IRS1* in CS1 cells (Table 3) and a significant effect on *IRS1* and a statistical trendwise effect on *INSULIN* mRNA expression in CDS11 cells (Table 4). For CS1 cells (Figure S4), post hoc tests showed significant DOX-mediated increases in *IGF1R* (Figure S4E) and *IGF2R* (Figure S4F) and BEZ- and DOX+BEZ-mediated increases in *IGF2R* (Figure S4F), as well as reductions in *IRS1* (Figure S4G). For CDS11 cells, the post hoc tests demonstrated significantly lower *INSULIN* (Figure S5A) and *IRS1* (Figure S5G) mRNA levels with BEZ or DOX+BEZ, and increased *IGF2R* associated with BEZ (Figure S5F).

Table 4. CDS11 multiplex RNA hybridization assay ANOVA test results.

Gene Code	Gene Name	F-Ratio	*p*-Value
ASPH	Aspartyl-Asparaginyl-β-Hydroxylase	66.64	0.0007
NOTCH1	Notch 1	0.8998	N.S.
JAG1	Jagged 1	9.533	0.0270
HES1	Hairy and enhancer of split-1	9.187	0.0288
HEY1	HES-related family bHLH transcription factor	0.5119	N.S.
HIF1α	Hypoxia-Inducible Factor 1-alpha	80.51	0.0005
INS	Insulin	*4.845*	*0.0808*
IGF1	Insulin-Like Growth Factor 1	0.5891	N.S.
IGF2	Insulin-Like Growth Factor 2	0.8381	N.S.
INSR	Insulin Receptor	0.4804	N.S.
IGF1R	Insulin-Like Growth Factor 1 Receptor	1.100	N.S.
IGF2R	Insulin-Like Growth Factor 2 Receptor	2.613	N.S.
IRS1	Insulin Receptor Substrate, Type 1	13.75	0.0142
IRS2	Insulin Receptor Substrate, Type 2	1.245	N.S.
IRS4	Insulin Receptor Substrate, Type 4	0.4039	N.S.

Molecular pathways of DOX-, BEZ-, and DOX+BEZ (D+B)-mediated CDS11 cytotoxicity via insulin, IGF, and IRS pathways. The mRNA levels were measured with a multiplex bead-based RNA hybridization protocol, with results normalized to Ribosomal Protein L13a (RPL). Inter-group comparisons were made by one-way ANOVA. See Figure S5 for the results of Tukey post hoc repeated measures tests for all inter-group comparisons. n = 4 cultures/group. Significant ($p \leq 0.05$) and statistical trendwise ($0.05 < p < 0.10$) (italics) *p*-values are displayed. N.S. = not significant.

3.8. Akt-mTOR Pathway Analysis

To examine the effects of SMI1182, DOX, and BEZ on insulin/IGF/IRS1 signaling through Akt pathways to mTOR, we measured total and phosphorylated proteins using multiplex bead-based ELISA platforms. In addition, to assess the effects on relative levels of phosphorylation, we calculated the ratios of phosphorylated to total protein immunoreactivity. Results were analyzed by ANOVA with repeated measures post hoc Tukey tests.

CS1 Cells: SMI1182 had no statistically significant or trendwise effects on the expression of upstream (Insulin-R, IGF1R, IRS1) (Figure 15), intermediary (Akt, GSK-3α, GSK-3β, PTEN) (Figure 17), or downstream (TSC2, mTOR, RPS6, p70S6K) (Figure 19) proteins, phosphoproteins, or the relative levels of protein phosphorylation (p/T) assessed in the panel. DOX significantly reduced the levels of both the IGF1R (Figure 15B), tyrosine phosphorylated IGF1R (Figure 15E), IRS-1 (Figure 15C), Akt (Figure 17A), GSK-3β (Figure 17C), p-GSK-3β (Figure 17G), PTEN (Figure 17D), p-PTEN (Figure 17H), TSC2 (Figure 19A), p-TSCE (Figure 19E), mTOR (Figure 19B), and RPS6 (Figure 19C), and increased p-GSK-3α (Figure 17F). The discordant or unbalanced inhibition of total versus phosphoprotein expression resulted in significant increases in relative phosphorylation of IGF1R (Figure 15H), IRS-1 (Figure 15I), Akt (Figure 17I), and GSK-3α (Figure 17J) in the DOX versus vehicle-treated CS1 cells. BEZ significantly reduced the levels of all proteins and phosphoproteins except for p-IGF1R and p-p70S6K, indicating that the dominant effect of BEZ was to broadly inhibit the expression of nearly all critical signaling molecules in the Akt-mTOR pathway in CS1 cells.

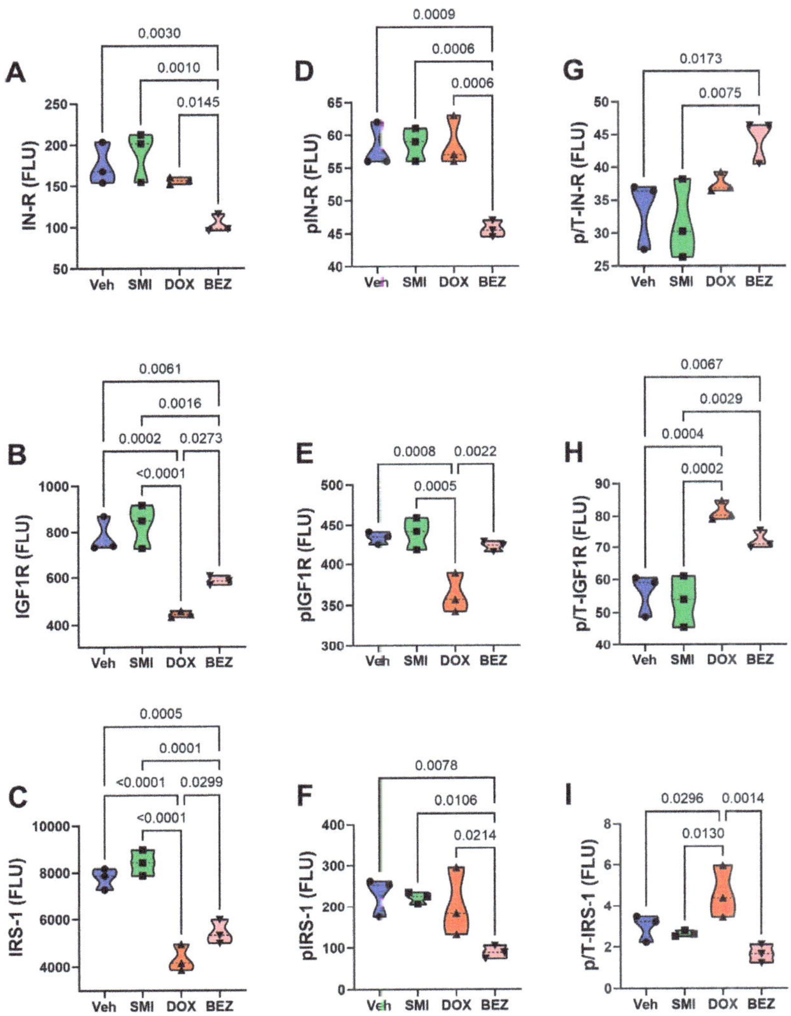

Figure 15. Effects of SMI1182 (SMI), DOX, and BEZ on the expression of upstream components of the Insulin/IGF1-Akt-mTOR pathway in CS1 cells. Commercial 11-plex magnetic bead-based ELISAs measured (**A**) Insulin receptor (IN-R), (**B**) IGF1R, (**C**) IRS-1, (**D**) pYpY1162/1163-IN-R, (**E**) pYpY1135/1136-IGF1R, and (**F**) pS636-IRS-1 and the calculated levels of relative phosphorylation of (**G**) p/T-IN-R, (**H**) p/T-IGF1R, and (**I**) p/T-IRS-1 in CS1 cells (n = 4/group) treated for 48 h with Veh, SMI, DOX, or BEZ. Values correspond to arbitrary fluorescent light units (FLU). Inter-group comparisons were made using two-way ANOVA tests (Table 5) with post -hoc Tukey tests. Software-generated p-values corresponding to significant differences ($p \leq 0.05$) are shown in the panels.

Table 5. CS grade and treatment effects on insulin/IGF1/IRS-Akt-mTOR signaling.

	Treatment	*p*-Value	CS Grade	*p*-Value	Treatment × CS Grade	*p*-Value
IN-R	13.26	0.0001	16.07	0.001	2.285	N.S.
p-IN-R	29.07	<0.0001	175.5	<0.0001	6.212	0.005
IGF1R	39.36	<0.0001	1.979	N.S.	2.400	N.S.
p-IGF1R	33.27	<0.0001	31.05	<0.0001	0.703	N.S.
IRS1	81.54	<0.0001	0.25	N.S.	4.209	0.0225
p-IRS1	7.535	0.0023	52.91	<0.0001	3.132	*0.054*
Akt	28.64	<0.0001	224.2	<0.0001	22.0	<0.0001
p-Akt	21.07	<0.0001	11.58	0.0036	0.829	N.S.
GSK-3α	34.43	<0.0001	0.863	N.S.	1.05	N.S.
p-GSK-3α	34.54	<0.0001	25.68	<0.0001	5.146	0.01
GSK-3β	90.35	<0.0001	4.49	0.05	0.785	N.S.
p-GSK-3β	132.8	<0.0001	0.284	N.S.	3.563	0.038
PTEN	91.46	<0.0001	3.718	*0.0718*	1.89	N.S.
p-PTEN	43.21	<0.0001	6.12	0.0249	0.073	N.S.
TSC	53.03	<0.0001	43.21	<0.0001	1.602	N.S.
p-TSC	23.81	<0.0001	35.48	<0.0001	0.447	N.S.
mTOR	24.57	<0.0001	12.28	0.0029	1.68	N.S.
p-mTOR	49.11	<0.0001	31.42	<0.0001	3.459	0.041
RPS6	23.97	<0.0001	0.764	N.S.	0.909	N.S.
p-RPS6	23.48	<0.0001	47.73	<0.0001	8.774	0.001
P70S6K	13.53	0.0001	1.191	N.S.	1.475	N.S.
p-P70S6K	5.049	0.012	68.16	<0.0001	4.346	0.02

Two-way ANOVA test results reflecting CS grade (CS1-Grade 3 versus CDS11 cells-Grade 2), treatment, and CS grade × treatment interactive effects on the expression of signaling and phosphorylated signaling molecules ($^{pYpY1162/1163}$-Insulin R, $^{pYpY1162/1163}$-IGF1R, ^{pS636}-IRS-1, ^{pS473}-Akt, ^{pS21}-GSK-3α, ^{pS9}-GSK-β, 3 ^{pS380}-PTEN, ^{pS939}-TSC2, ^{pS2448}-mTOR, $^{pS235/S236}$-RPS6, ^{pT412}-P70S6K) in the Akt/mTOR pathway. Significant differences ($p \leq 0.05$) and statistical trend ($0.05 < p < 0.10$; italic) results are displayed. N.S. = not significant. n = 4 cultures/group. See Figures 15–20 for corresponding graphs and post hoc test results. Abbreviations: R = receptor; IGF = insulin-like growth factor; IRS = insulin receptor substrate; GSK = glycogen synthase kinase; PTEN = phosphatase and tensin homolog; TSC = tuberous sclerosis complex 2; mTOR = mechanistic target of rapamycin; RPS6 = Ribosomal Protein S6; P70S6K = 70 kDa ribosomal protein S6 kinase.

CDS11 Cells: As observed in CS1 cells, SMI1182 monotherapy had no significant or trendwise effects on any of the signaling proteins or phosphoproteins analyzed with the multiplex panels, except for pAkt, which was significantly reduced relative to the vehicle control (Figure 16, Figure 18, and Figure 20). DOX significantly reduced Insulin R (Figure 16A), IGF1R (Figure 16B), IRS1 (Figure 16C), GSK-3β (Figure 18C), PTEN (Figure 18D), TSC2 (Figure 20A), mTOR (Figure 20B), RPS6 (Figure 20C), and p70S6K (Figure 20D), with limited concordant inhibitory effects on corresponding phosphoproteins, including p-Insulin R (Figure 14D), p-IGF1R (Figure 16E), p-GSK-3β (Figure 18G), p-PTEN (Figure 18H), and pTSC2 (Figure 20E). In addition, DOX had discordant inhibitory effects on p-Akt (Figure 18E). BEZ had broad inhibitory effects on all protein and phosphoprotein signaling molecules in the panel (Figure 16, Figure 18, and Figure 20).

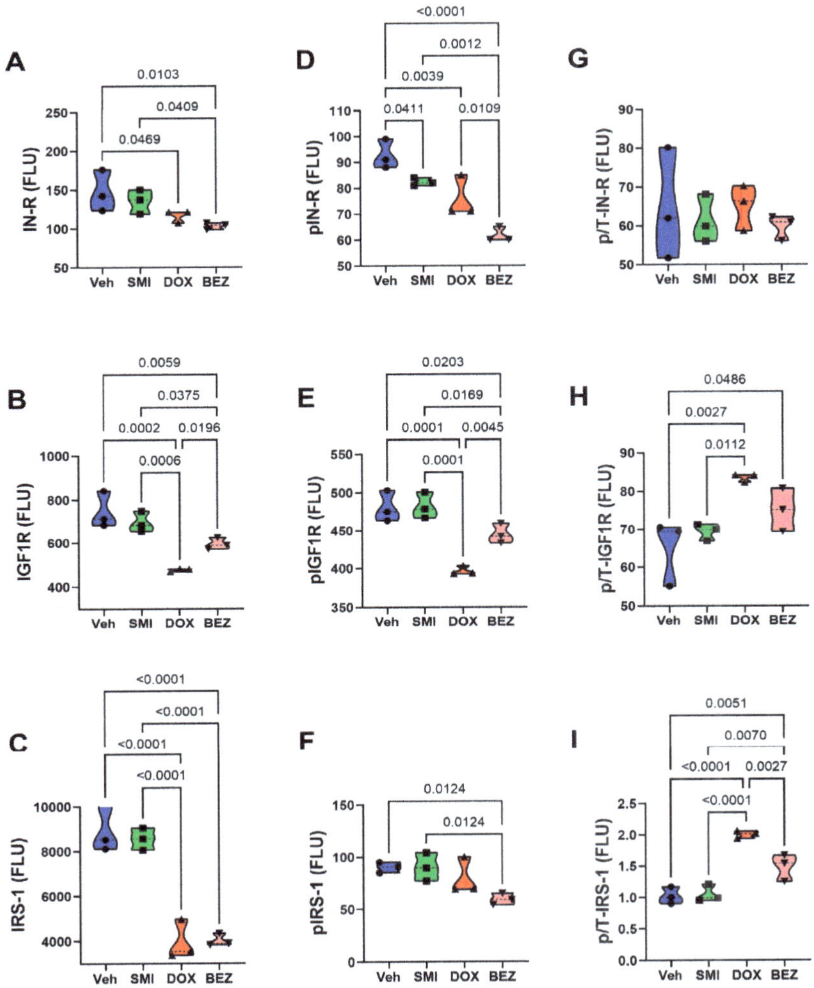

Figure 16. Effects of SMI1182 (SMI), DOX, and BEZ on the expression of upstream components of the Insulin/IGF1-Akt-mTOR pathway in CDS11 cells. Commercial 11-plex magnetic bead-based ELISAs measured (**A**) Insulin receptor (IN-R), (**B**) IGF1R, (**C**) IRS-1, (**D**) pYpY1162/1163-IN-R, (**E**) pYpY1135/1136-IGF1R, and (**F**) pS636-IRS-1 and the calculated levels of relative phosphorylation of (**G**) p/T-IN-R, (**H**) p/T-IGF1R, and (**I**) p/T-IRS-1 in CDS11 cells (n = 4/group) treated for 48h with Veh, SMI, DOX, or BEZ. Values correspond to arbitrary fluorescent light units (FLU). Inter-group comparisons were made using two-way ANOVA tests (Table 5) with post -hoc Tukey tests. Software-generated p-values corresponding to significant differences ($p \leq 0.05$) are shown in the panels.

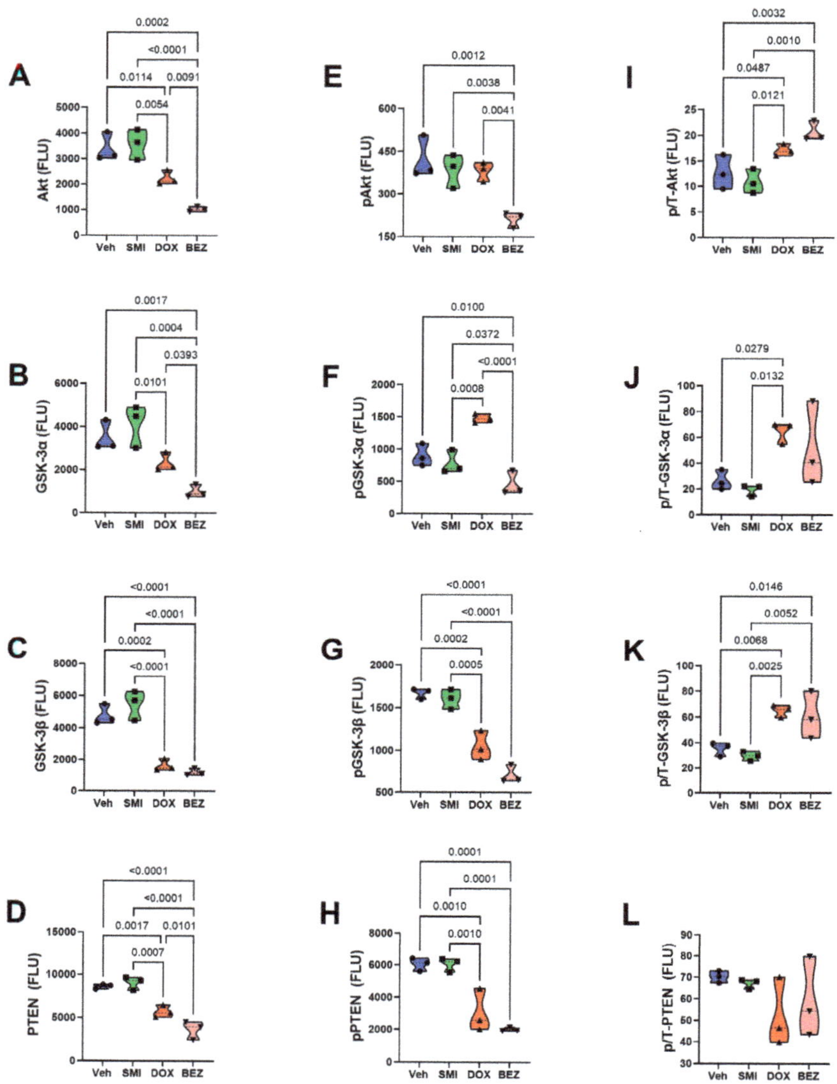

Figure 17. Effects of SMI1182 (SMI), DOX, and BEZ on the expression of mid-level components of the iInsulin/IGF1-Akt-mTOR pathway in CS1 cells. Magnetic bead-based protein and phosphoprotein 11-plex ELISAs measured frontal lobe levels of (**A**) Akt, (**B**) GSK-3α, (**C**) GSK-3β, (**D**) PTEN, (**E**) pS473-Akt, (**F**) pS21-GSK-3α, (**G**) pS9-GSK-3β, (**H**) pS380-PTEN, and the calculated levels of relative phosphorylation of (**I**) p/T-Akt, (**J**) p/T-GSK-3α, and (**K**) p/T-GSK-3β, and (**L**), and p/T-PTEN, (compared with total protein) in cells (n = 4/group) treated for 48 h with Veh, SMI, DOX, or BEZ. Graphed values correspond to arbitrary fluorescent light units (FLU). Inter-group comparisons were made using two-way ANOVA tests (Table 5) with post -hoc Tukey tests. Software-generated p-values corresponding to significant differences ($p \leq 0.05$) are shown in the panels.

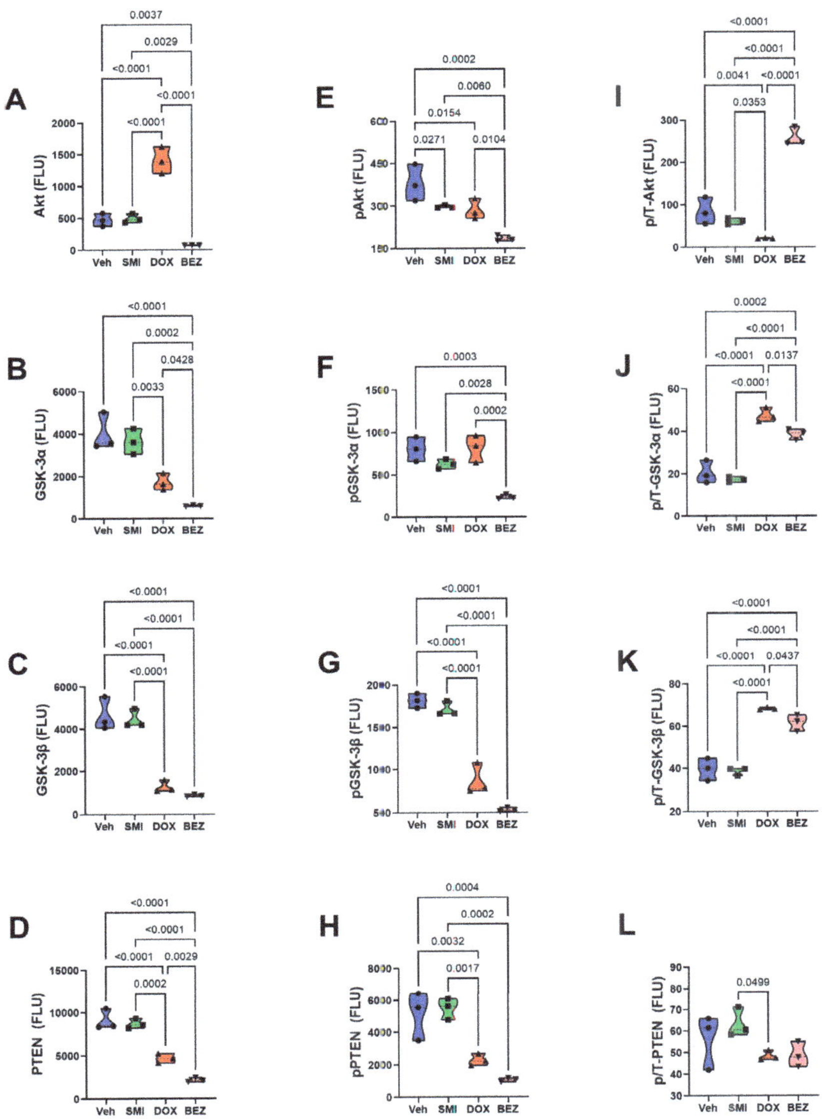

Figure 18. Effects of SMI1182 (SMI), DOX, and BEZ on the expression of mid-level components of the iInsulin/IGF1-Akt-mTOR pathway in CDS11 cells. Magnetic bead-based protein and phosphoprotein 11-plex ELISAs measured frontal lobe levels of **(A)** Akt, **(B)** GSK-3α, **(C)** GSK-3β, **(D)** PTEN, **(E)** pS473-Akt, **(F)** pS21-GSK-3α, **(G)** pS9-GSK-3β, **(H)** pS380-PTEN, and the calculated levels of relative phosphorylation of **(I)** p/T-Akt, **(J)** p/T-GSK-3α, and **(K)** p/T-GSK-3β, and **(L)**, and p/T-PTEN, (compared with total protein) in cells (n = 4/group) treated for 48 h with Veh, SMI, DOX, or BEZ. Graphed values correspond to arbitrary fluorescent light units (FLU). Inter-group comparisons were made using two-way ANOVA tests (Table 5) with post -hoc Tukey tests. Software-generated *p*-values corresponding to significant differences ($p \leq 0.05$) are shown in the panels.

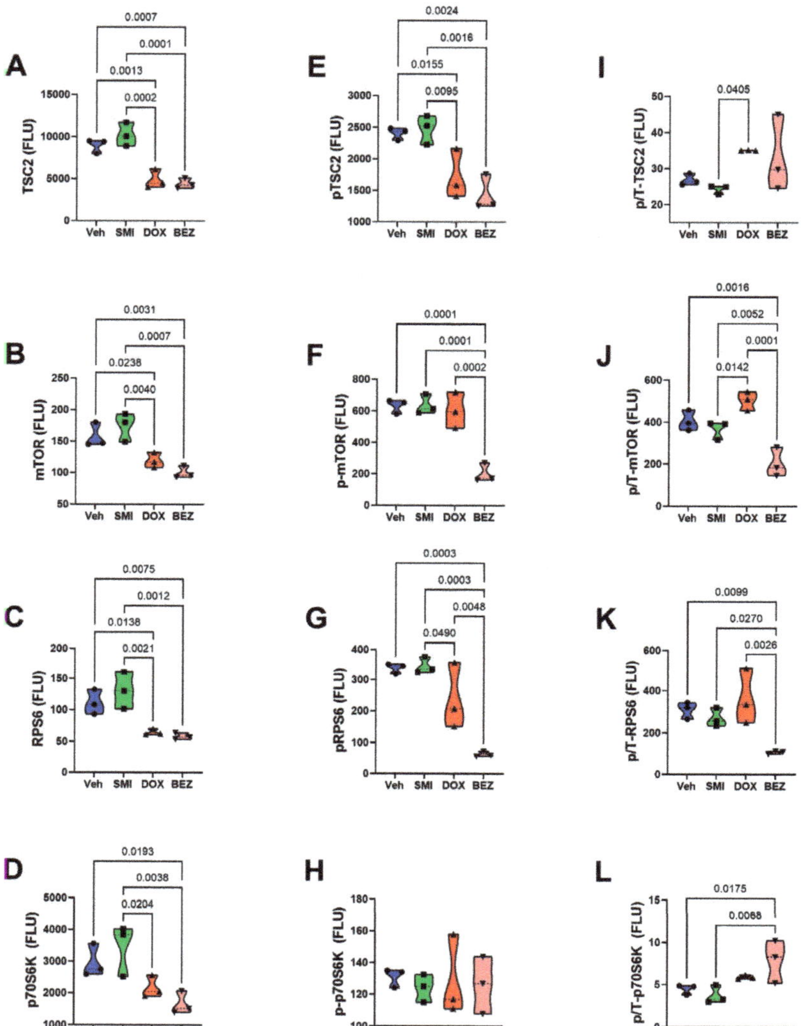

Figure 19. Effects of SMI1182 (SMI), DOX, and BEZ on the expression of downstream signaling through mTOR in CS1 cells. Magnetic bead-based protein and phosphoprotein 11-plex ELISAs measured immunoreactivity to (**A**) TSC2, (**B**) mTOR, (**C**) RPS6, (**D**) P70S6K, (**E**) pS939-TSC2, (**F**) pS2448-mTOR, (**G**) pS235/236-RPS6, (**H**) pT412-P70S6K, and the calculated relative levels of signaling molecule phosphorylation of (**I**) TSC, (**J**) mTOR, (**K**) RPS6, and (**L**) P70S6K in cells (n = 4 cultures/group) treated for 48 h with Veh, SMI, DOX, or BEZ. Graphed values correspond to arbitrary fluorescent light units (FLU). Inter-group comparisons were made by a two-way ANOVA (Table 5). Software-calculated p-values reflecting significant differences ($p \leq 0.05$) are shown in the panels.

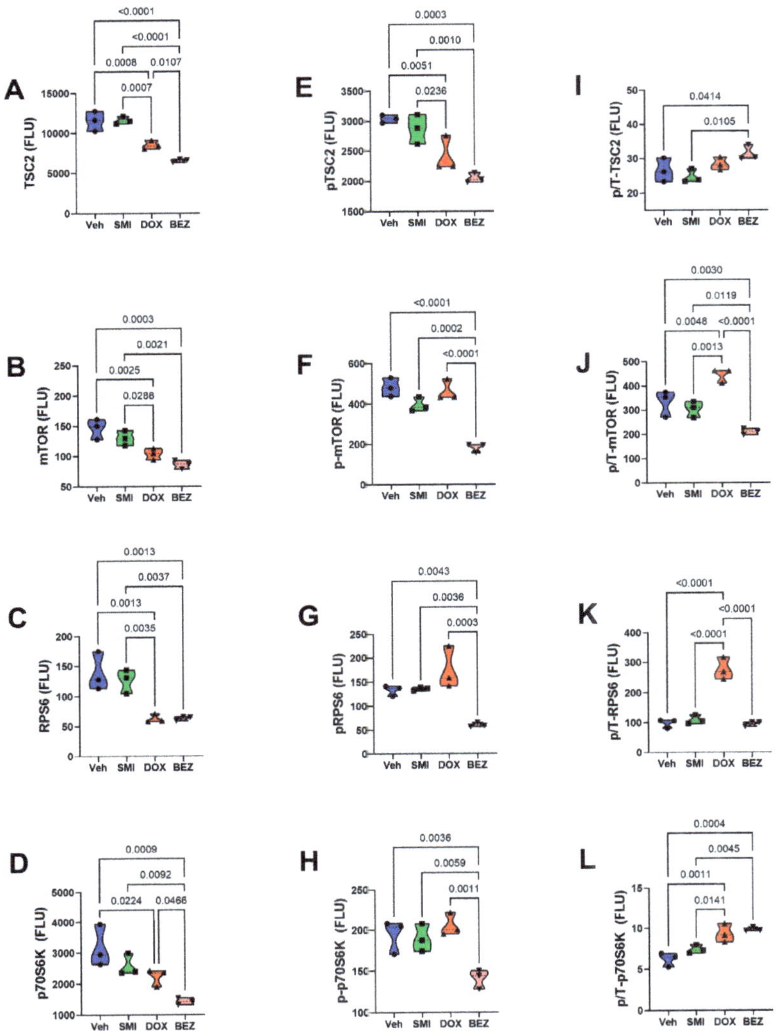

Figure 20. Effects of SMI1182 (SMI), DOX, and BEZ on the expression of downstream signaling through mTOR in CDS11 cells. Magnetic bead-based protein and phosphoprotein 11-plex ELISAs measured immunoreactivity to (**A**) TSC2, (**B**) mTOR, (**C**) RPS6, (**D**) P70S6K, (**E**) pS939-TSC2, (**F**) pS2448-mTOR, (**G**) pS235/236-RPS6, (**H**) pT412-P70S6K, and the calculated relative levels of signaling molecule phosphorylation of (**I**) TSC, (**J**) mTOR, (**K**) RPS6, and (**L**) P70S6K in cells (n = 4 cultures/group) treated for 48 h with Veh, SMI, DOX, or BEZ. Graphed values correspond to arbitrary fluorescent light units (FLU). Inter-group comparisons were made by a two-way ANOVA (Table 5). Software-calculated *p*-values reflecting significant differences ($p \leq 0.05$) are shown in the panels.

3.9. Comparative Effects of Treatment in Relation to CS Grade

Two-way ANOVA tests compared the effects of CS grade, treatment, and CS grade × treatment interactions on the expression of insulin/IGF/IRS-Akt-mTOR pathway proteins and phosphoproteins. SMI1182 results were excluded from these analyses because we did not detect any significant or statistical trendwise effects relative to the vehicle. Treatment significantly modulated the expression of all 22 signaling proteins and phosphoproteins (Table 5). CS grade (III-CS1; II-CDS11) significantly modulated the expression of 5 of the 11 (45.5%) signaling molecules and 10 of 11 (90.9%) phosphoproteins. Treatment × CS grade interactive effects were observed for two (18.1%) proteins and seven (63.6%) phosphoproteins. Importantly, these results indicate substantial effects of treatment modality and CS grade, as well as interactive effects of treatment and CS grade.

4. Discussion

The goals of this study were to characterize the expression and regulation of ASPH in moderate- versus high-grade human conventional CS cells and examine the differential effects of treatments designed to target molecular pathways that mediate ASPH's expression and function. ASPH stands out as an important therapeutic target in malignancy [26,78] due to its high levels of expression on cancer cell surfaces and very low or absent expression in most normal cells [34]. Identifying new therapeutic targets for CS is a priority set by the American Academy of Orthopaedic Surgeons. To this end, we extended our investigations beyond DOX, one of the standard chemotherapeutic agents used to treat post-surgical recurrent CS [79], by also examining the effects of SMI1182, a small molecule inhibitor of ASPH, and BEZ235, a potent inhibitor of PI3K/mTOR [80,81]. BEZ235 was studied because PI3K/mTOR signaling plays a critical role in sarcoma progression [10,82,83], and BEZ235's inhibitory effects on cancer cell growth are independent of EGFR status [84].

This study utilized two established human conventional CS cell lines regarded as histologic Grade 3 (CS1) or Grade 2 (CDS11). Efforts were made to delineate their similarities and differences in relation to ASPH expression and function, as well as their sensitivity to SMI1182, DOX, BEZ, and combination treatment. Corresponding with previous studies, higher levels of ASPH were detected in CS1 compared with CDS11 cells. CDS11 cells were further distinguished from CS1 by their higher level of Humbug (FB50-ASPH), which has been reported for malignancies such as those of gastrointestinal lineage [69,73,75]. Despite higher levels of ASPH in CS1 cells, the motility indices were like those of CDS11 cells, suggesting that factors other than ASPH levels govern CS motility and invasion potential. For example, the increased Humbug expression in CDS11 cells may be significant in relation to its Ca2+ binding effects in the ER [85]. Other findings reflective of CS1's higher tumor grade and more aggressive behavior compared with CDS11 cells include the significantly higher mRNA levels of *INSULINR*, *IGF2R*, *IRS1*, *NOTCH1*, *JAGGED1*, *HES1*, and *HIF1α*, which function to promote tumor cell growth and survival.

Various experiments compared the effects of treatment in CS1 and CDS11 cells to gain a better understanding of how therapeutic approaches can be optimized based on tumor grade. MTT and viability assays demonstrated that CS1 cells were relatively resistant to DOX but sensitive to BEZ, whereas CDS11 cells were sensitive to DOX but relatively resistant to BEZ. However, with a combination of DOX+BEZ treatment, both CS1 and CDS11 cells were effectively killed, indicating the importance of multi-targeting therapeutic approaches for both high- and intermediate-grade CS. Surprisingly, the impact of SMI1182 was inconsistent, as it primarily enhanced the BEZ killing of CDS11 cells and also exhibited small but significant anti-tumor effects in DOX-treated CS1 cells, but it produced cytoprotection in DOX-treated CDS11 cells. Notably, these conclusions were based on H33342 data, reflecting cell loss, since corresponding increases in MTT activity in CS1 cells co-administered BEZ and DOX or DOX+SMI1182 were likely related to oxidative stress. The SMI1182 dose–response and co-treatment studies confirmed that SMI1182 was largely ineffective as a monotherapy but suitable as an adjuvant for enhancing the chemotherapeutic effects of DOX or BEX. Correspondingly, preliminary studies suggest that while cytotoxic effects of SMI1182 monotherapy occur at doses above 10 μM CS1 and CDS11 cells, synergistic cytotoxicity with 0.5 μM DOX may occur with as little as 3.125 μM SMI1182.

Altogether, these findings highlight greater DOX monotherapy resistance in CS1 versus CDS11 cells and greater BEZ monotherapy resistance in CDS11 versus CS1 cells. However, the studies showed significant and largely similar benefits of DOX+BEZ or DOX+BEZ+SMI combination therapy for anti-tumor cell effects in high-grade (CS1) and intermediate-grade (CDS11) CS. The concerns revealed by these studies were that (1) although the SMI1182 ASPH inhibitor enhanced the DOX killing of CS1 cells, in CDS11 cells, it had opposite effects when administered in combination with DOX; (2) BEZ or BEZ+SMI combined with DOX effectively killed CS1 cells to greater extents than DOX only—this was not the case for CDS11 cells; (3) the BEZ killing of CS1 cells was incrementally increased by the addition of SMI, DOX, or SMI+DOX but strikingly increased by the co-treatment of CDS11 cells with BEZ+ SMI, DOX, or SMI+DOX. These differential responses to monotherapy and combination therapy underscore the importance of considering CS grade as a key variable in regulating the effectiveness of novel combination chemotherapy protocols. Correspondingly, several studies have demonstrated at least additive effects of dual treatment with BEZ235 and DOX for enhancing cancer cell killing and reducing tumor cell growth [86–88].

The cytomorphological effects of treatment reflected the impact of treatment on cell viability and cytotoxicity. Correspondingly, the chamber slide cultures showed that SMI1182 treatment yielded similar growth patterns and cellular morphology as with vehicle treatment, whereas the cytotoxic effects of BEZ or DOX were manifested by conspicuous cell loss, apoptosis, and cellular necrosis. Cytospin preparations provided better opportunities to assess treatment effects on cell shape and appreciate the degree to which BEZ induced cell cycle arrest and apoptosis. BEZ-associated surface membrane blebbing, i.e., irregular bulging of the plasma membrane, likely reflected the decoupling of the cytoskeleton from the plasma membrane. Physiologically, the response could occur during cell migration, such as cancer cell metastasis [89], or with apoptosis [90]. The prominent cell shrinkage with nuclear condensation associated with DOX or BEZ, with or without SMI1182, is consistent with known mechanisms of DOX- [91] and BEZ-mediated [84,92] cancer cell killing, including necrosis, cytotoxicity, autophagy, and apoptosis. In contrast, the cell swelling, pallor, nuclear loss, and fragmentation noted more in SMI1182-treated cells likely represented apoptosis as a prominent mechanism of cell death linked to Notch pathway inhibition [93–95].

Previous studies have shown that ASPH plays functional roles in cell motility and invasion through its catalytic activity, which hydroxylates Asp and Asn residues in the EGF-like domains of Notch and Jagged (Ince, 2000 #52). SMI's targeting ASPH's catalytic domain was shown to reduce malignant neoplastic cell motility by inhibiting the expression of Notch transcription factors rather than *ASPH*, *NOTCH*, or *JAGGED* [36,38]. Herein, the SMI1182 treatment of CS cells reduced ASPH-A85G6 expression, ASPH cell surface immunoreactivity, and the percentages of migrated adherent cells. BEZ also reduced ASPH immunoreactivity and its cell surface localization but did not impact motility relative to the vehicle, most likely due to its failure to reduce *NOTCH*, *HES1*, and *HEY1*. Apart from its selective inhibition of *ASPH* mRNA, DOX exhibited anti-therapeutic effects, including enhanced migration, Notch pathway signaling, and ASPH surface fibrillary staining in CS1 cells. Reduced surface ASPH immunoreactivity in SMI1182-treated cells may have been instrumental in impairing the hydroxylation of EGF-like domains of Notch and Jagged and thereby impairing cell motility, whereas increased surface fibrillar localization of ASPH immunoreactivity associated with DOX may have contributed to the increased CS cell motility.

The findings with respect to SMI1182 suggest that the inhibition of ASPH's catalytic activity, coupled with reduced ASPH surface localization, accounts for its inhibitory effects on cell motility. The additional findings that DOX-associated increases in *IGF1R*, *IGF2R*, and *IRS1* in CS1 cells, and BEZ and DOX+BEZ reductions in *IRS1* expression in CS1 and CDS11 cells, are significant because increased signaling through insulin/IGF/IRS corresponds with the observed increases in cell motility. Further studies designed to better understand the mechanistic differential treatment responses of CS1 versus CDS11 cells revealed that DOX treatment of CS1 cells mainly enhanced the expression of Notch pathway signaling molecules, whereas BEZ treatment inhibited the key mRNAs, including JAGGED, HES1, and HIF1α, which have important roles in cell migration and invasiveness [96,97]. The largely stimulatory effects of DOX on Notch pathway signaling in CS1 cells versus its limited effects in CDS11 cells may account for high-grade CS resistance to DOX monotherapy. The BEZ-mediated reduction in *HES1* mRNA suggests that BEZ can have inhibitory effects on Notch with attendant reductions in critical transcription factors [30,31,38]. These findings are consistent with previous reports that BEZ or a related PI3K/mTOR inhibitor can have significant anti-cancer/therapeutic effects in malignancies, including high-grade CS [98–100], and that BEZ235 can enhance the cytotoxic effects of DOX in malignancies [86,88], in part due to the BEZ-mediated increased accumulation of DOX [87]. However, there are caveats signaled by the finding that dual treatment with BEZ235 can also worsen outcomes by further increasing Notch pathway activity, as reported previously [101]. Similarly, the co-administration of DOX with cisplatin or knockdown of *Notch1* in osteosarcoma reduced the cytotoxic effects of DOX via the unanticipated upregulation of Notch1 and related target genes [102].

The final approach for characterizing molecular mediators of tumor cell responsiveness based on CS grade was to examine the effects of SMI1182, DOX, and BEZ on insulin/IGF/IRS signaling downstream through mTOR networks using multiplex ELISA platforms. DOX and BEZ significantly reduced the expression of most upstream (Insulin-R, IGF1R, and IRS1), intermediate (Akt, GSK-3, and PTEN), and downstream (TSC2, mTOR, RPS6, and p70S6K) signaling mediators in both CS1 and CDS11 cells. The few exceptions were that DOX had no significant effect on Insulin-R, GSK-3α, or p70S6K in CS1 cells, and it upregulated Akt in CDS11 cells. It is noteworthy that the reduced levels of IRS1 protein were concordant with the effects of DOX and BEZ on IRS1 mRNA, whereas the downward shifts in Insulin-R and IGF1R were detected by multiplex ELISA and not mRNA analysis, suggesting post-transcription mechanisms lead to the increased degradation of signaling molecules. The broadly parallel reductions in phosphoproteins correlated with the downregulated expression of specific signaling molecules and, therefore, were unlikely to have been caused by the selective disruption of kinase activities. In contrast, SMI1182 treatment had no significant effects on the expression levels of signaling molecules or their corresponding phosphoproteins, indicating that its anti-cancer effects were produced by alternative indirect mechanisms linked to ASPH's functions. In terms of distinguishing treatment responses between high-grade and intermediate-grade CS, the multiplex assays demonstrated DOX inhibition of Insulin-R and p-Insulin-R in CDS11 but not CS1 cells, and increased relative levels of GSK-3b phosphorylation (silencing) in CS1 but not CDS11 cells, suggesting that the failure of DOX to inhibit insulin signaling in CS1 cells, combined with the suppression of GSK-3β activity, contributes to DOX chemotherapy resistance in CS1 cells. On the other hand, the differential inhibitory effects of BEZ on IGF1R and p70S6K phosphorylation in CDS11 but not CS1 cells may have contributed to its sharp dose-dependent reductions in CDS11 cell viability.

5. Conclusions

This study characterized ASPH expression in human CS cells and demonstrated that high-grade CS cells expressed higher levels of ASPH, Notch pathway genes, and selected insulin/IGF/IRS pathway-related genes compared with intermediate-grade CS. The identified differences could account for the more aggressive behavior of high-grade versus intermediate-grade CS.

SMI1182, DOX, and BEZ independently inhibited CS cell viability and increased cytotoxicity in a dose-dependent manner. However, combination therapy with SMI1182+BEZ, SMI1182+DOX, or BEZ+DOX mainly had additive effects on CS cell cytotoxicity/killing compared with monotherapy.

The therapeutic effects of SMI1182, DOX, and BEZ were mediated in part by the inhibition of ASPH expression, Notch pathway genes, and signaling through the insulin/IGF/IRS-Akt-mTOR pathways. The greater inhibitory effects of BEZ versus DOX or SMI1182 were associated with broader impairments in signaling molecule expression, including in CDS11 (intermediate grade) versus CS1 (high grade) cells.

The SMI1182 small molecule inhibitor of ASPH functioned more effectively as a cofactor or adjuvant agent with DOX or BEZ for killing CS cells rather than as a monotherapy.

Combination therapy with DOX or BEZ plus SMI1182 could improve clinical outcomes from CS.

Supplementary Materials: The following supporting information can be downloaded at: https://www.mdpi.com/article/10.3390/cancers17101671/s1, Figure S1: Directional cell motility and adhesion in CS1 and CDS11 cells; Figure S2: Directional cell motility and adhesion in CS1 and CDS11 cells; Figure S3: Molecular pathways of DOX-, BEZ-, and DOX+BEZ (D+B)-mediated CS1 cytotoxicity via insulin, IGF, and IRS pathways; Figure S4: Molecular pathways of DOX-, BEZ-, and DOX+BEZ (D+B)-mediated CDS11 cytotoxicity via insulin, IGF, and IRS pathways. Figure S5: Molecular Pathways of DOX-, BEZ-, and DOX+BEZ (D+B)-Mediated CDS11 Cytotoxicity Via Insulin, IGF, and IRS Pathways.

Author Contributions: M.F. conducted most of the experiments, contributed to the development and refinement of the experimental protocols, contributed to data analysis, and wrote the initial draft of the manuscript. M.T. provided supportive efforts for data acquisition, execution of experiments, and data quality assurances; B.D. and P.C. chemically synthesized the SMI1182 used in this research. R.R. generated the CDS11 cells, contributed to the research design, and critically reviewed the science. R.I.C. provided administrative support and guidance to students participating in the research. S.M.d.l.M. directed the research, designed the experiments, performed data analysis, critically reviewed and interpreted results, wrote intermediate and final versions of the manuscript, and secured funding. All authors have read and agreed to the published version of the manuscript.

Funding: This research was funded by The National Institute on Alcohol Abuse and Alcoholism AA-011431, AA028408, AA032106, and the National Cancer Institute (CA-270795) at the National Institutes of Health. The APC was funded by SMdlM.

Institutional Review Board Statement: Not applicable.

Informed Consent Statement: Not applicable.

Data Availability Statement: Data will be made available upon reasonable request from the corresponding author.

Acknowledgments: During the initial stages of this research, RIC was supported by CA-270795.

Conflicts of Interest: The authors declare no conflicts of interest.

Abbreviations

The following abbreviations are used in this manuscript:

Abbreviation	Definition
Akt	Akt (Protein Kinase B)
ASPH	Aspartyl-asparaginyl-β-hydroxylase
AUC	Area-under-curve
BEZ	BEZ235; combination of everolimus and the imidazoquinoline derivative
CS	Chondrosarcoma
DMSO	Dimethyl sulfoxide
DOX	Doxorubicin
EDTA	Ethylenediaminetetraacetic acid
EIA	Enzyme immunoassay
ELISA	Enzyme-linked immunosorbent assay
GAPDH	Glyceraldehyde-3-phosphate dehydrogenase
GSK-3a	Glycogen Synthase Kinase 3a
GSK-3β	Glycogen Synthase Kinase 3β
HES1	Hairy and enhancer of split-1
HEY1	HES-related family bHLH transcription factor
HIF1α	Hypoxia-Inducible Factor 1-alpha
HRP	Horseradish peroxidase
IGF1	Insulin-Like Growth Factor 1
IGF1R	Insulin-Like Growth Factor 1 Receptor
IGF1R	Insulin-Like Growth Factor Receptor Type 1
IGF2	Insulin-Like Growth Factor 2
IGF2R	Insulin-Like Growth Factor 2 Receptor
INS	Insulin
INSR	Insulin receptor
Insulin-R	Insulin receptor
IRS1	Insulin receptor substrate, type 1
IRS1	Insulin receptor substrate, type 1
IRS2	Insulin receptor substrate, type 2
IRS4	Insulin receptor substrate, type 4
JAGGED1	Jagged 1
mTOR	Mechanistic target of rapamycin
MTT	3-(4,5-dimethylthiazol-2-yl)-2,5-diphenyltetrazolium bromide
NOTCH1	Notch 1
P70S6K	p70 Ribosomal S6 kinase
PBS	Phosphate-buffered saline
PTEN	Phosphatase and tensin homolog
RPS6	Ribosomal Protein S6
SMI	Small molecule inhibitor
TBS	Tris-buffered saline
TSC2	Tuberous sclerosis complex 2

References

1. Limaiem, F.; Davis, D.D.; Sticco, K.L. Chondrosarcoma. In *StatPearls*; StatPearls Publishing: Treasure Island, FL, USA, 2024.
2. van Praag Veroniek, V.M.; Rueten-Budde, A.J.; Ho, V.; Dijkstra, P.D.S.; Fiocco, M.; van de Sande, M.A.J. Incidence, outcomes and prognostic factors during 25 years of treatment of chondrosarcomas. *Surg. Oncol.* **2018**, *27*, 402–408. [CrossRef]
3. Thorkildsen, J.; Myklebust, T.A. The national incidence of chondrosarcoma of bone; a review. *Acta Oncol.* **2023**, *62*, 110–117. [CrossRef] [PubMed]
4. Tlemsani, C.; Larousserie, F.; De Percin, S.; Audard, V.; Hadjadj, D.; Chen, J.; Biau, D.; Anract, P.; Terris, B.; Goldwasser, F.; et al. Biology and Management of High-Grade Chondrosarcoma: An Update on Targets and Treatment Options. *Int. J. Mol. Sci.* **2023**, *24*, 1361. [CrossRef]

Frezza, A.M.; Cesari, M.; Baumhoer, D.; Biau, D.; Bielack, S.; Campanacci, D.A.; Casanova, J.; Esler, C.; Ferrari, S.; Funovics, P.T.; et al. Mesenchymal chondrosarcoma: Prognostic factors and outcome in 113 patients. A European Musculoskeletal Oncology Society study. *Eur. J. Cancer* **2015**, *51*, 374–381. [CrossRef] [PubMed]

Chawla, S.P.; Papai, Z.; Mukhametshina, G.; Sankhala, K.; Vasylyev, L.; Fedenko, A.; Khamly, K.; Ganjoo, K.; Nagarkar, R.; Wieland, S.; et al. First-Line Aldoxorubicin vs Doxorubicin in Metastatic or Locally Advanced Unresectable Soft-Tissue Sarcoma: A Phase 2b Randomized Clinical Trial. *JAMA Oncol.* **2015**, *1*, 1272–1280. [CrossRef] [PubMed]

Weinschenk, R.C.; Wang, W.L.; Lewis, V.O. Chondrosarcoma. *J. Am. Acad. Orthop. Surg.* **2021**, *29*, 553–562. [CrossRef]

Zajac, W.; Drozdz, J.; Kisielewska, W.; Karwowska, W.; Dudzisz-Sledz, M.; Zajac, A.E.; Borkowska, A.; Szumera-Cieckiewicz, A.; Szostakowski, B.; Rutkowski, P.; et al. Dedifferentiated Chondrosarcoma from Molecular Pathology to Current Treatment and Clinical Trials. *Cancers* **2023**, *15*, 3924. [CrossRef]

Andreou, D.; Ruppin, S.; Fehlberg, S.; Pink, D.; Werner, M.; Tunn, P.U. Survival and prognostic factors in chondrosarcoma: Results in 115 patients with long-term follow-up. *Acta Orthop.* **2011**, *82*, 749–755. [CrossRef]

Polychronidou, G.; Karavasilis, V.; Pollack, S.M.; Huang, P.H.; Lee, A.; Jones, R.L. Novel therapeutic approaches in chondrosarcoma. *Future Oncol.* **2017**, *13*, 637–648. [CrossRef]

Boehme, K.A.; Schleicher, S.B.; Traub, F.; Rolauffs, B. Chondrosarcoma: A Rare Misfortune in Aging Human Cartilage? The Role of Stem and Progenitor Cells in Proliferation, Malignant Degeneration and Therapeutic Resistance. *Int. J. Mol. Sci.* **2018**, *19*, 311. [CrossRef]

Rey, V.; Tornin, J.; Alba-Linares, J.J.; Robledo, C.; Murillo, D.; Rodriguez, A.; Gallego, B.; Huergo, C.; Viera, C.; Brana, A.; et al. A personalized medicine approach identifies enasidenib as an efficient treatment for IDH2 mutant chondrosarcoma. *EBioMedicine* **2024**, *102*, 105090. [CrossRef] [PubMed]

Bernstein-Molho, R.; Kollender, Y.; Issakov, J.; Bickels, J.; Dadia, S.; Flusser, G.; Meller, I.; Sagi-Eisenberg, R.; Merimsky, O. Clinical activity of mTOR inhibition in combination with cyclophosphamide in the treatment of recurrent unresectable chondrosarcomas. *Cancer Chemother. Pharmacol.* **2012**, *70*, 855–860. [CrossRef] [PubMed]

Miele, L. Notch signaling. *Clin. Cancer Res.* **2006**, *12*, 1074–1079. [CrossRef] [PubMed]

Leong, K.G.; Karsan, A. Recent insights into the role of Notch signaling in tumorigenesis. *Blood* **2006**, *107*, 2223–2233. [CrossRef]

Yuan, X.; Wu, H.; Xu, H.; Xiong, H.; Chu, Q.; Yu, S.; Wu, G.S.; Wu, K. Notch signaling: An emerging therapeutic target for cancer treatment. *Cancer Lett.* **2015**, *369*, 20–27. [CrossRef]

Purow, B. Notch inhibition as a promising new approach to cancer therapy. *Adv. Exp. Med. Biol.* **2012**, *727*, 305–319. [CrossRef]

Li, X.; Yan, X.; Wang, Y.; Kaur, B.; Han, H.; Yu, J. The Notch signaling pathway: A potential target for cancer immunotherapy. *J. Hematol. Oncol.* **2023**, *16*, 45. [CrossRef]

Siar, C.H.; Ha, K.O.; Aung, L.O.; Nakano, K.; Tsujigiwa, H.; Nagatsuka, H.; Ng, K.H.; Kawakami, T. Immunolocalization of notch signaling protein molecules in a maxillary chondrosarcoma and its recurrent tumor. *Eur. J. Med. Res.* **2010**, *15*, 456–460. [CrossRef]

Massard, C.; Azaro, A.; Soria, J.C.; Lassen, U.; Le Tourneau, C.; Sarker, D.; Smith, C.; Ohnmacht, U.; Oakley, G.; Patel, B.K.R.; et al. First-in-human study of LY3039478, an oral Notch signaling inhibitor in advanced or metastatic cancer. *Ann. Oncol.* **2018**, *29*, 1911–1917. [CrossRef]

Gounder, M.M.; Rosenbaum, E.; Wu, N.; Dickson, M.A.; Sheikh, T.N.; D'Angelo, S.P.; Chi, P.; Keohan, M.L.; Erinjeri, J.P.; Antonescu, C.R.; et al. A Phase Ib/II Randomized Study of RO4929097, a Gamma-Secretase or Notch Inhibitor with or without Vismodegib, a Hedgehog Inhibitor, in Advanced Sarcoma. *Clin. Cancer Res.* **2022**, *28*, 1586–1594. [CrossRef]

Collins, M.; Michot, J.M.; Bellanger, C.; Mussini, C.; Benhadji, K.; Massard, C.; Carbonnel, F. Notch inhibitors induce diarrhea, hypercrinia and secretory cell metaplasia in the human colon. *EXCLI J.* **2021**, *20*, 819–827. [CrossRef]

Mir, O.; Azaro, A.; Merchan, J.; Chugh, R.; Trent, J.; Rodon, J.; Ohnmacht, U.; Diener, J.T.; Smith, C.; Yuen, E.; et al. Notch pathway inhibition with LY3039478 in soft tissue sarcoma and gastrointestinal stromal tumours. *Eur. J. Cancer* **2018**, *103*, 88–97. [CrossRef] [PubMed]

Lopez-Nieva, P.; Gonzalez-Sanchez, L.; Cobos-Fernandez, M.A.; Cordoba, R.; Santos, J.; Fernandez-Piqueras, J. More Insights on the Use of gamma-Secretase Inhibitors in Cancer Treatment. *Oncologist* **2021**, *26*, e298–e305. [CrossRef] [PubMed]

Jia, S.; VanDusen, W.J.; Diehl, R.E.; Kohl, N.E.; Dixon, R.A.; Elliston, K.O.; Stern, A.M.; Friedman, P.A. cDNA cloning and expression of bovine aspartyl (asparaginyl) beta-hydroxylase. *J. Biol. Chem.* **1992**, *267*, 14322–14327. [CrossRef] [PubMed]

Greve, J.M.; Pinkham, A.M.; Cowan, J.A. Human aspartyl (asparaginyl) hydroxylase. A multifaceted enzyme with broad intra- and extra-cellular activity. *Metallomics* **2021**, *13*, mfab044. [CrossRef]

Lin, Q.; Chen, X.; Meng, F.; Ogawa, K.; Li, M.; Song, R.; Zhang, S.; Zhang, Z.; Kong, X.; Xu, Q.; et al. ASPH-notch Axis guided Exosomal delivery of Prometastatic Secretome renders breast Cancer multi-organ metastasis. *Mol. Cancer* **2019**, *18*, 156. [CrossRef]

Borgas, D.L.; Gao, J.S.; Tong, M.; Roper, N.; de la Monte, S.M. Regulation of Aspartyl-(Asparaginyl)-beta-Hydroxylase Protein Expression and Function by Phosphorylation in Hepatocellular Carcinoma Cells. *J. Nat. Sci.* **2015**, *1*, e84.

29. Cantarini, M.C.; de la Monte, S.M.; Pang, M.; Tong, M.; D'Errico, A.; Trevisani, F.; Wands, J.R. Aspartyl-asparagyl beta hydroxyla over-expression in human hepatoma is linked to activation of insulin-like growth factor and notch signaling mechanis *Hepatology* **2006**, *44*, 446–457. [CrossRef]
30. Barboro, P.; Benelli, R.; Tosetti, F.; Costa, D.; Capaia, M.; Astigiano, S.; Vene, R.; Poggi, A.; Ferrari, N. Aspartate beta-hydroxyla targeting in castration-resistant prostate cancer modulates the NOTCH/HIF1alpha/GSK3beta crosstalk. *Carcinogenesis* **2020**, 1246–1252. [CrossRef]
31. Zheng, W.; Wang, X.; Hu, J.; Bai, B.; Zhu, H. Diverse molecular functions of aspartate beta-hydroxylase in cancer (Review). *On Rep.* **2020**, *44*, 2364–2372. [CrossRef]
32. Engel, J. EGF-like domains in extracellular matrix proteins: Localized signals for growth and differentiation? *FEBS Lett.* **1989**, 2 1–7. [CrossRef] [PubMed]
33. Borgas, D.L.; Gao, J.S.; Tong, M.; de la Monte, S.M. Potential Role of Phosphorylation as a Regulator of Aspartyl-(asparagin beta-hydroxylase: Relevance to Infiltrative Spread of Human Hepatocellular Carcinoma. *Liver Cancer* **2015**, *4*, 139–153. [CrossR [PubMed]
34. Ince, N.; de la Monte, S.M.; Wands, J.R. Overexpression of human aspartyl (asparaginyl) beta-hydroxylase is associated w malignant transformation. *Cancer Res.* **2000**, *60*, 1261–1266. [PubMed]
35. Dong, X.; Lin, Q.; Aihara, A.; Li, Y.; Huang, C.K.; Chung, W.; Tang, Q.; Chen, X.; Carlson, R.; Nadolny, C.; et al. Asparta beta-Hydroxylase expression promotes a malignant pancreatic cellular phenotype. *Oncotarget* **2015**, *6*, 1231–1248. [CrossR [PubMed]
36. Sturla, L.M.; Tong, M.; Hebda, N.; Gao, J.; Thomas, J.M.; Olsen, M.; de la Monte, S.M. Aspartate-beta-hydroxylase (ASPH): potential therapeutic target in human malignant gliomas. *Heliyon* **2016**, *2*, e00203. [CrossRef]
37. Yang, H.; Song, K.; Xue, T.; Xue, X.P.; Huyan, T.; Wang, W.; Wang, H. The distribution and expression profiles of hum Aspartyl/Asparaginyl beta-hydroxylase in tumor cell lines and human tissues. *Oncol. Rep.* **2010**, *24*, 1257–1264. [CrossRef]
38. Aihara, A.; Huang, C.K.; Olsen, M.J.; Lin, Q.; Chung, W.; Tang, Q.; Dong, X.; Wands, J.R. A cell-surface beta-hydroxylase is biomarker and therapeutic target for hepatocellular carcinoma. *Hepatology* **2014**, *60*, 1302–1313. [CrossRef]
39. Wang, K.; Liu, J.; Yan, Z.L.; Li, J.; Shi, L.H.; Cong, W.M.; Xia, Y.; Zou, Q.F.; Xi, T.; Shen, F.; et al. Overexpression of aspart (asparaginyl)-beta-hydroxylase in hepatocellular carcinoma is associated with worse surgical outcome. *Hepatology* **2010**, 164–173. [CrossRef]
40. Sun, X.; Ramirez, J.; Hart, J.; Schwab, J.; Molino, J.; Dong, X.; Carlson, R.; Wands, J.R.; Terek, R. ASPH Targeted Therapy Inhib Chondrosarcoma Progression. In Proceedings of the Orthopaedic Research Society, Tampa, FL, USA, 4–8 February 2022; p. 12
41. Revskaya, E.; Jiang, Z.; Morgenstern, A.; Bruchertseifer, F.; Sesay, M.; Walker, S.; Fuller, S.; Lebowitz, M.S.; Gravekamp, Ghanbari, H.A.; et al. A Radiolabeled Fully Human Antibody to Human Aspartyl (Asparaginyl) beta-Hydroxylase Is a Promisi Agent for Imaging and Therapy of Metastatic Breast Cancer. *Cancer Biother. Radiopharm.* **2017**, *32*, 57–65. [CrossRef]
42. Xue, T.; Su, J.; Li, H.; Xue, X. Evaluation of HAAH/humbug quantitative detection in the diagnosis of hepatocellular carcinon *Oncol. Rep.* **2015**, *33*, 329–337. [CrossRef]
43. Xue, T.; Xue, X.P.; Huang, Q.S.; Wei, L.; Sun, K.; Xue, T. Monoclonal antibodies against human aspartyl (asparaginyl) be hydroxylase developed by DNA immunization. *Hybridoma* **2009**, *28*, 251–257. [CrossRef] [PubMed]
44. Yeung, Y.A.; Finney, A.H.; Koyrakh, I.A.; Lebowitz, M.S.; Ghanbari, H.A.; Wands, J.R.; Wittrup, K.D. Isolation and characterizati of human antibodies targeting human aspartyl (asparaginyl) beta-hydroxylase. *Hum. Antibodies* **2007**, *16*, 163–176. [CrossR [PubMed]
45. Lavanya, V.; Adil, M.; Ahmed, N.; Rishi, A.K.; Jamal, S. Small molecule inhibitors as emerging cancer therapeutics. *Integr. Canc Sci. Ther.* **2014**, *1*, 39–46. [CrossRef]
46. McGowan, P.M.; Mullooly, M.; Caiazza, F.; Sukor, S.; Madden, S.F.; Maguire, A.A.; Pierce, A.; McDermott, E.W.; Crown, O'Donovan, N.; et al. ADAM-17: A novel therapeutic target for triple negative breast cancer. *Ann. Oncol.* **2013**, *24*, 362–3 [CrossRef]
47. Moss, M.L.; Minond, D. Recent Advances in ADAM17 Research: A Promising Target for Cancer and Inflammation. *Mediato Inflamm.* **2017**, *2017*, 9673537. [CrossRef]
48. Huang, C.K.; Iwagami, Y.; Aihara, A.; Chung, W.; de la Monte, S.; Thomas, J.M.; Olsen, M.; Carlson, R.; Yu, T.; Dong, X.; et Anti-Tumor Effects of Second Generation beta-Hydroxylase Inhibitors on Cholangiocarcinoma Development and Progressi *PLoS ONE* **2016**, *11*, e0150336. [CrossRef]
49. Nagaoka, K.; Ogawa, K.; Ji, C.; Cao, K.Y.; Bai, X.; Mulla, J.; Cheng, Z.; Wands, J.R.; Huang, C.-K. Targeting Aspartate Be Hydroxylase with the Small Molecule Inhibitor MO-I-1182 Suppresses Cholangiocarcinoma Metastasis. *Dig. Dis. Sci.* **2021**, 1080–1089. [CrossRef]
50. Lin, Q.; Chen, X.; Meng, F.; Ogawa, K.; Li, M.; Song, R.; Zhang, S.; Zhang, Z.; Kong, X.; Xu, Q.; et al. Multi-organ metastasis destination for breast cancer cells guided by biomechanical architecture. *Am. J. Cancer Res.* **2021**, *11*, 2537–2567.

51. Susa, M.; Morii, T.; Yabe, H.; Horiuchi, K.; Toyama, Y.; Weissbach, L.; Hornicek, F.J.; Morioka, H. Alendronate inhibits growth of high-grade chondrosarcoma cells. *Anticancer Res.* **2009**, *29*, 1879–1888.
52. Rey, V.; Menendez, S.T.; Estupinan, O.; Rodriguez, A.; Santos, L.; Tornin, J.; Martinez-Cruzado, L.; Castillo, D.; Ordonez, G.R.; Costilla, S.; et al. New Chondrosarcoma Cell Lines with Preserved Stem Cell Properties to Study the Genomic Drift During In Vitro/In Vivo Growth. *J. Clin. Med.* **2019**, *8*, 455. [CrossRef]
53. de la Monte, S.M.; Tamaki, S.; Cantarini, M.C.; Ince, N.; Wiedmann, M.; Carter, J.J.; Lahousse, S.A.; Califano, S.; Maeda, T.; Ueno, T.; et al. Aspartyl-(asparaginyl)-beta-hydroxylase regulates hepatocellular carcinoma invasiveness. *J. Hepatol.* **2006**, *44*, 971–983. [CrossRef] [PubMed]
54. Eliasz, S.; Liang, S.; Chen, Y.; De Marco, M.A.; Machek, O.; Skucha, S.; Miele, L.; Bocchetta, M. Notch-1 stimulates survival of lung adenocarcinoma cells during hypoxia by activating the IGF-1R pathway. *Oncogene* **2010**, *29*, 2488–2498. [CrossRef] [PubMed]
55. Mi, L.; Chen, Y.; Zheng, X.; Li, Y.; Zhang, Q.; Mo, D.; Yang, G. MicroRNA-139-5p Suppresses 3T3-L1 Preadipocyte Differentiation Through Notch and IRS1/PI3K/Akt Insulin Signaling Pathways. *J. Cell Biochem.* **2015**, *116*, 1195–1204. [CrossRef]
56. Patravale, V.; Dandekar, P.; Jain, R. Nanotoxicology: Evaluating toxicity potential of drug-nanoparticles. In *Nanoparticulate Drug Delivery*; Patravale, V., Dandekar, P., Jain, R., Eds.; Woodhead Publishing: Cambridge, UK, 2012; pp. 123–155. [CrossRef]
57. Jain, A.K.; Singh, D.; Dubey, K.; Maurya, R.; Mittal, S.; Pandey, A.K. Models and Methods for In Vitro Toxicity. In *In Vitro Toxicology*; Dhawan, A., Kwon, S., Eds.; Academic Press: Cambridge, MA, USA, 2018; pp. 45–65. [CrossRef]
58. van Meerloo, J.; Kaspers, G.J.L.; Cloos, J. Cell Sensitivity Assays: The MTT Assay. In *Cancer Cell Culture: Methods and Protocols*; Cree, I.A., Ed.; Humana Press: Totowa, NJ, USA, 2011; pp. 237–245. [CrossRef]
59. Gundogan, F.; Elwood, G.; Longato, L.; Tong, M.; Feijoo, A.; Carlson, R.I.; Wands, J.R.; de la Monte, S.M. Impaired placentation in fetal alcohol syndrome. *Placenta* **2008**, *29*, 148–157. [CrossRef]
60. Tong, M.; Ziplow, J.L.; Mark, P.; de la Monte, S.M. Dietary Soy Prevents Alcohol-Mediated Neurocognitive Dysfunction and Associated Impairments in Brain Insulin Pathway Signaling in an Adolescent Rat Model. *Biomolecules* **2022**, *12*, 676. [CrossRef] [PubMed]
61. Yalcin, E.B.; Tong, M.; Delikkaya, B.; Pelit, W.; Yang, Y.; de la Monte, S.M. Differential effects of moderate chronic ethanol consumption on neurobehavior, white matter glial protein expression, and mTOR pathway signaling with adolescent brain maturation. *Am. J. Drug Alcohol Abuse* **2024**, *50*, 492–516. [CrossRef] [PubMed]
62. Yang, Y.; Tong, M.; de la Monte, S.M. Early-Stage Moderate Alcohol Feeding Dysregulates Insulin-Related Metabolic Hormone Expression in the Brain: Potential Links to Neurodegeneration Including Alzheimer's Disease. *J. Alzheimers Dis. Rep.* **2024**, *8*, 1211–1228. [CrossRef]
63. Alexander, M.C.; Lomanto, M.; Nasrin, N.; Ramaika, C. Insulin stimulates glyceraldehyde-3-phosphate dehydrogenase gene expression through cis-acting DNA sequences. *Proc. Natl. Acad. Sci. USA* **1988**, *85*, 5092–5096. [CrossRef]
64. Nicholls, C.; Li, H.; Liu, J.P. GAPDH: A common enzyme with uncommon functions. *Clin. Exp. Pharmacol. Physiol.* **2012**, *39*, 674–679. [CrossRef]
65. de la Monte, S.M.; Lahousse, S.A.; Carter, J.; Wands, J.R. ATP luminescence-based motility-invasion assay. *Biotechniques* **2002**, *33*, 98–100, 102, 104 passim. [CrossRef]
66. Kanwal, M.; Smahel, M.; Olsen, M.; Smahelova, J.; Tachezy, R. Aspartate beta-hydroxylase as a target for cancer therapy. *J. Exp. Clin. Cancer Res.* **2020**, *39*, 163. [CrossRef] [PubMed]
67. Lavaissiere, L.; Jia, S.; Nishiyama, M.; de la Monte, S.; Stern, A.M.; Wands, J.R.; Friedman, P.A. Overexpression of human aspartyl(asparaginyl)beta-hydroxylase in hepatocellular carcinoma and cholangiocarcinoma. *J. Clin. Investig.* **1996**, *98*, 1313–1323. [CrossRef] [PubMed]
68. de la Monte, S.M.; Tong, M.; Carlson, R.I.; Carter, J.J.; Longato, L.; Silbermann, E.; Wands, J.R. Ethanol inhibition of aspartyl-asparaginyl-beta-hydroxylase in fetal alcohol spectrum disorder: Potential link to the impairments in central nervous system neuronal migration. *Alcohol* **2009**, *43*, 225–240. [CrossRef]
69. Dinchuk, J.E.; Focht, R.J.; Kelley, J.A.; Henderson, N.L.; Zolotarjova, N.I.; Wynn, R.; Neff, N.T.; Link, J.; Huber, R.M.; Burn, T.C.; et al. Absence of post-translational aspartyl beta-hydroxylation of epidermal growth factor domains in mice leads to developmental defects and an increased incidence of intestinal neoplasia. *J. Biol. Chem.* **2002**, *277*, 12970–12977. [CrossRef]
70. Feriotto, G.; Finotti, A.; Breveglieri, G.; Treves, S.; Zorzato, F.; Gambari, R. Transcriptional activity and Sp 1/3 transcription factor binding to the P1 promoter sequences of the human AbetaH-J-J locus. *FEBS J.* **2007**, *274*, 4476–4490. [CrossRef]
71. Finotti, A.; Treves, S.; Zorzato, F.; Gambari, R.; Feriotto, G. Upstream stimulatory factors are involved in the P1 promoter directed transcription of the A beta H-J-J locus. *BMC Mol. Biol.* **2008**, *9*, 110. [CrossRef]
72. Bandyopadhyay, G.K.; Yu, J.G.; Ofrecio, J.; Olefsky, J.M. Increased p85/55/50 expression and decreased phosphotidylinositol 3-kinase activity in insulin-resistant human skeletal muscle. *Diabetes* **2005**, *54*, 2351–2359. [CrossRef]
73. Lee, J.H. Overexpression of humbug promotes malignant progression in human gastric cancer cells. *Oncol. Rep.* **2008**, *19*, 795–800. [CrossRef] [PubMed]

74. Luu, M.; Sabo, E.; de la Monte, S.M.; Greaves, W.; Wang, J.; Tavares, R.; Simao, L.; Wands, J.R.; Resnick, M.B.; Wang, L. Prognosti value of aspartyl (asparaginyl)-beta-hydroxylase/humbug expression in non-small cell lung carcinoma. *Hum. Pathol.* **2009**, *4* 639–644. [CrossRef]
75. Wang, J.; de la Monte, S.M.; Sabo, E.; Kethu, S.; Tavares, R.; Branda, M.; Simao, L.; Wands, J.R.; Resnick, M.B. Prognostic value o humbug gene overexpression in stage II colon cancer. *Hum. Pathol.* **2007**, *38*, 17–25. [CrossRef]
76. Qu, Y.; Qi, L.; Hao, L.; Zhu, J. Upregulation of circ-ASPH contributes to glioma cell proliferation and aggressiveness by targetin the miR-599/AR/SOCS2-AS1 signaling pathway. *Oncol. Lett.* **2021**, *21*, 388. [CrossRef] [PubMed]
77. Benelli, R.; Costa, D.; Mastracci, L.; Grillo, F.; Olsen, M.J.; Barboro, P.; Poggi, A.; Ferrari, N. Aspartate-beta-Hydroxylase: A Promising Target to Limit the Local Invasiveness of Colorectal Cancer. *Cancers* **2020**, *12*, 971. [CrossRef]
78. Kanwal, M.; Smahelova, J.; Ciharova, B.; Johari, S.D.; Nunvar, J.; Olsen, M.; Smahel, M. Aspartate beta-hydroxylase Regulate Expression of Ly6 Genes. *J. Cancer* **2024**, *15*, 1138–1152. [CrossRef]
79. Radaelli, S.; Stacchiotti, S.; Casali, P.G.; Gronchi, A. Emerging therapies for adult soft tissue sarcoma. *Expert Rev. Anticancer The* **2014**, *14*, 689–704. [CrossRef] [PubMed]
80. Maira, S.M.; Stauffer, F.; Brueggen, J.; Furet, P.; Schnell, C.; Fritsch, C.; Brachmann, S.; Chene, P.; De Pover, A.; Schoemaker, K et al. Identification and characterization of NVP-BEZ235, a new orally available dual phosphatidylinositol 3-kinase/mammalia target of rapamycin inhibitor with potent in vivo antitumor activity. *Mol. Cancer Ther.* **2008**, *7*, 1851–1863. [CrossRef]
81. Schnell, C.R.; Stauffer, F.; Allegrini, P.R.; O'Reilly, T.; McSheehy, P.M.; Dartois, C.; Stumm, M.; Cozens, R.; Littlewood-Evans A.; Garcia-Echeverria, C.; et al. Effects of the dual phosphatidylinositol 3-kinase/mammalian target of rapamycin inhibito NVP-BEZ235 on the tumor vasculature: Implications for clinical imaging. *Cancer Res.* **2008**, *68*, 6598–6607. [CrossRef] [PubMed]
82. Tzeng, H.E.; Tang, C.H.; Wu, S.H.; Chen, H.T.; Fong, Y.C.; Lu, Y.C.; Chen, W.C.; Huang, H.D.; Lin, C.Y.; Wang, S.W. CCN6 mediated MMP-9 activation enhances metastatic potential of human chondrosarcoma. *Cell Death Dis.* **2018**, *9*, 955. [CrossRef] [PubMed]
83. Mery, B.; Espenel, S.; Guy, J.B.; Rancoule, C.; Vallard, A.; Aloy, M.T.; Rodriguez-Lafrasse, C.; Magne, N. Biological aspects o chondrosarcoma: Leaps and hurdles. *Crit. Rev. Oncol. Hematol.* **2018**, *126*, 32–36. [CrossRef]
84. Wu, Y.Y.; Wu, H.C.; Wu, J.E.; Huang, K.Y.; Yang, S.C.; Chen, S.X.; Tsao, C.J.; Hsu, K.F.; Chen, Y.L.; Hong, T.M. The dua PI3K/mTOR inhibitor BEZ235 restricts the growth of lung cancer tumors regardless of EGFR status, as a potent accompanist ir combined therapeutic regimens. *J. Exp. Clin. Cancer Res.* **2019**, *38*, 282. [CrossRef]
85. Huyan, T.; Tang, R.; Li, J.; Li, Q.; Xue, X.; Yang, H. Optimized Expression and Purification of Humbug in Pichia pastoris and Its Monoclonal Antibody Preparation. *Iran. J. Public Health* **2015**, *44*, 1632–1642.
86. Durrant, D.E.; Das, A.; Dyer, S.; Kukreja, R.C. A dual PI3 kinase/mTOR inhibitor BEZ235 reverses doxorubicin resistance ir ABCB1 overexpressing ovarian and pancreatic cancer cell lines. *Biochim. Biophys. Acta Gen. Subj.* **2020**, *1864*, 129556. [CrossRef] [PubMed]
87. Durrant, D.E.; Das, A.; Dyer, S.; Tavallai, S.; Dent, P.; Kukreja, R.C. Targeted Inhibition of Phosphoinositide 3-Kinase/Mammaliar Target of Rapamycin Sensitizes Pancreatic Cancer Cells to Doxorubicin without Exacerbating Cardiac Toxicity. *Mol. Pharmacol* **2015**, *88*, 512–523. [CrossRef] [PubMed]
88. Durrant, D.E.; Dyer, S.; Das, A.; Kukreja, R.C. Abstract 16201: Combined Treatment of the PI3k/mTOR Inhibitor BEZ235 With Doxorubicin Does Not Increase Cardiotoxicity While Improving Efficacy in Breast Cancer Cells. *Circulation* **2014**, *130* (Suppl. 2) [CrossRef]
89. Ikenouchi, J.; Aoki, K. Membrane bleb: A seesaw game of two small GTPases. *Small GTPases* **2017**, *8*, 85–89. [CrossRef]
90. Aoki, K.; Satoi, S. Coordinated changes in cell membrane and cytoplasm during maturation of apoptotic bleb. *Mol. Biol. Cell* **2020** *31*, 833–844. [CrossRef]
91. Tacar, O.; Dass, C.R. Doxorubicin-induced death in tumour cells and cardiomyocytes: Is autophagy the key to improving future clinical outcomes? *J. Pharm. Pharmacol.* **2013**, *65*, 1577–1589. [CrossRef]
92. Brachmann, S.M.; Hofmann, I.; Schnell, C.; Fritsch, C.; Wee, S.; Lane, H.; Wang, S.; Garcia-Echeverria, C.; Maira, S.M. Specific apoptosis induction by the dual PI3K/mTor inhibitor NVP-BEZ235 in HER2 amplified and PIK3CA mutant breast cancer cells *Proc. Natl. Acad. Sci. USA* **2009**, *106*, 22299–22304. [CrossRef]
93. Mo, J.S.; Yoon, J.H.; Ann, E.J.; Ahn, J.S.; Baek, H.J.; Lee, H.J.; Kim, S.H.; Kim, Y.D.; Kim, M.Y.; Park, H.S. Notch1 modulates oxidative stress induced cell death through suppression of apoptosis signal-regulating kinase 1. *Proc. Natl. Acad. Sci. USA* **2013** *110*, 6865–6870. [CrossRef]
94. Tasca, A.; Helmstadter, M.; Brislinger, M.M.; Haas, M.; Mitchell, B.; Walentek, P. Notch signaling induces either apoptosis or cell fate change in multiciliated cells during mucociliary tissue remodeling. *Dev. Cell* **2021**, *56*, 525–539.e526. [CrossRef]
95. Wang, Z.; Zhang, Y.; Li, Y.; Banerjee, S.; Liao, J.; Sarkar, F.H. Down-regulation of Notch-1 contributes to cell growth inhibition and apoptosis in pancreatic cancer cells. *Mol. Cancer Ther.* **2006**, *5*, 483–493. [CrossRef]
96. Sahlgren, C.; Gustafsson, M.V.; Jin, S.; Poellinger, L.; Lendahl, U. Notch signaling mediates hypoxia-induced tumor cell migration and invasion. *Proc. Natl. Acad. Sci. USA* **2008**, *105*, 6392–6397. [CrossRef] [PubMed]

. Liu, L.; Zhang, L.; Zhao, S.; Zhao, X.Y.; Min, P.X.; Ma, Y.D.; Wang, Y.Y.; Chen, Y.; Tang, S.J.; Zhang, Y.J.; et al. Non-canonical Notch Signaling Regulates Actin Remodeling in Cell Migration by Activating PI3K/AKT/Cdc42 Pathway. *Front. Pharmacol.* **2019**, *10*, 370. [CrossRef]

. Yu, Y.; Yu, X.; Ma, J.; Tong, Y.; Yao, J. Effects of NVP-BEZ235 on the proliferation, migration, apoptosis and autophagy in HT-29 human colorectal adenocarcinoma cells. *Int. J. Oncol.* **2016**, *49*, 285–293. [CrossRef] [PubMed]

. Vazquez, N.; Lopez, A.; Cuello, V.; Persans, M.; Schuenzel, E.; Innis-Whitehouse, W.; Keniry, M. NVP-BEZ235 or JAKi Treatment leads to decreased survival of examined GBM and BBC cells. *Cancer Treat. Res. Commun.* **2021**, *27*, 100340. [CrossRef] [PubMed]

0. Zhang, Y.X.; van Oosterwijk, J.G.; Sicinska, E.; Moss, S.; Remillard, S.P.; van Wezel, T.; Buhnemann, C.; Hassan, A.B.; Demetri, G.D.; Bovee, J.V.; et al. Functional profiling of receptor tyrosine kinases and downstream signaling in human chondrosarcomas identifies pathways for rational targeted therapy. *Clin. Cancer Res.* **2013**, *19*, 3796–3807. [CrossRef]

1. Bhola, N.E.; Jansen, V.M.; Koch, J.P.; Li, H.; Formisano, L.; Williams, J.A.; Grandis, J.R.; Arteaga, C.L. Treatment of Triple-Negative Breast Cancer with TORC1/2 Inhibitors Sustains a Drug-Resistant and Notch-Dependent Cancer Stem Cell Population. *Cancer Res.* **2016**, *76*, 440–452. [CrossRef]

2. Zhdanovskaya, N.; Firrincieli, M.; Lazzari, S.; Pace, E.; Scribani Rossi, P.; Felli, M.P.; Talora, C.; Screpanti, I.; Palermo, R. Targeting Notch to Maximize Chemotherapeutic Benefits: Rationale, Advanced Strategies, and Future Perspectives. *Cancers* **2021**, *13*, 5106. [CrossRef]

cancers

MDPI

Article

FOLFIRINOX: The Best Adjuvant Treatment for Ampullary Adenocarcinoma? A Multicenter Study by the Turkish Oncology Group (TOG)

Ali Kalem [1,*], Tulay Kus [1], Taha Koray Sahin [2], Omer Dizdar [2], Safa Can Efil [3], Mehmet Ali Nahit Sendur [3], Talat Aykut [4], Murat Araz [4], Hatice Bolek [5], Yuksel Urun [5], Nadiye Sever [6], Ibrahim Vedat Bayoglu [6], Eyyup Cavdar [7], Muhammed Fatih Sagıroglu [8], Tugce Kubra Gunes [9], Melike Ozcelik [9], Nadide Demirel [10], Bulent Yıldız [10], Berkan Karabuga [11], Ulku Yalcıntas Arslan [11], Savas Gokcek [12], Ilkay Tugba Unek [12], Seray Saray [13], Ferit Aslan [14], Omer Acar [15], Atike Pınar Erdogan [15], Mustafa Seyyar [8], Gokmen Aktas [16] and Suayib Yalcın [2]

1 Department of Medical Oncology, School of Medicine, Gaziantep University, Gaziantep 27310, Turkey; tulaykus@gantep.edu.tr
2 Department of Medical Oncology, School of Medicine, Hacettepe University, Ankara 06100, Turkey; takorsah@gmail.com (T.K.S.); dromerdizdar@gmail.com (O.D.); suayibyalcin@gmail.com (S.Y.)
3 Department of Medical Oncology, School of Medicine, Ankara YıldırımBeyazıt University, Ankara 06110, Turkey; safacan.efil@saglik.gov.tr (S.C.E.); mansendur@aybu.edu.tr (M.A.N.S.)
4 Department of Medical Oncology, School of Medicine, Necmettin Erbakan University, Konya 42090, Turkey; talat_aykut@hotmail.com (T.A.); zaratarum@yahoo.com (M.A.)
5 Department of Medical Oncology, School of Medicine, Ankara University, Ankara 06230, Turkey; hati.kocc@gmail.com (H.B.); yukselurun@gmail.com (Y.U.)
6 Department of Medical Oncology, School of Medicine, Marmara University, İstanbul 34854, Turkey; nadiye.sever@marmara.edu.tr (N.S.); dr.vebay@gmail.com (I.V.B.)
7 Department of Medical Oncology, Adıyaman Education and Research Hospital, Adıyaman 02100, Turkey; ecavdar@nku.edu.tr
8 Department of Medical Oncology, Gaziantep City Hospital, Gaziantep 27470, Turkey; dr.mfsagiroglu@gmail.com (M.F.S.); mustafa.seyyar@saglik.gov.tr (M.S.)
9 Department of Medical Oncology, Ümraniye Education and Research Hospital, İstanbul 34764, Turkey; drtugcekubragunes@gmail.com (T.K.G.); drmelike.ozcelik@gmail.com (M.O.)
10 Department of Medical Oncology, School of Medicine, Eskisehir Osmangazi University, Eskisehir 26040, Turkey; nadide.demirel@saglik.gov.tr (N.D.); bulentyildiz@ogu.edu.tr (B.Y.)
11 Department of Medical Oncology, Dr. AbdurrahmanYurtaslan Oncology Education and Research Hospital, Ankara 06200, Turkey; drbkarabuga@gmail.com (B.K.); ulkuarslan63@gmail.com (U.A.Y.)
12 Department of Medical Oncology, School of Medicine, Dokuz Eylul University, Izmir 35220, Turkey; savas.gokcek@saglik.gov.tr (S.G.); ilkaytugbaunek@gmail.com (I.T.U.)
13 Department of Medical Oncology, Ataturk Education and Research Hospital, Balıkesir 10100, Turkey; seraysaray@saglik.gov.tr
14 Department of Medical Oncology, Medical Park Ankara Batıkent Hospital, Ankara 06680, Turkey; feritferhat21@gmail.com
15 Department of Medical Oncology, School of Medicine, Manisa Celal Bayar University, Manisa 45030, Turkey; dracaromer@gmail.com (O.A.); atike.erdogan@cbu.edu.tr (A.P.E.)
16 Gaziantep Medical Point Private Hospital, Gaziantep 27584, Turkey; gokmen.aktas@mph.com.tr
* Correspondence: kalemali88@gmail.com; Tel.: +90-342-360-60-60; Fax: +90-342-360-60-61

Simple Summary: Ampullary adenocarcinoma is a rare cancer with no universally accepted adjuvant treatment. Our multi-center retrospective study evaluated the role of modified FOLFIRINOX (mFOLFIRINOX) compared to other adjuvant regimens. We found that mFOLFIRINOX was associated with significantly improved disease-free survival and a favorable trend toward better overall survival, independent of histological subtype. While the regimen has a higher toxicity profile, it may be considered for fit, younger patients. This is the largest dataset on adjuvant FOLFIRINOX in ampullary adenocarcinoma and can help guide future clinical trials.

Abstract: Background: Ampullary adenocarcinoma is a rare cancer for which there are no standard adjuvant treatment recommendations due to the lack of randomized clinical trials. The primary aim of this analysis is to investigate the efficacy of adjuvant FOLFIRINOX treatment in patients with resected ampullary adenocarcinoma. Materials and Methods: This multicenter retrospective cohort study was conducted at 15 institutions in Turkey between August 2007 and January 2024, involving 211 patients with resected, non-metastatic ampullary adenocarcinoma receiving adjuvant chemotherapy with various chemotherapy regimens with or without chemoradiation. Clinicopathological and treatment-related parameters were recorded. Disease-free survival (DFS) and overall survival (OS) were analyzed by using Kaplan–Meier estimates. Cox proportional hazards regression was used to identify covariates associated with OS. Results: The median follow-up time was 52 months, and 116 patients (55.2%) were alive at the time of the analysis. The median age was 61 years (32–82). mFOLFIRINOX was administered to 16.6% of the patients (n = 35). The 3-year DFS rate was 79.41% in the FOLFIRINOX-treated arm and 53.9% in the other treatment arm (p = 0.034 for mDFS). The median OS was non-reached in patients receiving mFOLFIRINOX treatment, while it was 51 months in patients receiving other treatments (p = 0.071). While no statistically significant results were reached, a trend toward statistically significant survival times was observed in the FOLFIRINOX arm. After adjustment for other prognostic parameters, mFOLFIRINOX remained an independent statistically significant parameter for better OS (HR; 95% CI: 3.24; 1.02–10.9; p = 0.046). Conclusions: FOLFIRINOX treatment has shown efficacy in the adjuvant treatment of ampullary cancer, independent of histological subtype. The findings should be validated in large prospective trials.

Keywords: ampullary adenocarcinoma; FOLFIRINOX; adjuvant; radiotherapy; gemcitabine

1. Introduction

Ampullary cancers are defined as tumors originating from the ampulla of Vater, which comprises three anatomical components: the ampulla, the intraduodenal portion of the bile duct, and the intraduodenal portion of the pancreatic duct [1]. Although they are a single type of tumor, ampullary cancers can have different clinical courses. Kimura W et al. first proposed that ampullary tumors exhibit different characteristics and prognoses based on the epithelial cloaca from which they originate. Accordingly, ampullary tumors are histologically classified into two types: (1) an intestinal type that resembles tubular adenocarcinoma of the stomach or colon and (2) a pancreaticobiliary type characterized by papillary projections with scant fibrous cores. The age distributions of both types are similar; however, the prognosis of the intestinal type is much better than that of the pancreaticobiliary type [2,3]. Additionally, lymph node metastasis and a strong infiltrative tendency are more common in the pancreaticobiliary subtype, while the intestinal type tends to have little or no invasion into the surrounding interstitium [2,3]. This histological heterogeneity may be responsible for the differing clinical outcomes and treatment responses [4–7].

In addition to their histological heterogeneity, ampullary cancers are very rare tumors, accounting for only 0.2% of gastrointestinal malignancies [8]. Consequently, the results of clinical trials regarding ampullary adenocarcinoma are not definitive due to their rarity and are often analyzed as part of biliary tract cancers. There is no consensus on the optimal adjuvant treatment for patients with ampullary cancer. Recommendations for the management of ampullary carcinoma have only recently been included in the National

Comprehensive Cancer Network (NCCN) guidelines as a separate category [9]. Given the more favorable prognosis of ampullary cancer compared to biliary tract cancers and the lack of strong data from randomized trials proving a survival advantage, the level of evidence for adjuvant chemoradiotherapy or specific chemotherapy regimens is weak [9,10].

In our study, we aimed to analyze the effectiveness of mFOLFIRINOX chemotherapy, which has been shown to be effective in both pancreatic cancer and bowel cancer, in the adjuvant treatment of patients with ampullary cancer who underwent curative intent surgery. This analysis is based on histological subtype and is compared with other systemic treatments [11].

2. Materials and Methods

2.1. Patients and Study Design

This multicenter, retrospective observational study included patients from 15 centers across various regions of Turkey between August 2007 and January 2024. Patients aged at least 18 years who had received a histological and clinical diagnosis of ampullary adenocarcinoma and underwent curative intent Whipple procedures were screened. Patients who underwent R0 or R1 resection and received adjuvant chemotherapy or chemoradiotherapy were included.

Patients were followed up using computed tomography, routine blood tests, and tumor marker assays every 3–4 months for the first 2 years, then every 6 months for 5 years, and annually thereafter. Additionally, we meticulously examined the recurrence patterns (local or distant) among patients who experienced recurrence and documented the treatments they received during these recurrence episodes.

The study was conducted in accordance with the principles outlined in the Declaration of Helsinki and received approval from the ethics committees of the participating centers. The Ethics Committee of Gaziantep University Faculty of Medicine approved the study protocol, which included patients' clinical data, informed consent, and the multicenter design (Approval No. 2024/43). In addition, corresponding clinical information (e.g., age and gender) and pathological characteristics were collected from the hospital information management system. The datasets used and/or analyzed during the current study are available from the corresponding author upon reasonable request.

2.2. Patients and Disease Features

Baseline characteristics, treatment regimens, and treatment and recurrence dates were retrospectively examined from patient records and electronic systems. The following parameters were recorded: patient age, gender, comorbidities, alcohol use, smoking status, the Eastern Cooperative Oncology Group (ECOG) performance status (0–2), and disease pathological T stage (T1-4) and N stage (0–2) using the American Joint Committee on Cancer staging system (AJCC, 8th edition of the TNM classification). Additionally, histopathological subtype (pancreaticobiliary type or intestinal type), presence of lymphovascular invasion (LVI) and perineural invasion (PNI), resection margin (R0 or R1), presence of pancreatic invasion and biliary obstruction, lymph node ratio (LNR; positive lymph node/total harvested lymph nodes), and total harvested lymph nodes (THLN; <15 vs. ≥ 16) were recorded.

Laboratory parameters, including hemoglobin, neutrophil, lymphocyte, and platelet levels, detailed liver function tests, and tumor markers (CEA ng/mL, CA 19-9 U/mL) were also noted.

2.3. Adjuvant Treatments

Other treatments administered in addition to mFOLFIRINOX (oxaliplatin plus irinotecan with leucovorin and short-term infusional5-fluorouracil [FU]) chemotherapy included gemcitabine in combination with FU-based therapy or cisplatin; gemcitabine or FU monotherapy; chemoradiotherapy (with gemcitabine or FU/capecitabine); and FU-based doublets (XELOX, FOLFOX). Additionally, radiotherapy was administered to some patients outside of chemoradiotherapy.

All investigators used the Common Terminology Criteria for Adverse Events "CTCAE" version 5.0 to assess toxicity.

2.4. Statistical Analysis

The normal distribution of numerical variables was assessed using the Shapiro–Wilk test. Relationships between categorical variables were examined using the χ^2 test and Fisher's Exact Test, with categorical variables presented as counts (ns) and percentages (%s). Continuous variables were reported as mean ± standard deviation (SD). Disease-free survival (DFS) and overall survival (OS) were estimated using the Kaplan–Meier method, and the log-rank test was used to compare the effects of clinicopathological and laboratory parameters on survival in the univariate analysis. Univariate Cox regression was employed to estimate hazard ratios (HRs) with 95% confidence intervals (CIs) for OS and DFS. All potential predictive factors with a p-value <0.10 in the univariate analyses were included in the multivariate analysis. The Cox proportional hazards model was used for multivariate analysis to determine independent significant factors. Cox regression curve of overall survival of patients with intestinal-type ampullary carcinoma and pancreaticobiliary/mix type ampullary carcinoma was performed after adjustment for age, comorbidities, performance status, stage, lymphovascular and perineural invasion status, resection margin, pancreatic invasion, and CA 19-9.Cox regression curve was performed for DFS after adjustment for performance status, stage, lymphovascular and perineural invasion status, resection margin, pancreatic invasion, and CA 19-9. Next, Cox regression curve of the OS of patients was performed after adjustment for age, comorbidities, performance status, stage, lymphovascular and perineural invasion status, resection margin, pancreatic invasion, and CA 19-9.A p-value of $\leq$0.05 was considered statistically significant, and statistical analyses were conducted using the SPSS Statistics version 22.0 (IBM Corp., Armonk, NY, USA).

The primary endpoint was to assess the efficacy of FOLFIRINOX on DFS by comparing it with other treatment options. DFS was defined as the time from the start of adjuvant therapy to the occurrence of the first recurrence (local, regional, or distant metastasis), death from any cause, or the date of the last follow-up. OS was defined as the period from the initiation of adjuvant treatment to the date of death or the last follow-up.

3. Results

3.1. Patient Demographics and Tumor Characteristics

A total of 211 patients with ampullary adenocarcinoma who underwent curative intent surgery and received adjuvant treatment were included in this retrospective study. Patients with R2 resection, inadequate follow-up (<30 days), or those who did not receive adjuvant systemic chemotherapy were excluded. The median age was 61 years (range, 32–82 years); 123 patients (58.3%) were men, and the remaining were women. Seventy-two patients (34.1%) were 65 years or older, and 109 patients (51.7%) had comorbidities (such as coronary artery disease, diabetes, hypertension, or inflammatory diseases). The ECOG performance score was 0 or 1 for 93.9% of the patients, and the remaining patients had a score of 2 (Table 1). The most common histological subtype was pancreatobiliary or mixed intestinal and pancreatobiliary (n = 141; 76.2%), followed by intestinal (n = 44; 23.8%). Most patients

(n = 136; 63.4%) had stage III disease. All other potentially prognostic parameters are shown in Table 1.

Table 1. Baseline characteristics of patients with resected ampullary adenocarcinoma.

	mFOLFIRINOX N (%)	The Other Treatments N (%)	*p* Value
Age, years (median; range)			
<65	33 (9.4.3)	106 (60.2)	<0.001
≥65	2 (5.7)	70 (39.8)	
Gender			
Female	19 (54.3)	104 (59.1)	0.60
Male	16 (45.7)	72 (40.9)	
Comorbidity			
Present	17 (48.6)	85 (48.3)	0.97
Absent	18 (51.4)	91 (51.7)	
ECOG PS			
0	15 (42.9)	74 (42.0)	0.96
1	18 (51.4)	91 (51.7)	
2	2 (5.7)	11 (6.2)	
BMI			
<18.5, underweight	1 (3.1)	5 (3.8)	0.98
18.5–24.9, normal	17 (53.1)	73 (56.2)	
25–29.9, overweight	11 (34.4)	39 (30.0)	
≥30, obese	3 (9.4)	13 (10.0)	
Smoking	14 (40)	76 (43.2)	0.73
None	21 (60)	100 (56.8)	
Alcohol use	3 (8.6)	16 (9.1)	0.92
None	32 (91.4)	160 (90.9)	
Histological subtype			
Intestinal	7 (21.2)	37 (24.3)	0.70
Pancreaticobiliary/mix	26 (78.8)	115 (75.7)	
pT stage			
T1	3 (8.6)	21 (11.9)	0.66
T2	11 (31.4)	49 (27.8)	
T3	20 (57.1)	92 (52.3)	
T4	1 (2.9)	14 (8.0)	
pN stage			
N0	13 (38.2)	65 (37.1)	0.77
N1	16 (47.19)	91 (52.0)	
N2	5 (14.7)	19 (10.9)	
LNR			
0	14 (41.2)	61 (41.9)	0.84
<0.2	10 (29.4)	47 (29.4)	
0.2–0.4	5 (14.7)	30 (18.8)	
≥0.4	5 (14.7)	16 (10.0)	
THLN			
≥16	18 (52.9)	75 (48.1)	0.61
<16	16 (47.1)	81 (51.9)	

Table 1. *Cont.*

	mFOLFIRINOX N (%)	The Other Treatments N (%)	*p* Value
Surgical margin			
R0	33 (97.1)	160 (92.0)	0.29
R1	1 (2.9)	14 (8.0)	
Pancreatic invasion			
Present	21 (61.8)	98 (61.6)	0.98
Absent	13 (38.2)	61 (38.4)	
LVI			
Present	23 (67.6)	121 (70.8)	0.71
Absent	11 (32.4)	50 (29.2)	
PNI			
Present	19 (55.9)	87 (50.3)	0.55
Absent	15 (44.1)	86 (49.7)	
Biliary obstruction			
Present	16 (45.7)	82 (47.4)	0.86
Absent	19 854.3)	91 (52.6)	
CA 19-9 U/mL			
<150	30 (85.7)	129 (81.6)	0.57
≥150 (if bilirubin is high: 300)	5 (14.3)	29 (18.4)	
Albumi, g/dL			
<3.5	7 (20)	44 (27.3)	0.37
≥3.5	28 (80)	117 (72.7)	
Hemoglobin, g/dL			
<12 for women; 13 for men	15 (44.1)	83 (50.6)	0.49
≥3.5	19 (55.9)	81 (49.4)	
Neutrophils, $\times 10^9$/L			
≥7000	28 (84.3)	137 (82.0)	0.69
<7000	5 (15.2)	30 (18.0)	
Lymphocyte, $\times 10^9$/L			
<1500	8 (24.2)	57 (34.3)	0.26
≥1500	25 (75.3)	109 (65.7)	
Platelets, $\times 10^9$/L			
<500	34 (97.1)	156 (91.2)	0.23
≥500	1 (2.9)	15 (8.8)	
Radiotherapy			
Received	5 (14.3)	76 (43.2)	0.001
None	30 (85.7)	100 (56.8)	

LNR: lymph node ratio (positive lymph node/total harvested lymph nodes); THLN: total harvested lymph nodes; LVI: lymphovascular invasion; PNI: perineural invasion.

In terms of the clinicopathological characteristics of the patients, there was generally no statistically significant difference between the group receiving mFOLFIRINOX and the group receiving other treatments. However, it was observed that the mFOLFIRINOX treatment was less commonly recommended for patients over 65 years of age. Additionally, the rate of radiotherapy administered in the mFOLFIRINOX group was statistically significantly lower (Table 1). Nevertheless, neither age nor radiotherapy treatment was a statistically significant parameter for DFS ($p > 0.05$) (Table 2).

Table 2. Univariate and multivariate analyses of effect of clinicopathological and treatment-related parameters on disease-free survival.

	Univariate Analysis, Mo (95%CI)	p Value	Multivariate Analysis HR (95%CI)	p Value
Age, years (median; range)				
<65	31.0 (24.1–37.9)	0.58		
≥65	24.0 (8.8–39.2)			
Gender				
Female	31.0 (27.1–34.9)	0.56		
Male	23.0 (19.8–26.1)			
Comorbidity				
Present	24.0 (14.2–33.8)	0.90		
Absent	31.0 (22.5–39.4)			
ECOG PS				
0	33 (28.7–37.3)	0.088	1 (ref)	0.91
1	24 (15.1–32.9)		0.9 (0.6–1.5)	
2	11.0 (8.2–13.7)		1.1 (0.5–2.7)	
BMI				
<18.5, underweight	10.0 (3.0–24.0)	0.82		
18.5–24.9, normal	31.0 (22.1–39.9)			
25–29.9, overweight	24.0 (13.5–34.6)			
≥30, obese	21.0 (16.1–25.6)			
Smoking	27.0 (16.9–37.0)	0.72		
None	31.0 (22.6–39.1)			
Alcohol use	30 (23.9–36.0)	0.64		
None	14 (3.0–31.47)			
Histological subtype				
Pancreaticobiliary/mix	27.0 (20.8–33.1)	0.28		
Intestinal	34.0 (26.8–41.2)			
pT stage				
T1	34.0 (29.3–38.7)	0.059		
T2	37.0 (22.9–51.0)		-	
T3	23.0 (19.0–26.6)			
T4	10.0 (8.2–11.7)			
pN stage				
N0	43.0 (9.1–76.9)	<0.001	-	
N1	23.0 (18.4–27.6)			
N2	14.0 (4.8–23.2)			
Stage				
I (T1-2N0)	72 (Non-reached)		1 (ref)	
II (T3N0)	33.0 (28.3–37.7)	<0.001	1.1 (0.5–2.7)	0.26
IIIA (T 1-3N1)	24.0 (17.2–30.8)		1.5 (0.7–3.2)	
IIIB (T4 or N2)	13.0 (3.8–22.1)		2.0 (0.9–4.7)	
LNR				
0	43.0 (15.5–70.5)	0.007	-	
<0.2	21.0 (14.6–27.3)			
0.2–0.4	33.0 (7.7–58.2)			
≥0.4	17.0 (9.0–24.9)			
THLN				
≥16	24.0 (15.4–32.5)	0.49		
<16	30.0 (20.1–39.9)			

Table 2. *Cont.*

	Univariate Analysis, Mo (95%CI)	*p* Value	Multivariate Analysis HR (95%CI)	*p* Value
Surgical margin				
R0	32.0 (25.1–38.3)	0.001	2.1 (1.1–4.2)	0.035
R1	14.0 (6.4–21.6)			
Pancreatic invasion				
Present	19.0 (12.6–25.3)	<0.001	1.62 (0.91–2.8)	0.073
Absent	51.0 (14.6–87.4)			
LVI				
Present	23.0 (17.5–28.4)	<0.001	2.1 (1.2–3.8)	0.010
Absent	NR			
PNI				
Present	21.0 (16.6–25.4)	0.010	0.94 (0.58–1.50)	0.80
Absent	34.0 (22.6–45.4)			
Biliary obstruction				
Present	23.0 (13.9–32.1)	0.017		
Absent	33.0 (18.8–47.1)			
CA 19-9 U/mL				
<150	31.0 (23.9–38.0)	0.003	1.70 (0.9–3.1)	0.076
≥150 (if bilirubin is high: 300)	13.0 (8.3–17.7)			
Albumin, g/dL				
<3.5	24.0 (7.4–40.6)	0.20		
≥3.5	30.0 (24.4–35.5)			
Hemoglobin, g/dL				
<12 for women; 13 for men	24.0 (17.9–30.0)	0.15		
≥12–13	32.0 (15.4–48.5)			
Neutrophils, $\times 10^9$/L				
≥7000	24.0 (16.9–31.1)	0.15		
<7000	32.0 (22.9–41.0)			
Lymphocyte, $\times 10^9$/L				
<1500	27.0 (20.0–33.9)	0.68		
≥1500	31.0 (17.9–44.1)			
Platelets, $\times 10^9$/L				
<500	15.0 (2.0–31.6)	0.27		
≥500	30.0 (23.9–36.1)			
Radiotherapy				
Received	29.0 (21.5–36.5)	0.45		
None	31.0 (21.2–40.8)			
mFOLFIRINOX	NR	0.034	2.22 (1.1–4.6)	0.034
The other treatments	27.0 (20.7–33.3)			
mFOLFIRINOX	NR	0.064		
5-FU-doublet	16.0 (1.5–30.6)			
Gemcitabine doublet	24.0 (17.1–30.9)			
Gemcitabine or 5-FU	33.0 (19.2–46.7)			
CRT	34.0 (18.8–35.7)			

LNR: lymph node ratio (positive lymph node/total harvested lymph nodes); THLN: total harvested lymph nodes; LVI: lymphovascular invasion; PNI: perineural invasion.

3.2. Effect of Clinicopathologic, Inflammatory, and Treatment-Related Parameters on Disease-Free Survival

The median DFS was 30 (24.3–35.7) months. Of the 211 patients, 114 experienced a recurrence (54.3%), and the DFS data for one patient could not be accessed; only the overall survival data was available for that patient.

The most frequently administered regimen was gemcitabine in combination with FU-based or cisplatin (N = 62; 29.4%), followed by gemcitabine/FU monotherapy (N = 43, 20.4%), chemoradiotherapy (with gemcitabine or FU/capecitabine, N = 40; 19%), FU-based doublet (XELOX, FOLOX, n = 31; 14.7%), and mFOLFIRINOX (n = 35; 16.6%).

The median DFS was non-reached (NR) (NR; NR-NR) in patients treated with mFOLFIRINOX, while it was 27 (20.1–33.3) months in patients receiving other treatment modalities (p = 0.034). (The 3-year DFS rate was 79.41% in the FOLFIRINOX-treated arm and 53.9% in the other treatment arm.) As shown in Table 2, while age, gender, presence of comorbidities, BMI, smoking and alcohol use, histopathological subtype, and radiotherapy use were not effective parameters for DFS, better ECOG performance status, lower T and N stages, absence of pancreatic invasion and biliary obstruction, negative surgical margins, absence of lymphovascular and perineural invasion, low CA 19-9 level, and use of FOLFIRINOX adjuvant therapy were associated with better DFS in the univariate analysis (Table 2). The optimal total harvested lymph nodes value (≥16) was not predictive for DFS; however, the lymph node ratio (LNR; positive lymph nodes/total harvested lymph nodes) was a predictive parameter for DFS. Since LNR was highly correlated with lymph node stage, it was not included in the multivariate analysis. Additionally, there was a high correlation between the presence of biliary obstruction and the CA 19-9 level; therefore, the CA 19-9 level was chosen for the model. In the multivariable analysis, only negative surgical margins, absence of LVI, and mFOLFIRINOX treatment were associated with better DFS (HR 95%CI: 2.22:1.1–4.6; p = 0.034) (Table 2; Figure 1A).

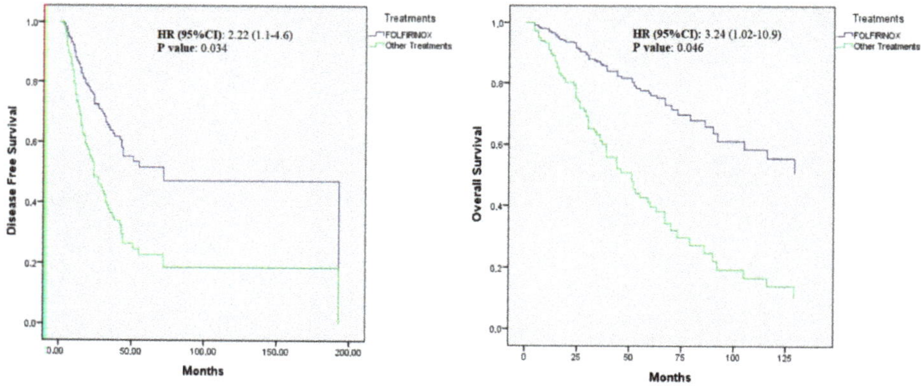

(A) (B)

Figure 1. (**A**): Cox regression curve of disease-free survival in the adjuvant treatments of patients with ampullary adenocarcinoma after adjustment for Eastern Cooperative Oncology Group performance status, stage, lymphovascular and perineural invasion status, resection margin, pancreatic invasion, and CA 19-9. (**B**): Cox regression curve of overall survival in the adjuvant treatments of patients with ampullary adenocarcinoma after adjustment for age, comorbidities, Eastern Cooperative Oncology Group performance status, stage, lymphovascular and perineural invasion status, resection margin, pancreatic invasion, and CA 19-9.

In patients who received radiotherapy, the rate of local recurrence was statistically significantly lower (40% vs. 20%; p = 0.034).

3.3. Effect of Clinicopathologic, Inflammatory, and Treatment-Related Parameters on Overall Survival

The median follow-up time was 52 months. At the end of the follow-up, 116 patients (55.2%) were alive. The median OS was NR in patients receiving mFOLFIRINOX treatment, while it was 51 months in patients receiving other treatments (p = 0.071). (The 5-year overall survival rate was 88.2% in the FOLFIRINOX treatment arm and 55.7% in the other treatment arm.) While no statistically significant result was reached, a trend toward statistically significant survival times was observed in the FOLFIRINOX arm. Since the p-value was below 0.10, it was included in the multivariate analysis. Accordingly, as shown in Table 3, while BMI, smoking and alcohol use, histopathological subtype, LNR, THLN, biliary obstruction, and radiotherapy use were not effective parameters for OS, younger age, absence of comorbidities, better ECOG performance status, lower T and N stages, absence of pancreatic invasion, negative surgical margins, absence of lymphovascular and perineural invasion, low CA 19-9 level, and use of mFOLFIRINOX were associated with better OS in the univariate analysis. In the multivariate analysis, mFOLFIRINOX treatment was an independent statistically significant parameter for better OS (HR; 95% CI: 3.24; 1.02–10.9; p = 0.046) (Table 3; Figure 1B).

Table 3. Univariate and multivariate analyses of effect of clinicopathological and treatment-related parameters on overall survival.

	Univariate Analysis, Mo (95%CI)	***p* Value**	**Multivariate Analysis HR (95%CI)**	***p* Value**
Age, years (median; range)				
<65	59.0 (41.0–76.9)	0.045	1.1 (0.63–1.85)	0.76
≥65	44.0 (28.7–59.3)			
Gender				
Female	51.0 (34.7–67.3)	0.98		
Male	55.0 (35.9–74.0)			
Comorbidity				
Present	44.0 (30.7–57.3)	0.007	1.57 (0.94–2.6)	0.084
Absent	70.0 (31.2–108.8)			
ECOG PS				
0	52.0 (24.9–79.1)	0.009	-	0.90
1	55.0 (36.0–73.9)			
2	11.0 (3.0–34.8)			
BMI				
<18.5, underweight	11.0 (3.0–33.2)	0.226		
18.5–24.9, normal	51.0 (39.6–62.4)			
25–29.9, overweight	90.0 (32.9–147.1)			
≥30, obese	42.0 (29.3–54.7)			
Smoking	53.0 (36.8–69.2)	0.93		
None	51.0 (34.3–67.7)			
Alcohol use	52 (38.7–65.3)	0.78		
None	32 (3.0–109)			
Histological subtype				
Pancreaticobiliary/mix	44.0 (34.0–53.9)	0.56		
Intestinal	63.0 (41.7–84.3)			

Table 3. *Cont.*

	Univariate Analysis, Mo (95%CI)	p Value	Multivariate Analysis HR (95%CI)	p Value
Stage				
I (T1-2N0)	92 (59.2–124.7)			
II (T3N0)	44.0 (10.0–77.8)	0.098	-	0.59
IIIA (T1-3N1)	48.0 (33.4–61.6)			
IIIB (T4 or N2)	44.0 (25.0–56.8)			
LNR				
0	67.0 (52.4–81.6)	0.42		
<0.2	44.0 (23.9–64.1)			
0.2–0.4	51.0 (43.9–58.1)			
≥0.4	36.0 (17.6–54.4)			
THLN				
≥16	51.0 (38.8–63.2)	0.51		
<16	52.0 (31.4–72.6)			
Surgical margin				
R0	55.0 (42.6–67.0)	0.008	1.87 (0.79–4.3)	0.15
R1	30.0 (27.7–32.3)			
Pancreatic invasion				
Present	36.0 (25.1–46.8)	<0.001	1.61 (0.90–2.86)	0.107
Absent	73.0 (46.1–99.9)			
LVI				
Present	44.0 (36.2–51.8)	0.020	1.33 (0.8–2.6)	0.23
Absent	92.0 (43.8–140.1)			
PNI				
Present	51.0 (37.1–64.9)	0.016	1.38 (0.82–2.34)	0.62
Absent	73.0 (42.3–103.6)			
Biliary obstruction				
Present	51.0 (37.6–64.4)	0.61		
Absent	63.0 (40.9–85.1)			
CA 19-9 U/mL				
<150	55.0 (41.4–68.6)	0.001	1.62 (0.83–3.2)	0.11
≥150 (if bilirubin is high: 300)	28.0 (20.8–35.1)			
Radiotherapy				
Received	52.0 (32.6–71.4)	0.607		
None	52.0 (40.3–63.7)			
mFOLFIRINOX	NR	0.071	3.24 (1.02–10.9)	0.046
The other treatments	51.0 (41.6–60.4)			

LNR: lymph node ratio (positive lymph node/total harvested lymph nodes); THLN: total harvested lymph nodes; LVI: lymphovascular invasion; PNI: perineural invasion.

As shown in Figure 2, the clinical efficacy of FOLFIRINOX was observed in both histopathological subtypes (log-rank *p*-value: 0.040).

3.4. Toxicity

In the FOLFIRINOX group, dose reduction was implemented in 41.2% of patients (n = 14), and treatment was temporarily interrupted due to toxicity in 31.4% of patients (n = 11). In the other treatment group, dose reduction occurred in 20.5% of patients (n = 36) (p = 0.009), and treatment was interrupted in 17% of patients (n = 30) (p = 0.50).

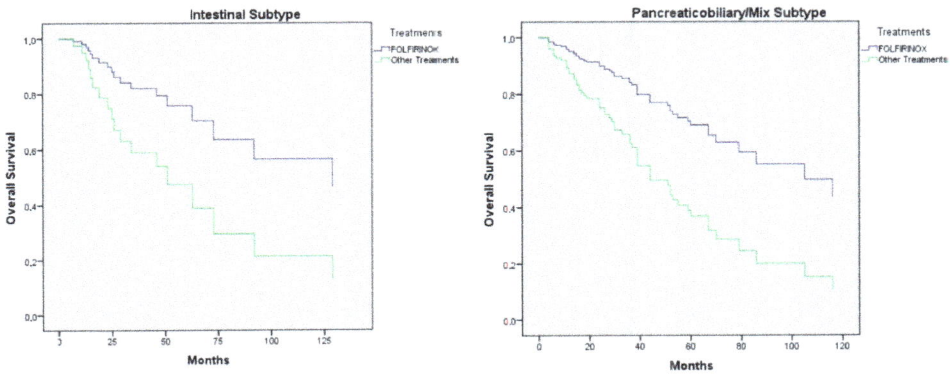

(A) (B)

Figure 2. (**A**): Cox regression curve of overall survival of patients with intestinal-type ampullary carcinoma. (**B**): Cox regression curve of overall survival of patients with pancreaticobiliary/mix type ampullary carcinoma after adjustment for age, comorbidities, Eastern Cooperative Oncology Group performance status, stage, lymphovascular and perineural invasion status, resection margin, pancreatic invasion, and CA 19-9.

In the FOLFIRINOX group, chemotherapy was discontinued due to toxicity in 5 patients (14.3%), while in the other treatment group, chemotherapy was discontinued in 12 patients (6.8%) ($p = 0.138$).

In the FOLFIRINOX group, grade 3–4 neutropenia occurred in 7 patients (20%) (Grade 1–2: $n = 11$; 31.4%), febrile neutropenia in 5 patients (14.7%), grade 3–4 anemia in 2 patients (5.9%) (Grade 1–2: $n = 19$; 55.9%), and grade 3–4 thrombocytopenia in 2 patients (5.8%) (Grade 1–2: $n = 10$; 29.4%). In the other treatment group, grade 3–4 neutropenia was observed in 25 patients (14.2%) (Grade 1–2: $n = 65$; 36.9%), febrile neutropenia in 6 patients (3.5%), grade 3–4 anemia in 3 patients (1.8%) (Grade 1–2: N = 92; 55.5%), and grade 3–4 thrombocytopenia in 4 patients (2.4%) (Grade 1–2: $n = 51$; 31.3%) ($p = 0.013$; $p = 0.008$; $p = 0.164$; $p = 0.234$, respectively).

Grade 3–4 fatigue was higher in the FOLFIRINOX group compared to the other treatment group (10.3% vs. 2.6% for grade 3–4; 75.8% vs. 62.1% for grade 1–2, respectively; $p = 0.043$). However, appetite loss was similar between the groups (7.1% vs. 2.0% for grade 3–4; 60.7% vs. 54.3% for grade 1–2, respectively; $p = 0.30$), and grade 1–2 diarrhea was more common in the FOLFIRINOX group (32.2% vs. 16.2%, respectively; $p = 0.034$; no grade ≥ 3). Stomatitis (grade 1–2: 25.9% vs. 20.2%; $p = 0.69$; no grade ≥ 3) and liver function test abnormalities were similar between the groups (0.0% vs. 1.3% for grade 3–4; 34.3% vs. 20.9%, respectively; $p = 0.37$; no grade ≥ 3) (Table 4).

Table 4. Treatment-related toxicities in patients receiving adjuvant therapy.

Adverse Event	FOLFIRINOX Grade 1–2 *n* (%)	FOLFIRINOX Grade 3–4 *n* (%)	Other Tx Grade 1–2 *n* (%)	Other Tx Grade 3–4 *n* (%)	*p*-Value
Neutropenia	11 (31.4)	7 (20.0)	65 (36.9)	25 (14.2)	0.013
Febrile neutropenia	-	5 (14.7)	-	6 (3.5)	0.008
Anemia	19 (55.9)	2 (5.9)	92 (55.5)	3 (1.8)	0.164
Thrombocytopenia	10 (29.4)	2 (5.8)	51 (31.3)	4 (2.4)	0.234
Fatigue	26 (75.8)	4 (10.3)	109(62.1)	4 (2.6)	0.043

Table 4. *Cont.*

Adverse Event	FOLFIRINOX Grade 1–2 *n* (%)	FOLFIRINOX Grade 3–4 *n* (%)	Other Tx Grade 1–2 *n* (%)	Other Tx Grade 3–4 *n* (%)	*p*-Value
Appetite loss	21 (60.7)	2 (7.1)	93 (54.3)	3 (2.0)	0.30
Diarrhea	11 (32.2)	0	27 (16.2)	0	0.034
Stomatitis	9 (25.9)	0	34 (20.2)	0	0.69
LFT abnormalities	12 (34.3)	0	36 (20.9)	2 (1.3)	0.37

4. Discussion

Our study is the largest database analysis to date examining the effectiveness of FOLFIRINOX in the adjuvant treatment of ampullary cancers. FOLFIRINOX has revolutionized survival times in the adjuvant and metastatic stages of pancreatic cancer [12,13] and has also shown survival benefits, particularly in certain subgroups of metastatic intestinal cancers [14]. Although there are no randomized controlled trials specifically evaluating FOLFIRINOX for the adjuvant treatment of ampullary cancers, it is now recommended by the current National Comprehensive Cancer Network (NCCN) guidelines for the adjuvant treatment of the pancreaticobiliary subtype of ampullary tumors [9]. Our study demonstrates that FOLFIRINOX provides superior disease-free survival (DFS) and a favorable trend in overall survival (OS) benefits in the adjuvant treatment of ampullary cancer compared to other adjuvant chemotherapy or chemoradiotherapy modalities. When other prognostic parameters were included in the multivariate analysis, the overall survival benefit of FOLFIRINOX treatments reached statistical significance [9–11,15,16].

In our study, well-established predictive parameters for recurrence, such as ECOG performance status, stage, positive surgical margins, lymphovascular invasion, perineural invasion, pancreatic invasion, development of jaundice, and elevated CA 19-9 levels, were confirmed as risk factors for recurrence. In the multivariate analysis, positive surgical margins and FOLFIRINOX chemotherapy emerged as independent predictive parameters (Table 2). The characteristics related to patients and tumors receiving FOLFIRINOX were similar to those receiving other treatments, as shown in Table 1. However, it was observed that FOLFIRINOX was less frequently administered to patients over 65 years of age, although age was not found to be a predictive factor for DFS. Patient age, gender, habits, and comorbidities were not predictive of recurrence. While the dissection of at least 16 lymph nodes is recommended [17], our study did not find a predictive parameter for recurrence based on the number of lymph nodes dissected, as both patients with $\geq$16 and <16 lymph nodes were compared. Although an increased positive lymph node ratio (NLR) has been associated with reduced survival, our study found that NLR, while related to early recurrence and poor survival, was less effective than nodal stage, which was therefore included in the model. Elevated CA 19-9 levels were found to be prognostic for both DFS and OS in the univariate analysis, consistent with existing literature [18].

Radiation therapy (RT) can be used in localized ampullary cancer, sometimes in combination with chemotherapy, as suggested by NCCN guidelines; however, there is no high-level evidence to support its use [19]. The Surveillance, Epidemiology, and End Results (SEER) database was used to assess the benefit of RT in patients with ampullary cancer. According to the propensity score matching analysis, adjuvant radiotherapy did not improve either OS ($p = 0.119$) or disease-specific survival ($p = 0.188$) [20]. Consistent with the literature, our study found that adding RT did not have a positive impact on DFS or OS. However, it was observed that the local recurrence rate was statistically significantly lower in patients who received RT compared to those who did not.

The European Study Group for Pancreatic Cancer (ESPAC)-3 trial was the first phase 3 randomized controlled trial to compare the survival benefits of adjuvant chemotherapy versus observation following resection for patients with periampullary carcinoma (including ampullary, intrapancreatic bile duct, nondescript, or periampullary duodenal cancer) and analyze the efficacy of fluorouracil plus folinic acid chemotherapy and gemcitabine alone [21]. Of the 428 patients included in the primary analysis, 297 had ampullary cancer, 96 had bile duct cancer, and 35 had other types of cancer. Among the 105 patients with ampullary cancer in the observation group, the median survival was 40.6 months (95% CI, 30.6–61.4 months); among 100 patients in the fluorouracil plus folinic acid group, it was 57.8 months (95% CI, 32.8–84.0 months); and among 92 patients in the gemcitabine group, it was 70.8 months (95% CI, 45.3–not reached months). There were no statistically significant differences in overall survival between patients with the pancreatobiliary subtype and those with the intestinal subtype of ampullary tumors. Multivariate analysis, adjusting for prognostic variables, demonstrated a statistically significant survival benefit with chemotherapy, specifically for gemcitabine compared to observation [19]. In our study, although the DFS and OS of the intestinal subtype were observed to be longer compared to the pancreatobiliary/mixed subtype, this difference was not statistically significant. FOLFIRINOX treatment was shown to be superior to other treatments in both subtypes (Figure 2). Our study also found that gemcitabine doublet or 5-FU doublet therapies did not show superiority over monotherapy in terms of DFS (Table 2) [9,15].

With the exception of the ESPAC-3 study, no treatment recommendations in the NCCN guidelines are based on phase II or phase III studies specifically designed for ampullary tumors. In the ABC-2 study, which included ampullary tumors among patients with metastatic biliary tract cancers, the survival advantage of adding cisplatin to gemcitabine was demonstrated, and it was observed that ampullary tumors also benefited from this combination treatment [22]. Another recommended combination therapy, XELOX, was shown to be an effective treatment in a phase II study that included metastatic small bowel cancers and ampullary cancers. Consequently, gemcitabine doublet and 5-FU doublet treatments, which were effective in the metastatic stage according to subgroup analyses, have been recommended as adjuvant treatments by guidelines [23]. Similarly, XELOX has been included in the adjuvant treatment of intestinal-type ampullary cancer based on its efficacy in the adjuvant treatment of colon cancer [24,25]. In our study, the FU-doublet chemotherapy regimen did not show superiority over 5-FU monotherapy; however, FOLFIRINOX treatment provided a better DFS advantage, including in the intestinal subtype.

Another combination therapy, gemcitabine and capecitabine, demonstrated effectiveness in the ESPAC-4 study, where it was compared with gemcitabine alone. This combination, which has been shown to be effective in the adjuvant treatment of pancreatic cancer, has become one of the recommended combinations for the adjuvant treatment of ampullary cancers based on this study [26]. In our study, this combination was evaluated within the gemcitabine-based chemotherapy group and showed a negative DFS outcome compared to FOLFIRINOX [10].

Although there are no phase studies in the literature, there are extensive series of adjuvant treatment studies for ampullary cancers. In a database study by Ecker BL. et al., similar to our study, it was observed that neither chemotherapy nor radiotherapy contributed to overall survival in patient populations receiving 5-FU monotherapy or combinations with oxaliplatin or irinotecan, as well as gemcitabine monotherapy and gemcitabine-based doublet chemotherapy [15]. A similar meta-analysis also yielded comparable results [27]. Given the superiority of FOLFIRINOX over other treatments in our study, it is suggested that the negative outcomes in these studies may be related to the fact that patients did not receive optimal adjuvant systemic chemotherapy [11].

In our study, higher rates of neutropenia and febrile neutropenia were observed in the mFOLFIRINOX group, with a greater frequency of dose reductions and treatment discontinuations. Additionally, the administration of mFOLFIRINOX was notably lower in patients aged 65 and older compared to other treatment groups. In this context, it is advisable to use mFOLFIRINOX primarily in fit patients younger than 65, to initiate treatment with dose reductions, and to routinely use growth factors to manage side effects.

5. Limitations

The primary limitations of this study are its retrospective nature and the relatively small number of patients in the mFOLFIRINOX group compared to the control group. Additionally, the follow-up duration for the mFOLFIRINOX group did not reach the median follow-up time, and there was data loss due to the retrospective nature of the study. One of the major limitations of the study is that the hazard ratio for OS (HR = 3.24) showed a wide confidence interval (1.02–10.9), indicating considerable variance.

6. Conclusions

This multi-institutional study represents the largest series of surgically resected ampullary adenocarcinoma patients treated with adjuvant chemotherapy, including mFOLFIRINOX, with detailed histopathological data. Therefore, we have reported significantly important parameters regarding the overall management of patients with ampullary adenocarcinoma. FOLFIRINOX provides better DFS and a favorable trend in OS benefits in the adjuvant treatment of ampullary cancer compared to other adjuvant chemotherapy or chemoradiotherapy modalities. When other prognostic parameters were included in the multivariate analysis, the OS benefit of FOLFIRINOX reached statistical significance, independent of histological subtype.

Given the high toxicity rate of FOLFIRINOX, its use should be carefully considered based on patients' performance status and comorbidities. It should also be noted that FOLFIRINOX treatment was not preferred in the elderly population in our database, and, therefore, the data is insufficient for this population. Although our study provides the highest level of patient data in the literature, confirmation of our data through prospective randomized trials is needed to recommend FOLFIRINOX treatment with a high level of evidence; therefore, our data can shed light on future planned studies [28].

Author Contributions: Conceptualization, A.K. and T.K.; methodology, A.K. and T.K.; software, A.K. and T.K.; validation, A.K. and T.K.; formal analysis, A.K. and T.K.; investigation, A.K. and T.K.; resources, A.K. and T.K.; data curation, A.K., T.K., T.K.S., O.D., S.C.E., M.A.N.S., T.A., M.A., H.B., Y.U., N.S., I.V.B., E.C., M.F.S., T.K.G., M.O., N.D., B.Y., B.K., U.A.Y., S.G., I.T.U., S.S., F.A., O.A., A.P.E., M.S., G.A. and S.Y.; writing—original draft preparation, A.K. and T.K.; writing—review and editing, A.K. and T.K.; visualization, A.K. and T.K.; supervision, S.Y.; project administration, A.K. and T.K.; funding acquisition, A.K. All authors have read and agreed to the published version of the manuscript.

Funding: This research received no external funding.

Institutional Review Board Statement: The study was conducted in accordance with the Declaration of Helsinki and approved by the Gaziantep University Faculty of Medicine Ethics Committee (No. 2024/43). The ethical approval was obtained in February 2024.

Informed Consent Statement: Informed consent was obtained from all subjects involved in the study.

Data Availability Statement: The datasets generated and/or analyzed during the current study are available from the corresponding author on reasonable request.

Conflicts of Interest: The authors declare no conflicts of interest.

eferences

Ahn, D.H.; Bekaii-Saab, T. Ampullary Cancer: An Overview. *Am. Soc. Clin. Oncol. Educ. Book* **2014**, *34*, 112–115. [CrossRef] [PubMed]

Kimura, W.; Futakawa, N.; Yamagata, S.; Wada, Y.; Kuroda, A.; Muto, T.; Esaki, Y. Different Clinicopathologic Findings in Two Histologic Types of Carcinoma of Papilla of Vater. *Jpn. J. Cancer Res.* **1994**, *85*, 161–166. [CrossRef] [PubMed]

Kohler, I.; Jacob, D.; Budzies, J.; Lehmann, A.; Weichert, W.; Schulz, S.; Neuhaus, P.; Röcken, C. Phenotypic and Genotypic Characterization of Carcinomas of the Papilla of Vater Has Prognostic and Putative Therapeutic Implications. *Am. J. Clin. Pathol.* **2011**, *135*, 202–211. [CrossRef] [PubMed]

Bonet, M.; Rodrigo, A.; Vázquez, S.; Carrizo, V.; Vilardell, F.; Mira, M. Adjuvant treatment in ampullary adenocarcinoma: A systematic review. *Clin. Transl. Oncol.* **2020**, *22*, 1407–1413. [CrossRef]

Kim, J.H.; Jeong, J.H.; Ryoo, B.Y.; Kim, K.P.; Chang, H.M.; Oh, D.; Song, T.J.; Lee, S.S.; Seo, D.W.; Lee, S.K.; et al. Adjuvant Chemotherapy for Resected Ampulla of Vater Carcinoma: Retrospective Analysis of 646 Patients. *Cancer Res. Treat.* **2021**, *53*, 424–435. [CrossRef]

Vining, C.C.; Schuitevoerder, D.; Turaga, K.K. Ampullary adenocarcinoma: The current state of adjuvant therapies. *Hepatobiliary Surg. Nutr.* **2020**, *9*, 477–489. [CrossRef]

Tsagkalidis, V.; Langan, R.C.; Ecker, B.L. Ampullary Adenocarcinoma: A Review of the Mutational Landscape and Implications for Treatment. *Cancers* **2023**, *15*, 5772. [CrossRef]

Westgaard, A.; Tafjord, S.; Farstad, I.N.; Cvancarova, M.; Eide, T.J.; Mathisen, O.; Clausen, O.P.F.; Gladhaug, I.P. Pancreatobiliary versus intestinal histologic type of differentiation is an independent prognostic factor in resected periampullary adenocarcinoma. *BMC Cancer* **2008**, *8*, 170. [CrossRef]

Chiorean, E.G.; Del Chiaro, M.; Tempero, M.A.; Malafa, M.P.; Benson, A.B.; Cardin, D.B.; Christensen, J.A.; Chung, V.; Czito, B.; Dillhoff, M.; et al. Ampullary Adenocarcinoma, Version 1.2023, NCCN Clinical Practice Guidelines in Oncology. *J. Natl. Compr. Cancer Netw.* **2023**, *21*, 753–782. [CrossRef]

. Siebenhüner, A.R.; Seifert, H.; Bachmann, H.; Seifert, B.; Winder, T.; Feilchenfeldt, J.; Breitenstein, S.; Clavien, P.-A.; Stupp, R.; Knuth, A.; et al. Adjuvant treatment of resectable biliary tract cancer with cisplatin plus gemcitabine: A prospective single center phase II study. *BMC Cancer* **2018**, *18*, 72. [CrossRef]

. ClinicalTrials.gov. ADAPTA Trial: Adjuvant CAPCX vs mFOLFIRINOX in Ampullary Adenocarcinoma by Histology. NCT06068023. Available online: https://clinicaltrials.gov/study/NCT06068023 (accessed on 20 August 2025).

. Conroy, T.; Hammel, P.; Hebbar, M.; Ben Abdelghani, M.; Wei, A.C.; Raoul, J.-L.; Choné, L.; Francois, E.; Artru, P.; Biagi, J.J.; et al. FOLFIRINOX or Gemcitabine as Adjuvant Therapy for Pancreatic Cancer. *N. Engl. J. Med.* **2018**, *379*, 2395–2406. [CrossRef] [PubMed]

. Stintzing, S.; Heinrich, K.; Tougeron, D.; Modest, D.P.; Schwaner, I.; Eucker, J.; Pihusch, R.; Stauch, M.; Kaiser, F.; Kahl, C.; et al. FOLFOXIRI Plus Cetuximab or Bevacizumab as First-Line Treatment of *BRAF*^{V600E}-Mutant Metastatic Colorectal Cancer: The Randomized Phase II FIRE-4.5 (AIO KRK0116) Study. *J. Clin. Oncol.* **2023**, *41*, 4143–4153. [CrossRef] [PubMed]

. Conroy, T.; Desseigne, F.; Ychou, M.; Bouché, O.; Guimbaud, R.; Bécouarn, Y.; Adenis, A.; Raoul, J.-L.; Gourgou-Bourgade, S.; De La Fouchardière, C.; et al. FOLFIRINOX versus Gemcitabine for Metastatic Pancreatic Cancer. *N. Engl. J. Med.* **2011**, *364*, 1817–1825. [CrossRef] [PubMed]

. Ecker, B.L.; Vollmer, C.M.; Behrman, S.W.; Allegrini, V.; Aversa, J.; Ball, C.G.; Barrows, C.E.; Berger, A.C.; Cagigas, M.N.; Christein, J.D.; et al. Role of Adjuvant Multimodality Therapy After Curative-Intent Resection of Ampullary Carcinoma. *JAMA Surg.* **2019**, *154*, 706–714. [CrossRef]

. Vo, N.P.; Nguyen, H.S.; Loh, E.W.; Tam, K.W. Efficacy and safety of adjuvant therapy after curative surgery for ampullary carcinoma: A systematic review and meta-analysis. *Surgery* **2021**, *170*, 1205–1214. [CrossRef]

. Falconi, M.; Crippa, S.; Domínguez, I.; Barugola, G.; Capelli, P.; Marcucci, S.; Beghelli, S.; Scarpa, A.; Bassi, C.; Pederzoli, P. Prognostic Relevance of Lymph Node Ratio and Number of Resected Nodes after Curative Resection of Ampulla of Vater Carcinoma. *Ann. Surg. Oncol.* **2008**, *15*, 3178–3186. [CrossRef]

. Smith, R.A.; Ghaneh, P.; Sutton, R.; Raraty, M.; Campbell, F.; Neoptolemos, J.P. Prognosis of Resected Ampullary Adenocarcinoma by Preoperative Serum CA19-9 Levels and Platelet-Lymphocyte Ratio. *J. Gastrointest. Surg.* **2008**, *12*, 1422–1428. [CrossRef]

. Klinkenbijl, J.H.; Jeekel, J.; Sahmoud, T.; van Pel, R.; Couvreur, M.L.; Veenhof, C.H.; Arnaud, J.P.; Gonzalez, D.G.; de Wit, L.T.; Hennipman, A.; et al. Adjuvant Radiotherapy and 5-Fluorouracil After Curative Resection of Cancer of the Pancreas and Periampullary Region. *Ann. Surg.* **1999**, *230*, 776. [CrossRef]

20. Zhao, W.; Wang, B.; Zhao, A.; Tian, Q.; Zhang, L.; Wang, L.; Zhao, X.; Yang, J.; Dong, D. The role of radiotherapy in patients w resected ampullary carcinoma: Findings based on the SEER database. *HPB* **2019**, *21*, 1535–1540. [CrossRef]
21. Neoptolemos, J.P.; Moore, M.J.; Cox, T.F.; Valle, J.W.; Palmer, D.H.; McDonald, A.C.; Carter, R.; Tebbutt, N.C.; Dervenis, C.; Smi D.; et al. Effect of Adjuvant Chemotherapy with Fluorouracil Plus Folinic Acid or Gemcitabine vs Observation on Survival Patients with Resected Periampullary Adenocarcinoma. *JAMA* **2012**, *308*, 147–156. [CrossRef]
22. Valle, J.; Wasan, H.; Palmer, D.H.; Cunningham, D.; Anthoney, A.; Maraveyas, A.; Madhusudan, S.; Iveson, T.; Hughes, Pereira, S.P. et al. Cisplatin plus Gemcitabine versus Gemcitabine for Biliary Tract Cancer. *N. Engl. J. Med.* **2010**, *362*, 1273–12 [CrossRef] [PubMed]
23. Overman, M.J.; Varadhachary, G.R.; Kopetz, S.; Adinin, R.; Lin, E.; Morris, J.S.; Eng, C.; Abbruzzese, J.L.; Wolff, R.A. Phase Study of Capecitabine and Oxaliplatin for Advanced Adenocarcinoma of the Small Bowel and Ampulla of Vater. *J. Clin. Onc* **2009**, *27*, 2598–2603. [CrossRef]
24. Adjuvant chemotherapy with oxaliplatin, in combination with fluorouracil plus leucovorin prolongs disease-free survival, b causes more adverse events in people with stage II or III colon cancer. Abstracted from: André T, Boni C, Mounedji-Boud L, et al. Multicenter international study of oxaliplatin/5-fluorouracil/leucovorin in the adjuvant treatment of colon canc (MOSAIC) investigators. *Cancer Treat Rev.* **2004**, *30*, 711–713. [CrossRef]
25. Haller, D.G.; Tabernero, J.; Maroun, J.; de Braud, F.; Price, T.; Van Cutsem, E.; Hill, M.; Gilberg, F.; Rittweger, K.; Schmoll, H Capecitabine Plus Oxaliplatin Compared with Fluorouracil and Folinic Acid As Adjuvant Therapy for Stage III Colon Canc *J. Clin. Oncol.* **2011**, *29*, 1465–1471. [CrossRef] [PubMed]
26. Neoptolemos, J.P.; Palmer, D.H.; Ghaneh, P.; Psarelli, E.E.; Valle, J.W.; Halloran, C.M.; Faluyi, O.; O'Reilly, D.A.; Cunningham, Wadsley, J.; et al. Comparison of adjuvant gemcitabine and capecitabine with gemcitabine monotherapy in patients with resect pancreatic cancer (ESPAC-4): A multicentre, open-label, randomised, phase 3 trial. *Lancet* **2017**, *389*, 1011–1024. [CrossR [PubMed]
27. Acharya, A.; Markar, S.R.; Sodergren, M.H.; Malietzis, G.; Darzi, A.; Athanasiou, T.; Khan, A.Z. Meta-analysis of adjuvant thera following curative surgery for periampullary adenocarcinoma. *Br. J. Surg.* **2017**, *104*, 814–822. [CrossRef]
28. Keane, F.; O'Connor, C.; Moss, D.; Chou, J.F.; Perry, M.A.; Crowley, F.; Saxena, P.; Chan, A.; Schoenfeld, J.D.; Singhal, A.; et Adjuvant modified FOLFIRINOX for resected pancreatic adenocarcinoma: Clinical insights and genomic features from a lar contemporary cohort. *J. Natl. Cancer Inst.* **2025**, *117*, 496–506. [CrossRef]

cancers

MDPI

Article

Targeted Hyaluronan Degradation Enhanced Tumor Growth Inhibition in Gastrointestinal Cancer Models

Fulai Zhou [1], Guangmao Mu [1], Honglei Bi [1], Limin Chen [1], Zhengxia Zha [1], Ying Jin [1] and Mark L. Chiu [1,2,*]

1 Research & Development Department, Tavotek Biotherapeutics, Suzhou 215000, China
2 Research & Development, Tavotek Biotherapeutics, Spring House, PA 19102, USA
* Correspondence: mark.chiu@tavotek.com

Simple Summary: Stronger tumor growth inhibition was demonstrated with targeted disruption of the hyaluronan (HA) barrier that surrounded solid tumor cells. Systemic delivery of hyaluronidase can break down HA but is not effective when used alone and can cause adverse effects. We demonstrated how an "antibody–enzyme complex" (AbEn) disrupted the HA barrier via homing to the tumor or cancer associated fibroblast cells. In mouse models of colorectal cancer, these targeted AbEn molecules inhibited tumor growth better and enhanced the effectiveness of chemotherapy, immunotherapy, and other cancer drugs. Likewise, the AbEn molecules allowed the immune cells to have stronger anti-tumor growth activity.

Abstract: Background. The dense hyaluronan (HA)-rich stroma in solid tumors can prevent effective tumor growth inhibition by hindering drug delivery and immune cell infiltration. However, the degradation of HA alone by systemic delivery of hyaluronidase has not shown significant improvement of tumor growth inhibition. **Objectives/Methods.** In this study, we targeted hyaluronan degradation by using antibody–enzyme (AbEn) molecules by fusing antibodies to a recombinant human hyaluronidase (HYAL). **Results.** The AbEn molecules were stable, retained both antigen-binding and enzymatic activities, and demonstrated a prolonged serum half-life of 132 h in rodent models. In the HA-rich colorectal cancer model, the cancer-associated fibroblast (CAF)-directed AbEn, TAVO423 (FAP × LRRC15 × HYAL trispecific antibody) achieved greater intratumoral HA depletion resulting in superior tumor growth inhibition compared to untargeted HYAL. Furthermore, the combination of TAVO423 in combination with other solid tumor cell targeting modalities such as 5-fluorouracil (5-FU), anti-PD-L1 monoclonal antibody, a PD-L1 × CD3 bispecific T-cell engager (TCE), and a CD318-targeting antibody–drug conjugate (ADC) all demonstrated enhanced tumor growth inhibition (TGI) values of 49–67% as compared to the respective monotherapy TGI values of 1–28%. In addition, TAVO423 improved the antitumor response of a 5T4 × CD3 TCE with an increase in TGI from 73% to 92% in an in vivo HA-rich pancreatic cancer model. The CAF-targeted HA degradation mediated by TAVO423 also reversed immune exclusion by increasing the density of $CD8^+$ tumor-infiltrating lymphocytes (TILs) by 6–9-fold and synergized with PD-1 blockade to enhance TGI from 33% to 51% in an in vivo immunocompetent EMT-6 breast cancer model. **Conclusions.** These findings demonstrated the broad potential of the modular AbEn platform for targeted HA degradation to overcome barrier entry in stromal HA-rich solid tumors.

Keywords: stromal hyaluronan; FAP; LRRC15; hyaluronidase; solid tumor microenvironment

1. Introduction

Tumor cells in the solid tumor is a part of a highly complex tumor microenvironment (TME), which includes the desmoplastic stroma, immune cells, blood vessels, cytokines, and other non-cellular components [1,2]. Each of these components participate in interactions that modulate tumor metabolism, growth, metastasis, immune evasion, as well as resistance to drug therapy [3]. In recent decades, anti-cancer therapies, such as targeted tumor cell therapies and immune checkpoint inhibitors (ICI), have emerged as effective treatments for certain solid tumors [4]. Unfortunately, many patients either do not respond to such therapies or experience relapses after having initial responses; concomitantly, the tumor cell resistance mechanisms remain poorly understood [5–7]. We hypothesized that better tumor control in the TME required control of the tumor cells and the stroma, a major component of the TME. The TME stroma includes the cancer-associated fibroblasts (CAFs) and the extracellular matrix (ECM) that act as physical barriers that hinder effective drug delivery and immune cell infiltration to make contact with the tumor cells. Additionally, the stroma can promote tumorigenesis, cancer progression, metastasis, and therapy resistance, thereby limiting the effectiveness of treatments for solid tumors [8]. We presented how disruption of these TME barriers could improve therapeutic outcomes.

CAFs are key elements of the tumor stroma that promote tumor growth, invasion, and metastasis [9]. The development of the CAF phenotype is associated with various CAF-related markers, most notably the increased levels of the fibroblast activation protein (FAP). FAP has been considered a promising therapeutic target since FAP positive ($FAP^{+)}$ CAFs are linked to immunosuppressive tumor microenvironments and correlate with worse prognoses in solid tumor cancers [10,11]. Various strategies targeting FAP, including small molecule inhibitors, monoclonal antibodies (mAbs), antibody–drug conjugates (ADCs), and chimeric antigen receptor T (CAR-T) cells, have demonstrated therapeutic effects in murine tumor models [12–14]. Another dominant CAF marker is the leucine-rich repeat-containing protein 15 (LRRC15), which has minimal expression in normal tissues. Elevated LRRC15 in stromal fibroblasts of multiple solid tumors, osteosarcoma, soft tissue sarcomas, and mesenchymal cancer cells are associated with higher tumor grades and worse outcomes in osteosarcoma and soft tissue sarcomas [15,16]. $LRRC15^{+}$ CAFs also promote tumor growth and suppress $CD8^{+}$ T cell function, indicating that therapies targeting LRRC15 may enhance patient survival and responses to immunotherapy [17]. Nonetheless, FAP and LRRC15-targeted therapies that focus on CAFs alone may not be sufficient for cancer therapy. Rather, additional mechanism of action strategies to dismantle the tumor stromal barriers could result in better tumor growth inhibition. Combining CAF-targeted therapies with agents that target the ECM could allow more effective inhibition of tumor cells in the tumor microenvironment.

Hyaluronic acid (HA) or hyaluronan is the major polysaccharide component of the ECM [18]. Beyond serving as a barrier, HA polymers in the ECM of solid tumors regulate cellular functions in a molecular-weight (MW)-dependent manner through interactions with HA receptors, particularly cluster of differentiation 44 (CD44) and the hyaluronan-mediated motility receptor (RHAMM) [19]. High-molecular-weight (HMW) HA exhibits anticancer effects by inhibiting cell proliferation and angiogenesis, while low-molecular-weight (LMW) HA promotes tumorigenesis [20,21]. The conversion of HMW-HA to LMW-HA is facilitated by hyaluronidase (HYAL) enzymes, including HYAL1, HYAL2, HYAL3, and PH20 (sperm adhesion molecule 1) [22]. PH20, the only neutral-active hyaluronidase, is commonly employed to hydrolyze HA in tumors or is co-administered with subcutaneously injected mAbs to enhance dispersion and absorption [23]. Since, PH20 HYAL has a short half-life, a pegylated recombinant human PH20 variant (PEGPH20) can remodel the stroma of HA-rich tumors by improving the intra-tumor distribution of anticancer drugs and enhancing their therapeutic efficacy without increasing toxicity [24–26]. However, despite promising results in phase I and II studies for tumors with high HA expression, the phase III study of PEGPH20 combined with chemotherapy for pancreatic cancer failed to extend patient survival [27–30]. Perhaps, the systemic delivery of PEGPH20 for HA depletion alone may not be sufficient for effective treatment.

To address these challenges, we hypothesized that a targeted hyaluronidase delivery platform that specifically degraded stromal HA within the TME could enhance treatment efficiency as compared to non-targeted HA depletion approaches. We also showed how targeted CAF delivery using two well-known CAF targets: FAP and LRRC15 [17,31]. TAVO423 is a FAP × LRRC15 × HYAL trispecific AbEn that effectively degraded HA, delayed tumor growth, and enhanced the efficacy of various therapeutics in an in vivo HA-rich colorectal cancer model. Furthermore, combining CAF-targeted HA degradation with PD-1/PD-L1 blockade significantly increased tumor-infiltrating T cells in a syngeneic breast cancer model, thereby amplifying the antitumor responses to immune checkpoint therapy. We demonstrated how CAF-targeted hyaluronidase could improve treatment outcomes in HA-rich solid tumors.

2. Materials and Methods

2.1. Cell Lines

Human Expi293F cells (Gibco, #A14527) were cultured in suspension using Expi293 expression medium (Gibco, #A1435101, Grand Island, NY, USA), while ExpiCHO-S cells (Gibco, #A29127) were cultured in suspension with ExpiCHO expression medium (Gibco, #A2910001). Both cell types were routinely cultured at a density of 3–5 × 10^6 cells per mL to maintain a healthy state, under conditions of 8% CO_2, ≥80% relative humidity, and a shaking speed of 125 rpm in a 37 °C incubator. Human RKO (#TCHu116) and HCC827 (#SCSP-538) cells were obtained from the National Collection of Authenticated Cell Cultures of the Chinese Academy of Sciences (Shanghai, China), while human U-87 MG (#HTB-14) cells were purchased from the American Type Culture Collection (ATCC, Manassas, VA, USA). RKO cells were cultured in Dulbecco's modified Eagle's medium (DMEM, Gibco, #10566016) supplemented with 10% (*v*/*v*) fetal bovine serum (FBS, Sigma, #F8318, St. Louis, MO, USA). HCC827 cells were cultured in RPMI-1640 medium (Corning, #10-040-CVRC, Corning, NY, USA) with 10% FBS, and U-87 MG cells were cultured in Eagle's Minimum Essential Medium (EMEM, Corning, #10-009-CV) with 10% (*v*/*v*) FBS. BxPC-3-WT (ATCC, CRL-1573) and BxPC-3-HAS3 stable cell lines (GeneChembio, Shanghai, China) were cultured in DMEM (Gibco, #10566016). All cancer cell lines were incubated in a humidified chamber with 5% CO2 at 37 °C. The EMT-6 cells (ATCC) were cultured in vitro using Waymouth's MB752/1 medium (Gibco) supplemented with 15% (*v*/*v*) FBS at 37 °C in a 5% CO_2 atmosphere. Cells in exponential growth phase were harvested, counted with a cell counter, and then used for tumor inoculation.

2.2. Constructs, Expression, and Purification

All codon-optimized genes were synthesized and cloned into the pcDNA3.4 vector (GenScript, Nanjing, China). To enhance protein expression and secretion, a signal peptide (MAWVWTLLFLMAAAQSIQA) was inserted at the N-termini of each DNA open reading frame. For purification purposes, the C-terminus of the human PH20 sequence (L36-Y482) was linked to a histidine 6 (His6) tag or to an IgG1 Fc domain via a flexible polypeptide linker (GGGGS). The antibody amino acid sequences were annotated according to the Kabat numbering scheme [32]. The antibodies used in this study were expressed in the Expi293F system using the Expifectamine293 transfection kit (Gibco, #A14527), in accordance with the manufacturer's recommendations. For the exemplary IgG1 monoclonal antibodies (mAbs), the transfection weight ratio of heavy chain (HC) to light chain (LC) plasmids was 2:3. The PH20-His6 and PH20-Fc proteins were expressed in ExpiCHO-S cells using electroporation (Etta Biotech, X-Porator F1, Suzhou, China), following the manufacturer's instructions. Six days post-transfection, the supernatants containing recombinant proteins and antibodies were harvested by centrifugation and filtration using a 0.2 μm filter.

For the purification of antibodies and PH20-Fc, the filtered supernatants were loaded onto a MabSelect SuRe column (Cytiva, #11003494, Marlborough, MA, USA) using an AKTA pure 25 system (Cytiva, #29018224). The column was washed with PBS, and the bound antibodies were eluted using an elution buffer containing 150 mM sodium citrate, pH 3.4 (prepared in-house) and collected in tubes with 1 M tris-HCl solution, pH 9.0 (Sangon Biotech, #8548128, Shanghai, China) for rapid neutralization. For the purification of PH20-His6 protein, the supernatant was incubated with TALON Metal Affinity Resin (Clontech, #635501, Mountain View, CA, USA) for 2 h at 4 °C. The resin was then packed and washed with 30 column volumes of wash buffer (PBS) before the protein was eluted in 150 mM imidazole.

The integrity and components of the purified protein were analyzed using sodium dodecyl sulfate polyacrylamide gel electrophoresis (SDS-PAGE) on precast SurePAGE 4–12% Bis-Tris gels (GenScript, #M00654) under reducing or non-reducing conditions, following the manufacturer's instructions, and stained with Coomassie brilliant blue (prepared in-house). Monomeric purity was assessed using high-performance liquid chromatography-size exclusion chromatography (HPLC-SEC) on an Agilent 1260 Infinity II LC system (Agilent Technologies, #G7167A, Santa Clara, CA, USA) equipped with an AdvanceBio SEC column (Agilent Technologies, #PL1180-5301). The running buffer consisted of 150 mM KH_2PO_4 (Sigma, #104873), pH 7.3, at room temperature. Proteins containing aggregates or fragments were concentrated using an Amicon Ultra Centrifugal Filter (MWCO: 30 kDa, Millipore, #UFC5030, Billerica, MA, USA) and further purified via a pre-equilibrated Superdex 200 Increase 10/300 GL 24 mL column (Cytiva, #28990944). Peak fractions were collected for subsequent characterization.

The AbEn molecules were generated by mixing equimolar solutions of the HYAL-Fc with the parental targeting arms in the presence of 2-MEA using the controlled Fab-arm exchange (cFAE) technique [33]. Briefly, equimolar solutions of HYAL-Fc and the parental antibody at concentrations of 1–10 mg/L were mixed at a 1:1 molar ratio. Reduction was initiated by adding 2-mercaptoethylamine (2-MEA) to a final concentration of 75 mM, followed by incubation for 5 h at 31 °C. The 2-MEA was then removed by dialysis against a large volume of PBS for 16–18 h at 4 °C to facilitate re-oxidation and bispecific molecule formation. The formation of the antibody–enzyme conjugates were confirmed using several chromatographic methods. HPLC-SEC analysis verified successful bispecific AbEn formation based on the distinct molecular weight differences between the parental molecules (HYAL-Fc or the parental antibody) and the medium conjugate. The bispecific AbEn exhibited an average molecular weight consistent with the sum of its two parental molecules. Monomeric and monodisperse purity were assessed using an Agilent 1260 Infinity II LC system equipped with an AdvanceBio SEC column. An isocratic method was employed with a 150 mM KH_2PO_4 buffer (pH 7.3) at room temperature and detection at 280 nm. The exchange efficiency was confirmed to be over 90%, as calculated by the monomeric peak area expressed as a percentage of the total injection area. The amino acid numbering followed the EU index for the Fc domain mutations.

2.3. Hyaluronidase Activity Assay

The activity of the hyaluronidase enzyme was measured using a 45 min turbidimetric assay, as previously described [34]. Briefly, enzymes were first diluted to 2–5 units per mL in a buffer containing 20 mM sodium phosphate (pH 7.0), 77 mM sodium chloride, and 0.01% (*w*/*v*) bovine serum albumin, and were equilibrated at 37 °C for 10 min. The enzyme dilution was then immediately mixed with 0.3 mg/mL HA (Sigma-Aldrich, #H5388, St. Louis, MO, USA), which was prepared in a buffer containing 300 mM sodium phosphate (pH 5.35) at 37 °C. The mixture was swirled and incubated at 37 °C for exactly 45 min to monitor HA degradation. Following incubation, each test and blank sample was mixed promptly with an acidic albumin solution consisting of 24 mM sodium acetate, 79 mM acetic acid, and 0.1% (*w*/*v*) bovine serum albumin (pH 3.75 at 25 °C). After mixing and waiting for an additional 10 min at room temperature, the absorbance values at 600 nm were recorded. One unit of hyaluronidase activity was defined as a change in absorbance (A600) of 0.330 per minute at pH 5.35 and 37 °C in a 2.0 mL reaction mixture. The units of activity were determined according to a standard curve of hyaluronidase activity generated from a solution of bovine testicular hyaluronidase (Sigma). The standard curve relating absorbance to enzyme activity units (0–6 units of enzyme per mL of working solution) was fit by linear regression in Prism 8 (GraphPad, Version 8.0.2, San Diego, CA, USA).

2.4. Flow Cytometry

RKO or HCC827 cells were digested with enzyme-free cell dissociation buffer (Gibco, #WHO621 K121) and washed twice with precooled buffer (PBS supplemented with 2% (*v*/*v*) FBS). A total of 1×10^5 cells and test articles at varying concentrations (starting from 33 nM with a series of 3-fold dilutions) were seeded into each well of a 96-well round-bottom plate and incubated on ice for 1 h to facilitate binding. After incubation, the cells were washed twice and stained with an Alexa Fluor 647 AffiniPure Goat Anti-Human IgG, Fcγ fragment-specific detection antibody (dilution 1:5000, Jackson ImmunoResearch, #109-605-190) at 4 °C for 30 min in the dark. Following this, two additional washing steps were performed, and the cells were resuspended for flow cytometric analysis on a CYTOFLEX flow cytometer (Beckman Coulter, Miami, FL, USA) using CytExpert 2.4.0.28 software. The EC50 and Emax values were calculated for each curve fitted using a four-parameter logistic equation in Prism 8 (GraphPad, Version 8.0.2).

2.5. T-Cell Cytotoxicity Assay

The T-cell-mediated cytotoxicity of PD-L1 × CD3 bispecific antibodies was assessed using PD-L1-expressing cancer cell line RKO and PBMCs. Briefly, target cells were collected and seeded into 96-well round-bottom plates at a density of 20,000 cells per well. Human PBMCs, obtained from healthy donors (SAILYBIO, #SC12282W), were rapidly thawed in a 37 °C water bath and added to the target cells at a 5:1 effector-to-target (E:T) cell ratio. Bispecific antibodies, at varying concentrations (starting at 10 nM and serially diluted 3-fold across the plate to generate an 8-point titration curve), were added to each well containing the target cells. The plates were then incubated under standard cell culture conditions for 72 h to allow cytotoxicity to occur. Control wells included cells alone, cells with PBMCs, and cells treated with 0.2% (*w*/*v*) Triton-X-100 (total lysis, Sigma, #T8787, St. Louis, MO, USA) to establish baseline values for calculating the percentage of killing. After incubation, 100 μL of supernatant from each well was transferred to a new flat-bottom 96-well plate for lactate dehydrogenase (LDH) detection (Roche, #11644793001), following the manufacturer's protocol.

2.6. In Vitro Cell Viability Assay

RKO cells were seeded into 96-well plates at 5000 cells/well. After overnight incubation, a serially diluted solution of CD318-MMAE ADC was added. Cell viability was evaluated after 4 days using the CellTiter-Glo® Luminescent Cell Viability kit (Promega, Madison, WI, USA) according to the manufacturer's instructions. The final luminescence signal was measured using a multiplate reader (Tecan). Concentration–response curves and half-maximal inhibitory concentration (IC_{50}) values were generated by nonlinear regression using a sigmoidal curve fit with variable slope in GraphPad Prism 8.0. Assays were performed in triplicate for each data point.

2.7. ELISA

A standard ELISA protocol was employed to measure the levels of FAP and LRRC15 antigen binding. Briefly, a 96-well microplate was coated with the target antigen diluted in coating buffer (1 µg/mL) and incubated overnight at 4 °C. After washing with PBST (PBS containing 0.05% (*w*/*v*) Tween-20), the plate was blocked with 5% (*w*/*v*) BSA or non-fat dry milk in PBS for 1–2 h at room temperature to prevent non-specific binding. Serial dilutions of the primary antibody were then added to the wells and incubated for 1–2 h, followed by washing and incubation with an enzyme-conjugated secondary antibody for 1 h. After a final wash, the substrate (e.g., TMB) was added, and the reaction was stopped with 1 M H_2SO_4. Absorbance was measured at 450 nm using a Tecan microplate reader.

For serum HA fragment measurement, Balb/c nude mice (n = 3 per group; 6–10 weeks old; Nanjing GemPharmatech Co., Ltd. Nanjing, China) were subcutaneously inoculated with 3×10^6 RKO tumor cells. Tumor-bearing mice were randomly grouped into treatment cohorts once mean tumors reached 200–250 mm^3. Following a single intravenous injection of 1 mg/kg HYAL-Fc IgG1 fusion protein, blood samples (20 µL) were collected at 2, 8, 24, 48, and 72 h post-administration. The HA levels in mice serum at different time points were quantified using the Hyaluronan Valukine ELISA Kit (R&D Systems, # VAL152), according to the manufacturer's instructions.

For intratumoral HA measurement, tumors from RKO xenograft models were harvested, weighed, and snap-frozen in liquid nitrogen. Frozen tissues were pulverized and homogenized on ice in a lysis buffer (100 mM Tris-HCl, pH 7.4, 1% (*w*/*v*) Triton X-100) supplemented with protease inhibitors. The homogenates were incubated at 4 °C for 2 h to extract HA, followed by centrifugation at 12,000× *g* to clarify the lysates. The concentration of HA in the supernatants was quantified using the Hyaluronan Quantikine ELISA Kit (DHYAL0, R&D Systems, Minneapolis, MN, USA) according to the manufacturer's instructions. Sample lysates were diluted as necessary in the provided calibrator diluent to fall within the assay's linear range. The intratumoral HA concentration was calculated from the standard curve of HA concentration with OD value, and normalized to the initial tumor weight, expressed as µg of HA per mg of tissue.

2.8. In Vivo Efficacy Studies

Mice were housed in a specific pathogen-free (SPF) facility in individual ventilated cage (5/cage) on corn cob bedding with crumpled tissue paper for enrichment. The SPF facility maintained a 12:12 light cycle (22 ± 1 °C, 60% humidity). Animals had ad libitum access to irradiated chow and autoclaved water. Health was monitored daily; no abnormalities were observed. For in vivo efficacy studies of the combination therapy involving TAVO423 with anti-PD-L1 mAb or PD-L1 × CD3 bispecific antibody in the RKO xenograft model or 5T4 x CD3 in the BxPC-3-HAS3 xenograft model, human PBMCs containing CD3-positive effector T cells were engrafted into immunocompromised mice five days prior to tumor cell inoculation. After human PBMC reconstitution, tumors were established by injecting 3×10^6 cells into the flanks of 6–10 weeks old female Balb/c nude mice. In other efficacy studies using the RKO xenograft model, Balb/c nude female mice received subcutaneous injections of 3×10^6 RKO tumor cells. Once tumors reached an average volume of 100–150 mm^3, mice were randomized (n = 6 or 8) and treated with isotype controls or test agents at specified dosages and regimens. 5-Fluorouracil (5-FU,

Shanghai Xudong Haipu Pharmaceutical Co., Ltd., # H31020593, Shanghai, China) and anti-PD-L1 mAb (Atezolizumab, MedChemExpress., # HY-P9904, Shanghai, China) were purchased from external suppliers. The CD318-MMAE ADC was prepared in-house via cysteine conjugation. Other antibodies used in the animal studies were also prepared in-house. The dose levels were selected based on prior in-house dose–response studies showing efficacy without toxicity. The primary outcome was tumor volume reduction. Tumor response was monitored twice weekly by measuring tumor size and mouse body weight. Tumor volumes were measured using calipers, with volumes calculated in mm^3 using the formula: $V = (L \times W \times W)/2$, where V is tumor volume, L is tumor length (longest dimension), and W is tumor width (perpendicular to L). At the study endpoint, mice were sacrificed, and tumors were excised and weighed to assess tumor shrinkage. For humane endpoints, mice were euthanized if body weight loss >20% or individual mouse with tumor volume exceeding 3000 mm^3. Daily monitoring included weight, tumor size, and activity. No severe adverse events (e.g., >20% weight loss) were observed in any group during the study. All procedures involving animal care, handling, and treatment complied with the guidelines of the Institutional Animal Care and Use Committee (IACUC) of Suzhou Charles River Accelerator and Development Lab (Approved Protocols: P202302160002-20240711 and P202302160002-20241219).

For the in vivo efficacy study of anti-mouse PD-1 antibody (anti-mouse PD-1 (CD279), BioXcell, #BP0146) in the syngeneic model, each mouse was subcutaneously inoculated in the right rear flank with 5×10^5 EMT-6 cells of 6–10 weeks old female Balb/c mice (Shanghai Lingchang Biotechnology Company Limited, Shanghai, China) to promote tumor growth. Randomization was initiated once the mean tumor size reached 80–100 mm^3. Mice ($n = 5$ per group) were then randomized and treated with either isotype controls or test agents at specified doses and regimens. The day of randomization was designated as day 0. Following tumor inoculation, animals were checked daily for morbidity and mortality. Routine monitoring included assessments of tumor growth effects and treatment-related impacts on behavior, such as mobility, food and water intake, body weight changes (measured two or three times per week post-randomization), eye and hair condition, and other abnormalities. Mortality and clinical signs were recorded in detail for each animal. At the end of the research, tumors were collected for IHC staining of tumor infiltrating lymphocytes. Tumor volumes were measured two or three times weekly using calipers, with volumes calculated in mm^3 using the formula: $V = (L \times W \times W)/2$, where V is tumor volume, L is tumor length (longest dimension), and W is tumor width (perpendicular to L). All dosing, tumor measurements, and body weight assessments were performed inside a laminar flow cabinet to maintain standard conditions. The efficacy study with EMT-6 model was performed by Crown Bioscience (Taicang, China) according to the IACUC-approved protocol (AN-2304-07-1441) and in accordance with the regulations of the Association for Assessment and Accreditation of Laboratory Animal Care (AAALAC).

2.9. *Pharmacokinetics*

Pharmacokinetics (PK) study was conducted on non-tumor-bearing female BALB/c nude mice. Mice received an intravenous injection of 1 mg/kg of human IgG1 isotype or HYAL-Fc IgG1 fusion protein. Serial blood samples were collected at 2, 8, 24, 48, 120, 192, 240, and 360 h post the single dose. Blood samples were allowed to clot for 2 h at room temperature, then centrifuged at 10,000 rpm for 10 min. The serum was separated and stored at −80 °C until analysis. Serum concentrations of human IgG1 isotype and HYAL-Fc IgG1 fusion protein were measured using the human IgG precoated ELISA kit (Dakewe, #1128162, Shenzhen, China). The half-life value was estimated using the noncompartmental analysis module in WinNonlin® (version 8.3; Certara, Mountain View, CA, USA).

2.10. Immunofluorescence Staining

IHC staining for FAP and LRRC15 was performed on formalin-fixed, paraffin-embedded (FFPE) sections from mouse xenograft samples using rabbit antibody clone (Solarbio, Cat# K110906P-50ul for FAP, and Cat# ab150376 for LRRC15, Beijing, China). HA staining was performed using Alcian Blue [35] or IHC with the HA-binding protein (Amsbio, Cat#AMS.HKD-BC41, Cambridge, MA, USA). Briefly, FFPE sections were deparaffinized in xylene and rehydrated through a graded ethanol series. Antigen retrieval was carried out using EDTA (pH 9.0) at 95–100 °C for 20 min. Endogenous peroxidase activity was quenched with 3% (v/v) hydrogen peroxide, and nonspecific binding was blocked with 5% (w/v) normal serum for 30 min. Sections were then incubated with primary antibodies (anti-mouse CD3, Abcam, #ab16669; anti-mouse CD8, Cell Signaling, #98941S; anti-mouse CD11b, Cell Signaling, #93169S; anti-mouse CD45, Cell Signaling, #70257S) overnight at 4 °C, followed by incubation with a horseradish peroxidase (HRP)-conjugated secondary antibody for 30 min at room temperature. Detection was performed using a DAB chromogen substrate, and sections were counterstained with hematoxylin, dehydrated, and mounted. Appropriate positive and negative controls were included to ensure staining specificity and accuracy. All images were acquired using the NanoZoomer-HT 2.0 Image system (Hamamatsu, Japan) at 40 times magnification (40×).

For image analysis for T cells infiltration in EMT-6 FFPE samples, all the images were analyzed with the HALO^{TM} platform. The whole slide image was analyzed, and big necrosis and stroma areas were excluded. Both the numbers of total cells and IHC positive cells were counted. IHC score was presented as the ratio of the positive cell counts against the total cell numbers of whole section.

2.11. Statistical Analyses

Statistical analyses were performed using GraphPad Prism (Version 8.0.2). The data were expressed as mean ± SEM values. For comparisons between two groups, an unpaired, two-tailed Student's t-test was used. For comparisons across more than two groups, the one-way ANOVA was applied. A p-value of less than 0.05 was considered statistically significant.

3. Results

3.1. Engineering an Antibody–Enzyme Conjugate

The AbEn was generated by fusing the human hyaluronidase PH20 domain (36–482) onto the hinge region of a human IgG1 Fc using a flexible GGGGS linker (Figure 1A, Supplementary Figure S1A). The resulting hyaluronidase-Fc (HYAL-Fc) construct had an average protein yield of 30 mg/L from the transient CHO expression system and demonstrated over 95% monomeric purity following a one-step protein A chromatography purification (Supplementary Figure S1B,C). In comparison to hyaluronidase with a C-terminal histidine tag (HYAL-His, 11,101 U/mg, 5,863 U/nM), the HYAL-Fc demonstrated comparable enzyme activity of 41,310 U/mg (6,352 U/nM) in a turbidity assay, confirming that the Fc region did not interfere with enzyme function (Supplementary Table S1).

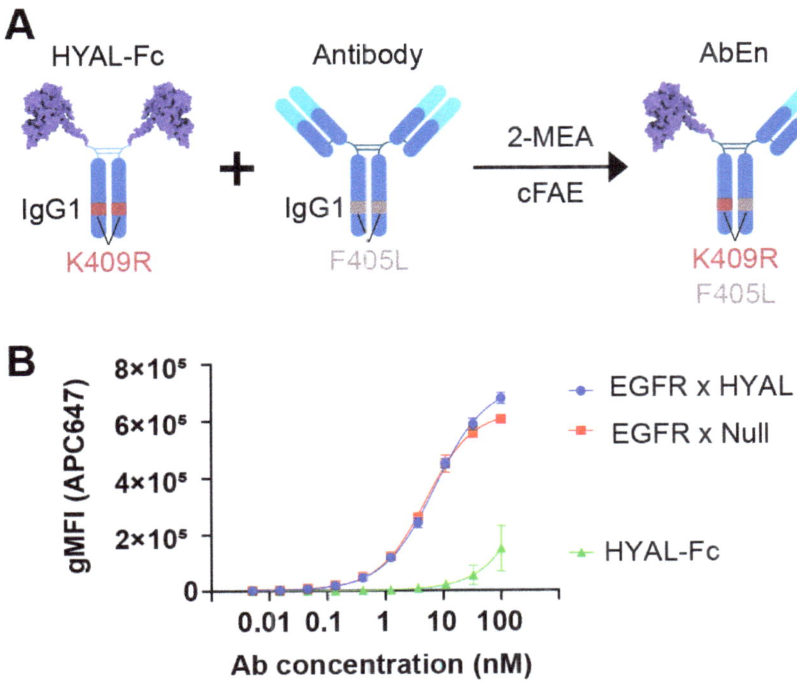

Figure 1. Engineering of an antibody–enzyme conjugate. (**A**) Schematic illustration of the generation of an antibody–enzyme conjugate (AbEn) via cFAE. The hyaluronidase-Fc fusion protein containing K409R mutation was incubated with an IgG1 molecule harboring matching F405L mutation in the presence of 2-MEA. (**B**) Concentration–response curves of the AbEn molecule binding to HCC827 cells were analyzed with Prism 8.0 using a four-parameter logistic equation. The data were presented as the mean ± SEM values of at least three independent experiments. The curve for EGFR × HYAL AbEn were shown in blue. The EGFR × Null antibody (red) and HYAL-Fc IgG1 fusion protein (green) served as the positive and negative controls, respectively. The abbreviations used: gMFI, geometrical mean fluorescent intensity; SEM, standard error of the mean; Fc, fragment crystallizable; AbEn, antibody–enzyme conjugate; cFAE, controlled Fab-arm exchange; 2-MEA, 2-mercaptoethylamine.

Subsequently, we generated an AbEn (Supplementary Figure S2A,B) by combining an anti-EGFR antibody with HYAL-Fc using the controlled Fab-arm exchange (cFAE) technique [33]. The bispecific EGFR x HYAL AbEn was confirmed to be greater than 95% (Supplementary Figure S2A,B). To evaluate whether the AbEn retained its target-binding capability, we conducted a flow cytometry-based cell binding assay using the EGFR-expressing cancer cell line HCC827. The HYAL-Fc fusion protein alone demonstrated no binding to HCC827 cells, while the EGFR × HYAL bispecific agent exhibited comparable binding potency (reflected by EC_{50}, 7.18 nM vs. 4.99 nM) and efficacy (reflected by E_{max}, 7.26×10^5 vs. 6.26×10^5) to that of the EGFR × Null control bispecific antibody (Figure 1B, Supplementary Table S2). Thus, the enzyme domain fusion did not interfere with cancer target binding. Furthermore, the bispecific EGFR × HYAL AbEn maintained a single enzyme activity of 2,047 U/nM (13,670 U/mg), nearly corresponding to half of that of the parental HYAL-Fc protein containing two enzyme domains (Supplementary Table S1). Therefore, neither the cFAE process nor the anti-EGFR Fab domain adversely affected the enzymatic function. Collectively, these results confirmed the dual functionality of this novel AbEn, which retained both antigen-binding and enzyme activity properties.

3.2. Targeted Hyaluronidase Demonstrated Effective Depletion of HA Both Ex Vivo and In Vivo

Hyaluronidase can effectively degrade HA within the tumor stroma, resulting in a delay in tumor growth in nonclinical models across various solid tumor types [26,36,37]. Furthermore, its antitumor activity corresponded to the extent of HA present in the tumor [38]. To characterize the expression patterns of HA in the stroma of tumor, Alcian blue staining [39] was conducted on various cell line-derived xenograft (CDX) models. Among them, a subset of cells including the colorectal cancer cell line RKO, the lung cancer cell line HCC827, the pancreatic cancer cell line AsPC-1, and the gastric cancer cell line N87 exhibited HA-rich staining (Supplementary Table S3).

Next, the enzyme activity of the AbEn in both the ex vivo and in vivo HA-rich RKO models were determined. The alcian blue staining on RKO tumor Formalin-Fixed Paraffin-Embedded (FFPE) section slides, with or without the pre-treatment of HYAL-Fc or EGFR × HYAL before the staining, demonstrated that both HYAL-Fc and EGFR × HYAL effectively degraded HA in the RKO tumor stromata (Figure 2A). Additionally, the in vivo enzyme activity of HYAL-Fc, via monitoring the time course of HA degradation, was evaluated following the injection into RKO tumor-bearing mice. Serum HA fragments were detectable as early as 2 h after HYAL-Fc treatment, increased until 48 h, and then sharply declined at 72 h. This pattern of dynamic homeostasis of HA degradation and clearance in vivo was shown in Figure 2B. This prolonged HA degradation activity was also corroborated by PEGPH20-mediated HA removal, which persisted for over 72 h in a prostate cancer xenograft model [40]. Moreover, a pharmacokinetic (PK) study to monitor the half-life of the HYAL-Fc fusion protein demonstrated a serum half-life of 132 h, which was comparable to that of the human IgG1 isotype control ($t_{1/2}$ = 157 h, Supplementary Figure S3). These results indicated that the enzyme domain did not affect FcRn binding and that the HYAL-Fc fusion protein was stable in circulation. In comparison to the pegylation strategy, which extended the serum half-life of PH20 from 3 min to approximately 10.3 h in mice [41], the Fc-fusion strategy produced a longer-acting hyaluronidase that could have advantages in facilitating sustained enzyme activity and reducing the frequency of administration.

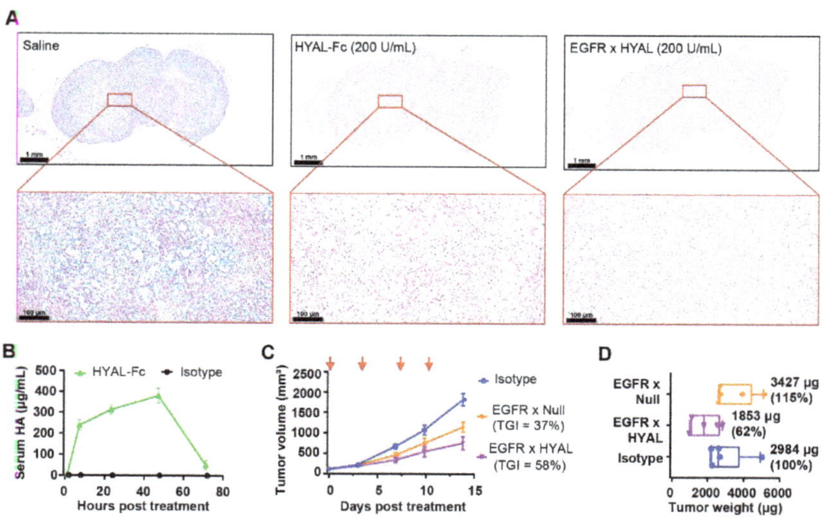

Figure 2. AbEn demonstrated effective depletion of HA both ex vivo and in vivo. (**A**) Alcian Blue staining of RKO tumor sections revealed changes in HA content following treatment with HYAL-Fc

IgG1 fusion protein and EGFR × HYAL AbEn at a concentration of 200 U/mL for 2 h at 37 °C. The saline treatment served as a negative control. Alcian Blue specifically stained HA in the tissues (blue staining). The scale bars spanned for the top images, 1 mm; and for the bottom images, 100 µm. (**B**) Time course of serum HA fragment levels after a single intratumoral injection of 1 mg/kg HYAL-Fc IgG1 fusion protein in mice bearing subcutaneous RKO tumors. Blood samples (20 µL) were collected at 2, 8, 24, 48, and 72 h post-administration, and serum HA fragments were quantified via ELISA. Each time point included the mean values of $n = 3$ mice. An IgG1 isotype control was used as a negative control. (**C**) Tumor growth curves of RKO xenografts treated with either IgG1 isotype control (blue), EGFR × Null (orange), or EGFR × HYAL (purple) at 5 mg/kg, administered twice weekly for a total of four doses. The red arrows indicated the specific dosing days. Tumor growth inhibition rates were calculated relative to the isotype control for each treatment group for comparison. $n = 5$ mice per group. (**D**) On day 14, the animals were sacrificed, and the respective tumor weights were measured. Tumor weights were normalized to the isotype control for comparison. $n = 5$ tumors per group. The abbreviations used: HA, hyaluronic acid; TGI: tumor growth inhibition; HYAL, hyaluronidase; EGFR, epidermal growth factor receptor; AbEn, antibody–enzyme conjugate.

To better distinguish the antitumor activity mediated by HA degradation from Fc effector functions, we introduced L234A and L235A (LALA) mutations into the CH2 domain of Fc to minimize the binding to Fc gamma receptors and C1q [42]. In HA-rich RKO tumors, intravenous administration of EGFR × HYAL effectively inhibited tumor growth, achieving a tumor growth inhibition (TGI) of 58% compared to 37% for EGFR × Null (Figure 2C), without affecting the mice's body weight (Supplementary Figure S4). Notably, a 38% reduction in tumor weight was observed following treatment with EGFR × HYAL, whereas no notable change was noted with EGFR × Null, highlighting the specific role of hyaluronidase in remodeling the tumor stroma (Figure 2D). Together, the AbEn was capable of effectively degrading HA and delaying tumor growth in HA-rich tumors. There was no difference in tumor growth inhibition between the WT Fc and the LALA IgG1 Fc.

3.3. Enhancing Anti-Tumor Activities Through Targeted Hyaluronidase Delivery

Having established that the AbEn could mediate enzymatic activity in vivo, we next explored whether antibody-mediated delivery of hyaluronidase enhanced HA degradation efficiency and antitumor efficacy compared to systemic hyaluronidase treatment. Given the low-EGFR and high-PD-L1 expression profiles of the RKO cell line, we generated a novel PD-L1 × HYAL AbEn using the cFAE technology to improve tumor specificity (Supplementary Figure S2C,D). As expected, the PD-L1 × HYAL AbEn exhibited superior binding efficacy to RKO cells by having a higher E_{max} value of 1.00×10^6 when compared to the EGFR × HYAL Aben with an E_{max} value of 1.42×10^4 (Supplementary Figure S5).

Moreover, RKO tumors exhibited a dose-dependent response to PD-L1 × HYAL treatment, showing TGI levels of 24% at 1 mg/kg and 53% at 6 mg/kg dosing, respectively (Figure 3A). In contrast, systemic administration of hyaluronidase resulted in lower TGI values for the HYAL × Null (34%) and PD-L1 × Null control (5%); both being dosed at 6 mg/kg, respectively (Figure 3A). Thus, the targeted delivery of hyaluronidase through a tumor-associated antigen (TAA) mediated more effective HA degradation and enhanced antitumor responses than the responses coming from systemic administration.

Our studies revealed that the AbEn facilitated HA degradation in the tumor stroma, highlighting its potential role in enhancing the therapeutic efficacy of oncological agents. To explore this, we selected TAVO412, a novel EGFR × cMet × VEGF trispecific antibody known for its potent tumor growth inhibition across multiple solid tumors [43–45]. In RKO tumor-bearing mice, PD-L1 × HYAL plus TAVO412 produced a stronger antitumor response (TGI = 66%) compared to either treatment alone: TAVO412 (TGI = 31%) or PD-L1 × HYAL (TGI = 38%) (Figure 3B). Tumor tissues from treated mice were collected at the time of sacrifice for intratumoral HA quantification, and tissue sections were also stained with Alcian blue. PD-L1 × HYAL administered alone or with TAVO412, effectively removed HA in RKO xenograft tissue (Supplementary Figure S6A, Figure 3C). In contrast, HA depletion was not observed with TAVO412 treatment alone, indicating that the enhanced antitumor responses from the combination therapy were linked to HA degradation (Supplementary Figure S6A, Figure 3C). Collectively, these findings demonstrated that combining oncological therapeutics with targeted HA degradation strategies could significantly improve their efficacy.

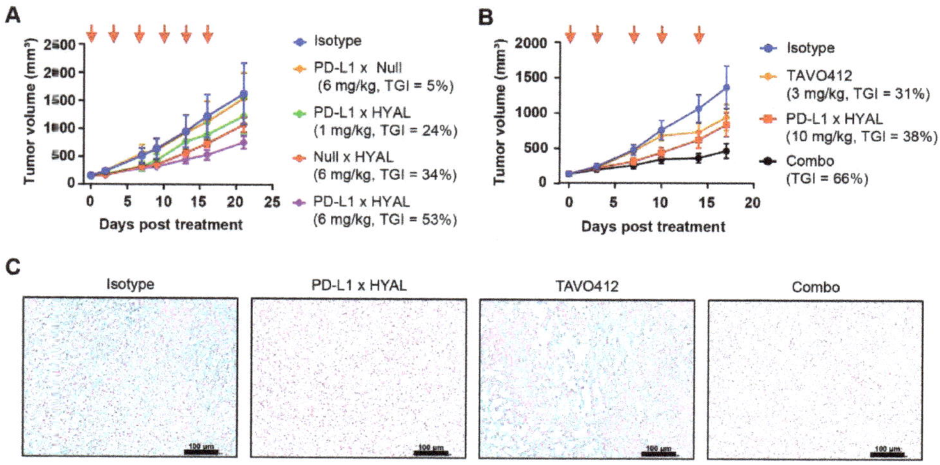

Figure 3. Targeted hyaluronidase enhanced the anti-tumor activities. (**A**) Tumor growth curves of RKO xenografts treated with IgG1 isotype control (blue), PD-L1 × Null (orange), or Null × HYAL (red) at 6 mg/kg, and PD-L1 × HYAL at 1 mg/kg (green) and 6 mg/kg (purple), administered twice weekly for six doses. Tumor growth inhibition rates were calculated relative to the isotype control for each treatment group. Each group included n = 5 mice. The red arrows indicated the specific dosing days. (**B**) Tumor growth curves of RKO xenografts treated with IgG1 isotype control (blue), TAVO412 (orange, 3 mg/kg for 4 doses and 5 mg/kg for fifth dose), PD-L1 × HYAL (red, 10 mg/kg), and the combination of PD-L1 × HYAL and TAVO412 (black), administered twice weekly for a total of five doses. The TGI rates were calculated relative to the isotype control for comparison. Each group included n = 5 mice. The red arrows indicated the specific dosing days. (**C**) Alcian Blue staining of RKO tumor sections demonstrated changes in hyaluronic acid (HA) content following treatment with IgG1 isotype control, PD-L1 × HYAL, TAVO412, the combination of PD-L1 × HYAL and TAVO412 (3 mg/kg). The scale bars spanned 100 µm. The abbreviations used: TGI, tumor growth inhibition; Combo, combination; HYAL, hyaluronidase.

3.4. *Enhancing Anti-Tumor Activities Through CAF-Targeted Hyaluronidase Delivery*

Given the enrichment of CAFs with the stroma in the TME, we hypothesized that delivering hyaluronidase via CAF targeting could achieve more effective HA degradation than targeting TAAs in solid tumor cells. Two CAF-specific targets, FAP and LRRC15, were selected to facilitate targeted delivery of hyaluronidase into the stroma. RKO tumor FFPE sections were stained with antibodies against FAP and LRRC15, along with HA-binding protein (HABP) to confirm the high levels of FAP and LRRC15 proteins in the tumor samples with a concomitant abundance of HA within the TME (Figure 4A).

To improve binding affinity to CAFs, we engineered a bispecific antibody by fusing tandem anti-LRRC15 VHH antibodies to the C-terminus of the heavy chain of an anti-FAP antibody. The anti-FAP antibody showed comparable binding affinity for human (EC_{50} = 0.08 nM) and mouse antigens (EC_{50} = 0.06 nM), while the anti-LRRC15 antibodies had slightly lower affinity for mouse LRRC15 (EC_{50} = 13.58 nM) compared to human LRRC15 (EC_{50} = 0.1326 nM); (Supplementary Figure S7, Supplementary Table S4). This cross-reactivity was crucial for evaluating targeted hyaluronidase delivery in mouse xenograft models, given that stromal components were of mouse origin. In flow cytometry assays using U-87 MG human glioma cells expressing both FAP and LRRC15, the dual-targeted antibody demonstrated increased binding (EC_{50} = 0.2645 nM vs. 12.68 nM and 0.8885 nM) compared to the FAP-only mAb (EC_{50} = 12.68 nM) and LRRC15-only mAb (EC_{50} = 0.8885 nM) controls (Supplementary Figure S7, Supplementary Table S4).

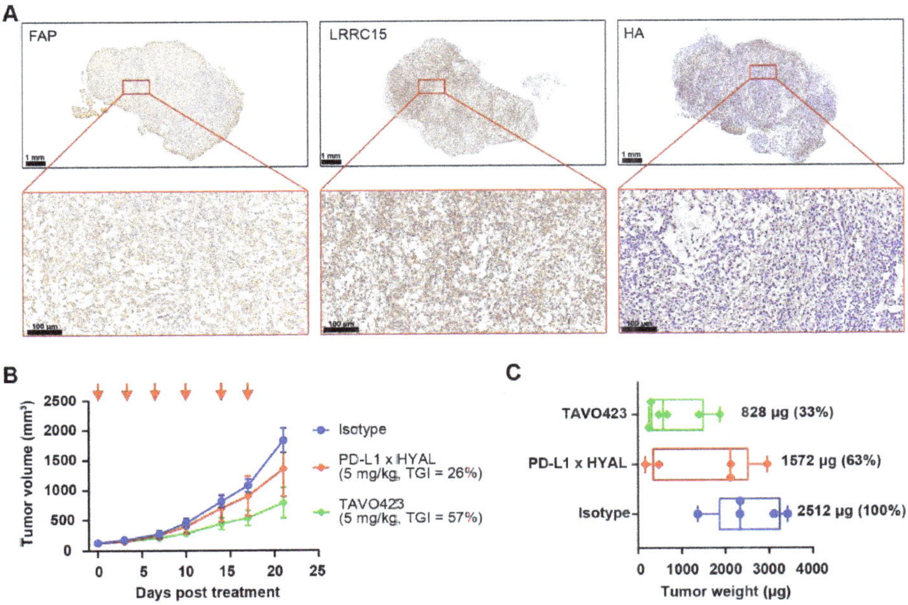

Figure 4. Targeted hyaluronidase delivery to stroma targets improved anti-tumor activity. (**A**) Representative immunohistochemistry images of RKO tumor sections showing FAP, LRRC15, and HA expression. The scale bar spans 100 µm. (**B**) Tumor growth curves of RKO xenografts treated with IgG1 isotype control (blue), PD-L1 × HYAL (red), or TAVO423 (green) at 5 mg/kg, administered twice weekly for six doses. Tumor growth inhibition rates were calculated relative to the isotype control. The data show an average of n = 5–6 mice per group. The red arrows indicate the specific dosing days. (**C**) On day 21, the animals were sacrificed and the respective tumor weights were measured. Tumor weights were normalized to the isotype control for comparison. Each group represents n = 5–6 mice per group. The abbreviations used: TGI, tumor growth inhibition; HYAL, hyaluronidase.

Subsequently, TAVO423, a trispecific FAP × LRRC15 × HYAL AbEn with an Fc-engineered IgG1 backbone carrying LALA mutations was generated using cFAE technology for in vivo proof of concept experiments (Supplementary Figure S8). In RKO tumor-bearing mice, TAVO423 demonstrated a superior antitumor effect (TGI = 57%) compared to PD-L1 × HYAL (TGI = 26%) when administered at the same dose (Figure 4B). Additionally, tumor tissues were collected at sacrifice to measure tumor weights, revealing a 67% reduction in tumor weight with TAVO423 treatment, compared to a 37% reduction with PD-L1 × HYAL treatment (Figure 4C). These results demonstrated a more pronounced tumor control effect, potentially driven by more extensive remodeling of the TME and alterations in tissue density and composition. Overall, the targeted hyaluronidase delivery via CAF antigens could disrupt the stromal barrier more effectively than strategies solely targeting TAAs, thereby enhancing antitumor responses.

3.5. CAF-Targeted HA Degradation Improves the Antitumor Responses of Diverse Antitumor Drugs in RKO Xenograft Model

Although being used as a standard of care drug for patients suffering from colorectal cancer (CRC), 5-fluorouracil (5-FU) chemotherapy has limited improvement in disease-free survival, largely due to TME-mediated chemoresistance and impaired drug penetration [46]. Similarly, while PD-1/PD-L1 inhibitors are approved for CRC, most patients experience poor responses or relapse due to TME-driven immunosuppression [47]. Recognizing HA accumulation as a common mediator of both chemoresistance and immunosuppression, we hypothesized that CAF-targeted HA degradation could potentiate existing therapies by inducing TME remodeling.

The in vivo models using RKO tumor-bearing mice revealed that while TAVO423 and 5-FU monotherapies achieved modest TGI of 28% and 14%, respectively, their combination synergistically enhanced TGI to 49% (Figure 5A, Supplementary Table S5). The combination therapy also induced rapid and significant HA depletion in the tumor stroma within 0.25 to 1 h of treatment (Supplementary Figure S6B). A similar readout was observed in the experiment involving immune checkpoint blockade. In the HA-rich RKO tumors, the anti-PD-L1 monotherapy (TGI = 14%) was substantially improved in the combination with TAVO423 (TGI = 56%) (Figure 5B, Supplementary Table S5). These findings showed how targeted HA degradation could improve current CRC therapies by disruption of stromal barriers.

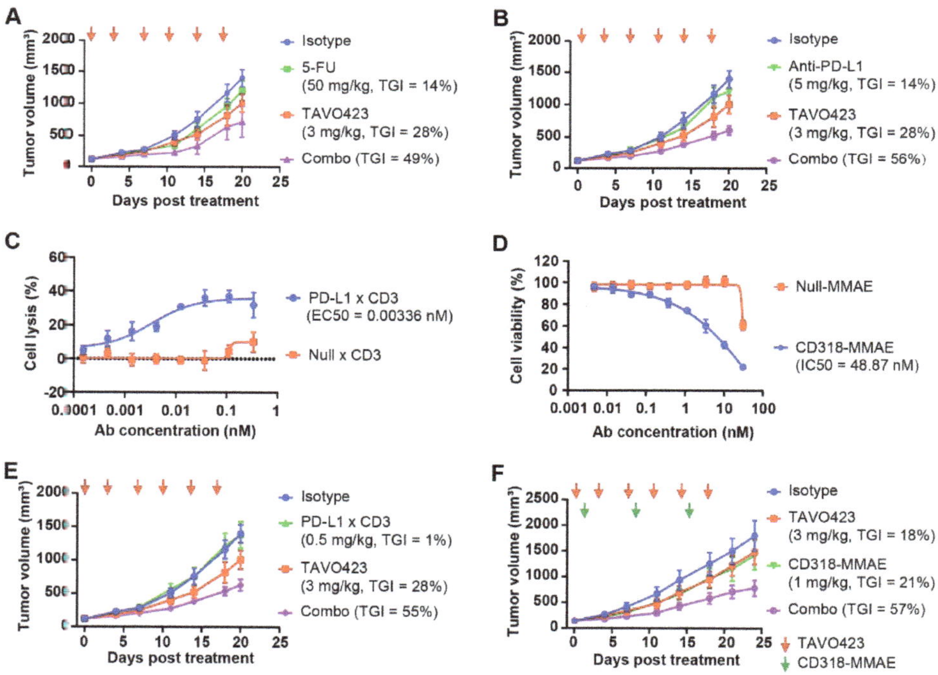

Figure 5. CAF-targeted HA degradation improved the antitumor responses of diverse anti-tumor drugs. (**A**) Tumor growth curves of RKO xenografts treated with IgG1 isotype control (3 mg/kg, blue), 5-FU (50 mg/kg, green), or TAVO423 (3 mg/kg, red), and a combination of 5-FU with TAVO423 (purple). Treatments were administered twice weekly for six doses. (**B**) Tumor growth curves of RKO xenografts in NOG mice reconstituted with human PBMCs, treated with IgG1 isotype control (3 mg/kg, blue), anti-PD-L1 monoclonal antibody (5 mg/kg, green), or TAVO423 (3 mg/kg, red), and

a combination of anti-PD-L1 mAb and TAVO423 (purple). Treatments were administered twice weekly for six doses. (**C**) Representative concentration–response curve demonstrating PD-L1 × CD3-induced lysis of RKO cells (E:T ratio = 5:1). Cytotoxicity was measured via LDH release after 72 h of cocultures of PBMCs and RKO cells at an E:T ratio of 5:1, compared to the 0.2% (*w*/*v*) Triton X-100 control. The Null × CD3 antibody served as negative control. EC_{50} values were determined by four-parameter logistic curve fitting. Data are mean ± SEM values from at least three independent experiments. (**D**) Representative concentration–response curve showing CD318-MMAE-induced proliferation inhibition of RKO cells. Cytotoxicity was assessed for 4 days post-treatment. Null-MMAE served as a negative control. IC_{50} values were calculated using four-parameter logistic curve fitting. The data represented the mean ± SEM values from at least three independent experiments. (**E**) Tumor growth curves of RKO xenografts in NOG mice reconstituted with human PBMCs, treated with IgG1 isotype control (3 mg/kg, blue), PD-L1 × CD3 (0.5 mg/kg, green), or TAVO423 (3 mg/kg, red), and a combination of PD-L1 × CD3 and TAVO423 (purple). Treatments were administered twice weekly for six doses. (**F**) Tumor growth curves of RKO xenografts treated with IgG1 isotype control (3 mg/kg, blue), CD318–MMAE (1 mg/kg, green), or TAVO423 (3 mg/kg, red), along with a combination of CD318–MMAE and TAVO423 (purple). Treatment with TAVO423 was administered thrice weekly for 9 doses, and CD318-MMAE was given weekly for 3 doses. For all animal studies, tumor growth inhibition (TGI) rates were calculated relative to the isotype control. The studies showed the average of n = 5–6 mice per group. The red and green arrows indicated the specific dosing days. The abbreviations used: 5-FU, 5-Fluorouracil; PBMCs, peripheral blood mononuclear cells; mAb, monoclonal antibody; TGI, tumor growth inhibition; Combo, combination; EC_{50}, median effective concentration; IC_{50}, half-maximal inhibitory concentration; E:T, effector to target ratio; SEM, standard error of the mean.

Since HA degradation has been shown to enhance drug penetration and immune infiltration [26,48], we expanded our investigation to include T-cell engagers (TCEs) and antibody–drug conjugates (ADCs). Both modalities could benefit from having stronger activity through more efficient tumor penetration and immune recruitment. Leveraging the expression profiles of the RKO cell line, we engineered a bispecific TCE targeting PD-L1 and CD3, as well as a novel ADC targeting CD318. CD318, also known as cub domain containing protein 1 (CDCP1), was overexpressed in many solid tumors and could be a promising target in treating colorectal tumors [49]. The PD-L1 × CD3 bispecific antibody comprised an anti-PD-L1 VHH and an anti-CD3 IgG1 domains. The CD318-MMAE ADC was created by conjugating an anti-CD318 IgG1 to the anti-mitotic toxin payload monomethyl auristatin E (MMAE), via a lysosomal cleavable dipeptide linker. Both constructs incorporated Fc-engineered IgG1 (LALA mutants) to reduce Fcγ receptor (FcγR) interactions.

To assess target engagement and efficacy prior to in vivo testing, the anti-tumor effects were assessed using two different in vitro cytotoxicity assay formats. The PD-L1 × CD3 TCE exhibited potent tumor killing (EC_{50} = 0.00336 nM) in RKO-PBMC cocultures (Figure 5C), while the CD318-MMAE ADC showed nanomolar growth inhibition (IC_{50} = 48.87 nM) against RKO monolayers (Figure 5D). In subsequent in vivo studies using PBMC-humanized RKO xenografts, administering PD-L1 × CD3 (0.5 mg/kg) or TAVO423 (3 mg/kg) alone twice weekly for six doses, showed limited activity with TGI values of 1% and 28%, respectively. However, the combination of TAVO423 and PD-L1 x CD3 achieved a TGI of 55% (Figure 5E, Supplementary Table S5). Similarly, in RKO xenografts, combining targeted HA degradation (3 mg/kg, administered thrice weekly for nine doses) with the CD318-MMAE ADC (1 mg/kg, weekly for three doses) resulted in superior efficacy, reaching 67% TGI versus 18–21% with either agent alone (Figure 5F, Supplementary Table S5). Together, these findings further supported how CAF-targeted HA degradation could remodel the TME to synergize TGI with various therapeutic modalities in HA-rich CRC tumors.

To further validate the broad applicability of CAF-targeted HA degradation across diverse tumor types, we assessed TAVO423 in a pancreatic cancer model employing BxPC-

3 cells engineered to overexpress hyaluronan synthase 3 (BxPC-3-HAS3). These cells exhibited a 17-fold increase in intratumoral HA accumulation compared to wild-type (WT) BxPC-3 controls (Supplementary Figure S9A,B). In this HA-high setting, TAVO423 monotherapy (5 mg/kg) achieved moderate tumor growth inhibition (TGI = 30%) and reduced intratumoral HA levels by 5-fold (Supplementary Figure S9B,C). While a suboptimal dose of 5T4 × CD3 TCE (0.1 or 0.5 mg/kg) alone showed substantial efficacy (TGI = 73%; Supplementary Figure S9C), the combination of TAVO423 with 5T4 × CD3 TCE led to superior antitumor activity (TGI = 92%), indicating that targeted HA degradation potently enhanced T-cell-mediated cytotoxicity in pancreatic tumors. All treatments were well tolerated in the rodent models as confirmed by minimal changes in the body weight throughout the study (Supplementary Figure S9D). Together, these results demonstrated the broad potential of CAF-directed HA degradation to augment diverse therapeutic modalities across multiple HA rich solid tumor types.

3.6. Targeted Hyaluronic Acid Degradation Sensitizes Tumors to Immune Checkpoint Blockade Therapy by Promoting the T Cells Infiltration

Compelling clinical evidence demonstrated that sufficient amounts of tumor-infiltrating lymphocytes (TILs) were essential for PD-1/PD-L1 blockade response and improved prognoses in many solid tumors [50]. Targeted HA degradation enhancement of ICIs efficacy by facilitating TIL recruitment was assessed by characterizing TAVO423 trispecific AbEn in combination with anti-PD-1 therapy in immunocompetent mice bearing EMT-6 syngeneic breast tumors. Unlike human tumor xenografts requiring immunocompromised hosts, these syngeneic models maintained fully functional immune systems so as to assess immunotherapy mechanisms and efficacy [51]. In addition, the constructs were fully cross-reactive with mouse target antigen counterparts, allowing for more relevant evaluation in these syngeneic models. IHC staining revealed substantial HA accumulation within the EMT-6 TME (Figure 6A), confirming the relevance of the model for hyaluronidase-based intervention. Treatment with anti-PD-1 monotherapy (3 mg/kg) demonstrated the expected antitumor activity (33% TGI; Figure 6B) in this immunotherapy-responsive model [52]. While TAVO423 alone (10 mg/kg) achieved modest efficacy (TGI = 16%), the combination regimen exhibited synergistic tumor control (TGI = 51%) while maintaining an excellent safety profile as demonstrated by stable body weight measured during the study (Figure 6C). These findings demonstrated both the preliminary therapeutic potential and favorable tolerability of this CAF-targeted hyaluronidase approach in immunocompetent hosts.

To determine if the enhanced antitumor activity of the combination of anti-PD-1 antibody and HA degradation mediated increased levels of TILs, we collected and analyzed tumor tissues post-treatment using IHC staining with various immune cell markers, including CD45 for leukocytes, CD11b for myeloid cells, and CD3 and CD8 for T cells (Figure 6D). While treatment regimens did not significantly alter overall leukocyte or myeloid cell infiltration, there were profound differences in the T cell populations (Figure 6D,E, Supplementary Table S6). Compared to the isotype control, the anti-PD-1 monotherapy induced a 3–4-fold expansion of both $CD3^+$ and $CD8^+$ T cells. The combination therapy of targeted HA degradation with anti-PD-1 had a 6–9-fold increase in T cell infiltration (Figure 6D,E; Supplementary Table S6). This robust infiltration of $CD8^+$ cytotoxic T lymphocytes (CTLs) could drive the improved tumor growth inhibition control. Notably, the TAVO423 monotherapy alone failed to increase TILs, consistent with published evidence that HA degradation strategy required the combination with T cell priming agents for optimal efficacy [53]. Thus, the combination of TAVO423 for targeted HA degradation and ICIs could result in remodeling of the TME to facilitate CTL infiltration thereby enhancing tumor growth inhibition.

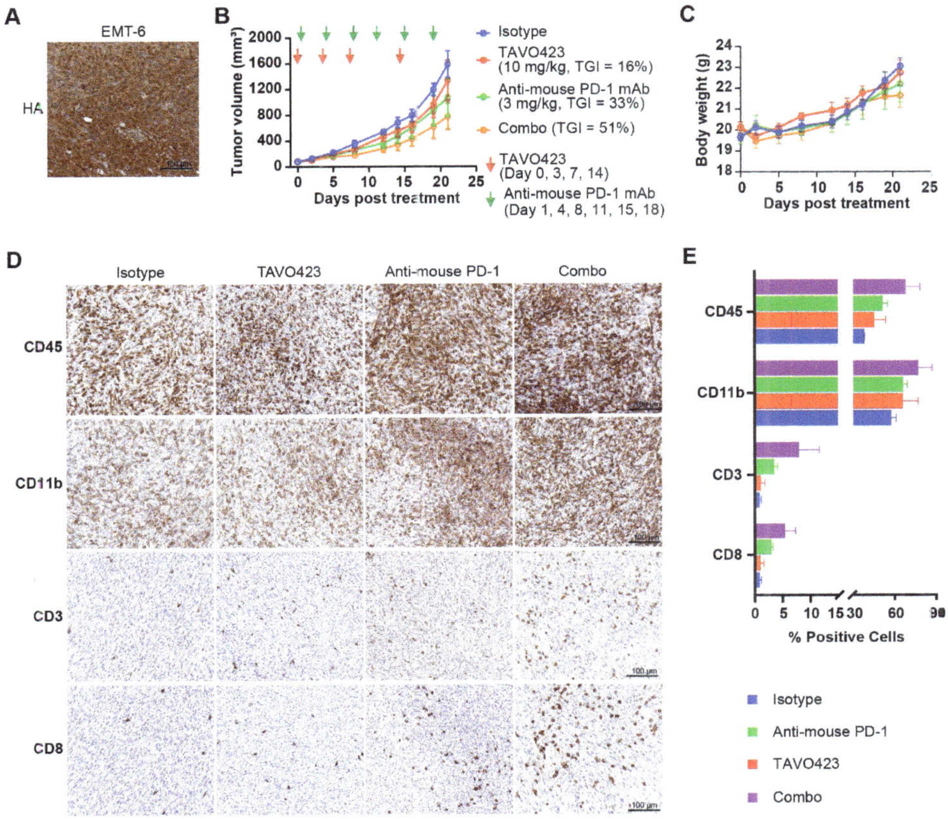

Figure 6. TAVO423 enhanced immune checkpoint blockade therapy by promoting T cell infiltration. (**A**) Representative immunohistochemistry images of EMT-6 tumor sections stained with HA-binding protein, demonstrating HA expression. Scale bars: 100 µm. (**B**,**C**) Balb/c mice were subcutaneously (s.c.) inoculated with 5×10^5 EMT-6 tumor cells and treated as follows: 10 mg/kg TAVO423 on days 0, 3, 7, and 14; 3 mg/kg anti-mouse PD-1 mAb twice weekly on days 1, 4, 8, 11, 15, and 18; and a combination of TAVO423 plus anti-mouse PD-1 mAb with the same dosage and regimen as monotherapy. The red and green arrows indicated the specific dosing days. Tumor growth curves (**B**) and mouse body weights (**C**) of EMT-6-bearing mice were shown for groups receiving the IgG1 isotype control (blue), TAVO423 (red), anti-mouse PD-1 mAb (green), and combination therapy (orange). Tumor growth inhibition rates were calculated relative to the isotype control. The studies were an average of $n = 5$ mice per group. (**D**,**E**) At day 21 post-treatment, mice were sacrificed, and tumor tissues were collected for immunohistochemical analysis of intratumoral levels of CD3, CD8, CD45, and CD11b. (**D**) Representative IHC images of EMT-6 tumor sections. Scale bar: 100 µm. (**E**) Quantitative analysis of the percentage of CD3, CD8, CD45, and CD11b positive cells in tumor samples. The X-axis showed a broken scale with a gap between 15% and 30% to accommodate the data range, while maintaining a consistent linear scale throughout. All bars were color-coded as in the legend (Isotype: blue; Anti-mouse PD-1: green; TAVO423: red; Combo: purple). Results from two independent experiments are shown as mean ± SEM values. The abbreviations used: mAb, monoclonal antibody; TGI, tumor growth inhibition; Combo, combination; IHC, immunohistochemistry; SEM, standard error of the mean. HA, hyaluronan acid.

4. Discussion

Tumor-targeting cell therapy often has limited efficacy because of the immunosuppressive and physically obstructive TME. Thus, there have been efforts to remodel the

TME to enhance anti-cancer responses from immunotherapies [54,55]. The HA-rich desmoplastic stroma in the solid tumors are barriers that impede drug penetration and immune cell infiltration. Although systemic hyaluronidase therapy has demonstrated promising preclinical success [30], we hypothesized that an antibody-mediated tumor-targeted delivery strategy could enhance HA degradation while mitigating off-tumor toxicity from systemic exposure. To this end, we developed a HYAL-Fc fusion protein to extend the hyaluronidase serum half-life and provide tissue targeting using a multifunctional AbEn using the cFAE technology. The AbEn molecules were engineered to preserve both target-binding specificity and enzymatic activity. In HA-rich colorectal cancer models, the AbEn molecules demonstrated superior therapeutic efficacy compared to conventional systemic delivery of hyaluronidase through more efficient HA degradation. Notably, the CAF-targeted AbEns exhibited greater therapeutic potency than tumor cell-targeted AbEns and synergistically improved outcomes via combination with other multiple treatment modalities. Furthermore, tumor-infiltrating lymphocyte augmentation was confirmed as a key driver of the enhanced anti-PD-1 response. This targeted HA degradation could overcome ECM-mediated inhibition of diverse therapies to treat patients suffering from growth of stroma-rich solid tumors.

The AbEn platform could overcome the lack of clinical responses from earlier attempts at HA-targeted therapies. Preclinical studies of PEGPH20 demonstrate effective HA degradation and TME remodeling effects by improving vascular patency thereby allowing enhanced chemotherapeutic delivery [26]. However, the clinical translation fail to meet the primary endpoint and suffered from dose-limiting toxicities [30]. These setbacks highlight two key challenges: (1) the narrow therapeutic window of systemic hyaluronidase therapy, and (2) the inability of HA depletion alone to sustain antitumor responses. Furthermore, PEGPH20 also show that disrupting the stromal barrier lead to HA fragmentation that can exacerbate tumor inflammation [30,56].

To circumvent such challenges, we developed the targeted HA AbEn to have two key features: (1) Fc fusion and heterodimerization for precise IgG1-scaffold assembly, (2) manipulation of tumor-targeting domain and the tunable valency to balance tumor penetration and enzyme delivery. As a result, compared to the untargeted PEGPH20 monotherapy approach, the AbEn platform offered four transformative advantages: (1) targeted HA degradation to enhance antitumor responses while minimizing systemic toxicity; (2) CAF-directed delivery to allow better immune cell infiltration; (3) the IgG1 scaffold enabled an extended half-life and concomitant immune effector functions (e.g., Depletion of CAFs mediated by Fc effector function); and (4) modular design permitted 'plug-and-play' integration of additional mechanisms of action, including antiangiogenic or immune-checkpoint inhibition.

Recent advances to localize HA degradation include mesothelin-specific CAR-T cells expressing PH20 and oncolytic adenoviruses that encode PH20 (e.g., VCN-01) to improve T-cell infiltration and stromal modulation in preclinical models [37,48]. However, these approaches may have two critical limitations. First, their therapeutic activities require successful tumor homing and proliferation of CAR-T cells or viral replication that can be compromised in densely fibrotic tumors. Second, continuous PH20 production by these living systems may lead to systemic enzyme dissemination which can raise safety concerns from off-tumor tissue damage. In contrast, our AbEn strategy did not depend on biological amplification, but rather allowed dose control for precise spatiotemporal control of HA degradation. This 'on-demand' mechanism offered distinct advantages including reduced systemic toxicity, streamlined manufacturing, and predictable pharmacokinetics to address key challenges in clinical translation.

Despite these promising findings, our study had several limitations that warrant consideration. First, although CAF-directed HA degradation demonstrated enhanced anti-tumor efficacy, we could not determine whether this resulted from CAF apoptosis or functional TME reprogramming. This distinction was critical because complete CAF elimination could unexpectedly accelerate tumor progression in certain contexts [57,58]. Future studies employing single-cell RNA sequencing and functional assays (e.g., contractility measurements) could elucidate these mechanistic details. Second, while co-expression of FAP and LRRC15 was common on pro-fibrotic CAFs, neither marker captured the full complexity of CAFs [17,59]. Third, although our PK data indicated prolonged serum stability of HYAL-Fc and efficacy studies in immunocompetent models demonstrated acceptable safety profiles, a comprehensive evaluation in non-human primates would be essential to assess potential off-tumor effects in HA-rich tissues (e.g., synovium, skin) and FAP (e.g., bone marrow) or LRRC15 (e.g., hair follicles) expressing sites. Finally, in the analysis of immune cell infiltration, the increase in $CD8^+$ T cells could not be enough for robust anti-tumor activity in clinical settings. We did not characterize other T cell subsets, such as regulatory T cells (Tregs), within the total $CD3^+$ population, nor did we dissect the heterogeneous $CD11b^+$ myeloid compartment into its pro- and anti-inflammatory subsets. A more detailed immune profiling in future studies could provide deeper insights into the immunomodulatory mechanisms of targeted HA degradation. Additionally, while our syngeneic models effectively demonstrated increased immune infiltration, they could not fully capture human stromal heterogeneity. More clinically relevant systems, such as patient-derived organoid/CAF cocultures, could improve translational predictability. Addressing these gaps would be critical for both fundamental understanding of biology and subsequent clinical development.

5. Conclusions

In conclusion, our study demonstrated that targeted HA degradation can overcome ECM-mediated treatment resistance to enhance tumor growth inhibition. The AbEn platform's modularity not only addressed key limitations of existing hyaluronidase base tumor growth inhibition approaches (e.g., short half-life, narrow therapeutic window) but also provided a modular platform for precision ECM remodeling to optimize combination regimens against stroma-rich malignancies. Notably, the PH20-based selective enzymatic activity for efficient HA degradation while minimizing off-target effects could enhance both safety and therapeutic efficacy and broaden the therapeutic index. Future studies exploring its application across diverse tumor types and stromal microenvironments could facilitate clinical implementation of this therapeutic paradigm. Further optimization of PH20 binding and catalytic efficiency could refine stromal targeting for both oncology, autoimmune, and inflammatory indications.

Supplementary Materials: The following supporting information can be downloaded at: https://www.mdpi.com/article/10.3390/cancers17213411/s1, Figure S1: Design and confirmation of the purified hyaluronidase-Fc fusion protein; Supplementary Figure S2: Generation of the TAA-targeted hyaluronidase AbEn molecules; Supplementary Figure S3: PK study of a hyaluronidase-Fc fusion protein in female BALB/c nude mice; Supplementary Figure S4: Effect of dosing on mouse body weights; Supplementary Figure S5: Comparison of PD-L1 binding with EGFR binding on the RKO cell line; Supplementary Figure S6: Quantification of intratumoral HA in RKO xenografts following treatment; Supplementary Figure S7: Binding of anti-FAP and anti-LRRC15 antibodies; Supplementary Figure S8: Engineering of a FAP x LRRC15 x HYAL AbEn, CAF-targeted hyaluronidase; Supplementary Figure S9: TAVO423 effects on the TGI efficacy of a 5T4 × CD3 TCE in a hyaluronan-rich pancreatic cancer model; Supplementary Table S1: Hyaluronidase enzyme activity of HYAL-Fc, HYAL-His, and EGFR × HYAL AbEn; Supplementary Table S2: Binding of EGFR x HYAL AbEn to the HCC827 cell line; Supplementary Table S3: Target Profile Screening of CDX models with HA-high stroma; Supplementary Table S4: Binding properties of anti-FAP x anti-LRRC15 bispecific antibodies;

Supplementary Table S5: Tumor growth inhibition values of TAVO423 combined with various antitumor drugs; Supplementary Table S6: Statistical analyses of immune cells populations of tumor samples via IHC staining; Supplementary Table S7: Profiles of the drugs and antibody constructs.

Author Contributions: F.Z.: Conceptualization, Formal Analysis, Investigation, Project administration, Resources, Supervision, Writing—original draft, Writing—review and editing. G.M.: Data curation, Formal Analysis, Methodology, Resources, Visualization, Writing—original draft. H.B.: Data curation, Formal Analysis, Methodology, Resources, Visualization, Writing—original draft. L.C.: Data curation, Formal Analysis, Methodology, Resources, Visualization, Writing—original draft. Z.Z.: Data curation, Formal Analysis, Methodology, Resources, Visualization, Writing—original draft. Y.J.: Conceptualization, Formal Analysis, Investigation, Data curation, Formal Analysis, Visualization, Writing—original draft. M.L.C.: Conceptualization, Funding acquisition, Investigation, Resources, Supervision, Writing—review and editing. All authors have read and agreed to the published version of the manuscript.

Funding: This research was funded by Tavotek Biotherapeutics. The APC was funded by Tavotek BioTherapeutics.

Institutional Review Board Statement: All procedures related to animal care, handling, and treatment were carried out in accordance with the guidelines of Suzhou Charles River accelerator and development lab's Institutional Animal Care and Use Committee (IACUC).

Informed Consent Statement: Not applicable.

Data Availability Statement: Data will be made available upon reasonable request from the corresponding author.

Acknowledgments: We acknowledge Yijun Zhou, Chao Han, Pu Pu, and Maria P. MacWilliams for helpful discussions and technical support.

Conflicts of Interest: All authors are employees of Tavotek Biotherapeutics, Inc., and declare no conflicts of interest.

Abbreviations

The following abbreviations are used in this manuscript:

Abbreviation	Definition
2-MEA	2-mercaptoethylamine
5-FU	5-Fluorouracil
ADCs	Antibody–drug conjugates
AbEn	Antibody–enzyme conjugate
ATCC	American Type Culture Collection
CAF	Cancer-associated fibroblast
CAR-T	Chimeric antigen receptor T
CD44	Cluster of differentiation 44
CDCP1	Cub domain containing protein 1
CDX	Cell line-derived xenograft
cFAE	Controlled Fab-arm exchange
CH2	Constant Heavy domain 2
CH3	Constant Heavy domain 3
CI	Confidence interval
Combo	Combination
CRC	Colorectal cancer
CTLs	Cytotoxic T lymphocytes
DMEM	Dulbecco's modified Eagle's medium
ET	Effector-to-target
EC50	Median effective concentration
ECM	Extracellular matrix

EGFR	Epidermal growth factor receptor
ELISA	Enzyme linked immunosorbent assay
EMEM	Eagle's Minimum Essential Medium
FAP	Fibroblast activation protein
FBS	Fetal bovine serum
Fc	Fragment crystallizable
FcγR	Fcγ receptor
FFPE	Formalin-fixed, paraffin-embedded
FITC	Fluorescein isothiocyanate
GGGGS	Glycine–Glycine–Glycine–Glycine–Serine
gMFI	Geometrical mean fluorescent intensity
HA	Hyaluronan or Hyaluronic acid
HABP	HA-binding protein
HAS3	Hyaluronan synthase 3
HC	Heavy chain
His	Histidine
HMW	High molecular weight
HPLC-SEC	High-performance liquid chromatography–size exclusion chromatography
HRP	Horseradish peroxidase
HYAL	Hyaluronidase
IACUC	Institutional Animal Care and Use Committee
IC50	Half-maximal inhibitory concentration
ICIs	Immune checkpoint inhibitors
IHC	Immunohistochemistry
LALA	L234A and L235A
LC	Light chain
LDH	Lactate dehydrogenase
LMW	Low molecular weight
LRRC15	Leucine-rich repeat-containing protein 15
mAbs	Monoclonal antibodies
mAU	Milliabsorbance
min	Minutes
MMAE	Monomethyl auristatin E
MW	Molecular weight
nM	Nanomolar
OD450	Optical density at 450 nm
PBMCs	Peripheral blood mononuclear cells
PEGPH20	Pegylated recombinant human PH20
PH20	Plasma membrane-associated Hyaluronidase 20, Sperm adhesion molecule 1
PK	Pharmacokinetics
RHAMM	Hyaluronan-mediated motility receptor
SDS-PAGE	Sodium dodecyl sulfate polyacrylamide gel electrophoresis
SEM	Standard error of the mean
TAA	Tumor-associated antigen
TCEs	T-cell engagers
TGI	Tumor growth inhibition
TILs	Tumor-infiltrating lymphocytes
TME	Tumor microenvironment
U	Units
UV280	Ultraviolet absorption at 280 nm
WT	Wild-type

References

1. Jin, M.-Z.; Jin, W.-L. The updated landscape of tumor microenvironment and drug repurposing. *Signal Transduct. Target. Ther.* **2020**, *5*, 166. [CrossRef] [PubMed]
2. de Visser, K.E.; Joyce, J.A. The evolving tumor microenvironment: From cancer initiation to metastatic outgrowth. *Cancer Cell* **2023**, *41*, 374–403. [CrossRef] [PubMed]
3. Giraldo, N.A.; Sanchez-Salas, R.; Peske, J.D.; Vano, Y.; Becht, E.; Petitprez, F.; Validire, P.; Ingels, A.; Cathelineau, X.; Fridman, W.H.; et al. The clinical role of the TME in solid cancer. *Br. J. Cancer* **2019**, *120*, 45–53. [CrossRef]
4. Liu, B.; Zhou, H.; Tan, L.; Siu, K.T.H.; Guan, X.-Y. Exploring treatment options in cancer: Tumor treatment strategies. *Signal Transduct. Target. Ther.* **2024**, *9*, 175. [CrossRef]
5. Meng, L.; Zheng, Y.; Liu, H.; Fan, D. The tumor microenvironment: A key player in multidrug resistance in cancer. *Oncologie* **2023**, *26*, 41–58. [CrossRef]
6. Cui, J.-W.; Li, Y.; Yang, Y.; Yang, H.-K.; Dong, J.-M.; Xiao, Z.-H.; He, X.; Guo, J.-H.; Wang, R.-Q.; Dai, B.; et al. Tumor immunotherapy resistance: Revealing the mechanism of PD-1/PD-L1-mediated tumor immune escape. *Biomed. Pharmacother.* **2024**, *171*, 116203. [CrossRef]
7. Lin, X.; Kang, K.; Chen, P.; Zeng, Z.; Li, G.; Xiong, W.; Yi, M.; Xiang, B. Regulatory mechanisms of PD-1/PD-L1 in cancers. *Mol. Cancer* **2024**, *23*, 108. [CrossRef]
8. Xu, M.; Zhang, T.; Xia, R.; Wei, Y.; Wei, X. Targeting the tumor stroma for cancer therapy. *Mol. Cancer* **2022**, *21*, 208. [CrossRef]
9. Biffi, G.; Tuveson, D.A. Diversity and biology of cancer-associated fibroblasts. *Physiol. Rev.* **2021**, *101*, 147–176. [CrossRef]
10. Zhang, H.E.; Hamson, E.J.; Koczorowska, M.M.; Tholen, S.; Chowdhury, S.; Bailey, C.G.; Lay, A.J.; Twigg, S.M.; Lee, Q.; Roediger, B.; et al. Identification of novel natural substrates of fibroblast activation protein-alpha by differential degradomics and proteomics. *Mol. Cell. Proteom.* **2019**, *18*, 65–85. [CrossRef]
11. Liu, R.; Li, H.; Liu, L.; Yu, J.; Ren, X. Fibroblast activation protein: A potential therapeutic target in cancer. *Cancer Biol. Ther.* **2012**, *13*, 123–129. [CrossRef] [PubMed]
12. Zhao, L.; Pang, Y.; Zhou, Y.; Chen, J.; Fu, H.; Guo, W.; Xu, W.; Xue, X.; Su, G.; Sun, L.; et al. Antitumor efficacy and potential mechanism of FAP-targeted radioligand therapy combined with immune checkpoint blockade. *Signal Transduct. Target. Ther.* **2024**, *9*, 142. [CrossRef] [PubMed]
13. Shahvali, S.; Rahiman, N.; Jaafari, M.R.; Arabi, L. Targeting fibroblast activation protein (FAP): Advances in CAR-T cell, antibody, and vaccine in cancer immunotherapy. *Drug Deliv. Transl. Res.* **2023**, *13*, 2041–2056. [CrossRef] [PubMed]
14. Bughda, R.; Dimou, P.; D'Souza, R.R.; Klampatsa, A. Fibroblast activation protein (FAP)-targeted CAR-T cells: Launching an attack on tumor stroma. *ImmunoTargets Ther.* **2021**, *10*, 313–323. [CrossRef]
15. Ray, U.; Pathoulas, C.L.; Thirusangu, P.; Purcell, J.W.; Kannan, N.; Shridhar, V. Exploiting LRRC15 as a novel therapeutic target in cancer. *Cancer Res.* **2022**, *82*, 1675–1681. [CrossRef]
16. Slemmons, K.K.; Mukherjee, S.; Meltzer, P.; Purcell, J.W.; Helman, L.J. LRRC15 antibody-drug conjugates show promise as osteosarcoma therapeutics in preclinical studies. *Pediatr. Blood Cancer* **2021**, *68*, e28771. [CrossRef]
17. Krishnamurty, A.T.; Shyer, J.A.; Thai, M.; Gandham, V.; Buechler, M.B.; Yang, Y.A.; Pradhan, R.N.; Wang, A.W.; Sanchez, P.L. Qu, Y.; et al. LRRC15$^+$ myofibroblasts dictate the stromal setpoint to suppress tumour immunity. *Nature* **2022**, *611*, 148–154. [CrossRef]
18. Kotla, N.G.; Bonam, S.R.; Rasala, S.; Wankar, J.; Bohara, R.A.; Bayry, J.; Rochev, Y.; Pandit, A. Recent advances and prospects of hyaluronan as a multifunctional therapeutic system. *J. Control. Release* **2021**, *336*, 598–620. [CrossRef]
19. Misra, S.; Hascall, V.C.; Markwald, R.R.; Ghatak, S. Interactions between hyaluronan and its receptors (CD44, RHAMM) regulate the activities of inflammation and cancer. *Front. Immunol.* **2015**, *6*, 201. [CrossRef]
20. Michalczyk, M.; Humeniuk, E.; Adamczuk, G.; Korga-Plewko, A. Hyaluronic acid as a modern approach in anticancer therapy-review. *Int. J. Mol. Sci.* **2022**, *24*, 103. [CrossRef]
21. Bhattacharyya, M.; Jariyal, H.; Srivastava, A. Hyaluronic acid: More than a carrier, having an overpowering extracellular and intracellular impact on cancer. *Carbohydr. Polym.* **2023**, *317*, 121081. [CrossRef]
22. Weber, G.C.; Buhren, B.A.; Schrumpf, H.; Wohlrab, J.; Gerber, P.A. Clinical applications of hyaluronidase. *Adv. Exp. Med. Biol.* **2019**, *1148*, 255–277. [CrossRef]
23. Locke, K.W.; Maneval, D.C.; LaBarre, M.J. ENHANZE® Drug Delivery Technology: A novel approach to subcutaneous administration using recombinant human hyaluronidase PH20. *Drug Deliv.* **2019**, *26*, 98–106. [CrossRef]
24. Wong, K.M.; Horton, K.J.; Coveler, A.L.; Hingorani, S.R.; Harris, W.P. Targeting the tumor stroma: The biology and clinical development of pegylated recombinant human hyaluronidase (PEGPH20). *Curr. Oncol. Rep.* **2017**, *19*, 47. [CrossRef] [PubMed]
25. Shepard, H.M. Breaching the castle walls: Hyaluronan depletion as a therapeutic approach to cancer therapy. *Front. Oncol.* **2015**, *5*, 192. [CrossRef] [PubMed]

Morosi, L.; Meroni, M.; Ubezio, P.; Fuso Nerini, I.; Minoli, L.; Porcu, L.; Panini, N.; Colombo, M.; Blouw, B.; Kang, D.W.; et al. Pegylated recombinant human hyaluronidase (PEGPH20) pre-treatment improves intra-tumour distribution and efficacy of paclitaxel in preclinical models. *J. Exp. Clin. Cancer Res.* **2021**, *40*, 286. [CrossRef] [PubMed]

Hingorani, S.R.; Harris, W.P.; Beck, J.T.; Berdov, B.A.; Wagner, S.A.; Pshevlotsky, E.M.; Tjulandin, S.A.; Gladkov, O.A.; Holcombe, R.F.; Korn, R.; et al. Phase Ib study of pegylated recombinant human hyaluronidase and gemcitabine in patients with advanced pancreatic cancer. *Clin. Cancer Res.* **2016**, *22*, 2848–2854. [CrossRef]

Van Cutsem, E.; Tempero, M.A.; Sigal, D.; Oh, D.Y.; Fazio, N.; Macarulla, T.; Hitre, E.; Hammel, P.; Hendifar, A.E.; Bates, S.E.; et al. Randomized phase III trial of pegvorhyaluronidase alfa with nab-paclitaxel plus gemcitabine for patients with hyaluronan-high metastatic pancreatic adenocarcinoma. *J. Clin. Oncol.* **2020**, *38*, 3185–3194. [CrossRef]

Infante, J.R.; Korn, R.L.; Rosen, L.S.; LoRusso, P.; Dychter, S.S.; Zhu, J.; Maneval, D.C.; Jiang, P.; Shepard, H.M.; Frost, G.; et al. Phase 1 trials of pegylated recombinant human hyaluronidase PH20 in patients with advanced solid tumours. *Br. J. Cancer* **2018**, *118*, 153–161. [CrossRef]

Hakim, N.; Patel, R.; Devoe, C.; Saif, M.W. Why HALO 301 failed and implications for treatment of pancreatic cancer. *Pancreas* **2019**, *3*, e1–e4. [CrossRef]

Xin, L.; Gao, J.; Zheng, Z.; Chen, Y.; Lv, S.; Zhao, Z.; Yu, C.; Yang, X.; Zhang, R. Fibroblast activation protein-α as a target in the bench-to-bedside diagnosis and treatment of tumors: A narrative review. *Front. Oncol.* **2021**, *11*, 648187. [CrossRef] [PubMed]

Kabat, E.A.; Wu, T.T. Attempts to locate complementarity-determining residues in the variable positions of light and heavy chains. *Ann. N. Y. Acad. Sci.* **1971**, *190*, 382–393. [CrossRef] [PubMed]

Labrijn, A.F.; Meesters, J.I.; Priem, P.; de Jong, R.N.; van den Bremer, E.T.J.; van Kampen, M.D.; Gerritsen, A.F.; Schuurman, J.; Parren, P.W.H.I. Controlled Fab-arm exchange for the generation of stable bispecific IgG1. *Nat. Protoc.* **2014**, *9*, 2450–2463. [CrossRef]

Dorfman, A.; Ott, M.L. A turbidimetric method for the assay of hyaluronidase. *J. Biol. Chem.* **1948**, *172*, 367–375. [CrossRef]

Reale, E.; Luciano, L.; Spitznas, M. Histochemical demonstration of hyaluronic acid molecules by alcian blue. *Histochem. J.* **1986**, *18*, 306–316. [CrossRef]

Xiong, X.; Xi, J.; Liu, Q.; Wang, C.; Jiang, Z.; Yue, S.Y.; Shi, L.; Rong, Y. Co-expression of IL-7 and PH20 promote anti-GPC3 CAR-T tumour suppressor activity in vivo and in vitro. *Liver Int.* **2021**, *41*, 1033–1043. [CrossRef]

Zhao, R.; Cui, Y.; Zheng, Y.; Li, S.; Lv, J.; Wu, Q.; Long, Y.; Wang, S.; Yao, Y.; Wei, W.; et al. Human hyaluronidase PH20 potentiates the antitumor activities of mesothelin-specific CAR-T cells against gastric cancer. *Front. Immunol.* **2021**, *12*, 660488. [CrossRef]

Jiang, P.; Li, X.; Thompson, C.B.; Huang, Z.; Araiza, F.; Osgood, R.; Wei, G.; Feldmann, M.; Frost, G.I.; Shepard, H.M. Effective targeting of the tumor microenvironment for cancer therapy. *Anticancer Res.* **2012**, *32*, 1203–1212.

Schofield, B.H.; Williams, B.R.; Doty, S.B. Alcian Blue staining of cartilage for electron microscopy. Application of the critical electrolyte concentation principle. *Histochem. J.* **1975**, *7*, 139–149. [CrossRef]

Kultti, A.; Li, X.; Jiang, P.; Thompson, C.B.; Frost, G.I.; Shepard, H.M. Therapeutic targeting of hyaluronan in the tumor stroma. *Cancers* **2012**, *4*, 873–903. [CrossRef]

Thompson, C.B.; Shepard, H.M.; O'Connor, P.M.; Kadhim, S.; Jiang, P.; Osgood, R.J.; Bookbinder, L.H.; Li, X.; Sugarman, B.J.; Connor, R.J.; et al. Enzymatic depletion of tumor hyaluronan induces antitumor responses in preclinical animal models. *Mol. Cancer Ther.* **2010**, *9*, 3052–3064. [CrossRef]

Wilkinson, I.; Anderson, S.; Fry, J.; Julien, L.A.; Neville, D.; Qureshi, O.; Watts, G.; Hale, G. Fc-engineered antibodies with immune effector functions completely abolished. *PLoS ONE* **2021**, *16*, e0260954. [CrossRef] [PubMed]

Chiu, M. Abstract 4494: Multi-specific cMet × EGFR × VEGF antibody for difficult to treat cancers. *Cancer Res.* **2023**, *83* (Suppl. S7), 4494. [CrossRef]

Jin, Y.; Chen, P.; Zhou, H.; Mu, G.; Wu, S.; Zha, Z.; Ma, B.; Han, C.; Chiu, M.L. Developing transcriptomic biomarkers for TAVO412 utilizing next generation sequencing analyses of preclinical tumor models. *Front. Immunol.* **2025**, *16*, 1505868. [CrossRef] [PubMed]

Jin, Y.; Sun, P.; Chen, P.; Xu, Y.; Mu, G.; Zha, Z.; Wu, S.; Fu, M.; Jiang, H.; Huang, S.; et al. A trispecific antibody targeting EGFR/cMET/VEGF-A demonstrates multiple mechanisms of action to inhibit wild-type and mutant NSCLC animal models. *Front. Oncol.* **2025**, *15*, 1533059. [CrossRef]

Cho, Y.-H.; Ro, E.J.; Yoon, J.-S.; Mizutani, T.; Kang, D.-W.; Park, J.-C.; Il Kim, T.; Clevers, H.; Choi, K.-Y. 5-FU promotes stemness of colorectal cancer via p53-mediated WNT/β-catenin pathway activation. *Nat. Commun.* **2020**, *11*, 5321. [CrossRef]

Chen, X.; Chen, L.-J.; Peng, X.-F.; Deng, L.; Wang, Y.; Li, J.-J.; Guo, D.-L.; Niu, X.-H. Anti-PD-1/PD-L1 therapy for colorectal cancer: Clinical implications and future considerations. *Transl. Oncol.* **2024**, *40*, 101851. [CrossRef]

Wang, S.; Li, Y.; Xu, C.; Dong, J.; Wei, J. An oncolytic vaccinia virus encoding hyaluronidase reshapes the extracellular matrix to enhance cancer chemotherapy and immunotherapy. *J. Immunother. Cancer* **2024**, *12*, e008431. [CrossRef]

Li, M.; Li, S.; Zhao, R.; Lv, J.; Zheng, D.; Qin, L.; Li, S.; Wu, Q.; Long, Y.; Tang, Z.; et al. CD318 is a target of chimeric antigen receptor T Cells for the treatment of colorectal cancer. *Clin. Exp. Med.* **2022**, *23*, 2409–2419. [CrossRef]

50. Lopez de Rodas, M.; Villalba-Esparza, M.; Sanmamed, M.F.; Chen, L.; Rimm, D.L.; Schalper, K.A. Biological and clinic significance of tumour-infiltrating lymphocytes in the era of immunotherapy: A multidimensional approach. *Nat. Rev. Cl Oncol.* **2025**, *22*, 163–181. [CrossRef]
51. Sanmamed, M.F.; Chester, C.; Melero, I.; Kohrt, H. Defining the optimal murine models to investigate immune checkpoi blockers and their combination with other immunotherapies. *Ann. Oncol.* **2016**, *27*, 1190–1198. [CrossRef] [PubMed]
52. Jin, Y.; An, X.; Mao, B.; Sun, R.; Kumari, R.; Chen, X.; Shan, Y.; Zang, M.; Xu, L.; Muntel, J.; et al. Different syngeneic tumc show distinctive intrinsic tumor-immunity and mechanisms of actions (MOA) of anti-PD-1 treatment. *Sci. Rep.* **2022**, *12*, 32 [CrossRef] [PubMed]
53. Martinez-Ordoñez, A.; Duran, A.; Ruiz-Martinez, M.; Cid-Diaz, T.; Zhang, X.; Han, Q.; Kinoshita, H.; Muta, Y.; Linares, J. Kasashima, H.; et al. Hyaluronan driven by epithelial aPKC deficiency remodels the microenvironment and creates a vulnerabili in mesenchymal colorectal cancer. *Cancer Cell* **2023**, *41*, 252–271.e9. [CrossRef] [PubMed]
54. Aghanejad, A.; Bonab, S.F.; Sepehri, M.; Haghighi, F.S.; Tarighatnia, A.; Kreiter, C.; Nader, N.D.; Tohidkia, M.R. A review targeting tumor microenvironment: The main paradigm shift in the mAb-based immunotherapy of solid tumors. *Int. J. Bi Macromol.* **2022**, *207*, 592–610. [CrossRef]
55. Xiang, X.; Wang, J.; Lu, D.; Xu, X. Targeting tumor-associated macrophages to synergize tumor immunotherapy. *Signal Transdu Target. Ther.* **2021**, *6*, 75. [CrossRef]
56. Hu, L.; Nomura, S.; Sato, Y.; Takagi, K.; Ishii, T.; Honma, Y.; Watanabe, K.; Mizukami, Y.; Muto, J. Anti-inflammatory effects differential molecular weight Hyaluronic acids on UVB-induced calprotectin-mediated keratinocyte inflammation. *J. Dermat Sci.* **2022**, *107*, 24–31. [CrossRef]
57. Chen, C.; Liu, J.; Lin, X.; Xiang, A.; Ye, Q.; Guo, J.; Rui, T.; Xu, J.; Hu, S. Crosstalk between cancer-associated fibroblasts ar regulated cell death in tumors: Insights into apoptosis, autophagy, ferroptosis, and pyroptosis. *Cell Death Discov.* **2024**, *10*, 1 [CrossRef]
58. Kwon, J.Y.; Vera, R.E.; Fernandez-Zapico, M.E. The multi-faceted roles of cancer-associated fibroblasts in pancreatic cancer. *C Signal.* **2025**, *127*, 111584. [CrossRef]
59. Mathieson, L.; Koppensteiner, L.; Dorward, D.A.; O'Connor, R.A.; Akram, A.R. Cancer-associated fibroblasts expressing fibrobla activation protein and podoplanin in Non-small cell lung cancer predict poor clinical outcome. *Br. J. Cancer* **2024**, *130*, 1758–17 [CrossRef]

cancers

MDPI

Review

FGFR Aberrations in Solid Tumors: Mechanistic Insights and Clinical Translation of Targeted Therapies

Zijie He [1,†], Yizhen Chen [1,†], Genglin Li [1], Jintao Wang [1], Yuxin Wang [1], Pengjie Tu [1], Yangyun Huang [1], Lilan Zhao [1], Xiaojie Pan [1], Hengrui Liu [2,*] and Wenshu Chen [1,*]

1 Department of Thoracic Surgery, Shengli Clinical Medical College of Fujian Medical University, Fujian Provincial Hospital, Fuzhou University Affiliated Provincial Hospital, 134 East Street, Fuzhou 350001, China; hezijie0824@fjmu.edu.cn (Z.H.); cyz765488791@126.com (Y.C.); lgl18065003992@fjmu.edu.cn (G.L.); wangjt690@fjmu.edu.cn (J.W.); 3211003371@stu.fjmu.edu.cn (Y.W.); tupengjie@tmu.com (P.T.); thoracichyy@163.com (Y.H.); zhaolilan@fzu.edu.cn (L.Z.); xyr8837924@126.com (X.P.)

2 Department of Oncology, University of Cambridge, Cambridge CB2 1SZ, UK

* Correspondence: doctorcws@fjmu.edu.cn (W.C.); hl546@cam.ac.uk (H.L.)

† These authors contributed equally to this work.

Simple Summary

Cancer treatment is increasingly guided by tumor DNA alterations. Changes in fibroblast growth factor receptors, a family of four related proteins, occur across many solid tumors and drive growth, spread, and treatment response by activating multiple signaling pathways. This review examines how these receptor changes—amplifications, mutations, and fusions—differ in their mechanisms and distribution across cancer types, and how they reshape the tumor environment, blood supply, and immune landscape. We summarize the clinical benefits and limitations of approved targeted drugs, explain why resistance develops through secondary mutations or alternative signaling routes, and describe how blood-based tests may detect resistance early. We also discuss next-generation strategies including combination with immunotherapy or anti-vessel therapy, antibody–drug conjugates, and nanotechnology-based delivery systems. By integrating molecular, translational, and clinical evidence, this review aims to guide patient selection, treatment monitoring, and the design of durable precision therapies targeting these receptor abnormalities.

Abstract

Aberrations in fibroblast growth factor receptors (FGFRs) constitute a key oncogenic mechanism across multiple solid tumors, influencing tumor initiation, therapeutic response, and clinical outcomes. This review synthesizes current knowledge on the molecular biology, signaling networks, and tumor-specific distribution of FGFR alterations, including amplifications, point mutations, and gene fusions. The mechanistic basis of FGFR-driven tumor progression is discussed, including activation of downstream signaling pathways, crosstalk with other receptor tyrosine kinases, and regulation of the tumor microenvironment, angiogenesis, and immune escape. Recent development of selective FGFR inhibitors—such as pemigatinib, erdafitinib, and futibatinib—has translated mechanistic insights into measurable clinical benefits in genomically defined patient populations. However, acquired resistance remains a major challenge, driven by secondary mutations, activation of bypass pathways, and intratumoral heterogeneity. Integration of multi-omics profiling, liquid biopsy monitoring, and biomarker-guided patient selection is essential to optimize therapeutic efficacy and overcome resistance. This review also highlights emerging therapeutic modalities, such as antibody–drug conjugates and nanotechnology-based delivery systems, which may improve target specificity and prolong therapeutic durability. By integrating molecular, translational, and clinical evidence, this review aims to establish a comprehensive framework for precision oncology strategies targeting FGFR-driven malignancies.

Keywords: FGFR aberrations; targeted therapy; receptor tyrosine kinases; oncogenic signaling; therapeutic resistance

1. Introduction

Malignant tumors represent the second leading cause of death worldwide. GLOBOCAN 2022 reported 20.1 million new cancer cases and 9.7 million cancer-related deaths globally in 2022. Projections indicate that new cancer cases will reach 28.4 million by 2040 [1]. Pan-cancer multi-omics studies have systematically identified widespread aberrations in key oncogenic drivers, including *TP53*, *KRAS*, and *PIK3CA* [2]. These discoveries have accelerated the development of precision oncology.

Cancer treatment has shifted from traditional surgery, radiotherapy, and chemotherapy toward comprehensive strategies guided by molecular biomarkers. The fibroblast growth factor receptor (FGFR) family comprises four receptor tyrosine kinases (FGFR1–4) that regulate cell proliferation, differentiation, and angiogenesis. Upon ligand binding, these receptors activate the RAS/MAPK, PI3K/AKT, and PLCγ signaling pathways. This activation drives tumor growth and remodels the tumor microenvironment [3]. Regulatory agencies have approved pemigatinib for cholangiocarcinoma and erdafitinib for urothelial carcinoma. Additionally, futibatinib has demonstrated promising selectivity and tolerability in clinical trials [4–6].

FGFR aberrations exhibit distinct distributions across cancer types. These aberrations include amplifications, mutations, fusions, and aberrant splicing. *FGFR2* fusions occur in approximately 10–16% of intrahepatic cholangiocarcinomas [7,8]. In contrast, *FGFR3* mutations are present in 17–20% of bladder cancers [9,10]. These genetic alterations drive oncogenic phenotypes. They also promote angiogenesis and facilitate immune evasion. Recent single-cell and spatial omics studies have revealed the critical role of FGFR signaling in tumor heterogeneity and clonal evolution [11].

Despite therapeutic advances, FGFR-targeted therapies face significant clinical challenges. Tumors develop acquired resistance through secondary mutations or bypass signaling activation. Patients experience dose-limiting toxicities, including hyperphosphatemia and cutaneous or ocular adverse events. Furthermore, clinicians face uncertainties in detecting FGFR aberrations across diverse cancer types and determining the druggability of rare variants [12–16]. Next-generation inhibitors, such as tinengotinib and rogaratinib, demonstrate improved selectivity and tolerability in preclinical and early clinical studies [17,18]. These agents may enhance clinical outcomes through optimized drug design and biomarker-driven patient selection.

This review systematically examines FGFR aberrations and oncogenic mechanisms in solid tumors, clinical development, and combination strategies, emphasizing resistance biology and biomarker advances. We propose multi-omics-based predictive frameworks to guide precision application and drug development of FGFR-targeted therapies.

2. Molecular Biology and Tumor Distribution of the FGFR Family

2.1. Structural and Functional Features of the FGFR Family

The FGFR family belongs to the receptor tyrosine kinase (RTK) superfamily and plays central roles in cell proliferation, differentiation, and metabolism. It comprises four highly conserved isoforms (FGFR1–4) that initiate complex signaling networks upon binding

 https://doi.org/10.3390/cancers18010089

to specific fibroblast growth factors (FGFs), thereby mediating diverse physiological and pathological processes.

2.1.1. Domain Architecture and Molecular Characteristics of FGFR1–4

FGFR1–4 share a highly conserved three-domain architecture: an extracellular domain, a transmembrane region, and an intracellular kinase domain [19]. The extracellular region contains three immunoglobulin-like domains (D1, D2, and D3). D1 and the acid box form an autoinhibitory module, while D2 and D3 constitute the primary ligand-binding region [20]. This structure maintains receptor quiescence in the resting state, with activation occurring only upon ligand binding and dimerization.

In addition to the classical kinase receptors FGFR1–4, the FGFR family includes an atypical member, fibroblast growth factor receptor-like 1 (FGFRL1, also known as FGFR5) [21]. The extracellular domain of FGFRL1 contains three immunoglobulin-like domains and a conserved FGF-binding site, while its intracellular region is limited to a short cytoplasmic tail, lacking the catalytic tyrosine kinase domain [22]. This structural feature classifies FGFRL1 as a "decoy receptor" or "pseudo-receptor." FGFRL1 binds multiple FGF ligands but does not initiate the classical kinase cascade, thereby modulating local FGF availability, influencing the signaling strength of other FGFRs, and playing a role in tissue morphogenesis and homeostasis [21].

Crystallographic studies reveal that FGFs bind primarily to the D2–D3 region, with heparan sulfate proteoglycans (HSPGs) enhancing binding affinity and promoting dimerization, thereby establishing the molecular basis for FGFR activation [23,24]. The intracellular kinase domain contains an ATP-binding site and multiple tyrosine residues; upon activation, autophosphorylation recruits downstream signaling proteins. Subtle differences in the kinase domain confer distinct substrate specificity and signaling intensity among isoforms [3].

2.1.2. Ligand-Binding Specificity and Signal Transduction Mechanisms

The FGF family comprises 22 members, classified into paracrine and endocrine subfamilies [25]. Paracrine FGFs, such as FGF1 and FGF2, require HSPGs as co-receptors, whereas endocrine FGFs, like FGF19, FGF21, and FGF23, interact specifically with Klotho proteins to modulate FGFR selectivity and activity [26,27]. Klotho proteins act as essential co-receptors that alter FGFR ligand specificity, particularly in tissues such as the liver and kidney. For instance, the FGF19-FGFR4-βKlotho complex enhances binding specificity by stabilizing receptor conformations that preferentially activate FGFR4 signaling in hepatocytes [24,28,29].

Structural studies have revealed that the presence of Klotho proteins influences the affinity of FGFs for FGFR isoforms. For example, FGF19 binding to FGFR4 is greatly enhanced by βKlotho, leading to a higher binding affinity compared to when βKlotho is absent. This alteration in ligand–receptor binding is essential for the regulation of key metabolic pathways such as bile acid synthesis and hepatic glucose metabolism [24,30,31]. On the other hand, Klotho proteins are absent in the binding of FGF1 and FGF2 to FGFRs, which exhibit a broader specificity across multiple FGFR isoforms [32].

Upon activation, FGFR signals predominantly through RAS/MAPK, PI3K/AKT, and PLCγ pathways [33]. FGFR substrate 2 (FRS2) serves as a key adaptor protein, recruiting GRB2 and SOS to initiate the RAS/MAPK pathway for proliferation and differentiation; the PI3K/AKT pathway promotes cell survival and metabolic reprogramming; PLCγ activation modulates intracellular calcium levels and protein kinase C activity [34].

Furthermore, ligand binding to the extracellular Ig-like domains of FGFR leads to receptor dimerization, which is further stabilized in the presence of Klotho proteins. This process is crucial for activating downstream signaling pathways, including RAS/MAPK and PI3K/AKT, which are involved in cell proliferation and survival. The altered specificity

https://doi.org/10.3390/cancers18010089

induced by Klotho proteins provides a mechanism for the tissue-specific regulation of FGFR signaling, which is pivotal for understanding both normal physiology and oncogenic transformations driven by FGFR aberrations [35,36].

2.1.3. Physiological Functions of the FGFR Family

FGFR signaling is indispensable in both embryonic development and adult homeostasis. During embryogenesis, FGFR1–3 participate in neural, skeletal, and organ formation: FGFR1 is critical for central nervous system and skeletal development, FGFR2 mediates epithelial–mesenchymal interactions, and FGFR3 regulates cartilage and bone growth [37,38].

Beyond the classical FGFRs, genetic and experimental evidence highlights a key role for FGFRL1 in organogenesis, particularly in embryonic kidney and diaphragm development. *Fgfrl1* knockout mice exhibit severe metanephric hypoplasia, often progressing to near-complete renal agenesis, accompanied by a thin, hypoplastic diaphragm that allows for liver herniation, resulting in respiratory failure and perinatal lethality [39,40]. In humans, germline loss-of-function or deleterious *FGFRL1* variants—arising from 4p16.3 terminal deletions or biallelic missense mutations—have been linked to congenital diaphragmatic hernia, though consistent extra-diaphragmatic features, such as cardiac or skeletal anomalies, remain sparsely documented [41]. Mechanistically, FGFRL1 interacts with FGF8 to modulate FGF–FGFR signaling during nephrogenesis, explaining the renal agenesis in null mice, while similar pathways in diaphragmatic myogenesis await elucidation. High FGFRL1 expression in musculoskeletal and renal tissues from embryonic to postnatal stages suggests potential contributions to adult tissue homeostasis and repair, a notion meriting further exploration via single-cell and spatial transcriptomics [39–41].

In adult tissues, FGFR pathways maintain homeostasis and participate in repair processes. For example, FGF2–FGFR1 promotes endothelial cell proliferation and migration to regulate angiogenesis; FGF7 stimulates epithelial cell proliferation and differentiation via FGFR2-IIIb, contributing to skin and lung repair [21,33]. Additionally, endocrine FGFs achieve tissue-specific functions in cooperation with Klotho family co-receptors: the FGF19–β-Klotho–FGFR4 axis regulates bile acid and hepatic glucose metabolism; FGF21 modulates lipid metabolism and insulin sensitivity; the FGF23–α-Klotho–FGFR1 axis maintains phosphate and vitamin D homeostasis. These findings suggest FGFR as a potential therapeutic target for metabolic disorders [26,27].

2.1.4. FGF-FGFR Interaction Networks and Regulatory Mechanisms

The complexity of the FGF–FGFR network arises not only from ligand–receptor binding diversity but also from multilayered regulation. HSPGs, serving as co-receptors for paracrine FGFs, enhance binding affinity and promote dimerization, thereby establishing local concentration gradients and shaping spatial signal distribution [23]. Endocrine FGFs depend on Klotho proteins to alter FGFR ligand selectivity, enabling precise regulation in specific tissues [42].

Beyond canonical ligand–receptor interactions, FGFRL1 acts as a molecular rheostat of FGF signaling. As a decoy receptor, FGFRL1 sequesters soluble FGF ligands and constrains the signaling range of select FGFs, thereby restricting signal amplitude and spatial spread. Under oncogenic conditions, dysregulated FGFRL1 expression or altered subcellular localization can reprogram FGF–FGFR signaling networks through context-dependent modulation of effector pathways such as PI3K–Akt and ERK1/2, potentially contributing to therapeutic resistance [43]. Thus, integrating FGFRL1 into comprehensive models of the FGF–FGFR network is imperative for understanding the dynamic regulation and plasticity of this signaling system.

 https://doi.org/10.3390/cancers1801008

Negative regulation is critical for maintaining spatiotemporal specificity: Sprouty (SPRY) proteins inhibit the GRB2/SOS–RAS axis to limit MAPK activation; SEF binds to receptor complexes to suppress kinase activity or promote endocytosis; dual-specificity phosphatases (DUSPs) terminate signaling by dephosphorylating MAPKs [44,45]. Furthermore, structural studies reveal that in the resting state, D1-acid box interactions maintain receptor autoinhibition; ligand binding induces receptor dimerization and conformational changes that relieve inhibition, thereby initiating downstream signaling [23,34].

2.2. Distribution Patterns of FGFR Aberrations in Solid Tumors

As a key receptor tyrosine kinase, FGFR exhibits structural and functional aberrations across multiple solid tumors. Large-scale genomic studies (e.g., TCGA, ICGC) have provided systematic evidence characterizing the frequency, types, and associations of FGFR aberrations with molecular subtypes, clinical features, and population differences.

2.2.1. Statistical Analysis of FGFR Aberrations in Solid Tumors Based on Large Databases

In a sequencing study of 4853 solid tumors, the overall prevalence of FGFR aberrations was 7.1%, comprising 66% amplifications, 26% mutations, and 8% fusions [46]. This provides a reference for understanding the overall landscape of FGFR aberrations.

Notably, reported pan-cancer prevalence estimates are strongly influenced by assay design and analytical sensitivity. DNA-based hybrid-capture panels and whole-exome sequencing often under-detect FGFR rearrangements compared with RNA-based NGS or anchored multiplex PCR, which provide higher sensitivity for in-frame and rare-partner fusions [47]. In contrast, assays with higher copy-number resolution—such as high-depth targeted NGS or array-CGH—tend to detect more low-level FGFR amplifications [48,49]. Conversely, cfDNA-based NGS may miss subclonal or low-allele-fraction FGFR events because detection relies on tumor DNA fraction and shedding dynamics [50,51].

Pan-cancer analyses further reveal significant heterogeneity in FGFR isoform distribution across cancer types [52]. A combined TCGA and ICGC whole-genome study demonstrated that each cancer harbors an average of 4–5 driver mutations, with *FGFR* aberrations representing an important component that appears across multiple cancer types, though their oncogenic contribution varies by tumor type [53,54]. Additionally, a large-scale study of Chinese patient cohorts found an overall FGFR aberration rate of 7.0%, consistent with international data, while showing certain ethnic differences in cancer type distribution [55].

2.2.2. Frequency and Type Distribution of FGFR Aberrations Across Cancer Types

FGFR aberrations exhibit marked tissue specificity and molecular preferences across tumor types. In urothelial carcinoma, *FGFR3* mutations are most prevalent, with an overall frequency of approximately 15–20%; detection rates reach 40–50% in non-muscle-invasive bladder cancer but decline to 10–15% in muscle-invasive or metastatic disease [56–58]. In intrahepatic cholangiocarcinoma (iCCA), *FGFR2* fusions are most common, occurring in approximately 10–16% of cases and correlating with distinct clinicopathological features [52,59]. Breast cancer harbors *FGFR* aberrations in 14–18% of cases, predominantly *FGFR1* amplifications [46,55]. Endometrial cancer frequently exhibits *FGFR2* mutations (e.g., S252W, N549K) in approximately 10–12% of cases, with established oncogenic activity [54,60]. While FGFR aberrations are generally rare in lung cancer, *FGFR1* amplification occurs in 15–20% of lung squamous cell carcinomas [46,61]. In glioblastoma, the overall prevalence of FGFR aberrations is approximately 2–8%, with *FGFR3* fusions being relatively common [62,63]. Additionally, certain rare tumors such as adenoid cystic carcinoma and ovarian malignant Brenner tumor display elevated proportions of FGFR fusions, typically associated with sensitivity to FGFR inhibitors [64].

https://doi.org/10.3390/cancers18010089

In contrast to FGFR1–4, the role of FGFRL1 in cancer research remains underexplored. However, large-scale cancer genomics analyses together with functional studies in several tumor types have suggested its potential importance in solid tumors. On the one hand, transcriptomic and clinicopathological data indicate that FGFRL1 expression in prostate cancer, esophageal cancer, and small-cell lung cancer is associated with tumor progression and, in some cohorts, with patient outcome, suggesting biological roles that are at least partly distinct from those of classical FGFR1–4 [65–67]. On the other hand, in esophageal cancer, increased FGFRL1 expression correlates with disease progression and is directly repressed by miR-107, whereas in small-cell lung cancer FGFRL1 promotes chemoresistance via an ENO1–PI3K/Akt axis, and in pancreatic cancer models exosomal miR-210 from cancer stem cells targets FGFRL1 to drive macrophage M2 polarization and gemcitabine resistance [68,69]. These findings suggest significant variation in FGFR signaling dependency across tumors of different tissue origins.

2.2.3. Association of FGFR Aberrations with Tumor Molecular Subtypes

FGFR aberrations display pronounced molecular subtype specificity across tumor types. In bladder cancer, *FGFR3* mutations predominantly occur in low-grade, non-muscle-invasive subtypes with papillary growth patterns and well-differentiated histology, suggesting they may define a distinct molecular subtype. Typical hotspot mutations (S249C, R248C, Y373C) involve extracellular cysteine substitutions that alter receptor conformation and confer ligand-independent activation [70–72]. In cholangiocarcinoma, *FGFR2* fusions and truncation variants (e.g., *FGFR2ΔE18*) represent mechanistically distinct driver events. *FGFR2* fusions typically act as solitary drivers activating RAS/MAPK and PI3K/AKT pathways to induce tumorigenesis, while *FGFR2ΔE18* causes receptor conformational stabilization and ligand-independent activation through structural deletion, indicating significant differences in signaling dependency and drug sensitivity [73,74]. In breast cancer, *FGFR1* amplification is enriched in hormone receptor-positive (ER+) subtypes, particularly Luminal B-like (high-proliferation HR+) tumors, and closely associates with endocrine therapy resistance and poor prognosis [75,76]. A large-scale functional study identified 71 newly confirmed oncogenic variants among 160 FGFR mutations; *FGFR2* mutations in endometrial cancer significantly correlate with molecular subtypes, showing markedly elevated frequency in microsatellite instability-high (MSI-H) tumors [77]. Table 1 summarizes the distribution patterns, frequency ranges, and molecular characteristics of FGFR aberrations in major solid tumors.

Table 1. FGFR Alterations Across Major Solid Tumor Types: Frequency, Predominant Alteration Type, and Molecular Characteristics.

Cancer Type	FGFR Alteration Frequency (%)	Predominant FGFR Alteration Type	Common Subtypes/Mutation Sites or Fusion Partners
Intrahepatic cholangiocarcinoma	10–16	FGFR2 fusion	BICC1, AHCYL1, PPHLN1
Bladder cancer/ Urothelial carcinoma	15–20	FGFR3 mutation	S249C, R248C, Y373C
Breast cancer	14–18	FGFR1 amplification	ER(+)
Endometrial cancer	10–12	FGFR2 mutation	S252W, N549K
Lung squamous cell carcinoma	15–20	FGFR1 amplification	NA
Glioblastoma	4–8	FGFR3 fusion	TACC3

Abbreviations: FGFR, fibroblast growth factor receptor; ER, estrogen receptor; NA, not applicable.

https://doi.org/10.3390/cancers18010089

2.2.4. Impact of Racial, Geographic, and Age Differences on FGFR Aberration Distribution

FGFR alterations also vary across populations. In a Chinese cohort study (n = 58), the frequency of FGFR alterations in colorectal cancer was 31%, significantly higher than in Western populations; in gastric cancer, the frequency was 16.8%, likewise indicating geographic variation. Regarding subtype distribution, *FGFR1* alterations were most common among Chinese patients (56.8%), followed by *FGFR3* (17.7%) and *FGFR2* (14.4%), differing from Western populations [55]. In Indian bladder cancer patients, the *FGFR3* mutation rate is 19%, comparable to the 15–50% range in European and American populations, but with differences in clinicopathological correlations [78]. Age also influences FGFR aberration distribution. In childhood and adolescent cancers, mutations, fusions, and amplifications of *FGFR1–4* are all relatively common, particularly in low-grade neuroepithelial tumors where they are significantly more prevalent than in adults, suggesting developmental stage-related differences in oncogenic mechanisms [79].

Taken together, available sequencing datasets indicate that the overall prevalence of FGFR alterations is broadly similar between Chinese and predominantly European/North American cohorts. By contrast, the relative contribution of individual tumor types differs by ethnicity and geography, with Chinese pan-cancer series enriched for FGFR-altered gastric and colorectal cancers, whereas Western datasets are more heavily weighted toward urothelial, breast, and endometrial tumors. Robust prevalence data from patients of African, South Asian, or Latin American ancestry remain scarce, underscoring the need for more ethnically diverse, globally representative genomic studies.

2.2.5. Aberrant Alternative Splicing of FGFR1–3 in Solid Tumors

Alternative splicing represents an important yet underappreciated source of functional diversity within the FGFR family. For FGFR1–3, the best-characterized splicing events occur in the third immunoglobulin-like domain (IgIII), giving rise to the epithelial-associated IIIb and mesenchymal-associated IIIc isoforms, which differ in ligand specificity and downstream signaling output. This splicing program is tightly regulated during development but is frequently dysregulated in cancer, contributing to altered tumor–stromal interactions and oncogenic signaling plasticity [80].

In FGFR1, expression of truncated splice variants such as FGFR1β, which lacks the first Ig-like domain, enhances ligand binding affinity and mitogenic signaling and has been associated with aggressive behavior in breast cancer and other solid tumors [81]. In FGFR2, a shift from the IIIb to the IIIc isoform is recurrently observed during carcinoma progression and has been mechanistically linked to epithelial–mesenchymal transition and increased invasiveness, partly through aberrant regulation by splicing factors such as PTBP1 [82,83]. FGFR3 also exhibits cancer-associated splice variants beyond the canonical IIIb/IIIc forms, including variants that alter C-terminal regulation of kinase signaling, leading to enhanced oncogenic activity and altered therapeutic sensitivity in prostate cancer models [84].

Collectively, aberrant alternative splicing of FGFR1–3 adds an additional layer of complexity to FGFR dysregulation in solid tumors, with implications for tumor biology, biomarker interpretation, and therapeutic targeting.

3. Molecular Mechanisms and Pathophysiological Roles of FGFR Signaling Pathways

3.1. FGFR Downstream Signaling Network

As key members of the receptor tyrosine kinase family, FGFRs activate multiple signaling cascades through intracellular kinase domain autophosphorylation upon ligand binding, thereby regulating core biological processes including cell proliferation, differentiation, survival, and migration.

https://doi.org/10.3390/cancers18010089

3.1.1. Activation Mechanisms of Classical Downstream Pathways: MAPK/ERK, PI3K/AKT, and PLCγ

FGFRs primarily function through three classical pathways: RAS-MAPK/ERK, PI3K-AKT, and PLCγ-PKC. The coordinated activation of these pathways constitutes the core signaling network [3]. These three signaling axes are spatially and temporally interconnected, collectively determining cell fate transitions and response patterns (Figure 1).

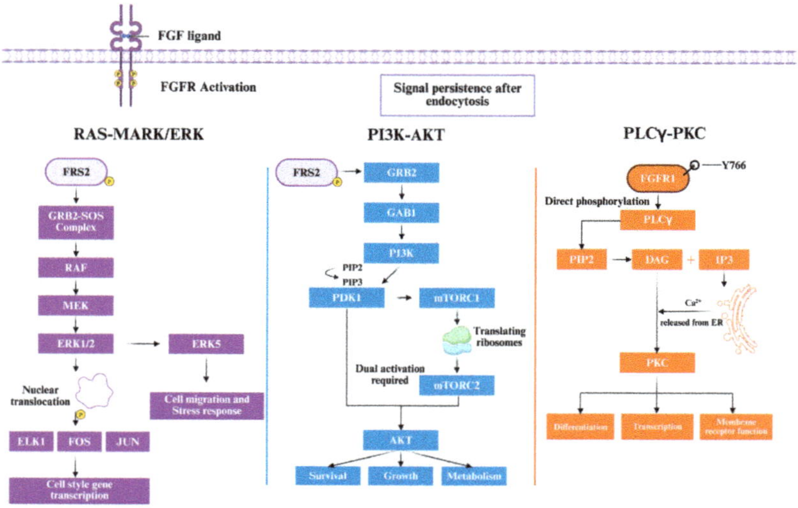

Figure 1. Major Signaling Cascades Downstream of FGFR Activation. A schematic representation of three major signaling cascades activated by FGFR: the RAS-MAPK/ERK pathway (purple) regulating transcription and mediating cell migration and stress responses, the PI3K-AKT pathway (blue) controlling cell survival, growth, and metabolism, and the PLCγ-PKC pathway (orange) mediating differentiation, transcription, and membrane receptor function. Detailed pathway descriptions and abbreviations are provided in the main text.

RAS-MAPK/ERK Pathway Activation Mechanism

Upon FGFR activation, the receptor substrate FRS2 is rapidly phosphorylated and recruits the GRB2-SOS complex, triggering the RAS-RAF-MEK-ERK cascade. Activated ERK1/2 translocates to the nucleus and phosphorylates transcription factors including ELK1, FOS, and JUN, driving transcription of cell cycle-related genes [85]. Additionally, the ERK5 (*MAPK7*) branch can also be activated by FGFR signaling, mediating cell migration and stress responses [86]. Recent studies reveal that the FGFR-FRS2-GRB2 complex maintains signaling persistence after receptor endocytosis, indicating spatiotemporal-specific regulation [87].

PI3K-AKT Pathway Activation Mechanism

PI3K-AKT pathway activation also depends on FRS2. Phosphorylated FRS2 recruits GAB1 (GRB2-associated binder 1), providing binding sites for PI3K. Activated PI3K generates PIP3, which further recruits and activates AKT and PDK1 [88]. Full AKT activation requires both PDK1 and mTORC2. Activated AKT regulates cell survival (BAD, FoxO3a, p53), growth (TSC2-mTOR pathway), metabolism (GSK3β), and cell cycle through phosphorylation of multiple substrates, playing critical roles in tumor cell survival, proliferation, and metabolic reprogramming [89].

https://doi.org/10.3390/cancers18010089

PLCγ-PKC Pathway Activation Mechanism

PLCγ pathway activation is relatively unique. PLCγ1 can directly bind to and be phosphorylated by specific sites on the FGFR activation loop, subsequently hydrolyzing PIP2 to generate IP3 and DAG. In FGFR1, this direct interaction is primarily mediated by autophosphorylation of the juxtamembrane tyrosine residue Y766, which serves as a canonical docking site for the SH2 domains of PLCγ1 and is essential for its recruitment and activation. IP3 promotes calcium release from the endoplasmic reticulum, elevating intracellular Ca^{2+}; DAG and Ca^{2+} jointly activate PKC [19]. PKC further regulates cell differentiation, transcription, and membrane receptor function. In tumors, aberrant PKC activation is closely associated with invasion, angiogenesis, and drug resistance [90].

Overall, these three classical pathways are both independent and tightly interconnected. FGFR signaling maintains cellular homeostasis through spatial restriction, negative feedback, and pathway crosstalk. Its aberrant sustained activation not only promotes tumor cell proliferation and anti-apoptosis but also provides a molecular basis for drug resistance development.

3.1.2. Crosstalk Between FGFR Signaling and Other Growth Factor Receptor Pathways

FGFR pathways exhibit extensive crosstalk with multiple RTKs, playing important roles in tumorigenesis and drug resistance. FGFR-EGFR interactions primarily depend on shared downstream adaptors (GRB2, GAB1) and common signaling nodes (RAS-MAPK, PI3K-AKT), generating synergistic oncogenic effects and mutual compensation in certain tumors, thereby affecting sensitivity to single-agent targeted therapy [91]. In non-small-cell lung cancer, co-occurrence of *FGFR1* amplification with *EGFR* alterations (amplification or mutations) has been observed in individual cases and small cohorts. The associated signaling is mediated through GRB2/GAB1 and may lead to downstream hyperactivation, though this remains a rare phenomenon overall, requiring larger studies to validate its clinical significance [92,93].

Mechanistic studies suggest that FGFRs can form complexes with EGFR and other RTKs at the membrane or in endosomes, or undergo transactivation, thereby altering signal output and promoting drug resistance. For example, studies report that FGFR4 can enhance EGFR oncogenic signaling and affect responses to EGFR inhibitors, suggesting potential targetable interaction mechanisms. However, the frequency and prevalence of such direct interactions remain unclear [94].

FGFR-MET interactions are also observed in triple-negative breast cancer, gastric cancer, and glioblastoma. Their co-expression or complementary activation is often associated with enhanced invasiveness and poor prognosis, with mechanisms frequently involving synergistic activation of the PI3K-AKT-mTOR axis, suggesting potential benefit from dual-target inhibition in selected patients [91,95].

In hormone receptor-positive breast cancer, FGFRs (especially FGFR2) exhibit functional interactions with estrogen receptor (ER): FGFR signaling promotes Erα phosphorylation and transcriptional activation via PI3K-AKT, and synergizes with EGFR/IGF-1R, thereby accelerating acquired endocrine resistance. This evidence indicates that comprehensive characterization of FGFR interactions with RTK/nuclear receptor networks holds significant value for optimizing combination or sequential therapy strategies and overcoming resistance [96].

3.1.3. Spatiotemporal Dynamics and Feedback Mechanisms of Signal Transduction

Precise transmission of FGFR signals depends on complex negative feedback mechanisms and spatiotemporal regulatory systems to ensure effective control of signal specificity, intensity, and duration.

https://doi.org/10.3390/cancers18010089

Multilevel Regulation by Negative Feedback Molecules

At the negative feedback level, the SPRY protein family serves as important inducible inhibitors of FGFR-mediated MAPK signaling. SPRY1–4 can be induced and translocate to the plasma membrane upon receptor stimulation, competitively blocking GRB2-SOS complex recruitment by binding GRB2, thereby inhibiting RAS activation. Additionally, SPRY can interfere at multiple levels of the MAPK axis, forming dual upstream/midstream inhibition to limit signal amplification. *DUSP6* (MKP-3), an ERK-specific dual-specificity phosphatase, rapidly terminates MAPK signaling by dephosphorylating activated ERK1/2. Its expression is positively regulated by ERK, forming a classical negative feedback loop that enables timely signal termination following rapid initiation [97–99]. Similar expression to FGF (SEF) inhibits FGFR activity through interaction with the receptor or receptor complexes and can competitively bind FRS2α to block downstream signal transmission. SEF can also alter signal output characteristics in different subcellular compartments by affecting receptor localization [44,100].

Precise Regulation of Receptor Endocytosis and Degradation

Receptor endocytosis and degradation represent another critical mechanism controlling signal intensity and duration. Ligand-activated FGFRs undergo clathrin-mediated endocytosis into endosomes, where they can either recycle to the plasma membrane for continued signaling or traffic to lysosomes for degradation. Receptor ubiquitination is mediated by specific E3 ubiquitin ligases (such as NEDD4/CBL), determining lysosomal sorting and affecting signal duration. Experimental evidence indicates that FGFR1 ubiquitination is not required for endocytosis but is essential for lysosomal trafficking and degradation; defective ubiquitination can increase receptor recycling and prolong signal transduction [101,102].

Spatiotemporal-Specific Signal Regulation Mechanisms

Regarding spatiotemporal dynamics, FGFR signaling exhibits significant spatial and temporal specificity: spatially, plasma membrane-localized receptor complexes are typically associated with rapid MAPK pathway activation, while endosomal receptor complexes more readily trigger metabolic pathways such as PI3K-AKT in certain cell types; temporally, FGFR-mediated signaling typically shows early MAPK dominance, followed by dynamic succession involving negative feedback inhibition and participation of other pathways [87,103]. This multilevel spatiotemporal regulation ensures orderly coordination of different biological processes according to temporal and spatial sequences.

3.1.4. Molecular Mechanisms of Aberrant Pathway Activation in Tumorigenesis and Progression

Aberrant activation of FGFR signaling pathways is a critical driver in various malignancies, commonly achieved through gene amplification, activating point mutations, and chromosomal translocations/fusions, leading to sustained downstream signaling that promotes tumor proliferation, survival, invasion, and therapeutic resistance [52].

Pathway Hyperactivation by Gene Amplification

Gene amplification causes receptor overexpression and amplifies signal intensity. *FGFR1* amplification is relatively enriched in lung squamous cell carcinoma, with multiple reports indicating an incidence of approximately 15–20%. Amplified FGFR1 synergistically drives tumorigenesis and progression through sustained activation of RAS-MAPK (upregulating cell cycle regulators such as *CCND1/CCNE1*), PI3K-AKT-mTOR (promoting metabolic reprogramming and survival), and PLCγ-PKC (affecting cytoskeleton and migration) pathways [93,104,105]. FGFR2 amplification occurs at low overall frequencies in gastric cancer but is relatively prominent in certain diffuse-type or specific cohorts. It promotes epithelial–

https://doi.org/10.3390/cancers18010089

mesenchymal transition, invasion, and tumor growth by activating PI3K-AKT and MAPK pathways to upregulate epithelial–mesenchymal transition (EMT) transcription factors (such as Snail/Slug) and induce angiogenic factors (such as VEGF) [106].

Oncogenic Mechanisms of Activating Mutations

FGFR3 mutations play important oncogenic roles in bladder cancer. Common mutations such as S249C, R248C, and Y373C, located in the receptor dimerization domain, cause constitutive receptor activation in the absence of ligand [107]. These mutations promote tumorigenesis through multiple mechanisms: sustained MAPK-ERK pathway activation phosphorylates and activates transcription factors ELK1 and AP-1, upregulating proliferation-related genes; meanwhile, aberrant PI3K-AKT pathway activation inhibits apoptosis by phosphorylating BAD and FoxO3a, enhancing tumor cell survival [108,109]. Activating *FGFR2* mutations commonly found in endometrial cancer (such as S252W, N549K) similarly enhance receptor kinase activity, thereby activating the PI3K-AKT-mTOR pathway to promote cell proliferation and metabolic reprogramming, while regulating cell cycle-related gene expression through the MAPK pathway [110].

Aberrant Activation Mechanisms of Fusion Genes

FGFR2 fusions occur in approximately 10–16% of cholangiocarcinomas, with common fusion partners including *BICC1*, *AHCYL1*, and *PPHLN1* [111,112]. These fusion proteins typically retain the FGFR2 kinase domain but lose the ligand-binding domain and autoinhibitory region, resulting in constitutive kinase activity. Oncogenic mechanisms include sustained MAPK pathway activation upregulating oncogenes such as MYC and CCND1 to promote cell proliferation; PI3K-AKT-mTOR pathway promotion of protein synthesis and lipid metabolism; and STAT3 pathway activation to regulate the tumor microenvironment and immune evasion [113,114].

Pathway Reactivation Associated with Therapeutic Resistance

During FGFR inhibitor therapy, tumor cells can reactivate downstream signaling through multiple mechanisms to acquire resistance. Major mechanisms include the following: ① Secondary kinase domain mutations: In *FGFR2* fusion-positive patients, polyclonal mutations frequently emerge at progression, with hotspot sites including the "molecular brake" (N549/N550) and gatekeeper residue (V564), these alterations significantly reduce inhibitor binding and drive resistance; ② Bypass pathway activation: Compensatory activation of RTKs such as EGFR, MET, and IGF1R can maintain MAPK and PI3K-AKT signaling, thereby circumventing FGFR dependency; ③ Downstream molecular aberrations: Mutations or amplifications in *RAS*, *PI3K*, or *AKT* can directly activate key signaling axes, enabling tumor cells to escape inhibition [12,115].

Pathway Activation Through Tumor Microenvironment Modulation

Aberrant FGFR signaling activation also promotes tumor progression by remodeling the tumor microenvironment. On one hand, FGFR signaling upregulates angiogenic factors such as VEGF and PDGF, promoting neovascularization; on the other hand, FGFR signaling enhances expression of cytokines including TGF-β and IL-6, regulating cancer-associated fibroblast and immune cell functions to establish a pro-tumorigenic microenvironment. Additionally, FGFR pathways can promote extracellular matrix remodeling by upregulating matrix metalloproteinases (MMPs), thereby enhancing tumor invasion and metastatic potential [116–119].

Available phosphoproteomic and immunoblot analyses demonstrate that FGFR amplifications, activating mutations, and oncogenic fusions consistently engage MAPK/ERK and PI3K/AKT signaling relative to FGFR-wild-type controls [87,120,121]. However, quantitative studies across different isoforms, fusion partners, and cellular contexts show sub-

https://doi.org/10.3390/cancers18010089

stantial variability in the magnitude of downstream activation, indicating that no uniform fold-change estimate can be applied across aberration classes or tumor types [122,123].

3.2. Interactions Between FGFR and the Tumor Microenvironment

3.2.1. Regulatory Mechanisms of FGFR Signaling in Tumor Angiogenesis

Angiogenesis is a key feature of the tumor microenvironment, providing nutrients and oxygen for tumor growth and metastasis. FGFR signaling plays a central role in vessel formation and maintenance, with FGF2 being the prototypical pro-angiogenic factor. Upon binding to endothelial cell-surface FGFR1 and its co-receptor HSPG, FGF2 activates the PI3K-AKT and MAPK pathways, driving endothelial cell proliferation, migration, and lumen formation, and promotes basement membrane degradation and cellular infiltration by inducing MMPs [124]. This regulatory mechanism is detailed in (Figure 2). In the tumor microenvironment, FGF/FGFR signaling not only acts directly on endothelial cells but also indirectly promotes angiogenesis by modulating immune and stromal cells. For example, tumor-derived FGF2 enhances the survival and migration of tumor-associated macrophages (TAMs) and skews them toward an immunosuppressive phenotype, thereby establishing a microenvironment conducive to angiogenesis and tumor progression [125].

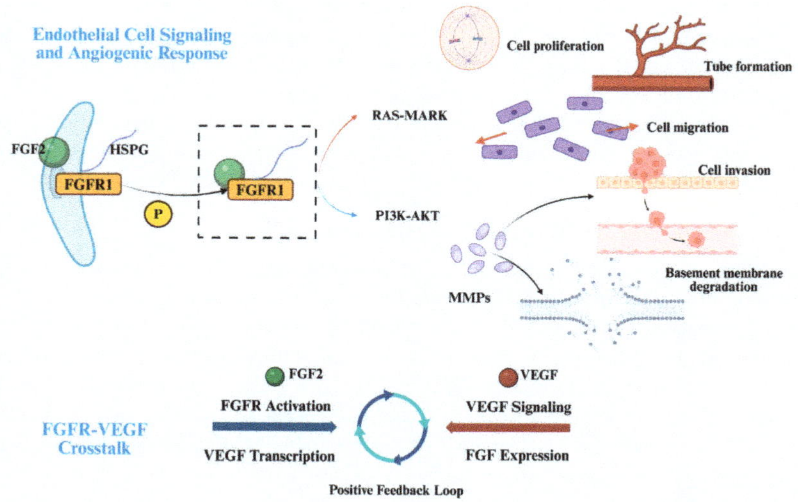

Figure 2. FGFR Signaling Regulates Tumor Angiogenesis Through Endothelial Cell Activation and VEGF Crosstalk. Upper panel: Tumor-derived FGF2 (green sphere) binds to FGFR1 (orange) and HSPG co-receptor (blue glycan chains) on endothelial cells, inducing receptor dimerization and autophosphorylation (marked as P). Activated FGFR1 triggers RAS-MAPK-ERK (red) and PI3K-AKT (blue) pathways, coordinately driving pro-angiogenic effects. Lower panel: FGFR-VEGF positive feedback loop amplifies angiogenic signaling. FGF2 (left, green sphere) activates FGFR to upregulate VEGF transcription and secretion. Secreted VEGF (right, red sphere) activates VEGFR, reciprocally enhancing FGF expression.

FGFR and VEGF signaling pathways exhibit significant complementarity and positive feedback regulation in tumor angiogenesis. FGFR activation can upregulate VEGF transcription and secretion, while VEGF signaling in turn enhances FGF expression, forming a bidirectional regulatory loop that cooperatively drives neovascularization under specific microenvironmental conditions (Figure 2). This crosstalk not only strengthens the angiogenic

https://doi.org/10.3390/cancers1801008

capacity of tumors but is also considered a key mechanism underlying compensatory resistance to anti-VEGF therapy. To disrupt such signaling interplay, multi-target tyrosine kinase inhibitors (TKIs), such as nintedanib, which simultaneously target VEGFR, FGFR, and PDGFR, have been developed to attenuate pathway cross-activation. Translational studies have shown that nintedanib effectively suppresses FGFR/VEGFR-mediated angiogenic signaling, improves tumor vascular structure and function, and enhances responsiveness to chemotherapy and immunotherapy. For instance, in non-small-cell lung cancer (NSCLC) xenograft models, nintedanib concurrently inhibits VEGFR and FGFR activities, significantly reduces tumor angiogenesis, and, when combined with immunotherapy, promotes $CD8^+$ T cell infiltration, thereby augmenting antitumor immune responses [126–128].

It is noteworthy that aberrant FGFR activation not only Increases the number of blood vessels but also induces structural disorganization and elevated vascular permeability, which together promote tumor cell dissemination and affect drug delivery. Based on the concept of "vascular normalization," moderate inhibition of FGFR signaling can help restore vascular architecture and enhance the efficacy of chemotherapy and immunotherapy [129,130]. In addition to FGF2, ligands such as FGF9 also play roles in microcirculatory regulation, influencing metastatic potential and tumor dynamics, further underscoring the multifaceted role of FGFR signaling in angiogenesis [131].

3.2.2. Relationship Between FGFR and Immune Cell Infiltration and Immune Evasion

FGFR signaling not only regulates angiogenesis but also reshapes the immune microenvironment to influence immune infiltration and immune evasion. The underlying mechanisms involve cancer-associated fibroblasts (CAFs), myeloid cells, and cytokine networks, which collectively alter immune cell recruitment and function, ultimately affecting the efficacy of immune checkpoint inhibitors (ICIs) [132].

Different FGFR subtypes exert distinct immunomodulatory effects. In gastric cancer, FGFR1 expression correlates positively with $CD8^+$, $CD4^+$, and dendritic cell infiltration, whereas FGFR4 expression shows a negative correlation with lymphocyte infiltration. Functional studies reveal that FGFR1 overexpression enhances the response to PD-1 blockade, while FGFR4 overexpression weakens it, suggesting that different FGFR subtypes shape immune lineages through distinct mechanisms [133].

Mechanistic studies indicate that FGFR inhibition can remodel the tumor microenvironment to enhance effector T cell infiltration. In triple-negative breast cancer models, FGFR inhibition alters CAF secretory profiles and matrix remodeling, markedly increasing $CD8^+$ T cell infiltration and suppressing tumor growth, supporting the rationale for FGFR-TKI and ICI combination therapy [134]. Additionally, FGFR inhibition reduces infiltration of myeloid-derived suppressor cells (MDSCs), decreases the incidence of distant metastases, and further potentiates antitumor immune responses [135].

A key mechanism involves the IFN-γ $\rightarrow$ JAK/STAT1/IRF1 axis. FGFR activation broadly attenuates IFN-γ-induced transcriptional programs, whereas FGFR inhibition restores them and enhances ICI responsiveness [136]. Tumor-type-specific datasets support this conserved inhibitory effect: in renal cell carcinoma and multiple solid-tumor models, constitutive FGFR signaling diminishes IFN-γ-induced STAT1 phosphorylation and downstream genes such as IRF1, CXCL10, B2M, and PD-L1, while FGFR blockade reinstates this pathway [136]. Similar suppression of interferon-stimulated genes by FGF/FGFR activation has been documented across cellular systems, suggesting a shared biological mechanism rather than a model-restricted phenomenon [137]. In bladder cancer, *FGFR3* activation further intersects with IFN-γ signaling by modulating PD-L1 stability through NEDD4-dependent ubiquitination or by altering IFN-γ-induced PD-L1 turnover in *FGFR3-TACC3* fusion models [138,139]. By contrast, evidence from FGFR2-fusion intrahepatic

https://doi.org/10.3390/cancers18010089

cholangiocarcinoma is limited and sometimes inconsistent, underscoring the need for tumor-specific interpretation of FGFR–interferon crosstalk [140,141].

On the other hand, in hepatocellular carcinoma, FGFR4 inhibition promotes PD-L1 ubiquitin–proteasome degradation via GSK3β activation and suppresses IL-2–induced STAT5 phosphorylation to limit Treg differentiation, thereby reducing PD-L1 levels and immunosuppression and enhancing anti-PD-1 efficacy [142]. These mechanisms are not contradictory but rather context-specific, depending on tumor type, FGFR subtype, and molecular background. This highlights the importance of considering tumor-specific contexts when interpreting FGFR–ICI interactions.

3.2.3. Role of FGFR in Cancer-Associated Fibroblasts and Its Crosstalk with Tumor Cells

CAFs are the most abundant stromal cells in the tumor microenvironment and profoundly influence tumor progression by secreting growth factors, cytokines, and extracellular matrix (ECM) components. FGF/FGFR signaling plays a central role in CAF activation and functional maintenance, forming an interactive network among CAFs, immune cells, and tumor cells [143,144].

In gastric cancer, CAF-derived IGFBP7 enhances the FGF2/FGFR1–PI3K/AKT pathway, promoting TAM recruitment and polarization, thereby establishing a pro-tumorigenic CAF–TAM–tumor cell axis [145]. In addition, FGF2 sustains CAF activation via autocrine and paracrine mechanisms. Activated CAFs upregulate MMPs and their regulators (such as TIMPs) through FGFR-related pathways, driving ECM remodeling that both creates invasion channels for tumor dissemination and releases matrix-sequestered pro-tumorigenic factors to amplify signaling [146,147].

The heterogeneity of CAFs has garnered considerable attention. Single-cell sequencing and functional analyses reveal transcriptional and functional differences among CAF subtypes. Differential expression patterns of FGFR isoforms have been observed in some contexts, though such "CAF subtype–FGFR" correspondence is not universal and requires validation across tumor types [148,149].

Recent spatial and single-cell transcriptomic studies further contextualize this heterogeneity. In pancreatic ductal adenocarcinoma (PDAC), inflammatory CAF (iCAF), myofibroblastic CAF (myCAF), and antigen-presenting CAF (apCAF) programs occupy discrete spatial niches and show characteristic proximity to tumor or immune compartments, as demonstrated by integrated spatial and single-cell profiling [150,151]. Similar integrated approaches in breast, colorectal, lung, and skin cancers have delineated conserved CAF subclasses (e.g., iCAF- and myCAF-like populations) while revealing tumor-specific spatial organization and neighborhood interactions. For instance, in breast cancer, spatial transcriptomics combined with scRNA-seq identifies CAF subpopulations with distinct proximity to immune infiltrates and tumor cells, implicating them in immune evasion, tumor progression, and potentially modulating responses to immunotherapy [152]. In colorectal cancer, spatially resolved profiling uncovers multicellular immune hubs involving CAFs that organize immunosuppressive niches [153]. In non-small-cell lung cancer, similar analyses underscore the plasticity of stromal and myeloid compartments, while in cutaneous squamous cell carcinoma and melanoma, CAF-associated stromal remodeling at the tumor interface supports the existence of conserved yet context-dependent stromal states [154–156]. Collectively, these integrated datasets refine CAF subset annotation and strengthen evidence for their specialized roles within distinct tumor microenvironments.

FGFR inhibition can partially reverse the pro-tumorigenic effects of CAFs. In triple-negative breast cancer, FGFR blockade remodels CAF phenotypes, enhances $CD8^+$ T cell infiltration, and boosts immune responses [134]. In lung adenocarcinoma, inhibition of the FGF/FGFR pathway suppresses CAF activation and reduces secretion of proinflammatory

https://doi.org/10.3390/cancers18010089

and prometastatic factors, thereby restraining tumor growth [157]. Furthermore, in prostate cancer, downregulation of miR-15/16 in CAFs relieves repression of the FGF2/FGFR1 axis, further promoting tumor growth and metastasis, highlighting the molecular complexity of CAF–tumor cell interactions [158].

3.2.4. Molecular Mechanisms of FGFR Signaling in Tumor Invasion and Metastasis

FGFR signaling plays a pivotal role in multiple stages of tumor invasion and metastasis by regulating cell adhesion, motility, angiogenesis, and the immune microenvironment [159]. Among these, EMT is a key mechanism that endows tumor cells with migratory and invasive capabilities. Numerous reviews and experimental studies have demonstrated that FGF/FGFR signaling can initiate or potentiate EMT-related molecular programs, typically accompanied by altered expression of EMT-associated transcription factors (such as Snail, Slug, and Twist) and epithelial markers (such as E-cadherin), thereby promoting mesenchymal phenotypes and stem-like properties in tumor cells [130,160]. This transition not only enhances cellular motility and invasiveness but also confers stemness and therapeutic resistance.

In addition, FGFR signaling promotes ECM degradation by regulating the expression and activity of MMPs. Activated FGFR upregulates MMP-2, MMP-9, and MMP-14 via the MAPK and PI3K-AKT pathways. These proteases degrade key ECM components such as collagen, fibronectin, and laminin, thus creating invasion corridors for tumor dissemination. Meanwhile, MMPs can also activate latent growth factors and cytokines, further amplifying pro-metastatic signaling cascades [161].

During hematogenous metastasis, tumor cells must traverse the vascular endothelial barrier. FGFR signaling facilitates this process by regulating endothelial permeability and the expression of adhesion molecules, thereby promoting tumor cell intravasation and extravasation [162]. Studies have shown that FGF2 increases the permeability of endothelial tight junctions and upregulates ICAM-1 and VCAM-1 expression, providing favorable conditions for circulating tumor cell adhesion and transendothelial migration [163,164].

Furthermore, FGFR signaling plays a crucial role in the formation and maintenance of distant metastatic niches. In colorectal cancer liver metastasis, tumor-derived FGF19 acts on FGFR4 expressed in hepatic stellate cells, activating the JAK2–STAT3 pathway and inducing their differentiation into iCAFs. These iCAFs secrete proinflammatory mediators such as IL-1, complement C5a, and various chemokines, which recruit neutrophils and promote the formation of neutrophil extracellular traps (NETs). NETs capture circulating tumor cells and release growth-promoting factors, enhancing tumor cell adhesion, survival, and proliferation in the liver. This "soil conditioning" mechanism enables metastatic cells to adapt and thrive within the new tissue microenvironment (Figure 3) [165,166].

Consistent with these mechanistic insights, preclinical CRC liver metastasis models provide causal evidence supporting this FGF19–FGFR4–iCAF–NET axis. Genetic silencing of *FGF19* or pharmacologic inhibition of FGFR4 or JAK2 markedly reduces iCAF polarization and hepatic NET deposition [165,167]. Similarly, neutrophil depletion or NET disruption—such as DNase I administration or the use of peptidylarginine deiminase 4 (*Padi4*)-deficient mice—significantly suppresses metastatic outgrowth. Functionally, NETs not only trap circulating tumor cells but also deliver proliferative and metabolic cues that facilitate micrometastatic expansion, in line with broader evidence implicating NETs in distant seeding and reactivation of dormant cancer cells [167–169].

In addition, FGFR signaling regulates the chemokine system; for instance, it upregulates CXCR4 expression, thereby enhancing tumor cell chemotaxis toward CXCL12. In metastatic breast cancer, blockade of CXCR4 signaling not only directly suppresses metastasis but also increases T cell infiltration and improves responses to immunotherapy [170,171].

https://doi.org/10.3390/cancers18010089

These findings suggest that the FGFR–CXCR4 signaling axis may represent a promising therapeutic target for anti-metastatic interventions.

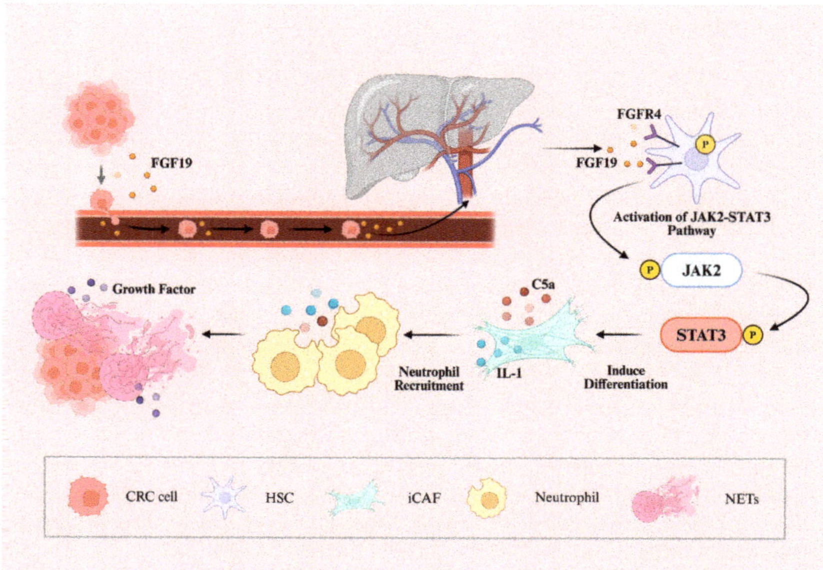

Figure 3. FGF19-FGFR4 axis promotes colorectal cancer liver metastasis through neutrophil extracellular trap formation. Tumor-derived FGF19 (orange sphere) activates FGFR4 on hepatic stellate cells (HSCs), triggering JAK2-STAT3 signaling and inducing HSC differentiation into inflammatory cancer-associated fibroblasts (iCAFs). iCAFs secrete IL-1 (blue sphere), complement C5a (red sphere), and chemokines to recruit neutrophils and promote NET formation. NETs facilitate colorectal cancer liver metastasis by capturing circulating tumor cells (CTCs) and releasing growth-promoting factors (purple sphere), thereby enhancing tumor cell adhesion, survival, and proliferation in the liver. CRC, colorectal cancer; HSC, hepatic stellate cell; iCAF, inflammatory cancer-associated fibroblast; NETs, neutrophil extracellular traps; CTCs, circulating tumor cells.

While other signaling axes such as HGF/MET and CXCL12/CXCR4 are also critical for metastatic niche formation, FGFR signaling exhibits distinct yet cooperative functional characteristics. The HGF/MET pathway is classically recognized as a primary initiator of "invasive growth" and EMT driven by stromal paracrine signaling, whereas the CXCL12/CXCR4 axis predominantly functions as a "chemotactic compass" guiding tumor cell homing to specific organs like the bone and lung [172,173]. In contrast to single-pathway chemotactic axes such as CXCR4, FGFR signaling operates as both an invasion driver and a microenvironmental "soil remodeler." FGFs can recruit neutrophils and reshape innate immune composition [174], while FGF2 potently modulates endothelial function and angiogenic programs, including VEGF induction, thereby facilitating intravasation and metastatic seeding [175,176]. FGFR activity also intersects with chemokine-driven trafficking, as FGF2 mobilizes stromal precursors that are subsequently recruited through the CXCL12–CXCR4 axis [177]. In parallel, extensive RTK crosstalk—particularly with MET—reinforces survival signaling and contributes to adaptive resistance to targeted therapies [178,179]. Collectively, these immune, vascular, and cooperative RTK effects define FGFR-driven niches as integrative hubs that strengthen metastatic establishment.

https://doi.org/10.3390/cancers1801008

3.3. Phenotypic Diversity Driven by Aberrant FGFR Activation

Aberrant activation of the FGFR signaling pathway is a major driver of tumorigenesis and progression across various solid tumors. Distinct types of FGFR alterations exhibit significant differences at the molecular, cellular, and metabolic levels, leading to diverse tumor phenotypes. This has been well supported by large-scale sequencing analyses and comprehensive reviews [53,130].

3.3.1. Distinct Oncogenic Mechanisms of Different Types of FGFR Alterations

Point mutations often cause constitutive receptor activation by altering receptor conformation or promoting aberrant dimerization. In urothelial carcinoma of the bladder, recurrent *FGFR3* hotspot mutations such as R248C, S249C and G370C cluster within the extracellular Ig-like domains or in the linker between the IgII and IgIII domains [181]. Biochemical analyses of the corresponding germline mutations that cause thanatophoric dysplasia type I have shown that substitution of these residues by cysteine introduces unpaired cysteines in the IgII–IgIII linker or juxtamembrane region, thereby promoting the formation of abnormal intermolecular disulfide bonds and covalent receptor dimers with ligand-independent kinase activation [182]. Together with crystallographic and biophysical studies that define how the D1–acid box module and HSPG–mediated FGF–FGFR assembly normally enforce an autoinhibited, ligand-dependent state [20,34], these data support a model in which Cys-substituting *FGFR3* mutations found in both bladder cancer and skeletal dysplasias drive constitutive signaling by stabilizing a covalent homodimer that at least partially bypasses D1–acid box-mediated autoinhibition and the strict requirement for HSPG-bridged dimerization, resulting in sustained trans-phosphorylation of the intracellular kinase domains

The protein levels of FGFR3 are also associated with mutation status and histological differentiation [181,183]. In contrast, in endometrial carcinoma, *FGFR2* kinase-domain mutations (such as N550K and K660E) directly enhance kinase activity and alter receptor conformation. Structural studies indicate that these mutations often occur within the ATP-binding pocket or activation loop, disrupting the "molecular brake" mechanism and maintaining constitutive activation [184,185].

Gene amplification primarily enhances physiological signaling via receptor overexpression. Among such alterations, *FGFR1* amplification is most prominent in breast cancer. It increases cellular sensitivity to FGF, alters signaling strength and spatiotemporal dynamics, and frequently co-occurs with amplification of the 8p11–12 region harboring multiple oncogenic drivers, thereby influencing prognosis. *FGFR1* amplification is also strongly correlated with endocrine therapy resistance. Notably, copy number gain does not always correspond to protein overexpression across tumor subtypes, suggesting a critical role for post-transcriptional regulation in modulating receptor activity [75,96,186]. This genomic instability further contributes to the complexity of amplification-driven oncogenesis.

Gene fusions remodel receptor function through domain rearrangement. In cholangiocarcinoma, the recurrent FGFR2–BICC1 fusion retains the kinase domain but lacks the autoinhibitory region, leading to constitutive activation and altered subcellular localization. This fusion represents a key molecular rationale for FGFR2-targeted therapy.

Emerging evidence indicates that the oncogenicity of *FGFR2* fusion proteins is shaped not only by the preserved kinase domain but also by fusion-partner-encoded modules that govern oligomerization and intracellular trafficking [187]. Many recurrent partners, including *TACC3* and *BICC1*, provide coiled-coil, Sterile Alpha Motif, or related multimerization motifs that drive ligand-independent dimerization or higher-order clustering, thereby stabilizing constitutive FGFR activation [188]. For example, in *FGFR3–TACC3*, the TACC3 coiled-coil domain promotes aberrant oligomerization and redirects the fusion

 https://doi.org/10.3390/cancers18010089

protein to mitotic spindle and centrosomal structures, producing ectopic signaling outputs and mitotic defects [189]. Similarly, BICC1 contains domains that facilitate spontaneous multimerization and influence membrane–endosomal or centrosomal localization, ultimately altering receptor trafficking dynamics [190]. These trafficking alterations—shifting the balance between recycling and lysosomal degradation, or redistributing receptors across plasma membrane, endosomal, or pericentrosomal compartments—can prolong kinase activity and reshape downstream signaling amplitude and duration. Consistent observations across engineered cell systems and functional genomic studies support these mechanisms and highlight disruption of fusion-driven oligomerization or trafficking as a potential therapeutic strategy complementary to kinase inhibition.

Another prototypical example is the FGFR3–TACC3 fusion, which was first identified and remains most prevalent in glioblastoma, with an incidence of approximately 2–5%. Beyond driving constitutive activation, its fusion partner TACC3 regulates cytoskeletal organization, spindle assembly, and cell division, thereby promoting malignant phenotypes. Moreover, this fusion induces distinct transcriptional and metabolic programs depending on the histological and cellular context [191–193].

Taken together, different types of FGFR alterations define distinct oncogenic mechanisms and downstream signaling networks: point mutations activate receptors through conformational changes or aberrant dimerization, amplifications enhance signal amplitude via a dosage effect, and fusions induce structural and localization remodeling. This molecular diversity underlies the heterogeneity of clinical phenotypes and therapeutic responses, highlighting the necessity of tailoring FGFR-targeted strategies according to specific alteration types and accompanying genomic contexts.

3.3.2. The Role of FGFR Alterations in Cancer Stem Cell Maintenance and Differentiation

FGFR signaling plays a central role in the self-renewal and pluripotency maintenance of cancer stem cells (CSCs). In glioblastoma, for instance, FGFR1 sustains CSC stemness and EMT-related phenotypes through an ADAMDEC1–FGF2 autocrine loop that activates ERK1/2 signaling and upregulates transcription factors such as ZEB1 [194]. Moreover, in other solid tumors, such as non-small-cell lung cancer with FGFR1 amplification, the same alteration has been shown to induce "stem cell-like" characteristics, suggesting that this pathway exerts a conserved role in maintaining stemness across different tissue contexts [195].

This regulation extends beyond transcriptional control and involves chromatin remodeling mechanisms. Nuclear-localized FGFR1 can bind chromatin and colocalize with active histone marks, as supported by ChIP-seq and transcriptomic analyses, thereby reorganizing the transcriptional program of stemness-associated genes and influencing therapeutic sensitivity [196].

In hematologic malignancies—particularly myeloid or mixed-phenotype leukemias harboring *FGFR1* fusions—the constitutive kinase activity of fusion proteins interferes with myeloid differentiation, leading to differentiation blockade, accumulation of immature progenitors, and clonal expansion that confers enhanced chemoresistance [197–199]. Patient-derived models and single-cell sequencing analyses have further demonstrated the heterogeneity in differentiation states driven by FGFR alterations. *FGFR1* amplification or overexpression often results in disrupted differentiation hierarchies: some cells retain CSC-like features, while others display aberrant differentiation-related transcriptional profiles. The coexistence of these subpopulations is closely associated with treatment response variability [200,201].

The CSC niche, composed of CAFs, vascular endothelial cells, and other stromal components, sustains CSC activity through the secretion of exogenous FGFs [119,202]. In colorectal and other solid tumor models, stromal cell-derived or membrane-bound FGFs (such as FGF2) activate downstream FGFR signaling to promote tumor cell migration, invasion, and stem-like phenotypes. Recent studies have further demonstrated that FGFR

 https://doi.org/10.3390/cancers18010089

signaling cooperates with multiple downstream pathways, including MAPK, PI3K, STATs, and SRC family kinases, to orchestrate these processes [203,204]. Furthermore, CSCs often localize within specialized microenvironmental niches, such as perivascular regions or tumor-invasive fronts, where local FGFR overexpression correlates positively with CSC enrichment. This suggests a spatially coordinated interaction between FGFR signaling and microenvironmental cues in shaping CSC heterogeneity and site-specific behavior [205,206].

Overall, FGFR abnormalities maintain CSC phenotypes and shape differentiation trajectories through multilayered mechanisms—including transcriptional regulation, chromatin remodeling, disruption of differentiation programs, and microenvironmental interactions—collectively driving tumor resistance, recurrence, and invasion. However, the precise molecular consequences vary significantly among tumor types and distinct *FGFR* alteration forms.

3.3.3. Effects of FGFR Signaling on Tumor Metabolic Reprogramming

FGFR signaling plays a pivotal role in tumor metabolic reprogramming, particularly in glycolytic and lipid metabolic pathways. In lung squamous cell carcinoma models, FGFR1 signaling activates the PI3K–AKT–mTOR axis, leading to HIF-1α accumulation and upregulation of GLUT1, thereby enhancing glucose uptake and glycolysis—the hallmark Warburg effect. In addition, FGFR1 regulates glucose metabolism through its interaction with PKM2, modulating its phosphorylation status and subcellular localization. Metabolomic analyses have further revealed that FGFR aberrations promote glycolytic bias by altering lactate dehydrogenase (LDH) isoenzyme composition—upregulating LDHA while downregulating LDHB—which collectively supports tumor growth and metabolic flexibility [207–210].

In terms of lipid metabolism, FGFR signaling has been shown to activate lipogenic pathways across multiple tumor models, particularly in *FGFR3*-altered bladder cancer. It upregulates SREBP1 downstream targets, such as SCD1 and FASN, thereby promoting monounsaturated fatty acid synthesis and membrane biogenesis. These findings indicate that FGFR aberrations can drive lipid metabolic reprogramming. Lipidomic studies have also demonstrated a close association between the RTK/PI3K signaling axis and lipid metabolism, whereby FGFR–PI3K signaling supports downstream transduction by modulating levels of phospholipid precursors, such as PIP_2, ultimately influencing membrane composition and signal propagation [211–213].

3.3.4. Single-Cell Heterogeneity of FGFR Expression and Its Biological Significance

Single-cell sequencing has revealed a high degree of heterogeneity in FGFR expression within tumor tissues. Multiple single-cell and spatial transcriptomic studies have shown that FGFR expression varies markedly across cellular subpopulations, and FGFR-high cells are frequently colocalized with invasive, stem-like phenotypes or microenvironmental niches enriched in angiogenic and immunosuppressive features. This phenomenon has been observed, to varying extents, across several solid tumors, including pancreatic cancer, head and neck squamous cell carcinoma, and lung cancer [214–216].

Pseudotime trajectory and clonal dynamics analyses further demonstrate that FGFR expression correlates with cell differentiation states and clonal selection. For example, in head and neck squamous cell carcinoma and breast cancer, FGFR-high subpopulations typically exhibit transcriptional signatures resembling stem/progenitor cells, whereas FGFR-low or -negative subgroups show a trend toward differentiation lineages. Under therapeutic pressure, rare resistant or "refuge" subclones may expand and drive tumor recurrence—a dynamic process directly observed in longitudinal studies of FGFR-driven cholangiocarcinoma [13,217].

In addition, single-cell ligand–receptor interaction analyses and spatial transcriptomic tools have been applied to dissect the roles of FGFR-mediated signaling circuits among

 https://doi.org/10.3390/cancers18010089

tumor cells, CAFs, and immune cells. These integrative analyses provide fresh insights into therapeutic response and resistance mechanisms and suggest that FGFR-targeted therapies should account for cooperative and compensatory interactions within the tumor microenvironment [218].

4. FGFR-Targeted Therapy: Drug Development and Clinical Applications

4.1. *Classification and Mechanisms of FGFR Inhibitors*

The development of FGFR inhibitors has evolved from non-selective multi-target inhibition toward highly selective and precise targeting. With deeper insights into FGFR structural biology and advances in rational drug design, a variety of strategies have emerged, including small-molecule TKIs, monoclonal antibodies, ADCs and novel delivery platforms. Understanding the design principles and clinical profiles of different inhibitor classes is essential for precision oncology and the development of next-generation therapeutics.

4.1.1. Design Concepts and Clinical Performance of Selective vs. Non-Selective Inhibitors

Early FGFR inhibitors were predominantly non-selective multi-target TKIs, such as nintedanib and dovitinib. Nintedanib inhibits VEGFR, FGFR, and PDGFR simultaneously and demonstrated moderate efficacy in NSCLC but was frequently associated with diarrhea, hepatotoxicity, and vascular-related adverse events [128,219]. Dovitinib exhibits inhibitory activity against FGFR1–3, VEGFR, FLT3, and c-KIT, yet clinical trials in breast and renal cancers reported limited efficacy and substantial toxicity, restricting its clinical utility [220,221]. These early attempts highlighted a key trade-off: while non-selective inhibitors benefit from broad target coverage and potential multi-target synergism, their lack of specificity often leads to off-target toxicity and suboptimal therapeutic precision.

With advances in structural biology, the development of selective FGFR inhibitors has become increasingly sophisticated. These agents are designed to exploit the unique structural features of FGFR kinase domains, thereby improving target specificity and minimizing off-target effects [222]. Erdafitinib, the first selective FGFR inhibitor approved by the U.S. FDA, demonstrates high affinity toward FGFR1–4 (IC_{50}: 1.8–10.8 nM). In patients with *FGFR3*-altered urothelial carcinoma, erdafitinib achieved an objective response rate (ORR) of 40%, markedly outperforming chemotherapy, which yielded an ORR of 13% [5,223].

Pemigatinib, a reversible ATP-competitive inhibitor, binds to the ATP-binding pocket of FGFR1–3, effectively suppressing their kinase activities. In the FIGHT-207 basket trial, pemigatinib demonstrated an ORR of 26.5% and a median progression-free survival (PFS) of 4.5 months among patients harboring FGFR fusions or rearrangements [224].

Futibatinib, a next-generation covalent FGFR inhibitor, exhibits potent antitumor activity, particularly in tumors harboring *FGFR2* and *FGFR3* mutations. Through optimized scaffold design, futibatinib overcomes several resistance-associated mutations and shows strong promise in both preclinical and early-phase clinical studies. Its molecular configuration enhances binding affinity for FGFR kinases while reducing nonspecific inhibition of other kinases. This highly selective targeting of FGFR2 and FGFR3 translates into superior efficacy and improved tolerability in genetically defined tumor subtypes [225].

4.1.2. Development Strategies for Monoclonal Antibodies and Antibody–Drug Conjugates

Monoclonal antibodies (mAbs) inhibit FGFR signaling by blocking ligand–receptor interactions or preventing receptor dimerization, representing a mechanism distinct from that of small-molecule TKIs. Bemarituzumab is a humanized monoclonal antibody targeting FGFR2b, which inhibits ligand binding of FGF7 and FGF10. In a phase II clinical trial involving patients with gastric and gastroesophageal junction cancer, bemarituzumab in

https://doi.org/10.3390/cancers1801008

combination with chemotherapy significantly improved progression-free survival (PFS) (9.5 vs. 7.4 months) and demonstrated a favorable efficacy profile [226].

Vofatamab (B-701) is a monoclonal antibody targeting FGFR3, which acts through allosteric binding to the extracellular domain to block receptor activation. In early clinical studies, vofatamab combined with pembrolizumab in patients with metastatic urothelial carcinoma refractory to platinum-based therapy showed acceptable safety and preliminary signals of efficacy, suggesting that this combination approach warrants further clinical investigation [227].

ADCs combine the target specificity of monoclonal antibodies with the cytotoxic potency of chemotherapeutic payloads, earning them the nickname "biological missiles". However, the development of FGFR-targeted ADCs faces multiple challenges, including the optimization of antibody selection, linker–payload design, and management of FGFR-specific toxicities [228].

Aprutumab ixadotin (BAY 1187982) was among the first FGFR2-targeting ADCs. Although it demonstrated selective cytotoxic activity against FGFR2-overexpressing tumors in preclinical models, its phase I clinical trial was terminated due to severe proteinuria and hematologic toxicities. Subsequent reviews have cited this case as an example to illustrate linker-associated toxicity and the unique risk profile of FGFR-targeted ADCs [229–231]. A novel tetravalent anti-FGFR1 ADC has recently been developed to enhance binding avidity and internalization efficiency. In vitro studies showed 5–10-fold greater cytotoxic potency compared to traditional bivalent ADCs, and superior antitumor efficacy was further validated in xenograft models [232,233].

The design of the ADC linker is a critical determinant of both therapeutic efficacy and safety. Cleavable linkers, such as the valine–citrulline linker, release their cytotoxic payload upon lysosomal enzymatic cleavage, while non-cleavable linkers require complete degradation of the antibody to liberate the drug [234]. Most FGFR-targeted ADCs employ cleavable linkers to ensure efficient intracellular drug release within target cells [235]. Emerging pH-sensitive or glutathione-responsive linkers exploit the unique biochemical characteristics of the tumor microenvironment to achieve precise site-specific drug release, potentially improving the therapeutic index and mitigating the impact of FGFR expression heterogeneity on treatment outcomes [236,237].

4.1.3. Advances in Novel Targeted Delivery Systems and Nanomedicine Carriers

The efficient internalization properties of FGFR confer a unique advantage for developing targeted delivery systems [187]. Upon ligand binding, FGFR undergoes clathrin-mediated endocytosis, a process that has been leveraged to engineer diverse FGFR-targeted carriers. For instance, constructs that use FGF1 as the recognition module and are conjugated to fluorescent oligomers can integrate targeting, tracking, and toxin loading. A representative molecule, 3xGFPp_FGF1–MMAE, exhibits nanomolar affinity and marked cytotoxicity toward FGFR1-high cells in vitro, while showing minimal effects on FGFR1-negative cells, indicating potential for imaging-guided and precision delivery applications [238].

Nanotechnology provides a platform for improving the stability, targetability, and therapeutic index of FGFR inhibitors, including liposomes, polymeric nanoparticles, and inorganic nanocarriers, but still faces challenges in clinical translation such as Enhanced Permeability and Retention effect (EPR) heterogeneity, in vivo clearance, and scaled-up manufacturing [239,240]. Notably, the robust EPR effect observed in rapidly growing murine xenografts is substantially weaker and more heterogeneous in human solid tumors [241]. Clinical and preclinical studies using quantitative imaging of radiolabeled or MRI/PET-tracked nanomedicines indicate that, in patients, only a limited and highly variable fraction of the injected dose accumulates in tumor tissue, with variability influenced by tumor type, anatomical site, stromal architecture, vascular normalization status,

https://doi.org/10.3390/cancers18010089

and prior therapies [242,243]. This mismatch between preclinical and clinical EPR behavior is thought to be one important factor contributing to the observation that many EPR-dependent nanomedicines with impressive efficacy in mouse models have achieved only modest, if any, overall survival benefit in late-phase clinical trials [244,245].

To mitigate this translational gap, current strategies increasingly explore functional "EPR imaging" (e.g., dynamic contrast-enhanced MRI or nanocarrier-based MRI/PET) for patient and lesion stratification, transient modulation of tumor vasculature in rational combinations to improve perfusion, and the development of actively targeted or stimuli-responsive carriers designed to be less reliant on passive EPR for intratumoral delivery [246–249].

Targeted liposomes typically achieve active targeting through surface modification with anti-FGFR antibodies or ligands, enabling prolonged circulation time and accumulation in tumor tissues. Preclinical studies have demonstrated that encapsulating FGFR-active compounds in targeted liposomes or other nanocarriers can achieve superior antitumor efficacy compared to free drug and a more controllable systemic toxicity profile in animal models [250,251].

Polymer–drug conjugates (PDCs) offer an additional option for delivering FGFR inhibitors. By tethering drugs to polymers via biodegradable linkages, PDCs enable controlled release and targeted delivery. PDC formulations of FGFR inhibitors have shown improved pharmacokinetic profiles and enhanced antitumor activity in preclinical and early-phase studies; however, further evidence is required to establish their clinical utility [252–254].

In addition, smart stimuli-responsive delivery systems leverage features of the tumor microenvironment to achieve precise release. For example, pH-responsive nanocarriers accelerate drug release under acidic conditions, whereas enzyme-responsive carriers rely on tumor-overexpressed proteases such as MMPs to cleave linkers and liberate the payload. These "microenvironment-responsive" strategies have demonstrated promising controlled-release behavior and targeting potential in vitro and in selected in vivo models, yet a meaningful gap to clinical translation remains [255–257].

4.2. Clinical Efficacy Evaluation of Approved FGFR Inhibitors

The evidence-based evaluation of approved FGFR inhibitors provides a scientific foundation for precision oncology strategies. Table 2 summarizes the key trial designs and efficacy outcomes of major FGFR inhibitors across different cancer types. With the regulatory approvals of pemigatinib, erdafitinib, and futibatinib in multiple countries, the body of clinical data regarding their efficacy and safety has expanded, while real-world studies continue to supplement findings from pivotal registration trials.

Table 2. Clinical Efficacy and Trial Design of FGFR Inhibitors Across Different Tumor Types and Molecular Alterations.

Drug	Cancer Type	Trial Name	Study Design	Key Efficacy Outcomes	References
Pemigatinib	Cholangiocarcinoma with FGFR2 fusion/rearrangement	FIGHT-202	Multicenter, open-label, phase II	ORR 37.0%; mPFS 7.0 months; mOS 17.5 months	[4,258]
	Solid tumors with FGFR alterations (multiple cancer types)	FIGHT-207	Phase II basket trial with cohorts stratified by molecular alteration type	ORR 26.5% in FGFR fusion patients; only 9.4% in non-kinase domain mutations	[224]
	Advanced cholangiocarcinoma with FGFR2 rearrangement	FIGHT-302 (ongoing)	Multicenter, open-label, randomized, phase III	Primary endpoint: PFS; secondary endpoints: OS, ORR, safety, etc.	[259]

https://doi.org/10.3390/cancers18010089

Table 2. *Cont.*

Drug	Cancer Type	Trial Name	Study Design	Key Efficacy Outcomes	References
Erdafitinib	Urothelial carcinoma with FGFR alterations	BLC2001	Single-arm, multicenter, phase II	ORR 40%; mDOR 5.6 months; mPFS 5.5 months	[5,260]
	Urothelial carcinoma with FGFR alterations	THOR	Multicenter, open-label, randomized controlled, phase III	mOS 12.1 months vs. 7.8 months (p = 0.005)	[261]
	Solid tumors with FGFR alterations (multiple cancer types)	RAGNAR	Multicenter, single-arm basket, phase II	ORR 29% in overall population; mDOR 6.9 months	[262]
Futibatinib	Cholangiocarcinoma with FGFR2 fusion/rearrangement	FOENIX-CCA2	Multicenter, single-arm, phase II	ORR 42%; mPFS 9.0 months; mOS 21.7 months	[191]
Lirafugratinib (RLY-4008)	Solid tumors with FGFR2 alterations	ReFocus (ongoing)	Multicenter, open-label, phase I/II	Primary endpoints: safety/tolerability, ORR; secondary endpoints: PFS, OS, DCR, DOR	[263,264]
Bemarituzumab	FGFR2b-positive gastric cancer/ GEJ cancer	FIGHT	Multicenter, randomized, double-blind, phase II	mPFS 9.5 months vs. 7.4 months (p = 0.07)	[226]
Vofatamab (B-701)	Metastatic urothelial carcinoma	FIERCE-22	Multicenter, open-label, phase Ib/II	ORR 32%; mPFS 4.7 months	[227]
Aprutumab ixadotin	Solid tumors with high FGFR2 expression	NCT02368951	Open-label, multicenter, phase I	Terminated due to severe proteinuria/ hematologic toxicity	[229]

Abbreviations: ORR, objective response rate; mPFS, median progression-free survival; mOS, median overall survival; mDOR, median duration of response; DCR, disease control rate; DOR, duration of response; GEJ, gastroesophageal junction.

4.2.1. Clinical Trial Outcomes of Pemigatinib in Cholangiocarcinoma

The FIGHT-202 trial (Phase II, multicenter, open-label) demonstrated the antitumor efficacy of pemigatinib in previously treated patients with *FGFR2* fusion or rearrangement–positive cholangiocarcinoma. The initial analysis reported an ORR of approximately 35.5%, a disease control rate (DCR) of about 82%, a median PFS of 6.9 months, and a median overall survival (OS) of 21.1 months [4]. An independent review committee (IRC)-based long-term follow-up analysis further validated these results, showing an ORR of 37.0%, median duration of response (DOR) of 9.1 months, median PFS of 7.0 months, and median OS of 17.5 months [258].

In the broader FIGHT-207 trial (Phase II, basket study encompassing multiple tumor types and FGFR alterations), the ORR was 26.5% among patients with FGFR fusions or rearrangements, while those harboring activating non-kinase domain mutations exhibited an ORR of only 9.4%, underscoring a marked difference in pemigatinib sensitivity across distinct *FGFR* alteration types, and providing direct evidence to support molecular stratification in clinical practice [224].

The ongoing FIGHT-302 study (Phase III, randomized controlled trial; pemigatinib vs. gemcitabine + cisplatin, 2:1 randomization; primary endpoint: PFS) is designed to evaluate the efficacy and safety of pemigatinib as first-line therapy for advanced *FGFR2*-rearranged cholangiocarcinoma. The trial has not yet reported final endpoint data [259].

https://doi.org/10.3390/cancers18010089

In terms of safety, both clinical trials and subsequent pooled analyses have identified hyperphosphatemia as the most common treatment-related adverse event, with a relatively high incidence rate. Effective management—including dose adjustment, low-phosphate diets, and use of phosphate binders—is recommended to maintain treatment adherence and optimize long-term benefit. Real-world data generally confirm efficacy and safety outcomes consistent with clinical trial findings but highlight the critical role of toxicity management in improving clinical outcomes [4,258].

4.2.2. Efficacy and Safety of Erdafitinib in Bladder Cancer

The BLC2001 trial, a pivotal phase II registration study, evaluated erdafitinib in 99 patients with advanced urothelial carcinoma harboring FGFR alterations. The trial reported an ORR of 40%, a median DOR of 5.6 months, and a median PFS of 5.5 months. Subgroup analysis revealed a higher ORR among patients with *FGFR3* mutations (49%) compared with those with FGFR fusions (33%), indicating that the type of FGFR alteration may influence therapeutic response [5,260].

The phase III THOR trial, which enrolled 266 patients with FGFR-altered advanced urothelial carcinoma, compared erdafitinib with standard chemotherapy. Erdafitinib significantly improved OS, achieving a median OS of 12.1 months versus 7.8 months with chemotherapy ($p = 0.005$), thereby establishing its role as the standard of care for patients with FGFR-altered urothelial carcinoma [261].

The RAGNAR trial, an international, single-arm phase II study, was designed to systematically evaluate the efficacy and safety of erdafitinib in patients with advanced or metastatic solid tumors harboring *FGFR1–4* alterations. In the overall study population, erdafitinib achieved an ORR of 29% and a median DOR of 6.9 months, supporting the tumor-agnostic druggability of FGFR-driven malignancies. Notably, in the cholangiocarcinoma cohort, erdafitinib demonstrated robust antitumor activity, with an ORR of approximately 60% and a median PFS of around 8.5 months [262].

With respect to safety, pooled analyses identified the most common treatment-related adverse events as dry skin (74%), fatigue (36%), diarrhea (32%), and hyperphosphatemia (>50%) [265]. Hyperphosphatemia represents a mechanism-related, on-target toxicity, while ocular adverse events—including central serous retinopathy—occurred in about 25% of patients, were largely reversible, and required regular ophthalmologic evaluation and serum phosphate monitoring to optimize the balance between efficacy and tolerability [266].

4.2.3. Clinical Advances of Futibatinib and Next-Generation FGFR Inhibitors

The FOENIX CCA2 trial, a pivotal phase II registration study, enrolled 103 patients with advanced iCCA harboring *FGFR2* fusions or rearrangements. The study reported an ORR of 42%, a median PFS of 9.0 months, and a median OS of 21.7 months, demonstrating durable and clinically meaningful benefit in this molecularly defined population [191].

Exploratory genomic analyses indicated no significant efficacy difference between *BICC1* and non *BICC1* fusion partners (ORR = 41.7% vs. 44.6%). Moreover, a rapid decline in circulating tumor DNA (ctDNA) after treatment correlated with improved clinical outcomes, underscoring the potential utility of liquid biopsy for response monitoring and early detection of resistance. Mechanistic studies revealed that secondary *FGFR2* kinase domain mutations represent the predominant mechanism of acquired resistance, with specific sites such as L617 reducing inhibitor sensitivity. The irreversible binding mode of futibatinib confers activity against certain resistant mutants; however, polyclonal secondary mutations remain a major therapeutic challenge [267].

Among emerging agents, RLY 4008 (lirafugratinib), a highly selective FGFR2 inhibitor, has demonstrated exceptional target specificity and broad activity against resistance mu-

https://doi.org/10.3390/cancers1801008

tations in early clinical evaluation. Preliminary results from the phase I ReFocus study showed encouraging antitumor activity at the recommended dose in *FGFR2* driven malignancies. These findings suggest that sequential use of more selective FGFR inhibitors may offer a novel strategy to extend the durability of FGFR targeted therapy [263,264].

4.2.4. Real World Evidence on the Efficacy and Safety of FGFR Inhibitors

A large-scale pharmacovigilance analysis based on the FDA Adverse Event Reporting System (FAERS) evaluated real world safety profiles of erdafitinib, pemigatinib, and futibatinib. Overall, the safety spectrum in real world settings was consistent with findings from registration trials, although certain rare but serious adverse events appeared at slightly higher frequencies, likely reflecting patient heterogeneity and reporting bias [268]. For instance, pemigatinib associated calciphylaxis has been reported in real world cases but was not observed in clinical trials [269].

In terms of clinical effectiveness, real world outcomes generally aligned with pivotal trial data but demonstrated modest variations due to differences in patient selection and response assessment. In a real-world cohort of 120 pemigatinib treated patients, the real-world ORR was 59.2%, and the median real world PFS was 7.4 months, exceeding those reported in the FIGHT 202 trial [270]. This improvement may reflect selection bias toward patients with more favorable baseline characteristics and non-uniform response assessment methods.

Real world populations also exhibited greater clinical complexity, including a higher prevalence of hepatic dysfunction and heavily pretreated disease, which likely contributed to outcome variability and underscored the limitations of real-world evidence studies [258,271]. Furthermore, inconsistent quality of FGFR genomic testing markedly influenced efficacy evaluations. Patients assessed using standardized next-generation sequencing (NGS) platforms showed significantly better outcomes than those tested by non-standardized methods [73]. Thus, establishing uniform FGFR testing standards is essential for enhancing clinical reproducibility and improving the reliability of real-world evidence.

5. Mechanisms of Resistance and Therapeutic Strategies

5.1. Molecular Basis of Resistance to FGFR Targeted Therapy

Resistance to FGFR targeted therapy represents a major barrier to sustaining durable clinical benefit, arising from highly diverse and complex molecular mechanisms. With the expanding clinical use of FGFR inhibitors, therapeutic resistance has become increasingly apparent. Detailed elucidation of the underlying mechanisms provides crucial insights into treatment failure and informs the design of next-generation inhibitors and predictive biomarkers [52]. Current evidence suggests that FGFR targeted resistance primarily results from acquired FGFR structural mutations, activation of compensatory bypass pathways, and clonal evolution driven by tumor heterogeneity, reflecting the multifaceted nature of therapeutic escape.

5.1.1. Molecular Characteristics and Structural Basis of Acquired Resistance Mutations

Acquired mutations constitute the most prevalent mechanism of resistance to FGFR inhibitors, with gatekeeper mutations being the archetypal example. These mutations occur within the ATP binding pocket of the kinase domain, where substitution of small residues (e.g., valine) with bulkier amino acids (e.g., methionine or phenylalanine) introduces steric hindrance that impedes inhibitor access or proper orientation, thereby diminishing drug affinity. Canonical examples include *FGFR1* V561M, *FGFR2* V564F, and *FGFR3* V555M [272–274]. Structural studies reveal that such mutations disrupt key hydrophobic interactions, promote stabilization of the active kinase conformation, and may enhance

https://doi.org/10.3390/cancers18010089

catalytic turnover, while also altering the kinetic stability of the inhibitor–protein complex rather than merely increasing ATP affinity [275].

Beyond gatekeeper substitutions, additional secondary mutations within the FGFR kinase domain have been implicated in clinical resistance. In patients with *FGFR2* fusion–positive cholangiocarcinoma, approximately 60% develop one or more secondary *FGFR2* kinase domain alterations upon progression, most frequently at N550 and V565 [12]. These mutations exhibit heterogeneous modes of action: N550 variants disrupt hydrogen bond networks within the hinge/molecular brake region, destabilizing the autoinhibited conformation, whereas gatekeeper like mutations primarily confer resistance via steric interference with inhibitor binding [276,277].

Molecular dynamics simulations further demonstrate that V564F/V565F mutations markedly reduce the structural stability of complexes formed between reversible ATP competitive inhibitors (e.g., pemigatinib and infigratinib) and FGFR2, providing a mechanistic explanation for the distinct resistance profiles observed across clinical and preclinical settings [275,278].

5.1.2. Bypass Signaling Activation as a Mechanism of Resistance

Upon pharmacologic blockade of FGFR signaling, tumor cells frequently exploit alternative RTKs and their downstream cascades to sustain proliferation and survival. Multiple RTKs—including EGFR, MET, ErbB3, and EphB3—have been validated across diverse preclinical models as key mediators of acquired resistance to FGFR inhibitors [279–281]. Their activation is driven by multifactorial mechanisms, encompassing transcriptional upregulation, enhanced protein stability, and increased availability of extracellular ligands, collectively enabling tumor cells to circumvent FGFR dependency and restore oncogenic signaling.

PI3K–AKT signaling represents a pivotal downstream effector of FGFR and constitutes a major bypass mechanism underlying therapeutic resistance. Loss or functional impairment of PTEN results in sustained accumulation of PIP3, thereby maintaining AKT activation even under FGFR inhibition and consequently diminishing inhibitor sensitivity [282–284]. In addition, reduced expression or inactivation of pleckstrin homology-like domain family A member 1 (PHLDA1)—a critical negative regulator of PI3K–AKT signaling—can unleash AKT activity and promote a resistant phenotype, a mechanism functionally validated in multiple RTK-driven tumor models [285,286].

Reactivation of the RAS–MAPK cascade represents another prevalent route to escape FGFR-targeted therapy. Acquired NRAS/KRAS amplifications or gain-of-function mutations have been detected in resistant tumor specimens, while downregulation of DUSP6—a phosphatase specifically targeting ERK—can sustain ERK phosphorylation and propagate proliferative output. Such convergent rewiring of the MAPK pathway highlights the inherent vulnerability of monotherapy against a single RTK, which is frequently bypassed by resilient downstream oncogenic signaling [287,288].

Beyond canonical oncogenic cascades, metabolic rewiring and autophagy have emerged as increasingly critical drivers of therapeutic resistance. In gastric cancer models, FGFR inhibition has been shown to activate the TAK1/AMPK axis and enhance autophagic flux, thereby enabling resistant cells to withstand metabolic and pharmacologic stress; concurrently, transcription factor EB (TFEB), a lysosome–autophagy master regulator, undergoes nuclear translocation to upregulate lysosomal and autophagy-associated gene programs, further strengthening cellular stress tolerance [289,290]. These findings collectively illustrate that, when classical RTK pathways are blocked, tumor cells can acquire alternative survival routes through metabolic adaptability and autophagy activation, ultimately inducing or sustaining resistance. However, substantial heterogeneity has been observed across distinct tumor types and molecular subgroups, suggesting that rational combinations or sequential strategies

https://doi.org/10.3390/cancers18010089

co-targeting downstream MAPK/PI3K signaling or metabolic vulnerabilities may outperform FGFR monotherapy in delaying or overcoming resistance.

5.1.3. The Role of Tumor Heterogeneity and Clonal Evolution in Resistance

Tumor heterogeneity represents a fundamental driver of resistance to FGFR-targeted therapies, manifesting across both temporal and spatial dimensions. Multi-omics profiling and rapid autopsy analyses have demonstrated that FGFR inhibitors can profoundly reshape the clonal architecture of cholangiocarcinoma: dominant FGFR-altered clones prior to treatment diminish following FGFR blockade, whereas minor subclones harboring secondary resistance mutations expand under therapeutic pressure and ultimately gain dominance [115,217]. Single-cell and high-resolution omics studies further reveal that these resistant subpopulations may pre-exist at low frequency before therapy, with drug exposure promoting their accelerated expansion. Such competitive or replacement clonal dynamics ultimately select for fitter clones, which govern the emergence of resistant phenotypes [291].

ctDNA has emerged as a powerful tool for real-time tracking of clonal evolution. Longitudinal analyses demonstrate that resistance-associated mutations can arise prior to radiographic progression, indicating opportunities for earlier therapeutic intervention. Moreover, ctDNA profiling reveals the prevalent nature of polyclonal resistance: diverse resistant mutations may coexist within the same patient, underscoring the tumor's ability to deploy multiple escape strategies simultaneously. This complexity explains why next-generation inhibitors alone often fail to fully eradicate resistance and highlights the necessity of rational combination therapies [12,292].

Clinical and multi-omics investigations suggest that resistant and sensitive clones may engage in competitive, coexistent, or even cooperative interactions. However, current evidence largely stems from observational molecular profiling, with limited real-time tumor ecology or functional validation. Future studies integrating high-resolution longitudinal omics, dynamic ctDNA monitoring, and ecological modeling will be essential to elucidate the evolutionary trajectories of resistance and optimize personalized therapeutic strategies [115].

5.1.4. Epigenetic Alterations Shaping Sensitivity to FGFR-Targeted Therapy

Epigenetic dysregulation plays a multilayered role in modulating resistance to FGFR-targeted therapies. Aberrations in chromatin-remodeling complexes—particularly loss of function of the SWI/SNF complex—represent an important molecular dimension influencing therapeutic response and have been directly implicated in adaptive resistance to FGFR inhibition in selected models [293]. In lung cancer and other solid tumors, deficiency of key SWI/SNF subunits (such as SMARCA4/BRG1) profoundly reorganizes chromatin architecture and transcriptional circuitry, thereby reshaping cellular sensitivity to multiple anticancer agents.

Mechanistically, SMARCA4 loss promotes heterochromatin accumulation and elevates replication stress, rendering tumor cells increasingly dependent on ATR signaling and consequently more vulnerable to ATR inhibitors in vitro and in vivo [294,295]. Moreover, SMARCA4/2 deficiency drives enhancer reprogramming and global transcriptional rewiring, activating MYC-linked or lineage plasticity programs and perturbing the DNA damage response (DDR), which in certain contexts confers potential susceptibility to PARP inhibition or ATR/PARP combinatorial strategies [296,297]. In addition, SMARCA4 defects attenuate expression of apoptosis-related programs—including mechanisms associated with impaired ER-to-mitochondria Ca^{2+} transfer—resulting in diminished responses to chemotherapy and other targeted agents (Figure 4) [298]. Although these molecular axes provide strong biological rationale for exploiting drug vulnerabilities in SMARCA4-

https://doi.org/10.3390/cancers18010089

deficient contexts, whether SMARCA4 directly drives clinical resistance to FGFR inhibitors remains to be validated through larger multi-omics datasets and longitudinal cohorts.

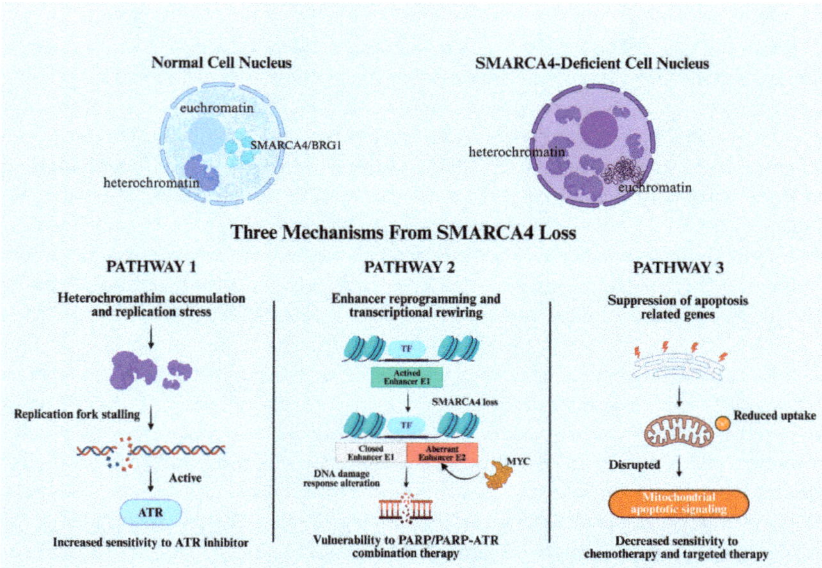

Figure 4. Molecular Mechanisms Underlying SMARCA4/BRG1 Loss-Mediated Therapeutic Response in Cancer. Upper panel: Chromatin organization in normal versus SMARCA4-deficient cell nuclei. The SMARCA4/BRG1 complex maintains euchromatin-heterochromatin balance in normal cells (**left**). SMARCA4 loss (**right**) causes global heterochromatin accumulation and reduced chromatin accessibility. Pathway 1 (**left**): Heterochromatin accumulation and replication stress. SMARCA4 deficiency promotes excessive heterochromatin formation (purple clusters), causing replication fork stalling and DNA damage. Persistent replication stress activates ATR signaling, creating oncogenic addiction to ATR-mediated DNA damage response and conferring sensitivity to ATR inhibitors. Pathway 2 (**center**): Enhancer reprogramming and transcriptional rewiring. SMARCA4 loss induces enhancer reprogramming, converting active enhancers to closed or aberrantly activated states. This activates MYC signaling and alters DNA damage response (DDR) gene expression, conferring vulnerability to PARP inhibitors. Pathway 3 (**right**): Suppression of apoptosis-related genes. SMARCA4 deficiency downregulates pro-apoptotic genes, disrupting ER-to-mitochondria Ca^{2+} transfer. Impaired mitochondrial Ca^{2+} uptake compromises apoptotic signaling, reducing sensitivity to chemotherapy and FGFR-targeted therapy.

Abnormalities in histone deacetylases (HDACs) may also contribute to resistance against FGFR-targeted therapies. Emerging evidence suggests that members such as HDAC2 promote tumor cell survival and drug-refractory phenotypes by modulating cell-cycle regulators, apoptosis pathways, EMT and cancer stemness features [299,300]. Although these findings have been supported in vitro and in animal models, causal evidence within the specific context of FGFR inhibition remains limited, underscoring the need for further functional and translational studies to define their actionable relevance.

DNA methylation represents another critical epigenetic determinant. FGFR-altered tumors frequently demonstrate unique methylation landscapes associated with transcriptional states and therapeutic responsiveness; however, systematic evidence remains insufficient, particularly regarding whether FGFR inhibitor treatment induces widespread remodeling of methylation programs involving DNA repair and cell death genes [301]. In glioblastoma, *MGMT* promoter methylation is strongly associated with enhanced sen-

https://doi.org/10.3390/cancers1801008

sitivity to alkylating chemotherapy, suggesting that impaired DNA repair capacity may modulate treatment response. Yet, whether similar principles apply to FGFR inhibition remains hypothetical, lacking direct clinical validation [302,303].

Non-coding RNAs, particularly microRNAs (miRNAs), play pivotal roles in therapeutic resistance regulation. Accumulating evidence demonstrates that the miR-100 family orchestrates FGFR3 expression through direct targeting of its 3′ untranslated region (3′UTR). In urothelial carcinoma, reduced miR-100 expression correlates with aberrant FGFR3 activation and enhanced downstream signaling, implicating the miR-100-associated non-coding RNA regulatory networks in modulating drug sensitivity and acquired resistance in FGFR-driven malignancies [304,305].

Emerging insights further reveal that lysosomal biology constitutes an underappreciated mechanism of resistance to small-molecule targeted therapies. Under therapeutic stress, the TFEB undergoes nuclear translocation and induces the expression of lysosomal and autophagy-related genes, thereby augmenting lysosomal biogenesis and functional capacity [290]. Certain small-molecule inhibitors—particularly lipophilic weak bases—undergo lysosomal sequestration, which substantially diminishes their effective intracellular concentrations at target sites. Moreover, alterations in vacuolar-type H^+-ATPase (V-ATPase) function or expression modulate lysosomal acidification status, potentiating this resistance phenotype [306–308]. While these mechanisms have been validated across diverse experimental models and provide a compelling framework for understanding intracellular drug distribution and resistance, direct clinical evidence specific to FGFR inhibitors remains limited, warranting rigorous investigation.

5.2. Prediction, Monitoring, and Early Intervention Strategies for Therapeutic Resistance

As FGFR-targeted therapies gain widespread clinical adoption, acquired resistance has emerged as the principal barrier limiting durable patient benefit. Establishing a systematic framework for resistance management necessitates not only mechanistic elucidation of resistance pathways but also the development of robust predictive tools, implementation of dynamic surveillance systems, and formulation of evidence-based early intervention protocols. The convergence of multi-omics-integrated prediction models, real-time ctDNA monitoring platforms, and artificial intelligence-driven personalized treatment algorithms has ushered in unprecedented opportunities for comprehensive resistance management in FGFR-driven malignancies. This multidimensional approach to resistance mitigation holds promise not only for enhancing therapeutic efficacy but also for postponing resistance emergence, ultimately improving patient clinical outcomes and long-term survival.

5.2.1. Resistance Prediction Models Integrating FGFR Genomic Alterations and Proteomic Profiling

Integrative prediction models incorporating FGFR genomic alterations and proteomic profiling have emerged as a pivotal approach for precision resistance prognostication. Multi-omics frameworks integrating genomic and proteomic signatures have demonstrated superior predictive performance relative to single-biomarker approaches across multiple cohort studies. However, prospective clinical validation remains nascent, with existing evidence predominantly derived from retrospective cohort analyses and exploratory early-phase trial investigations [309,310].

Proteomics and phosphoproteomics investigations have illuminated complex nonlinear relationships between genetic alterations and functional phenotypes, providing a rational foundation for patient-specific therapeutic target prioritization [311]. In cholangiocarcinoma, patients harboring identical *FGFR2* fusions frequently exhibit marked heterogeneity in molecular profiles and downstream signaling cascades. Converging evidence indicates that aberrant activation of compensatory pathways closely associates with intrinsic or

 https://doi.org/10.3390/cancers18010089

acquired resistance, manifesting distinct patient-specific patterns [191,287]. Consequently, elucidating resistance mechanisms typically necessitates serial biopsies or longitudinal ctDNA profiling to guide subsequent targeted strategies or sequential treatment selection. Recent investigations and comprehensive reviews consistently support the clinical utility of ctDNA in detecting resistance clones and informing therapeutic decision-making in *FGFR2* fusion-positive iCCA [312–314].

Beyond driver mutations and signaling pathway activation, genomic instability may also influence therapeutic efficacy. In certain solid tumors, homologous recombination deficiency correlates with treatment response; however, evidence supporting its role as a predictive biomarker for FGFR therapeutic efficacy remains insufficient. High tumor mutational burden (TMB) is typically accompanied by multi-pathway abnormalities and complex signaling networks, which may increase resistance risk. The relationship between TMB and FGFR inhibitor efficacy requires further validation through prospective studies [13,52].

Identification and functional validation of proteomic biomarkers are essential for establishing reliable prediction models. Large-scale screening of various FGFR-aberrant tumors has identified multiple candidate proteins associated with resistance [311,315]. For instance, low PHLDA1 expression is closely associated with resistance in RTK-driven tumors, including models related to the FGFR pathway. PHLDA1 negatively regulates AKT signaling; its loss leads to excessive PI3K-AKT activation, thereby promoting resistance [286,316]. These findings highlight the important value of proteomics in elucidating resistance mechanisms and constructing prediction models, laying the foundation for developing individualized resistance prediction tools based on functional parameters such as protein expression and phosphorylation status.

5.2.2. The Value of ctDNA Dynamic Monitoring of FGFR Secondary Mutations in Early Resistance Detection

ctDNA, as a core tool of liquid biopsy, has demonstrated unique advantages in monitoring *FGFR* secondary mutations. Compared with traditional tissue biopsy, ctDNA detection is characterized by its non-invasive, real-time, and repeatable nature, making it particularly suitable for dynamic assessment during treatment. In cholangiocarcinoma, ultra-high-sensitivity digital PCR technology has achieved detection of *FGFR2* secondary mutations at frequencies as low as 0.1% [115]. However, its application still faces several challenges, including the risk of false negatives when tumor burden is low or ctDNA shedding is limited, as well as difficulties in distinguishing tumor-derived mutations from clonal hematopoiesis-associated mutations. Additionally, the relatively short half-life of ctDNA (approximately 1.5–2 h) imposes stringent requirements on the timeliness of sample collection and processing [317–319].

Clinical studies have validated the practical value of ctDNA dynamic monitoring. Prospective investigations of *FGFR2* fusion-positive cholangiocarcinoma have established standardized protocols: baseline detection followed by initial assessment at 4 weeks of treatment, with subsequent monitoring every 8 weeks. Results demonstrate that the detection of emergent *FGFR2* secondary mutations in ctDNA precedes radiographic progression by an average of 10–14 weeks, providing a critical window for early intervention [292]. Furthermore, ctDNA analysis has revealed the polyclonal nature of resistance: individual patients may harbor multiple mutations, including *V564F*, *N549H*, and *E565A*, at different timepoints, indicating that tumors evade therapy through multiple pathways. This complexity suggests that single next-generation inhibitors are unlikely to fully overcome resistance, and combination therapies may offer superior efficacy [115].

ctDNA and imaging assessment demonstrate clear complementarity. In targeted therapy for urothelial carcinoma, ctDNA can reflect molecular-level resistance or response earlier, while CT/MRI retains irreplaceable advantages in tumor burden quantification and

https://doi.org/10.3390/cancers18010089

local control assessment [13]. Therefore, their integration can form a stratified management strategy: when ctDNA indicates resistance but imaging shows no progression, follow-up intervals should be shortened and biopsy considered; when both indicate progression, treatment strategies should be adjusted promptly. However, large-scale systematic studies to validate this combined monitoring approach are lacking, and future research should focus on this area.

5.2.3. Application Prospects of Artificial Intelligence in FGFR Resistance Prediction and Individualized Decision-Making

Artificial intelligence (AI) has demonstrated significant potential in predicting targeted drug resistance. By integrating genomic, transcriptomic, proteomic, and clinical data, researchers have constructed various machine learning models to assess resistance risk. In multiple pan-cancer studies, algorithms such as random forest and Bayesian multi-kernel learning have significantly improved prediction accuracy by integrating genetic features including point mutations, copy number variations, and gene fusions, along with clinical variables such as age, sex, stage, and prior treatment history, providing a methodological foundation for FGFR-related resistance risk assessment [320–322]. Deep learning has further advanced model performance, convolutional neural networks (CNNs) can identify complex genomic variation patterns. In cholangiocarcinoma research, CNN-based models have revealed multiple copy number and structural variation features associated with FGFR2 resistance [191,323].

Multimodal data fusion is considered a key development direction for AI in FGFR resistance prediction. By integrating ctDNA, imaging, pathology, and clinical information, complex interactions between molecular features and clinical phenotypes can be captured. Multiple studies have shown that multimodal models outperform single-modality models in precision medicine tasks [324,325]. For example, Vanguri et al. integrated radiomics, pathomics, and genomic features to predict immunotherapy response in 247 patients with advanced NSCLC, achieving an AUC of 0.80, significantly superior to single-modality models [326].

However, multimodal AI research targeting FGFR aberrations remains in early stages, predominantly consisting of small-sample or single-center explorations that lack external validation and multi-center consistency assessment [327,328]. Additionally, cross-modal data differences in dimensionality, distribution, and noise characteristics, along with data missingness and quality heterogeneity, as well as issues of model interpretability and reproducibility, all limit clinical translation. Future research requires interdisciplinary collaboration to establish standardized data collection and annotation systems, promote large-scale multi-center studies, and incorporate dynamic ctDNA monitoring information to enhance the verifiability and clinical applicability of FGFR resistance prediction models.

5.3. *Therapeutic Strategies for Overcoming Resistance*

The emergence of FGFR-targeted therapy resistance represents a significant challenge in clinical practice, necessitating innovative therapeutic strategies to overcome this obstacle. Strategies for reversing resistance must be based not only on in-depth understanding of resistance mechanisms, but also require the development of next-generation therapeutic tools and combination treatment regimens through multidisciplinary collaboration. From next-generation FGFR inhibitors designed at the molecular structural level, to exploration of synergistic mechanisms in immunotherapy combination strategies, and to formulation of precision treatment regimens based on individual resistance profiles, these innovative solutions offer new hope and avenues for overcoming FGFR-targeted therapy resistance. With deepening understanding of resistance mechanisms and continued advances in therapeutic technologies, individualized resistance reversal strategies will become an important development direction for FGFR-targeted therapy.

https://doi.org/10.3390/cancers18010089

5.3.1. Design Principles and Preclinical Research of Next-Generation FGFR Inhibitors

The core objective of next-generation FGFR inhibitors is to overcome acquired resistance mutations, particularly the steric barriers imposed by gatekeeper mutations. Irreversible inhibitors exemplified by futibatinib can achieve sustained inhibition by forming covalent bonds with cysteine residues in the FGFR kinase domain, thereby overcoming gatekeeper and other acquired mutations to a certain extent [225]. Structural studies indicate that utilizing electrophilic warheads such as acrylamide to form covalent linkages with cysteine can enhance inhibitory activity against gatekeeper mutations and prolong duration of action; however, the specific binding mode, selectivity, and pharmacokinetic effects remain dependent on molecular design [329].

The development of multi-targeted inhibitors aims to block bypass resistance. AZD4547 is a multi-targeted TKI primarily inhibiting FGFR1/2/3, with additional activity against VEGFR2/VEGFR3 and CSF1R. It can directly block FGFR-driven proliferative signaling in tumor cells while possessing anti-angiogenic potential through inhibition of FGFR1/2 and VEGFR2/3 signaling in endothelial cells. Consequently, it demonstrates potential advantages in tumors with active angiogenesis and/or synergistic FGF/VEGF signaling [130]. Another direction involves combined inhibition of FGFR and cell cycle pathways. Since FGFR signaling regulates the cell cycle through cyclin D1, preclinical studies have shown that in NSCLC models with FGF3/4/19/CCND1 amplification, LY2874455 and the CDK4/6 inhibitor abemaciclib exhibit synergistic effects and can reverse resistance; in estrogen receptor-positive breast cancer, both theoretical rationale and early clinical studies support the validity of triple combination therapy targeting ER, CDK4/6, and FGFR [330,331].

Novel drug delivery strategies also provide expanded avenues for FGFR-targeted therapy. Anti-FGFR3 antibody–drug conjugates (such as LY3076226) have demonstrated tolerable dose ranges and preliminary antitumor activity in phase I clinical studies, but require larger-scale validation [332]. Additionally, nanoparticle or lipid carriers (such as LNPs, PLGA nanoparticles, and liposomes) can significantly enhance tumor tissue accumulation of FGFR inhibitors and improve the therapeutic index in preclinical studies, suggesting their potential in drug delivery optimization; however, their clinical translation remains dependent on further evaluation of pharmacokinetics and toxicity profiles [333,334].

5.3.2. Synergistic Potential and Mechanistic Basis of FGFR Inhibition and Immunotherapy

FGFR signaling plays a central regulatory role in shaping the tumor immune microenvironment. Beyond promoting tumor cell proliferation and survival, aberrant FGFR activation can induce expression of immunosuppressive factors through multiple pathways, enhance the activity of regulatory T cells (Tregs) and myeloid suppressor cells, thereby attenuating antitumor immune responses [132,335]. Multiple murine model studies have demonstrated that FGFR inhibitors can partially reverse the immunosuppressive state, manifested by increased $CD8^+$ T cell infiltration and decreased proportions of immunosuppressive cells, thereby creating a more favorable immune environment for the action of ICIs [134,336].

Based on this rationale, the combination of FGFR inhibitors with PD-1/PD-L1 checkpoint inhibitors is considered to have potential synergistic effects. However, this effect exhibits significant tumor type and molecular dependency. For example, erdafitinib can reduce PD-L1 expression on tumor cells and enhance T cell activation in certain models, thereby improving the efficacy of anti-PD-1 antibodies; conversely, in FGFR3-activated bladder cancer models, FGFR3 inhibition can upregulate PD-L1 levels through modulation of the E3 ubiquitin ligase NEDD4, paradoxically attenuating immune responses [138,139,337]. These discrepancies suggest that complex bidirectional regulatory mechanisms exist between the FGFR pathway and immune checkpoint signaling, and combination strate-

https://doi.org/10.3390/cancers18010089

gies should be optimized in conjunction with specific molecular contexts and tumor ecosystem features.

In recent studies, FGFR has also been explored as a novel target for cell-based therapy. CAR-T cells targeting FGFR4 have demonstrated significant antitumor activity in multiple solid tumor models, providing feasibility evidence for FGFR-targeted cellular immunotherapy [338,339]. In bladder cancer, FGFR3 can regulate PD-L1 ubiquitination and stability through phosphorylation of NEDD4, thereby affecting $CD8^+$ T cell function [72,138]. Furthermore, preclinical evidence indicates that combination of FGFR inhibitors with CAR-T or ICIs can enhance therapeutic efficacy through dual mechanisms: on one hand, suppressing immunosuppressive signaling, and on the other hand, improving T cell infiltration and effector function. However, direct preclinical or clinical evidence remains limited at present, and further validation in standardized models and multi-center studies is needed in the future [340].

5.3.3. Formulation of Individualized Treatment Strategies Based on Resistance Mechanisms

The core of personalized therapy lies in the precise identification of patients' molecular characteristics. Different types of FGFR aberrations exhibit differential drug sensitivity: some studies suggest that patients with FGFR gene amplification or overexpression are more sensitive to ATP-competitive inhibitors, whereas those with fusions or specific activating mutations demonstrate more heterogeneous responses [52]. Monitoring of resistance mutation spectra is particularly critical. For patients harboring gatekeeper mutations, irreversible FGFR inhibitors may be effective [191], whereas if resistance mechanisms involve bypass pathway activation, a combination of FGFR inhibitors with pathway-specific inhibitors may hold greater promise [341]. Through ctDNA monitoring technology, the emergence of resistance mutations can be tracked in real time, providing evidence for timely adjustment of treatment regimens.

Beyond molecular characteristics, phenotype-based individualized strategies are equally important. In highly vascularized FGFR-aberrant tumors, FGFR inhibitors combined with anti-angiogenic agents have demonstrated synergistic effects in preclinical models and warrant further exploration in certain solid tumors [132,342]; in tumors with high FGFR4/KLB expression, selective FGFR4 inhibitors have shown feasibility in early studies, and preliminary results of their combination with PD-1 inhibitors suggest potential benefits [343]. Immune phenotype also serves as a key reference: in immunologically "hot" tumors, combination of FGFR inhibitors with checkpoint inhibitors may produce more significant therapeutic efficacy; whereas in immunologically "cold" tumors, it may be necessary to first remodel the microenvironment through FGFR inhibitors or other strategies before introducing immunotherapy [222].

Furthermore, dynamic and adaptive treatment strategies further expand the scope of individualized management. In early treatment phases, the most effective FGFR inhibitors can be prioritized to achieve rapid disease control; when resistance signals emerge, combination or sequential therapy should be promptly implemented. Through intermittent dosing or dose adjustment, selection pressure on resistant clones can be reduced, thereby delaying resistance development. In prostate cancer androgen deprivation therapy, intermittent treatment has been proven to delay resistance and improve quality of life, and similar strategies warrant exploration in FGFR-targeted therapy [344–346].

6. Combination Treatment Strategies and Multidisciplinary Integration

6.1. Combination Modalities of FGFR Inhibitors with Conventional Therapies

As an emerging targeted therapy, FGFR inhibitors have demonstrated application potential in various solid tumors. However, their single-agent efficacy is often limited by re-

 https://doi.org/10.3390/cancers18010089

sistance and tumor heterogeneity. Therefore, combining them with conventional treatments (chemotherapy, radiotherapy, and surgery) has become a critical direction for optimizing existing therapeutic paradigms. This section focuses on exploring combination modalities of FGFR inhibitors with conventional therapies, encompassing temporal sequencing, dose optimization, mechanisms of action, clinical validation, and toxicity management. By integrating the latest clinical and preclinical data, we evaluate their contribution to the therapeutic paradigm of solid tumors.

6.1.1. Studies on Temporal Sequencing and Dose Optimization in Combination with Chemotherapy

In cholangiocarcinoma, the combination of FGFR inhibitors (represented by pemigatinib) with chemotherapy (such as gemcitabine, GEM) has emerged as an important exploratory direction. Preclinical studies have demonstrated that GEM can induce downstream pathway activation (including phosphorylation of FRS2 and ERK) in certain FGF-activated tumor models, accompanied by alterations in cell cycle and mitotic markers; in cell lines and xenograft models harboring *FGFR2* fusions, the combination of GEM with pemigatinib produced synergistic antitumor effects both in vitro and in vivo, significantly enhancing apoptosis through inhibition of FGFR-mediated survival signaling, suggesting potential clinical value for concurrent combination in chemotherapy-naive patients with FGF-activated backgrounds [347]. Given that other receptor tyrosine kinases have shown schedule-dependent effects, systematic comparison of three strategies—"targeted therapy followed by chemotherapy," "chemotherapy followed by targeted therapy," and "concurrent administration"—is still needed in the future to confirm the optimal timing of FGFR inhibitors with chemotherapy [348].

Regarding dose optimization, FIGHT-101 established and recommended pemigatinib at 13.5 mg QD (intermittent dosing, 2 weeks on/1 week off) as the dose for subsequent studies. FIGHT-202 was conducted under this regimen and validated clinical activity in cholangiocarcinoma with *FGFR2* fusions/rearrangements. In FIGHT-202, the ORR in patients with *FGFR2* fusions/rearrangements was approximately 37.0%, supporting the clinical significance of dose and patient selection [4,349]. However, clinical evidence regarding combination of pemigatinib with standard chemotherapy (such as gemcitabine/cisplatin) remains limited, currently consisting primarily of preclinical studies, early-phase/small-sample clinical reports, and reviews, with a lack of large-scale randomized controlled trials to definitively establish optimal combination doses and administration schedules [350]. Additionally, chemotherapy dose intensity (clinically used GEM doses typically around 1000 mg/m^2, with variations of 800–1250 mg/m^2 in different regimens) is closely associated with toxicity risk, requiring careful balance between efficacy and safety in clinical practice [351–353].

6.1.2. Molecular Mechanisms of Radiosensitization and Clinical Translation Prospects

The radiosensitization effect of FGFR inhibitors in bladder cancer may depend on their intervention in DNA damage response and cell cycle regulation. FGFR/FGF signaling can affect multiple cell cycle checkpoints, commonly at the G1/S phase, and in specific contexts can also act on G2/M, thereby exhibiting context-dependent radiosensitivity [354,355]. Preclinical studies have shown that FGFR inhibitors can impair homologous recombination repair capacity, for instance, by prolonging the duration of γH2AX signaling and reducing RAD51 expression, thereby exacerbating radiation-induced DNA double-strand break damage and enhancing apoptosis [356,357].

Regarding clinical evidence, large-scale randomized trials confirming FGFR inhibitors as universal radiosensitizers are currently lacking. The phase III THOR study (cohort 2) compared the efficacy of erdafitinib versus pembrolizumab, with results showing no significant difference in OS (10.9 vs. 11.1 months, $p = 0.18$); PFS was numerically superior

 https://doi.org/10.3390/cancers18010089

(4.4 vs. 2.7 months) but did not reach statistical significance; however, ORR was significantly improved (40.0% vs. 21.6%) [358]. Nevertheless, this study did not include radiotherapy-related controls or subgroup analyses, and thus is insufficient to validate the radiosensitization effect of FGFR inhibitors. Recent systematic reviews have also emphasized that current evidence is inadequate to support widespread combination of FGFR inhibitors with radiotherapy, and future studies will need to rely on prospective randomized trials and systematic meta-analyses to clarify their clinical value [58].

6.1.3. Translational Prospects of FGFR2 Fusions in Postoperative Risk Stratification and Individualized Treatment of Cholangiocarcinoma

FGFR2 fusions have gradually demonstrated potential value in postoperative prognostic assessment and surgical decision-making. A retrospective single-center study revealed that *FGFR2* fusion-positive patients exhibited significantly superior long-term survival compared to fusion-negative patients following curative resection, and *FGFR2* fusion was identified as an independent protective factor in multivariate analysis, suggesting its utility as a molecular biomarker for preoperative or intraoperative risk stratification [359]. Furthermore, related cohort studies have found that patients with *FGFR2* fusions or rearrangements often exhibit reprogramming of immune-related pathways and differential immune cell infiltration patterns, which are closely associated with improved clinical outcomes, further reinforcing their potential application value in individualized treatment strategies [140].

At the mechanistic level, *FGFR2* fusions can constitutively activate the RAS/MAPK and PI3K/AKT signaling pathways, promoting tumor cell proliferation and invasion. Targeted intervention of these signaling axes may alter tumor biological behavior, providing a theoretical rationale for perioperative treatment [360]. Clinical studies have demonstrated that FGFR inhibitors can partially reverse the associated phenotypes. Notably, the irreversible inhibitor futibatinib has exhibited significant efficacy in patients with *FGFR2* fusion-positive iCCA who are resistant to ATP-competitive inhibitors, suggesting that perioperative treatment regimens should be individualized and optimized based on resistance mutation profiles [191,361].

However, existing evidence predominantly originates from retrospective or small-sample studies, and large-scale prospective validation remains lacking. Molecular detection methodology also represents a critical limiting factor for clinical translation. Different detection platforms exhibit variations in coverage, sensitivity, and fusion partner identification capability: studies have shown that DNA-only NGS has a higher false-negative rate compared to RNA sequencing or FISH, and concordance across different study series is limited. Therefore, clinical practice should adopt multimodal detection or repeat validation strategies to enhance accuracy [192,362]. Overall, *FGFR2* fusions demonstrate significant value in surgical prognostic assessment and individualized treatment, but their clinical application still requires further confirmation through multicenter large-scale studies.

6.1.4. Mechanisms of Action and Translational Evidence for FGFR Inhibitors Combined with Chemotherapy

Studies have demonstrated that *FGFR1* amplification or overexpression is closely associated with the EMT program and invasive/metastatic phenotypes. In multiple lung cancer models, inhibition of FGFR signaling can suppress or reverse EMT features and partially restore sensitivity to certain antitumor agents [363,364]. In the *KRAS*-mutant context, FGFR1 can serve as a compensatory pathway following inhibition of downstream KRAS signaling; therefore, the combination of FGFR inhibitors with chemotherapy or downstream pathway inhibitors has demonstrated synergistic effects in preclinical models under this context [365,366].

https://doi.org/10.3390/cancers18010089

Another potential synergistic mechanism involves c-Myc regulation: multiple models have shown that FGFR inhibitors can downregulate c-Myc and promote tumor cell apoptosis. Taxanes can also reduce c-Myc expression in certain systems, suggesting that the two have biological synergistic potential in regulating c-Myc and related proliferation/apoptosis networks. A recent review on the association between taxanes and c-Myc has summarized supporting evidence from different models, indicating that this synergistic hypothesis warrants further validation in in vivo and clinical trials [367,368].

5.1.5. Toxicity and Management Strategies for FGFR Inhibitor Combination Therapy

Multimodal combinations (FGFR inhibitor + chemotherapy ± radiotherapy) are frequently associated with on-target toxicities, including hyperphosphatemia, ocular toxicity, and fatigue. Hyperphosphatemia is the most common, with management measures including low-phosphate diet, phosphate binders, and dose adjustment or treatment discontinuation when necessary. Ocular toxicity (such as corneal epithelial changes and retinopathy) requires regular monitoring before and during treatment to enable early detection and intervention [14,369].

Clinical experience and guidelines indicate that combination therapy has a higher incidence of toxicity compared to monotherapy, but most adverse events can be controlled through supportive care and dose adjustment. Current toxicity management is largely experience-based and lacks large-scale prospective validation. Multiple reviews have emphasized that establishing standardized management frameworks (early monitoring, multidisciplinary collaboration, and timely intervention) can significantly reduce treatment discontinuation rates and improve adherence, though specific outcomes vary across studies [370,371]. Therefore, toxicity management of combination therapy requires multidisciplinary collaboration to ensure a balance between efficacy and safety and to improve patient outcomes.

6.2. *Synergistic Mechanisms of FGFR-Targeted Therapy and Immunotherapy*

The combination of FGFR-targeted therapy and immunotherapy is considered a cutting-edge strategy that can overcome the limitations of monotherapy by dually modulating the tumor microenvironment and enhancing antitumor immune responses. Current research focuses include: the mechanistic role of FGFR signaling in immune regulation, evidence from preclinical combination studies, exploration of adoptive cell therapy and cancer vaccines, and synergistic effects of gene regulation in specific cancer types. Based on multi-omics data and preclinical models, these strategies hold promise for improving therapeutic efficacy and reversing resistance.

6.2.1. Regulatory Mechanisms of FGFR Signaling on the Immune Microenvironment

FGFR signaling plays a pivotal regulatory role in shaping the immune microenvironment in bladder cancer. *FGFR3* activation or mutation is particularly common in bladder cancer, with an incidence of 50–70% in non-muscle-invasive cases and approximately 10–15% in muscle-invasive tumors, with an overall reported rate of 15–20% [78,372]. Mechanistic studies have revealed that FGFR3 can remodel the immune microenvironment through regulation of PD-L1 stability and the IFN-γ signaling pathway. Jing et al. found that FGFR3 activation promotes NEDD4-mediated ubiquitination and degradation of PD-L1, leading to PD-L1 downregulation and reduced CD8^{+} T cell infiltration; whereas FGFR3 inhibition can block this process, upregulate PD-L1, and enhance anti-PD-1 therapeutic efficacy in mouse models, providing a molecular rationale for combining FGFR inhibitors with immune checkpoint inhibitors [138]. Additionally, FGFR signaling activation can suppress the IFN-γ–JAK/STAT response, limiting cytokine and chemokine expression, thereby attenuating the recruitment and function of dendritic cells and natural

https://doi.org/10.3390/cancers1801008

killer cells; whereas FGFR inhibition can restore IFN-γ pathway activity and enhance immune activation [132,136]. These findings reveal the bidirectional role of FGFR signaling in immune regulation, although most remain based on in vitro and animal studies and await clinical validation.

Retrospective analyses have indicated that *FGFR3* alterations are typically associated with a "non-T cell-inflamed" tumor phenotype, characterized by insufficient $CD8^+$ T cell infiltration and poor response to ICI monotherapy. However, this association exhibits significant heterogeneity: in low-grade or well-differentiated lesions, *FGFR3* mutations are sometimes associated with better prognosis, whereas in high-grade tumors or those harboring co-mutations such as *TP53* and *PIK3CA*, they are frequently associated with immune evasion and poor ICI response [108,372,373]. These findings underscore that the immune regulatory effects of FGFR3 are jointly influenced by pathological grade, molecular context, and immune ecology.

In summary, FGFR3 influences tumor immune phenotype through regulation of PD-L1 stability and the IFN-γ signaling pathway, serving as an important molecular hub determining the immune microenvironment in bladder cancer. Its biological characteristics provide a theoretical foundation for FGFR status-based combination immunotherapy, but further validation in prospective clinical studies with clearly defined molecular stratification is still needed to clarify its translational potential and optimal combination strategies.

6.2.2. FGFR Inhibition-Mediated Immune Remodeling and Its Translational Potential in Combination Immunotherapy

Extensive preclinical studies have demonstrated that FGF/FGFR pathway inhibition can remodel the tumor immune microenvironment, thereby enhancing the antitumor efficacy of ICIs. Mechanistically, FGFR blockade can restore or enhance IFN-γ signaling pathway activity, upregulate antitumor immunity-related genes, and promote antigen presentation; additionally, it can induce vascular normalization, reduce immunosuppressive macrophage infiltration, and reactivate the IFN-γ pathway, thereby generating synergistic effects with anti-PD-1 or anti-PD-L1 therapy [136,336]. It should be noted that some agents used in combination immunotherapy (such as lenvatinib) are multi-target inhibitors, and their immunomodulatory effects are not solely mediated by FGFR inhibition. Recent clinical and real-world studies both suggest that careful distinction of the contributions of different targets is required when interpreting the efficacy of lenvatinib-ICI combinations [337,374].

Multiple animal model studies have further validated this synergistic effect: FGFR inhibitors combined with PD-1/PD-L1 inhibitors can significantly increase $CD8^+$ T cell infiltration, reduce immunosuppressive cell populations, and promote TCR clonal expansion, thereby achieving superior antitumor activity and survival benefit compared to monotherapy. For example, in *FGFR2*-driven lung cancer and *FGFR3*-driven urothelial carcinoma models, erdafitinib or other FGFR inhibitors combined with anti-PD-1 all demonstrated significant synergistic effects [132,336]. However, efficacy varied significantly across different models, and some mouse studies also observed increased combination therapy-related toxicity, suggesting that clinical translation requires thorough evaluation of safety and tolerability.

Clinically, the FIGHT-207 study demonstrated that pemigatinib achieved a monotherapy ORR of 26.5% in select solid tumors with *FGFR* alterations, suggesting that FGFR inhibition has definitive antitumor activity. However, clinical data on combinations with ICIs remain limited at present. Future studies should systematically evaluate the synergistic potential of FGFR inhibition and immunotherapy through stratified early- to mid-phase clinical trials, and validate their long-term impact on PFS, OS, and safety in randomized phase III trials [224,336].

https://doi.org/10.3390/cancers18010089

6.2.3. Exploratory Combination of Cellular Therapies Such as CAR-T and TCR-T with FGFR-Targeted Therapy

In cholangiocarcinoma, *FGFR2* fusions not only provide a target for small molecule inhibitors but also lay the foundation for neoantigen-directed cell therapies such as TCR-T. Research by White et al. demonstrated that TILs isolated from patients with metastatic cholangiocarcinoma can specifically recognize *FGFR2* fusion peptides, directly demonstrating that these sites possess natural immune accessibility and providing critical evidence for TCR-T design [375]. Currently, FGFR-targeted cell therapy research remains largely in preclinical or early clinical phases. Previous studies have indicated that FGFR-targeted CAR-T can be developed and have demonstrated potential in some solid tumor models; however, due to differences in tissue distribution and expression profiles among different subtypes of the FGFR family, *FGFR2* fusion-directed CAR-T development faces greater challenges [338,376,377].

Regarding combination strategies, theoretical and preliminary studies suggest that FGFR-TKIs can enhance CAR-T/TCR-T through the following pathways: first, by weakening tumor-dependent pathways and exposing more antigens; second, by remodeling the tumor microenvironment (reducing MDSCs/TAMs and improving vascular permeability), thereby promoting exogenous T cell infiltration and persistence; and third, by achieving precise elimination through fusion site-specific TCRs. Although these mechanisms are supported in vitro and in limited animal models, systematic preclinical data on pemigatinib or other selective FGFR-TKIs combined with CAR-T/TCR-T remain insufficient, and existing efficacy evidence should be cautiously interpreted [375,378].

Although this strategy possesses potential advantages through targeting of specific fusion proteins and improvement in the immune microenvironment, its limitations are also significant: first, fusion partner diversity leads to antigenic heterogeneity; second, cell therapy in solid tumors still faces challenges of insufficient infiltration, T cell exhaustion, and immunosuppressive barriers; third, safety concerns may compound the metabolic and electrolyte toxicities of FGFR-TKIs. Future efforts require establishment of a systematic preclinical validation framework and comprehensive evaluation of safety and durable efficacy through phased clinical trials progressing from small to expanded sample sizes [338,379].

6.2.4. Application Prospects of Tumor Vaccines in FGFR-Positive Tumors

Tumor vaccines can provide novel strategies for the treatment of FGFR-positive tumors by inducing specific immune responses. Animal model studies have demonstrated that vaccines based on xenogeneic FGFR1 (pxFR1) can significantly inhibit tumor growth and exert antitumor effects through activation of $CD8^+$ T cells and antibody responses [380]. Furthermore, FGFR1 as an immune target can be combined with FGFR inhibitors to enhance T cell responses and memory effects [381]. These findings suggest that FGFR-related antigens possess certain immunogenicity, providing a theoretical basis for the development of FGFR-targeted tumor vaccines.

However, compared to vaccine research for molecular targets such as HER2, EGFR, and NY-ESO-1, FGFR vaccine development is notably lagging. HER2 vaccines have entered multiple phase I/II clinical trials in breast cancer and demonstrated certain immunological activity; EGFR vaccines have also been explored in non-small-cell lung cancer, and although efficacy remains limited, translational experience has been accumulated; while cancer testis antigen vaccines such as NY-ESO-1 and MUC1 have demonstrated durable T cell responses in various solid tumors [382–384]. In contrast, FGFR vaccines still lack systematic preclinical and clinical validation, with existing evidence remaining primarily at the level of animal experiments.

 https://doi.org/10.3390/cancers18010089

Overall, vaccine development for FGFR-positive tumors remains in its nascent stages, with significant research gaps. Its potential value is more likely to be reflected in adjuvant therapy or recurrence prevention rather than monotherapy, and requires interdisciplinary collaboration and multicenter clinical trials to further advance the field.

7. Conclusions

Although extensive research has revealed the biological diversity of FGFR aberrations in various solid tumors and their feasibility as potential drug targets, there remains a lack of systematic, pan-cancer evidence integration that comprehensively compares different FGFR aberration types, resistance profiles, and optimal clinical management pathways. To address this gap, this review systematically reviews and synthesizes key evidence from recent years regarding the molecular mechanisms, clinical development progress, and translational challenges of FGFR aberration-driven solid tumors. Multi-omics analyses and functional studies consistently demonstrate that in several cancer types, FGFR aberrations can serve as definitive oncogenic driver events and significantly influence the heterogeneity in pharmacological sensitivity and clinical benefit to FGFR-targeted strategies. With the successive approval of targeted drugs such as pemigatinib, erdafitinib, and futibatinib and their demonstration of considerable efficacy in clinical trials, FGFR-targeted therapy has transitioned from concept to partial clinical benefit. However, the heterogeneity in real-world efficacy, complex resistance mechanisms, and lack of biomarkers indicate that substantial work remains in the clinical translation of this field.

First, different types of FGFR aberrations exhibit significant differences in signal dependency and drug sensitivity. Fusions and rearrangements typically lead to ligand-independent activation, whereas amplifications or point mutations display varying degrees of dependency. Therefore, precise stratification based on aberration type, fusion partners, and co-mutation profiles is critical for achieving personalized therapy. Although qualitative insights into pathway hyperactivation are presented here, a quantitative meta-analysis of signaling intensity remains unfeasible due to the lack of standardized data. Existing evidence relies largely on disparate assays (e.g., Western blot) with varying baseline conditions, preventing a unified comparison of MAPK/ERK or PI3K/AKT activation across different FGFR genotypes. Future efforts must therefore prioritize standardized phosphoproteomic quantification to bridge the gap between mechanistic characterization and clinical stratification.

Second, acquired resistance is widespread. Secondary kinase domain mutations, compensatory activation of downstream pathways or other RTKs represent the major mechanisms. Irreversible inhibitors and multi-target combination strategies developed specifically for these mechanisms have shown partial potential to overcome resistance, but polyclonal resistance and heterogeneity remain to be addressed. Liquid biopsy-assisted dynamic monitoring and sequential therapy will become future directions.

Additionally, FGFR inhibition exerts dual regulatory effects on the tumor immune microenvironment. It can enhance antitumor immune responses but may also upregulate PD-L1 expression in certain contexts. Precision combination designs based on molecular and immune features are expected to improve response rates to immune checkpoint inhibitors, while FGF-VEGF signaling crosstalk also suggests potential value in combining with antiangiogenic therapy.

Future research should focus on three aspects: ① establishing a standardized biomarker system that integrates tissue, ctDNA, and multi-omics data; ② optimizing combination and sequential strategies to delay resistance development; ③ promoting the development of new inhibitors with high selectivity or covalent binding properties. The

https://doi.org/10.3390/cancers18010089

introduction of artificial intelligence and multimodal data analysis is expected to further enhance the predictive and monitoring precision of FGFR-targeted therapy.

Overall, research on *FGFR* aberration-driven tumors is transitioning from mechanistic exploration toward clinical translation. Through molecular stratification, dynamic monitoring, and combination innovation, FGFR-targeted therapy is expected to achieve reproducible clinical benefit in a broader range of cancer types.

Author Contributions: Z.H.: Conceptualization, Writing—original draft preparation, Visualization, Methodology. G.L., J.W. and Y.W.: Data curation. P.T., Y.H. and X.P.: Visualization, Supervision. L.Z.: Visualization, Funding acquisition. Y.C., H.L. and W.C.: Conceptualization, Supervision, Project administration, Writing—review & editing, Funding acquisition. All authors have read and agreed to the published version of the manuscript.

Funding: The study was funded by the National Natural Science Foundation of China (82102981), the Natural Science Foundation of Fujian Province (2022J05073), Joint Funds for the Innovation of Science and Technology, Fujian Province (2023Y9352).

Institutional Review Board Statement: Not applicable.

Informed Consent Statement: Not applicable.

Data Availability Statement: No new data were generated or analyzed in this study. All data discussed in this review are derived from previously published studies and publicly available databases cited within the manuscript.

Acknowledgments: During the preparation of this work the authors used ChatGPT (version 5.1, OpenAI) in order to improve the language and readability of the manuscript. After using this tool, the authors reviewed and edited the content as needed and take full responsibility for the content of the publication.

Conflicts of Interest: The authors declare no conflicts of interest and the funders had no role in the design of the study; in the collection, analyses, or interpretation of data; in the writing of the manuscript; or in the decision to publish the results.

Abbreviations

The following abbreviations are used in this manuscript:

FGFRs	fibroblast growth factor receptors
FGFR	fibroblast growth factor receptor
RTK	receptor tyrosine kinase
FGFs	fibroblast growth factors
HSPGs	heparan sulfate proteoglycans
SPRY	Sprouty
DUSPs	dual-specificity phosphatases
iCCA	intrahepatic cholangiocarcinoma
GAB1	GRB2-associated binder 1
ER	estrogen receptor
SEF	similar expression to FGF
MMPs	matrix metalloproteinases
TAMs	tumor-associated macrophages
TKIs	tyrosine kinase inhibitors
NSCLC	non-small-cell lung cancer
CAFs	cancer-associated fibroblasts
ICIs	immune checkpoint inhibitors
MDSCs	myeloid-derived suppressor cells
ECM	extracellular matrix
EMT	epithelial–mesenchymal transition

https://doi.org/10.3390/cancers1801008

iCAFs	inflammatory cancer-associated fibroblasts
NETs	neutrophil extracellular traps
CSCs	cancer stem cells
LDH	lactate dehydrogenase
ADCs	antibody–drug conjugates
mAbs	monoclonal antibodies
PFS	progression-free survival
PDCs	polymer–drug conjugates
DCR	disease control rate
OS	overall survival
IRC	independent review committee
DOR	duration of response
ORR	objective response rate
ctDNA	circulating tumor DNA
FAERS	the FDA Adverse Event Reporting System
NGS	next-generation sequencing
PHLDA1	pleckstrin homology-like domain family A member 1
TFEB	transcription factor EB
DDR	DNA damage response
HDACs	histone deacetylases
miRNAs	microRNAs
3′UTR	3′ untranslated region
V-ATPase	vacuolar-type H^+-ATPase
TMB	tumor mutational burden
AI	Artificial intelligence
CNNs	convolutional neural networks
Tregs	regulatory T cells
GEM	gemcitabine
NA	not applicable
mPFS	median progression-free survival
mOS	median overall survival
mDOR	median duration of response
GEJ	gastroesophageal junction
EPR	Enhanced Permeability and Retention effect

References

1. Bray, F.; Laversanne, M.; Sung, H.; Ferlay, J.; Siegel, R.L.; Soerjomataram, I.; Jemal, A. Global cancer statistics 2022: GLOBOCAN estimates of incidence and mortality worldwide for 36 cancers in 185 countries. *CA Cancer J. Clin.* **2024**, *74*, 229–263. [CrossRef]
2. Hutter, C.; Zenklusen, J.C. The Cancer Genome Atlas: Creating Lasting Value beyond Its Data. *Cell* **2018**, *173*, 283–285. [CrossRef]
3. Zhang, P.; Yue, L.; Leng, Q.; Chang, C.; Gan, C.; Ye, T.; Cao, D. Targeting FGFR for cancer therapy. *J. Hematol. Oncol.* **2024**, *17*, 39. [CrossRef]
4. Abou-Alfa, G.K.; Sahai, V.; Hollebecque, A.; Vaccaro, G.; Melisi, D.; Al-Rajabi, R.; Paulson, A.S.; Borad, M.J.; Gallinson, D.; Murphy, A.G.; et al. Pemigatinib for previously treated, locally advanced or metastatic cholangiocarcinoma: A multicentre, open-label, phase 2 study. *Lancet Oncol.* **2020**, *21*, 671–684. [CrossRef]
5. Loriot, Y.; Necchi, A.; Park, S.H.; Garcia-Donas, J.; Huddart, R.; Burgess, E.; Fleming, M.; Rezazadeh, A.; Mellado, B.; Varlamov, S.; et al. Erdafitinib in Locally Advanced or Metastatic Urothelial Carcinoma. *N. Engl. J. Med.* **2019**, *381*, 338–348. [CrossRef]
6. Bahleda, R.; Meric-Bernstam, F.; Goyal, L.; Tran, B.; He, Y.; Yamamiya, I.; Benhadji, K.A.; Matos, I.; Arkenau, H.T. Phase I, first-in-human study of futibatinib, a highly selective, irreversible FGFR1-4 inhibitor in patients with advanced solid tumors. *Ann. Oncol.* **2020**, *31*, 1405–1412. [CrossRef]
7. Choudhury, N.J.; Harada, G.; Whiting, K.; Shen, R.; Yang, S.-R.; Solit, D.B.; Berger, M.F.; Dinapoli, S.; Abou-Alfa, G.K.; Iyer, G.; et al. Genomic characterization of *FGFR*-altered solid tumors using a clinically annotated pan-cancer repository. *J. Clin. Oncol.* **2023**, *41*, 3074. [CrossRef]
8. Zuo, W.; He, Y.; Li, W.; Wu, H.; Zhao, Z.; Zhang, Y.; Chen, S.; Yin, Y. Landscape of FGF/FGFR Alterations in 12,372 Chinese Cancer Patients. *J. Cancer* **2020**, *11*, 6695–6699. [CrossRef]

https://doi.org/10.3390/cancers18010089

9. Bou Zerdan, M.; Bratslavsky, G.; Jacob, J.; Ross, J.; Huang, R.; Basnet, A. Urothelial Bladder Cancer: Genomic Alterations in Fibroblast Growth Factor Receptor. *Mol. Diagn. Ther.* **2023**, *27*, 475–485. [CrossRef]
10. Li, J.; Hu, K.; Huang, J.; Zhou, L.; Yan, Y.; Xu, Z. Insights of fibroblast growth factor receptor 3 aberrations in pan-cancer and their roles in potential clinical treatment. *Aging* **2021**, *13*, 16541–16566. [CrossRef]
11. Wu, F.; Fan, J.; He, Y.; Xiong, A.; Yu, J.; Li, Y.; Zhang, Y.; Zhao, W.; Zhou, F.; Li, W.; et al. Single-cell profiling of tumor heterogeneity and the microenvironment in advanced non-small cell lung cancer. *Nat. Commun.* **2021**, *12*, 2540. [CrossRef]
12. Wu, Q.; Ellis, H.; Siravegna, G.; Michel, A.G.; Norden, B.L.; Fece de la Cruz, F.; Balasooriya, E.R.; Zhen, Y.; Silveira, V.S.; Che, J. et al. Landscape of Clinical Resistance Mechanisms to FGFR Inhibitors in FGFR2-Altered Cholangiocarcinoma. *Clin. Cancer Res.* **2024**, *30*, 198–208. [CrossRef] [PubMed]
13. Facchinetti, F.; Hollebecque, A.; Braye, F.; Vasseur, D.; Pradat, Y.; Bahleda, R.; Pobel, C.; Bigot, L.; Deas, O.; Florez Arango, J.D. et al. Resistance to Selective FGFR Inhibitors in FGFR-Driven Urothelial Cancer. *Cancer Discov.* **2023**, *13*, 1998–2011. [CrossRef] [PubMed]
14. Subbiah, V.; Verstovsek, S. Clinical development and management of adverse events associated with FGFR inhibitors. *Cell Rep. Med.* **2023**, *4*, 101204. [CrossRef] [PubMed]
15. Hsu, J.; Francis, J.H.; Ahmad, S. Ocular toxicities of fibroblast growth factor receptor inhibitors: A review. *Surv. Ophthalmol.* **2024**, *69*, 34–41. [CrossRef]
16. Gu, W.; Yang, J.; Wang, Y.; Xu, J.; Wang, X.; Du, F.; Hu, X.; Guo, H.; Song, C.; Tao, R.; et al. Comprehensive identification of FGFR1-4 alterations in 5557 Chinese patients with solid tumors by next-generation sequencing. *Am. J. Cancer Res.* **2021**, *11*, 3893–3906.
17. Piha-Paul, S.A.; Xu, B.; Dumbrava, E.E.; Fu, S.; Karp, D.D.; Meric-Bernstam, F.; Hong, D.S.; Rodon, J.A.; Tsimberidou, A.M. Raghav, K.; et al. First-In-Human Phase I Study of Tinengotinib (TT-00420), a Multiple Kinase Inhibitor, as a Single Agent in Patients with Advanced Solid Tumors. *Oncologist* **2024**, *29*, e514–e525. [CrossRef]
18. Schuler, M.; Cho, B.C.; Sayehli, C.M.; Navarro, A.; Soo, R.A.; Richly, H.; Cassier, P.A.; Tai, D.; Penel, N.; Nogova, L.; et al. Rogaratinib in patients with advanced cancers selected by FGFR mRNA expression: A phase 1 dose-escalation and dose expansion study. *Lancet Oncol.* **2019**, *20*, 1454–1466. [CrossRef]
19. Xie, Y.; Su, N.; Yang, J.; Tan, Q.; Huang, S.; Jin, M.; Ni, Z.; Zhang, B.; Zhang, D.; Luo, F.; et al. FGF/FGFR signaling in health and disease. *Signal Transduct. Target. Ther.* **2020**, *5*, 181. [CrossRef]
20. Olsen, S.K.; Ibrahimi, O.A.; Raucci, A.; Zhang, F.; Eliseenkova, A.V.; Yayon, A.; Basilico, C.; Linhardt, R.J.; Schlessinger, J. Mohammadi, M. Insights into the molecular basis for fibroblast growth factor receptor autoinhibition and ligand-binding promiscuity. *Proc. Natl. Acad. Sci. USA* **2004**, *101*, 935–940. [CrossRef]
21. Liu, Q.; Huang, J.; Yan, W.; Liu, Z.; Liu, S.; Fang, W. FGFR families: Biological functions and therapeutic interventions in tumors. *MedComm* **2023**, *4*, e367. [CrossRef]
22. Aprajita; Sharma, R. Comprehending fibroblast growth factor receptor like 1: Oncogene or tumor suppressor? *Cancer Treat. Res. Commun.* **2021**, *29*, 100472. [CrossRef]
23. Schlessinger, J.; Plotnikov, A.N.; Ibrahimi, O.A.; Eliseenkova, A.V.; Yeh, B.K.; Yayon, A.; Linhardt, R.J.; Mohammadi, M. Crystal structure of a ternary FGF-FGFR-heparin complex reveals a dual role for heparin in FGFR binding and dimerization. *Mol. Cell* **2000**, *6*, 743–750. [CrossRef]
24. Chen, L.; Fu, L.; Sun, J.; Huang, Z.; Fang, M.; Zinkle, A.; Liu, X.; Lu, J.; Pan, Z.; Wang, Y.; et al. Structural basis for FGF hormone signalling. *Nature* **2023**, *618*, 862–870. [CrossRef]
25. Goetz, R.; Mohammadi, M. Exploring mechanisms of FGF signalling through the lens of structural biology. *Nat. Rev. Mol. Cell Biol.* **2013**, *14*, 166–180. [CrossRef]
26. Aaldijk, A.S.; Verzijl, C.R.C.; Jonker, J.W.; Struik, D. Biological and pharmacological functions of the FGF19- and FGF21-coreceptor beta klotho. *Front. Endocrinol.* **2023**, *14*, 1150222. [CrossRef]
27. Phan, P.; Ternier, G.; Edirisinghe, O.; Kumar, T.K.S. Exploring endocrine FGFs—Structures, functions and biomedical applications. *Int. J. Biochem. Mol. Biol.* **2024**, *15*, 68–99. [CrossRef]
28. Dolegowska, K.; Marchelek-Mysliwiec, M.; Nowosiad-Magda, M.; Slawinski, M.; Dolegowska, B. FGF19 subfamily members FGF19 and FGF21. *J. Physiol. Biochem.* **2019**, *75*, 229–240. [CrossRef]
29. Liu, Y.; Cao, M.; Cai, Y.; Li, X.; Zhao, C.; Cui, R. Dissecting the Role of the FGF19-FGFR4 Signaling Pathway in Cancer Development and Progression. *Front. Cell Dev. Biol.* **2020**, *8*, 95. [CrossRef]
30. Beenken, A.; Mohammadi, M. The structural biology of the FGF19 subfamily. *Adv. Exp. Med. Biol.* **2012**, *728*, 1–24. [CrossRef]
31. Lin, B.C.; Wang, M.; Blackmore, C.; Desnoyers, L.R. Liver-specific activities of FGF19 require Klotho beta. *J. Biol. Chem.* **2007**, *282*, 27277–27284. [CrossRef]
32. Goetz, R.; Beenken, A.; Ibrahimi, O.A.; Kalinina, J.; Olsen, S.K.; Eliseenkova, A.V.; Xu, C.; Neubert, T.A.; Zhang, F.; Linhardt, R.J. et al. Molecular insights into the klotho-dependent, endocrine mode of action of fibroblast growth factor 19 subfamily members *Mol. Cell. Biol.* **2007**, *27*, 3417–3428. [CrossRef] [PubMed]

https://doi.org/10.3390/cancers18010089

Edirisinghe, O.; Ternier, G.; Alraawi, Z.; Suresh Kumar, T.K. Decoding FGF/FGFR Signaling: Insights into Biological Functions and Disease Relevance. *Biomolecules* **2024**, *14*, 1622. [CrossRef]

Mohammadi, M.; Olsen, S.K.; Ibrahimi, O.A. Structural basis for fibroblast growth factor receptor activation. *Cytokine Growth Factor Rev.* **2005**, *16*, 107–137. [CrossRef]

Farrell, B.; Breeze, A.L. Structure, activation and dysregulation of fibroblast growth factor receptor kinases: Perspectives for clinical targeting. *Biochem. Soc. Trans.* **2018**, *46*, 1753–1770. [CrossRef]

Park, J.S.; Choi, J.; Cao, L.; Mohanty, J.; Suzuki, Y.; Park, A.; Baker, D.; Schlessinger, J.; Lee, S. Isoform-specific inhibition of FGFR signaling achieved by a de-novo-designed mini-protein. *Cell Rep.* **2022**, *41*, 111545. [CrossRef]

Ornitz, D.M.; Itoh, N. New developments in the biology of fibroblast growth factors. *WIREs Mech. Dis.* **2022**, *14*, e1549. [CrossRef]

Kumar, V.; Goutam, R.S.; Park, S.; Lee, U.; Kim, J. Functional Roles of FGF Signaling in Early Development of Vertebrate Embryos. *Cells* **2021**, *10*, 2148. [CrossRef]

Gerber, S.D.; Beauchamp, P.; Zhuang, L.; Villiger, P.M.; Trueb, B. Functional domains of the FgfrL1 receptor. *Dev. Biol.* **2020**, *461*, 43–54. [CrossRef]

Zhuang, L.; Vogel, M.; Villiger, P.M.; Trueb, B. Dissecting the Interaction of FGF8 with Receptor FGFRL1. *Biomolecules* **2020**, *10*, 1399. [CrossRef]

Gofin, Y.; Mackay, L.P.; Machol, K.; Keswani, S.; Potocki, L.; Di Gregorio, E.; Naretto, V.G.; Brusco, A.; Hernandez-Garcia, A.; Scott, D.A. Evidence that FGFRL1 contributes to congenital diaphragmatic hernia development in humans. *Am. J. Med Genet. A* **2021**, *185*, 836–840. [CrossRef] [PubMed]

Eswarakumar, V.P.; Lax, I.; Schlessinger, J. Cellular signaling by fibroblast growth factor receptors. *Cytokine Growth Factor. Rev.* **2005**, *16*, 139–149. [CrossRef] [PubMed]

Guan, L.; Feng, L.; Wang, C.; Xie, Y. FGFRL1: Structure, Molecular Function, and Involvement in Human Disease. *Curr. Issues Mol. Biol.* **2025**, *47*, 286. [CrossRef]

Szybowska, P.; Kostas, M.; Wesche, J.; Haugsten, E.M.; Wiedlocha, A. Negative Regulation of FGFR (Fibroblast Growth Factor Receptor) Signaling. *Cells* **2021**, *10*, 1342. [CrossRef]

Relav, L.; Price, C.A. Regulation of dual specificity phosphatases by fibroblast growth factor signaling pathways in bovine granulosa cells. *Reproduction* **2021**, *162*, 367–374. [CrossRef]

Helsten, T.; Elkin, S.; Arthur, E.; Tomson, B.N.; Carter, J.; Kurzrock, R. The FGFR Landscape in Cancer: Analysis of 4853 Tumors by Next-Generation Sequencing. *Clin. Cancer Res.* **2016**, *22*, 259–267. [CrossRef] [PubMed]

Benayed, R.; Offin, M.; Mullaney, K.; Sukhadia, P.; Rios, K.; Desmeules, P.; Ptashkin, R.; Won, H.; Chang, J.; Halpenny, D.; et al. High Yield of RNA Sequencing for Targetable Kinase Fusions in Lung Adenocarcinomas with No Mitogenic Driver Alteration Detected by DNA Sequencing and Low Tumor Mutation Burden. *Clin. Cancer Res.* **2019**, *25*, 4712–4722. [CrossRef]

Hieronymus, H.; Murali, R.; Tin, A.; Yadav, K.; Abida, W.; Moller, H.; Berney, D.; Scher, H.; Carver, B.; Scardino, P.; et al. Tumor copy number alteration burden is a pan-cancer prognostic factor associated with recurrence and death. *Elife* **2018**, *7*, e37294. [CrossRef]

Boone, P.M.; Bacino, C.A.; Shaw, C.A.; Eng, P.A.; Hixson, P.M.; Pursley, A.N.; Kang, S.H.; Yang, Y.; Wiszniewska, J.; Nowakowska, B.A.; et al. Detection of clinically relevant exonic copy-number changes by array CGH. *Hum. Mutat.* **2010**, *31*, 1326–1342. [CrossRef]

Merker, J.D.; Oxnard, G.R.; Compton, C.; Diehn, M.; Hurley, P.; Lazar, A.J.; Lindeman, N.; Lockwood, C.M.; Rai, A.J.; Schilsky, R.L.; et al. Circulating Tumor DNA Analysis in Patients With Cancer: American Society of Clinical Oncology and College of American Pathologists Joint Review. *J. Clin. Oncol.* **2018**, *36*, 1631–1641. [CrossRef]

Aggarwal, C.; Thompson, J.C.; Black, T.A.; Katz, S.I.; Fan, R.; Yee, S.S.; Chien, A.L.; Evans, T.L.; Bauml, J.M.; Alley, E.W.; et al. Clinical Implications of Plasma-Based Genotyping With the Delivery of Personalized Therapy in Metastatic Non-Small Cell Lung Cancer. *JAMA Oncol.* **2019**, *5*, 173–180. [CrossRef]

Krook, M.A.; Reeser, J.W.; Ernst, G.; Barker, H.; Wilberding, M.; Li, G.; Chen, H.Z.; Roychowdhury, S. Fibroblast growth factor receptors in cancer: Genetic alterations, diagnostics, therapeutic targets and mechanisms of resistance. *Br. J. Cancer* **2021**, *124*, 880–892. [CrossRef]

The ICGC/TCGA Pan-Cancer Analysis of Whole Genomes Consortium; Campbell, P.J.; Getz, G. Pan-cancer analysis of whole genomes. *Nature* **2020**, *578*, 82–93. [CrossRef] [PubMed]

Li, J.; Hu, K.; Huang, J.; Zhou, L.; Yan, Y.; Xu, Z. A Pancancer Analysis of the Expression Landscape and Clinical Relevance of Fibroblast Growth Factor Receptor 2 in Human Cancers. *Front. Oncol.* **2021**, *11*, 644854. [CrossRef] [PubMed]

Sun, Y.; Li, G.; Zhu, W.; He, Q.; Liu, Y.; Chen, X.; Liu, J.; Lin, J.; Han-Zhang, H.; Yang, Z.; et al. A comprehensive pan-cancer study of fibroblast growth factor receptor aberrations in Chinese cancer patients. *Ann. Transl. Med.* **2020**, *8*, 1290. [CrossRef]

Noeraparast, M.; Krajina, K.; Pichler, R.; Niedersuss-Beke, D.; Shariat, S.F.; Grunwald, V.; Ahyai, S.; Pichler, M. FGFR3 alterations in bladder cancer: Sensitivity and resistance to targeted therapies. *Cancer Commun.* **2024**, *44*, 1189–1208. [CrossRef]

https://doi.org/10.3390/cancers18010089

57. Teo, M.Y.; Mota, J.M.; Whiting, K.A.; Li, H.A.; Funt, S.A.; Lee, C.H.; Solit, D.B.; Al-Ahmadie, H.; Milowsky, M.I.; Balar, A.V.; et Fibroblast Growth Factor Receptor 3 Alteration Status is Associated with Differential Sensitivity to Platinum-based Chemothera in Locally Advanced and Metastatic Urothelial Carcinoma. *Eur. Urol.* **2020**, *78*, 907–915. [CrossRef]
58. Li, R.; Linscott, J.; Catto, J.W.F.; Daneshmand, S.; Faltas, B.M.; Kamat, A.M.; Meeks, J.J.; Necchi, A.; Pradere, B.; Ross, J.S.; et FGFR Inhibition in Urothelial Carcinoma. *Eur. Urol.* **2025**, *87*, 110–122. [CrossRef]
59. Jain, A.; Borad, M.J.; Kelley, R.K.; Wang, Y.; Abdel-Wahab, R.; Meric-Bernstam, F.; Baggerly, K.A.; Kaseb, A.O.; Al-Shamsi, H.C Ahn, D.H.; et al. Cholangiocarcinoma With FGFR Genetic Aberrations: A Unique Clinical Phenotype. *JCO Precis. Oncol.* **201** 2, 1–12. [CrossRef]
60. Du, S.; Zhang, Y.; Xu, J. Current progress in cancer treatment by targeting FGFR signaling. *Cancer Biol. Med.* **2023**, *20*, 490–4 [CrossRef]
61. Makinen, N.; Meyerson, M. Genomic insights into the mechanisms of FGFR1 dependency in squamous cell lung cancer. *J. Cl Investig.* **2023**, *133*, e174171. [CrossRef]
62. Lassman, A.B.; Sepulveda-Sanchez, J.M.; Cloughesy, T.F.; Gil-Gil, M.J.; Puduvalli, V.K.; Raizer, J.J.; De Vos, F.Y.F.; Wen, P. Butowski, N.A.; Clement, P.M.J.; et al. Infigratinib in Patients with Recurrent Gliomas and FGFR Alterations: A Multicenter Pha II Study. *Clin. Cancer Res.* **2022**, *28*, 2270–2277. [CrossRef]
63. Marastoni, E.; Mulone, D.; Barresi, V. Diffuse Gliomas with FGFR3::TACC3 Fusion: Morphological and Molecular Features ar Classification Challenges. *Cancers* **2024**, *16*, 1644. [CrossRef]
64. Masago, K.; Kuroda, H.; Sasaki, E.; Fujita, Y.; Fujita, S.; Horio, Y.; Endo, M.; Ishihara, H.; Hanai, N.; Matsushita, H. Novel ger fusions in human oropharyngeal carcinoma. *Cancer Genet.* **2024**, *286–287*, 29–34. [CrossRef] [PubMed]
65. Sharma, P.; Kaushik, V.; Saraya, A.; Sharma, R. Aberrant Expression of FGFRL1 in Esophageal Cancer and Its Regulation miR-107. *Asian Pac. J. Cancer Prev.* **2023**, *24*, 1331–1341. [CrossRef] [PubMed]
66. Chen, R.; Li, D.; Zheng, M.; Chen, B.; Wei, T.; Wang, Y.; Li, M.; Huang, W.; Tong, Q.; Wang, Q.; et al. FGFRL1 affec chemoresistance of small-cell lung cancer by modulating the PI3K/Akt pathway via ENO1. *J. Cell. Mol. Med.* **2020**, *24*, 2123–21 [CrossRef]
67. Yu, L.; Toriseva, M.; Afshan, S.; Cangiano, M.; Fey, V.; Erickson, A.; Seikkula, H.; Alanen, K.; Taimen, P.; Ettala, O.; et al. Increase Expression and Altered Cellular Localization of Fibroblast Growth Factor Receptor-Like 1 (FGFRL1) Are Associated with Prosta Cancer Progression. *Cancers* **2022**, *14*, 278. [CrossRef]
68. Wang, L.; He, J.; Hu, H.; Tu, L.; Sun, Z.; Liu, Y.; Luo, F. Lung CSC-derived exosomal miR-210-3p contributes to a pro-metastat phenotype in lung cancer by targeting FGFRL1. *J. Cell. Mol. Med.* **2020**, *24*, 6324–6339. [CrossRef]
69. Guo, Y.; Cui, J.; Liang, X.; Chen, T.; Lu, C.; Peng, T. Pancreatic cancer stem cell-derived exosomal miR-210 mediates macropha M2 polarization and promotes gemcitabine resistance by targeting FGFRL1. *Int. Immunopharmacol.* **2024**, *127*, 111407. [CrossRe
70. Bannier, P.A.; Saillard, C.; Mann, P.; Touzot, M.; Maussion, C.; Matek, C.; Klumper, N.; Breyer, J.; Wirtz, R.; Sikic, D.; et al. allows pre-screening of FGFR3 mutational status using routine histology slides of muscle-invasive bladder cancer. *Nat. Commu* **2024**, *15*, 10914. [CrossRef] [PubMed]
71. Guercio, B.J.; Sarfaty, M.; Teo, M.Y.; Ratna, N.; Duzgol, C.; Funt, S.A.; Lee, C.H.; Aggen, D.H.; Regazzi, A.M.; Chen, et al. Clinical and Genomic Landscape of FGFR3-Altered Urothelial Carcinoma and Treatment Outcomes with Erdafitinib: Real-World Experience. *Clin. Cancer Res.* **2023**, *29*, 4586–4595. [CrossRef]
72. Wang, Z.; Muthusamy, V.; Petrylak, D.P.; Anderson, K.S. Tackling FGFR3-driven bladder cancer with a promising synergist FGFR/HDAC targeted therapy. *NPJ Precis. Oncol.* **2023**, *7*, 70. [CrossRef] [PubMed]
73. Silverman, I.M.; Hollebecque, A.; Friboulet, L.; Owens, S.; Newton, R.C.; Zhen, H.; Feliz, L.; Zecchetto, C.; Melisi, D.; Bur T.C. Clinicogenomic Analysis of FGFR2-Rearranged Cholangiocarcinoma Identifies Correlates of Response and Mechanisms Resistance to Pemigatinib. *Cancer Discov.* **2021**, *11*, 326–339. [CrossRef]
74. Zingg, D.; Bhin, J.; Yemelyanenko, J.; Kas, S.M.; Rolfs, F.; Lutz, C.; Lee, J.K.; Klarenbeek, S.; Silverman, I.M.; Annunziato, S.; et Truncated FGFR2 is a clinically actionable oncogene in multiple cancers. *Nature* **2022**, *608*, 609–617. [CrossRef]
75. Marin, A.; Morales, F.; Walbaum, B. Fibroblast growth factor receptor signaling in estrogen receptor-positive breast canc Mechanisms and role in endocrine resistance. *Front. Oncol.* **2024**, *14*, 1406951. [CrossRef]
76. Elbauomy Elsheikh, S.; Green, A.R.; Lambros, M.B.; Turner, N.C.; Grainge, M.J.; Powe, D.; Ellis, I.O.; Reis-Filho, J.S. FGF amplification in breast carcinomas: A chromogenic in situ hybridisation analysis. *Breast Cancer Res.* **2007**, *9*, R23. [CrossRe [PubMed]
77. Nakamura, I.T.; Kohsaka, S.; Ikegami, M.; Ikeuchi, H.; Ueno, T.; Li, K.; Beyett, T.S.; Koyama, T.; Shimizu, T.; Yamamoto, N.; et Comprehensive functional evaluation of variants of fibroblast growth factor receptor genes in cancer. *npj Precis. Oncol.* **2021**, *5*, [CrossRef]
78. Ahmad, F.; Mahal, V.; Verma, G.; Bhatia, S.; Das, B.R. Molecular investigation of FGFR3 gene mutation and its correlation wi clinicopathological findings in Indian bladder cancer patients. *Cancer Rep.* **2018**, *1*, e1130. [CrossRef] [PubMed]

https://doi.org/10.3390/cancers180100

79. Brown, L.M.; Ekert, P.G.; Fleuren, E.D.G. Biological and clinical implications of FGFR aberrations in paediatric and young adult cancers. *Oncogene* **2023**, *42*, 1875–1888. [CrossRef]
80. Kelleher, F.C.; O'Sullivan, H.; Smyth, E.; McDermott, R.; Viterbo, A. Fibroblast growth factor receptors, developmental corruption and malignant disease. *Carcinogenesis* **2013**, *34*, 2198–2205. [CrossRef]
81. Wendt, M.K.; Taylor, M.A.; Schiemann, B.J.; Sossey-Alaoui, K.; Schiemann, W.P. Fibroblast growth factor receptor splice variants are stable markers of oncogenic transforming growth factor beta1 signaling in metastatic breast cancers. *Breast Cancer Res.* **2014**, *16*, R24. [CrossRef]
82. Ishiwata, T. Role of fibroblast growth factor receptor-2 splicing in normal and cancer cells. *Front. Biosci.* **2018**, *23*, 626–639. [CrossRef]
83. Chen, Y.Y.; Zhang, Q.; Gui, M.H.; Feng, L.; Cao, P.B.; Zhou, G.Q. PTBP1 promotes the progression of hepatocellular carcinoma by enhancing the oncogenic splicing switch of FGFR2. *Yi Chuan* **2024**, *46*, 46–62. [CrossRef] [PubMed]
84. Olender, J.; Wang, B.D.; Ching, T.; Garmire, L.X.; Garofano, K.; Ji, Y.; Knox, T.; Latham, P.; Nguyen, K.; Rhim, J.; et al. A Novel FGFR3 Splice Variant Preferentially Expressed in African American Prostate Cancer Drives Aggressive Phenotypes and Docetaxel Resistance. *Mol. Cancer Res.* **2019**, *17*, 2115–2125. [CrossRef] [PubMed]
85. Turner, N.; Grose, R. Fibroblast growth factor signalling: From development to cancer. *Nat. Rev. Cancer* **2010**, *10*, 116–129. [CrossRef] [PubMed]
86. Song, C.; Zhang, Z.; Leng, D.; He, Z.; Wang, X.; Liu, W.; Zhang, W.; Wu, Q.; Zhao, Q.; Chen, G. ERK5 promotes autocrine expression to sustain mitogenic balance for cell fate specification in human pluripotent stem cells. *Stem Cell Rep.* **2024**, *19*, 1320–1335. [CrossRef]
87. Watson, J.; Ferguson, H.R.; Brady, R.M.; Ferguson, J.; Fullwood, P.; Mo, H.; Bexley, K.H.; Knight, D.; Howell, G.; Schwartz, J.M.; et al. Spatially resolved phosphoproteomics reveals fibroblast growth factor receptor recycling-driven regulation of autophagy and survival. *Nat. Commun.* **2022**, *13*, 6589. [CrossRef]
88. He, Y.; Sun, M.M.; Zhang, G.G.; Yang, J.; Chen, K.S.; Xu, W.W.; Li, B. Targeting PI3K/Akt signal transduction for cancer therapy. *Signal Transduct. Target. Ther.* **2021**, *6*, 425. [CrossRef]
89. Glaviano, A.; Foo, A.S.C.; Lam, H.Y.; Yap, K.C.H.; Jacot, W.; Jones, R.H.; Eng, H.; Nair, M.G.; Makvandi, P.; Geoerger, B.; et al. PI3K/AKT/mTOR signaling transduction pathway and targeted therapies in cancer. *Mol. Cancer* **2023**, *22*, 138. [CrossRef]
90. Zheng, J.; Zhang, W.; Li, L.; He, Y.; Wei, Y.; Dang, Y.; Nie, S.; Guo, Z. Signaling Pathway and Small-Molecule Drug Discovery of FGFR: A Comprehensive Review. *Front. Chem.* **2022**, *10*, 860985. [CrossRef]
91. Ferguson, H.R.; Smith, M.P.; Francavilla, C. Fibroblast Growth Factor Receptors (FGFRs) and Noncanonical Partners in Cancer Signaling. *Cells* **2021**, *10*, 1201. [CrossRef]
92. Peng, M.; Deng, J.; Li, X. Clinical advances and challenges in targeting FGF/FGFR signaling in lung cancer. *Mol. Cancer* **2024**, *23*, 256. [CrossRef]
93. Homayounfar, S.; Ayatollahi, H.; Ferns, G.; Hemmatan Attarbashi, R.; Gharib, M.; Sheikhi, M.; Khoshnegah, Z.; Siyadat, P.; Jafarian, A. Fibroblast growth factor receptor 1 gene (FGFR1) amplification in non-small cell lung cancer (NSCLC) by real-time PCR. *Caspian J. Intern. Med.* **2025**, *16*, 66–72. [CrossRef]
94. Hong, C.S.; Sun, E.G.; Choi, J.N.; Kim, D.H.; Kim, J.H.; Ryu, K.H.; Shim, H.J.; Hwang, J.E.; Bae, W.K.; Kim, H.R.; et al. Fibroblast growth factor receptor 4 increases epidermal growth factor receptor (EGFR) signaling by inducing amphiregulin expression and attenuates response to EGFR inhibitors in colon cancer. *Cancer Sci.* **2020**, *111*, 3268–3278. [CrossRef]
95. Sung, V.Y.C.; Knight, J.F.; Johnson, R.M.; Stern, Y.E.; Saleh, S.M.; Savage, P.; Monast, A.; Zuo, D.; Duhamel, S.; Park, M. Co-dependency for MET and FGFR1 in basal triple-negative breast cancers. *npj Breast Cancer* **2021**, *7*, 36. [CrossRef]
96. Servetto, A.; Formisano, L.; Arteaga, C.L. FGFR signaling and endocrine resistance in breast cancer: Challenges for the clinical development of FGFR inhibitors. *Biochim. Biophys. Acta Rev. Cancer* **2021**, *1876*, 188595. [CrossRef]
97. Hanafusa, H.; Torii, S.; Yasunaga, T.; Nishida, E. Sprouty1 and Sprouty2 provide a control mechanism for the Ras/MAPK signalling pathway. *Nat. Cell Biol.* **2002**, *4*, 850–858. [CrossRef]
98. Mason, J.M.; Morrison, D.J.; Basson, M.A.; Licht, J.D. Sprouty proteins: Multifaceted negative-feedback regulators of receptor tyrosine kinase signaling. *Trends Cell Biol.* **2006**, *16*, 45–54. [CrossRef]
99. Yoo, D.H.; Im, Y.S.; Oh, J.Y.; Gil, D.; Kim, Y.O. DUSP6 is a memory retention feedback regulator of ERK signaling for cellular resilience of human pluripotent stem cells in response to dissociation. *Sci. Rep.* **2023**, *13*, 5683. [CrossRef]
100. Cabrita, M.A.; Christofori, G. Sprouty proteins, masterminds of receptor tyrosine kinase signaling. *Angiogenesis* **2008**, *11*, 53–62. [CrossRef]
101. Haugsten, E.M.; Malecki, J.; Bjorklund, S.M.; Olsnes, S.; Wesche, J. Ubiquitination of fibroblast growth factor receptor 1 is required for its intracellular sorting but not for its endocytosis. *Mol. Biol. Cell* **2008**, *19*, 3390–3403. [CrossRef]
102. Tang, R.; Langdon, W.Y.; Zhang, J. Negative regulation of receptor tyrosine kinases by ubiquitination: Key roles of the Cbl family of E3 ubiquitin ligases. *Front. Endocrinol.* **2022**, *13*, 971162. [CrossRef] [PubMed]
103. Clark, J.F.; Soriano, P. Diverse Fgfr1 signaling pathways and endocytic trafficking regulate mesoderm development. *Genes Dev.* **2024**, *38*, 393–414. [CrossRef]

https://doi.org/10.3390/cancers18010089

104. Moes-Sosnowska, J.; Chorostowska-Wynimko, J. Fibroblast Growth Factor Receptor 1-4 Genetic Aberrations as Clinically Relevant Biomarkers in Squamous Cell Lung Cancer. *Front. Oncol.* **2022**, *12*, 780650. [CrossRef]
105. Weiss, J.; Sos, M.L.; Seidel, D.; Peifer, M.; Zander, T.; Heuckmann, J.M.; Ullrich, R.T.; Menon, R.; Maier, S.; Soltermann, A.; et al. Frequent and focal FGFR1 amplification associates with therapeutically tractable FGFR1 dependency in squamous cell lung cancer. *Sci. Transl. Med.* **2010**, *2*, 62ra93. [CrossRef]
106. Schrumpf, T.; Behrens, H.M.; Haag, J.; Kruger, S.; Rocken, C. FGFR2 overexpression and compromised survival in diffuse-type gastric cancer in a large central European cohort. *PLoS ONE* **2022**, *17*, e0264011. [CrossRef]
107. di Martino, E.; L'Hote, C.G.; Kennedy, W.; Tomlinson, D.C.; Knowles, M.A. Mutant fibroblast growth factor receptor 3 induces intracellular signaling and cellular transformation in a cell type- and mutation-specific manner. *Oncogene* **2009**, *28*, 4306–4316. [CrossRef] [PubMed]
108. Komura, K.; Hirosuna, K.; Tokushige, S.; Tsujino, T.; Nishimura, K.; Ishida, M.; Hayashi, T.; Ura, A.; Ohno, T.; Yamazaki, S.; et al. The Impact of FGFR3 Alterations on the Tumor Microenvironment and the Efficacy of Immune Checkpoint Inhibitors in Bladder Cancer. *Mol. Cancer* **2023**, *22*, 185. [CrossRef]
109. Yu, S.H.; Kim, S.S.; Kim, S.; Lee, H.; Kang, T.W. FGFR3 Mutations in Urothelial Carcinoma: A Single-Center Study Using Next-Generation Sequencing. *J. Clin. Med.* **2024**, *13*, 1305. [CrossRef]
110. Porta, R.; Borea, R.; Coelho, A.; Khan, S.; Araujo, A.; Reclusa, P.; Franchina, T.; Van Der Steen, N.; Van Dam, P.; Ferri, J.; et al. FGFR a promising druggable target in cancer: Molecular biology and new drugs. *Crit. Rev. Oncol. Hematol.* **2017**, *113*, 256–267. [CrossRef] [PubMed]
111. Graham, R.P.; Barr Fritcher, E.G.; Pestova, E.; Schulz, J.; Sitailo, L.A.; Vasmatzis, G.; Murphy, S.J.; McWilliams, R.R.; Hart, S.N.; Halling, K.C.; et al. Fibroblast growth factor receptor 2 translocations in intrahepatic cholangiocarcinoma. *Hum. Pathol.* **2014**, *45*, 1630–1638. [CrossRef] [PubMed]
112. Yin, L.; Han, Z.; Feng, M.; Wang, J.; Xie, Z.; Yu, W.; Fu, X.; Shen, N.; Wang, X.; Duan, A.; et al. Chimeric transcripts observed in non-canonical FGFR2 fusions with partner genes' breakpoint located in intergenic region in intrahepatic cholangiocarcinoma. *Cancer Genet.* **2022**, *266–267*, 39–43. [CrossRef] [PubMed]
113. Arai, Y.; Totoki, Y.; Hosoda, F.; Shirota, T.; Hama, N.; Nakamura, H.; Ojima, H.; Furuta, K.; Shimada, K.; Okusaka, T.; et al. Fibroblast growth factor receptor 2 tyrosine kinase fusions define a unique molecular subtype of cholangiocarcinoma. *Hepatology* **2014**, *59*, 1427–1434. [CrossRef]
114. Sia, D.; Losic, B.; Moeini, A.; Cabellos, L.; Hao, K.; Revill, K.; Bonal, D.; Miltiadous, O.; Zhang, Z.; Hoshida, Y.; et al. Massive parallel sequencing uncovers actionable FGFR2-PPHLN1 fusion and ARAF mutations in intrahepatic cholangiocarcinoma. *Nat. Commun.* **2015**, *6*, 6087. [CrossRef]
115. Goyal, L.; Saha, S.K.; Liu, L.Y.; Siravegna, G.; Leshchiner, I.; Ahronian, L.G.; Lennerz, J.K.; Vu, P.; Deshpande, V.; Kambadakone, A.; et al. Polyclonal Secondary FGFR2 Mutations Drive Acquired Resistance to FGFR Inhibition in Patients with FGFR2 Fusion-Positive Cholangiocarcinoma. *Cancer Discov.* **2017**, *7*, 252–263. [CrossRef]
116. Boichuk, S.; Dunaev, P.; Galembikova, A.; Valeeva, E. Fibroblast Growth Factor 2 (FGF2) Activates Vascular Endothelial Growth Factor (VEGF) Signaling in Gastrointestinal Stromal Tumors (GIST): An Autocrine Mechanism Contributing to Imatinib Mesylate (IM) Resistance. *Cancers* **2024**, *16*, 3103. [CrossRef]
117. Liu, G.; Chen, T.; Ding, Z.; Wang, Y.; Wei, Y.; Wei, X. Inhibition of FGF-FGFR and VEGF-VEGFR signalling in cancer treatment. *Cell Prolif.* **2021**, *54*, e13009. [CrossRef]
118. Thuault, S.; Ghossoub, R.; David, G.; Zimmermann, P. A Journey on Extracellular Vesicles for Matrix Metalloproteinases: A Mechanistic Perspective. *Front. Cell Dev. Biol.* **2022**, *10*, 886381. [CrossRef]
119. Nguyen, A.L.; Facey, C.O.B.; Boman, B.M. The Complexity and Significance of Fibroblast Growth Factor (FGF) Signaling for FGF-Targeted Cancer Therapies. *Cancers* **2024**, *17*, 82. [CrossRef] [PubMed]
120. Chapman, J.R.; Katsara, O.; Ruoff, R.; Morgenstern, D.; Nayak, S.; Basilico, C.; Ueberheide, B.; Kolupaeva, V. Phosphoproteomics of Fibroblast Growth Factor 1 (FGF1) Signaling in Chondrocytes: Identifying the Signature of Inhibitory Response. *Mol. Cell Proteom.* **2017**, *16*, 1126–1137. [CrossRef]
121. De Luca, A.; Esposito Abate, R.; Rachiglio, A.M.; Maiello, M.R.; Esposito, C.; Schettino, C.; Izzo, F.; Nasti, G.; Normanno, N. FGFR Fusions in Cancer: From Diagnostic Approaches to Therapeutic Intervention. *Int. J. Mol. Sci.* **2020**, *21*, 6856. [CrossRef]
122. Cunningham, D.L.; Sarhan, A.R.; Creese, A.J.; Larkins, K.P.B.; Zhao, H.; Ferguson, H.R.; Brookes, K.; Marusiak, A.A.; Cooper, H.J.; Heath, J.K. Differential responses to kinase inhibition in FGFR2-addicted triple negative breast cancer cells: A quantitative phosphoproteomics study. *Sci. Rep.* **2020**, *10*, 7950. [CrossRef] [PubMed]
123. Terp, M.G.; Jacobsen, K.; Molina, M.A.; Karachaliou, N.; Beck, H.C.; Bertran-Alamillo, J.; Gimenez-Capitan, A.; Cardona, A.F.; Rosell, R.; Ditzel, H.J. Combined FGFR and Akt pathway inhibition abrogates growth of FGFR1 overexpressing EGFR-TKI-resistant NSCLC cells. *npj Precis. Oncol.* **2021**, *5*, 65. [CrossRef] [PubMed]
124. Liu, Z.L.; Chen, H.H.; Zheng, L.L.; Sun, L.P.; Shi, L. Angiogenic signaling pathways and anti-angiogenic therapy for cancer. *Signal Transduct. Target. Ther.* **2023**, *8*, 198. [CrossRef]

5. Shen, L.; Li, Y.; Zhao, H. Fibroblast growth factor signaling in macrophage polarization: Impact on health and diseases. *Front. Immunol.* **2024**, *15*, 1390453. [CrossRef] [PubMed]
6. Tu, J.; Xu, H.; Ma, L.; Li, C.; Qin, W.; Chen, X.; Yi, M.; Sun, L.; Liu, B.; Yuan, X. Nintedanib enhances the efficacy of PD-L1 blockade by upregulating MHC-I and PD-L1 expression in tumor cells. *Theranostics* **2022**, *12*, 747–766. [CrossRef]
7. Liu, J.; Gao, J.; Wang, A.; Jiang, Z.; Qi, S.; Qi, Z.; Liu, F.; Yu, K.; Cao, J.; Chen, C.; et al. Nintedanib overcomes drug resistance from upregulation of FGFR signalling and imatinib-induced KIT mutations in gastrointestinal stromal tumours. *Mol. Oncol.* **2022**, *16*, 1761–1774. [CrossRef]
8. Yan, S.; Xue, S.; Wang, T.; Gao, R.; Zeng, H.; Wang, Q.; Jia, X. Efficacy and safety of nintedanib in patients with non-small cell lung cancer, and novel insights in radiation-induced lung toxicity. *Front. Oncol.* **2023**, *13*, 1086214. [CrossRef]
9. Jain, R.K. Normalization of tumor vasculature: An emerging concept in antiangiogenic therapy. *Science* **2005**, *307*, 58–62. [CrossRef]
0. Katoh, M. FGFR inhibitors: Effects on cancer cells, tumor microenvironment and whole-body homeostasis (Review). *Int. J. Mol. Med.* **2016**, *38*, 3–15. [CrossRef]
1. Yin, H.; Frontini, M.J.; Arpino, J.M.; Nong, Z.; O'Neil, C.; Xu, Y.; Balint, B.; Ward, A.D.; Chakrabarti, S.; Ellis, C.G.; et al. Fibroblast Growth Factor 9 Imparts Hierarchy and Vasoreactivity to the Microcirculation of Renal Tumors and Suppresses Metastases. *J. Biol. Chem.* **2015**, *290*, 22127–22142. [CrossRef]
2. Ruan, R.; Li, L.; Li, X.; Huang, C.; Zhang, Z.; Zhong, H.; Zeng, S.; Shi, Q.; Xia, Y.; Zeng, Q.; et al. Unleashing the potential of combining FGFR inhibitor and immune checkpoint blockade for FGF/FGFR signaling in tumor microenvironment. *Mol. Cancer* **2023**, *22*, 60. [CrossRef]
3. Yang, C.; Song, D.; Zhao, F.; Wu, J.; Zhang, B.; Ren, H.; Sun, Q.; Qin, S. Comprehensive analysis of the prognostic value and immune infiltration of FGFR family members in gastric cancer. *Front. Oncol.* **2022**, *12*, 936952. [CrossRef]
4. Wu, Y.; Yi, Z.; Li, J.; Wei, Y.; Feng, R.; Liu, J.; Huang, J.; Chen, Y.; Wang, X.; Sun, J.; et al. FGFR blockade boosts T cell infiltration into triple-negative breast cancer by regulating cancer-associated fibroblasts. *Theranostics* **2022**, *12*, 4564–4580. [CrossRef]
5. Liu, L.; Ye, T.H.; Han, Y.P.; Song, H.; Zhang, Y.K.; Xia, Y.; Wang, N.Y.; Xiong, Y.; Song, X.J.; Zhu, Y.X.; et al. Reductions in myeloid-derived suppressor cells and lung metastases using AZD4547 treatment of a metastatic murine breast tumor model. *Cell. Physiol. Biochem.* **2014**, *33*, 633–645. [CrossRef]
6. Adachi, Y.; Kamiyama, H.; Ichikawa, K.; Fukushima, S.; Ozawa, Y.; Yamaguchi, S.; Goda, S.; Kimura, T.; Kodama, K.; Matsuki, M.; et al. Inhibition of FGFR Reactivates IFNgamma Signaling in Tumor Cells to Enhance the Combined Antitumor Activity of Lenvatinib with Anti-PD-1 Antibodies. *Cancer Res.* **2022**, *82*, 292–306. [CrossRef]
7. Maddaluno, L.; Urwyler, C.; Rauschendorfer, T.; Meyer, M.; Stefanova, D.; Sporri, R.; Wietecha, M.; Ferrarese, L.; Stoycheva, D.; Bender, D.; et al. Antagonism of interferon signaling by fibroblast growth factors promotes viral replication. *EMBO Mol. Med.* **2020**, *12*, e11793. [CrossRef]
8. Jing, W.; Wang, G.; Cui, Z.; Xiong, G.; Jiang, X.; Li, Y.; Li, W.; Han, B.; Chen, S.; Shi, B. FGFR3 Destabilizes PD-L1 via NEDD4 to Control T-cell-Mediated Bladder Cancer Immune Surveillance. *Cancer Res.* **2022**, *82*, 114–129. [CrossRef]
9. Lin, Y.C.; Chu, C.Y.; Hsieh, T.H.; Lin, B.J.; Liou, J.P.; Yen, Y.; Chen, C.H. FGFR inhibitors promote the autophagic degradation of IFN-gamma-induced PD-L1 and alleviate the PD-L1-mediated transcriptional suppression of FGFR3-TACC3 in non-muscle-invasive bladder cancer. *Cell Death Dis.* **2025**, *16*, 485. [CrossRef]
0. Liu, S.; Weng, J.; Cao, M.; Zhou, Q.; Xu, M.; Xu, W.; Hu, Z.; Xu, M.; Dong, Q.; Sheng, X.; et al. FGFR2 fusion/rearrangement is associated with favorable prognosis and immunoactivation in patients with intrahepatic cholangiocarcinoma. *Oncologist* **2024**, *29*, e1734–e1747. [CrossRef]
41. Dong, L.; Lu, D.; Chen, R.; Lin, Y.; Zhu, H.; Zhang, Z.; Cai, S.; Cui, P.; Song, G.; Rao, D.; et al. Proteogenomic characterization identifies clinically relevant subgroups of intrahepatic cholangiocarcinoma. *Cancer Cell* **2022**, *40*, 70–87.e15. [CrossRef] [PubMed]
42. Yi, C.; Chen, L.; Lin, Z.; Liu, L.; Shao, W.; Zhang, R.; Lin, J.; Zhang, J.; Zhu, W.; Jia, H.; et al. Lenvatinib Targets FGF Receptor 4 to Enhance Antitumor Immune Response of Anti-Programmed Cell Death-1 in HCC. *Hepatology* **2021**, *74*, 2544–2560. [CrossRef] [PubMed]
43. Maia, A.; Schollhorn, A.; Schuhmacher, J.; Gouttefangeas, C. CAF-immune cell crosstalk and its impact in immunotherapy. *Semin. Immunopathol.* **2023**, *45*, 203–214. [CrossRef] [PubMed]
44. Wright, K.; Ly, T.; Kriet, M.; Czirok, A.; Thomas, S.M. Cancer-Associated Fibroblasts: Master Tumor Microenvironment Modifiers. *Cancers* **2023**, *15*, 1899. [CrossRef]
45. Li, D.; Xia, L.; Huang, P.; Wang, Z.; Guo, Q.; Huang, C.; Leng, W.; Qin, S. Cancer-associated fibroblast-secreted IGFBP7 promotes gastric cancer by enhancing tumor associated macrophage infiltration via FGF2/FGFR1/PI3K/AKT axis. *Cell Death Discov.* **2023**, *9*, 17. [CrossRef]
46. Ardizzone, A.; Bova, V.; Casili, G.; Repici, A.; Lanza, M.; Giuffrida, R.; Colarossi, C.; Mare, M.; Cuzzocrea, S.; Esposito, E.; et al. Role of Basic Fibroblast Growth Factor in Cancer: Biological Activity, Targeted Therapies, and Prognostic Value. *Cells* **2023**, *12*, 1002. [CrossRef]

 https://doi.org/10.3390/cancers18010089

147. Yuan, Z.; Li, Y.; Zhang, S.; Wang, X.; Dou, H.; Yu, X.; Zhang, Z.; Yang, S.; Xiao, M. Extracellular matrix remodeling in tum progression and immune escape: From mechanisms to treatments. *Mol. Cancer* **2023**, *22*, 48. [CrossRef]
148. Lujano Olazaba, O.; Farrow, J.; Monkkonen, T. Fibroblast heterogeneity and functions: Insights from single-cell sequencing wound healing, breast cancer, ovarian cancer and melanoma. *Front. Genet.* **2024**, *15*, 1304853. [CrossRef]
149. Solek, J.; Braun, M.; Sadej, R.; Romanska, H.M. FGFR-related phenotypic and functional profile of CAFs in prognostication breast cancer (Review). *Int. J. Oncol.* **2024**, *65*, 94. [CrossRef]
150. Elyada, E.; Bolisetty, M.; Laise, P.; Flynn, W.F.; Courtois, E.T.; Burkhart, R.A.; Teinor, J.A.; Belleau, P.; Biffi, G.; Lucito, M.; et al. Cross-Species Single-Cell Analysis of Pancreatic Ductal Adenocarcinoma Reveals Antigen-Presenting Cancer-Associate Fibroblasts. *Cancer Discov.* **2019**, *9*, 1102–1123. [CrossRef]
151. Croft, W.; Pearce, H.; Margielewska-Davies, S.; Lim, L.; Nicol, S.M.; Zayou, F.; Blakeway, D.; Marcon, F.; Powell-Brett, S.; Maho B.; et al. Spatial determination and prognostic impact of the fibroblast transcriptome in pancreatic ductal adenocarcinoma. *El* **2023**, *12*, e86125. [CrossRef]
152. Wu, S.Z.; Al-Eryani, G.; Roden, D.L.; Junankar, S.; Harvey, K.; Andersson, A.; Thennavan, A.; Wang, C.; Torpy, J.R.; Bartonicek, N et al. A single-cell and spatially resolved atlas of human breast cancers. *Nat. Genet.* **2021**, *53*, 1334–1347. [CrossRef] [PubMed]
153. Pelka, K.; Hofree, M.; Chen, J.H.; Sarkizova, S.; Pirl, J.D.; Jorgji, V.; Bejnood, A.; Dionne, D.; Ge, W.H.; Xu, K.H.; et al. Spatial organized multicellular immune hubs in human colorectal cancer. *Cell* **2021**, *184*, 4734–4752.e4720. [CrossRef]
154. Salcher, S.; Sturm, G.; Horvath, L.; Untergasser, G.; Kuempers, C.; Fotakis, G.; Panizzolo, E.; Martowicz, A.; Trebo, M.; Pall, C et al. High-resolution single-cell atlas reveals diversity and plasticity of tissue-resident neutrophils in non-small cell lung canc *Cancer Cell* **2022**, *40*, 1503–1520 e1508. [CrossRef]
155. Ji, A.L.; Rubin, A.J.; Thrane, K.; Jiang, S.; Reynolds, D.L.; Meyers, R.M.; Guo, M.G.; George, B.M.; Mollbrink, A.; Bergenstrahl J.; et al. Multimodal Analysis of Composition and Spatial Architecture in Human Squamous Cell Carcinoma. *Cell* **202** *182*, 497–514.e422. [CrossRef] [PubMed]
156. Thrane, K.; Eriksson, H.; Maaskola, J.; Hansson, J.; Lundeberg, J. Spatially Resolved Transcriptomics Enables Dissection of Genet Heterogeneity in Stage III Cutaneous Malignant Melanoma. *Cancer Res.* **2018**, *78*, 5970–5979. [CrossRef] [PubMed]
157. Hegab, A.E.; Ozaki, M.; Kameyama, N.; Gao, J.; Kagawa, S.; Yasuda, H.; Soejima, K.; Yin, Y.; Guzy, R.D.; Nakamura, Y.; et a Effect of FGF/FGFR pathway blocking on lung adenocarcinoma and its cancer-associated fibroblasts. *J. Pathol.* **2019**, *249*, 193–2([CrossRef]
158. Musumeci, M.; Coppola, V.; Addario, A.; Patrizii, M.; Maugeri-Sacca, M.; Memeo, L.; Colarossi, C.; Francescangeli, F.; Biffoni, M Collura, D.; et al. Control of tumor and microenvironment cross-talk by miR-15a and miR-16 in prostate cancer. *Oncogene* **201** *30*, 4231–4242. [CrossRef]
159. Facchinetti, F.; Hollebecque, A.; Bahleda, R.; Loriot, Y.; Olaussen, K.A.; Massard, C.; Friboulet, L. Facts and New Hopes o Selective FGFR Inhibitors in Solid Tumors. *Clin. Cancer Res.* **2020**, *26*, 764–774. [CrossRef]
160. Szymczyk, J.; Sluzalska, K.D.; Materla, I.; Opalinski, L.; Otlewski, J.; Zakrzewska, M. FGF/FGFR-Dependent Molecul Mechanisms Underlying Anti-Cancer Drug Resistance. *Cancers* **2021**, *13*, 5796. [CrossRef]
161. Nomura, S.; Yoshitomi, H.; Takano, S.; Shida, T.; Kobayashi, S.; Ohtsuka, M.; Kimura, F.; Shimizu, H.; Yoshidome, H.; Kato, A et al. FGF10/FGFR2 signal induces cell migration and invasion in pancreatic cancer. *Br. J. Cancer* **2008**, *99*, 305–313. [CrossRef
162. Psaila, B.; Kaplan, R.N.; Port, E.R.; Lyden, D. Priming the 'soil' for breast cancer metastasis: The pre-metastatic niche. *Breast D* **2006**, *26*, 65–74. [CrossRef]
163. Murakami, M.; Nguyen, L.T.; Zhuang, Z.W.; Moodie, K.L.; Carmeliet, P.; Stan, R.V.; Simons, M. The FGF system has a key role i regulating vascular integrity. *J. Clin. Investig.* **2008**, *118*, 3355–3366. [CrossRef]
164. Ribatti, D. The role of endothelial junctions in the regulation of the extravasation of tumor cells. A historical reappraisal. *Fror Oncol.* **2024**, *14*, 1415601. [CrossRef]
165. Li, C.; Chen, T.; Liu, J.; Wang, Y.; Zhang, C.; Guo, L.; Shi, D.; Zhang, T.; Wang, X.; Li, J. FGF19-Induced Inflammatory CA Promoted Neutrophil Extracellular Trap Formation in the Liver Metastasis of Colorectal Cancer. *Adv. Sci.* **2023**, *10*, e230261 [CrossRef] [PubMed]
166. Xie, M.; Lin, Z.; Ji, X.; Luo, X.; Zhang, Z.; Sun, M.; Chen, X.; Zhang, B.; Liang, H.; Liu, D.; et al. FGF19/FGFR4-mediated elevatio of ETV4 facilitates hepatocellular carcinoma metastasis by upregulating PD-L1 and CCL2. *J. Hepatol.* **2023**, *79*, 109–125. [CrossRe [PubMed]
167. Li, N.; Yang, S.; Hu, C.; Zheng, J.; Yang, H.; Jiang, T.; Li, X.; Han, Q.; Zhang, X.; Yang, Y.; et al. Neutrophil extracellular trap regulate LDHA expression to promote colorectal cancer liver metastasis. *J. Transl. Med.* **2025**, *23*, 1208. [CrossRef]
168. Okamoto, M.; Mizuno, R.; Kawada, K.; Itatani, Y.; Kiyasu, Y.; Hanada, K.; Hirata, W.; Nishikawa, Y.; Masui, H.; Sugimoto, N et al. Neutrophil Extracellular Traps Promote Metastases of Colorectal Cancers through Activation of ERK Signaling by Releasin Neutrophil Elastase. *Int. J. Mol. Sci.* **2023**, *24*, 1118. [CrossRef] [PubMed]
169. Ronchetti, L.; Boubaker, N.S.; Barba, M.; Vici, P.; Gurtner, A.; Piaggio, G. Neutrophil extracellular traps in cancer: Not onl catching microbes. *J. Exp. Clin. Cancer Res.* **2021**, *40*, 231. [CrossRef]

https://doi.org/10.3390/cancers1801008

70. Chen, I.X.; Chauhan, V.P.; Posada, J.; Ng, M.R.; Wu, M.W.; Adstamongkonkul, P.; Huang, P.; Lindeman, N.; Langer, R.; Jain, R.K. Blocking CXCR4 alleviates desmoplasia, increases T-lymphocyte infiltration, and improves immunotherapy in metastatic breast cancer. *Proc. Natl. Acad. Sci. USA* **2019**, *116*, 4558–4566. [CrossRef]
71. Shi, Y.; Riese, D.J., 2nd; Shen, J. The Role of the CXCL12/CXCR4/CXCR7 Chemokine Axis in Cancer. *Front. Pharmacol.* **2020**, *11*, 574667. [CrossRef]
72. Gherardi, E.; Birchmeier, W.; Birchmeier, C.; Vande Woude, G. Targeting MET in cancer: Rationale and progress. *Nat. Rev. Cancer* **2012**, *12*, 89–103. [CrossRef] [PubMed]
73. Muller, A.; Homey, B.; Soto, H.; Ge, N.; Catron, D.; Buchanan, M.E.; McClanahan, T.; Murphy, E.; Yuan, W.; Wagner, S.N.; et al. Involvement of chemokine receptors in breast cancer metastasis. *Nature* **2001**, *410*, 50–56. [CrossRef]
74. Haddad, L.E.; Khzam, L.B.; Hajjar, F.; Merhi, Y.; Sirois, M.G. Characterization of FGF receptor expression in human neutrophils and their contribution to chemotaxis. *Am. J. Physiol. Cell Physiol.* **2011**, *301*, C1036–C1045. [CrossRef] [PubMed]
75. Seghezzi, G.; Patel, S.; Ren, C.J.; Gualandris, A.; Pintucci, G.; Robbins, E.S.; Shapiro, R.L.; Galloway, A.C.; Rifkin, D.B.; Mignatti, P. Fibroblast growth factor-2 (FGF-2) induces vascular endothelial growth factor (VEGF) expression in the endothelial cells of forming capillaries: An autocrine mechanism contributing to angiogenesis. *J. Cell Biol.* **1998**, *141*, 1659–1673. [CrossRef]
76. Murakami, M.; Simons, M. Fibroblast growth factor regulation of neovascularization. *Curr. Opin. Hematol.* **2008**, *15*, 215–220. [CrossRef] [PubMed]
77. Qiu, H.; Zhang, X.; Qi, J.; Zhang, J.; Tong, Y.; Li, L.; Fu, L.; Qin, Y.R.; Guan, X.; Zhang, L. Identification and characterization of FGFR2(+) hematopoietic stem cell-derived fibrocytes as precursors of cancer-associated fibroblasts induced by esophageal squamous cell carcinoma. *J. Exp. Clin. Cancer Res.* **2022**, *41*, 240. [CrossRef]
78. Kim, S.M.; Kim, H.; Yun, M.R.; Kang, H.N.; Pyo, K.H.; Park, H.J.; Lee, J.M.; Choi, H.M.; Ellinghaus, P.; Ocker, M.; et al. Activation of the Met kinase confers acquired drug resistance in FGFR-targeted lung cancer therapy. *Oncogenesis* **2016**, *5*, e241. [CrossRef]
79. Kim, B.; Wang, S.; Lee, J.M.; Jeong, Y.; Ahn, T.; Son, D.S.; Park, H.W.; Yoo, H.S.; Song, Y.J.; Lee, E.; et al. Synthetic lethal screening reveals FGFR as one of the combinatorial targets to overcome resistance to Met-targeted therapy. *Oncogene* **2015**, *34*, 1083–1093. [CrossRef]
180. Touat, M.; Ileana, E.; Postel-Vinay, S.; Andre, F.; Soria, J.C. Targeting FGFR Signaling in Cancer. *Clin. Cancer Res.* **2015**, *21*, 2684–2694. [CrossRef]
181. Cappellen, D.; De Oliveira, C.; Ricol, D.; de Medina, S.; Bourdin, J.; Sastre-Garau, X.; Chopin, D.; Thiery, J.P.; Radvanyi, F. Frequent activating mutations of FGFR3 in human bladder and cervix carcinomas. *Nat. Genet.* **1999**, *23*, 18–20. [CrossRef] [PubMed]
182. d'Avis, P.Y.; Robertson, S.C.; Meyer, A.N.; Bardwell, W.M.; Webster, M.K.; Donoghue, D.J. Constitutive activation of fibroblast growth factor receptor 3 by mutations responsible for the lethal skeletal dysplasia thanatophoric dysplasia type I. *Cell Growth Differ.* **1998**, *9*, 71–78.
183. Tomlinson, D.C.; Baldo, O.; Harnden, P.; Knowles, M.A. FGFR3 protein expression and its relationship to mutation status and prognostic variables in bladder cancer. *J. Pathol.* **2007**, *213*, 91–98. [CrossRef]
184. Byron, S.A.; Gartside, M.; Powell, M.A.; Wellens, C.L.; Gao, F.; Mutch, D.G.; Goodfellow, P.J.; Pollock, P.M. FGFR2 point mutations in 466 endometrioid endometrial tumors: Relationship with MSI, KRAS, PIK3CA, CTNNB1 mutations and clinicopathological features. *PLoS ONE* **2012**, *7*, e30801. [CrossRef]
185. Chen, H.; Ma, J.; Li, W.; Eliseenkova, A.V.; Xu, C.; Neubert, T.A.; Miller, W.T.; Mohammadi, M. A molecular brake in the kinase hinge region regulates the activity of receptor tyrosine kinases. *Mol. Cell* **2007**, *27*, 717–730. [CrossRef]
186. Goncalves, E.; Poulos, R.C.; Cai, Z.; Barthorpe, S.; Manda, S.S.; Lucas, N.; Beck, A.; Bucio-Noble, D.; Dausmann, M.; Hall, C.; et al. Pan-cancer proteomic map of 949 human cell lines. *Cancer Cell* **2022**, *40*, 835–849.e8. [CrossRef]
187. Porebska, N.; Latko, M.; Kucinska, M.; Zakrzewska, M.; Otlewski, J.; Opalinski, L. Targeting Cellular Trafficking of Fibroblast Growth Factor Receptors as a Strategy for Selective Cancer Treatment. *J. Clin. Med.* **2018**, *8*, 7. [CrossRef]
188. Li, F.; Peiris, M.N.; Donoghue, D.J. Functions of FGFR2 corrupted by translocations in intrahepatic cholangiocarcinoma. *Cytokine Growth Factor Rev.* **2020**, *52*, 56–67. [CrossRef] [PubMed]
189. Nita, A.; Abraham, S.P.; Krejci, P.; Bosakova, M. Oncogenic FGFR Fusions Produce Centrosome and Cilia Defects by Ectopic Signaling. *Cells* **2021**, *10*, 1445. [CrossRef] [PubMed]
190. Reicher, A.; Harris, A.L.; Prinz, F.; Kiesslich, T.; Wei, M.; Ollinger, R.; Rad, R.; Pichler, M.; Kwong, L.N. Generation of An Endogenous FGFR2-BICC1 Gene Fusion/58 Megabase Inversion Using Single-Plasmid CRISPR/Cas9 Editing in Biliary Cells. *Int. J. Mol. Sci.* **2020**, *21*, 2460. [CrossRef]
191. Goyal, L.; Meric-Bernstam, F.; Hollebecque, A.; Valle, J.W.; Morizane, C.; Karasic, T.B.; Abrams, T.A.; Furuse, J.; Kelley, R.K.; Cassier, P.A.; et al. Futibatinib for FGFR2-Rearranged Intrahepatic Cholangiocarcinoma. *N. Engl. J. Med.* **2023**, *388*, 228–239. [CrossRef]
192. Zou, Y.; Zhu, K.; Pang, Y.; Han, J.; Zhang, X.; Jiang, Z.; Huang, Y.; Gu, W.; Ji, Y. Molecular Detection of FGFR2 Rearrangements in Resected Intrahepatic Cholangiocarcinomas: FISH Could Be An Ideal Method in Patients with Histological Small Duct Subtype. *J. Clin. Transl. Hepatol.* **2023**, *11*, 1355–1367. [CrossRef]

https://doi.org/10.3390/cancers18010089

193. Gott, H.; Uhl, E. FGFR3-TACCs3 Fusions and Their Clinical Relevance in Human Glioblastoma. *Int. J. Mol. Sci.* **2022**, *23*, 8675. [CrossRef]
194. Jimenez-Pascual, A.; Hale, J.S.; Kordowski, A.; Pugh, J.; Silver, D.J.; Bayik, D.; Roversi, G.; Alban, T.J.; Rao, S.; Chen, R.; et al. ADAMDEC1 Maintains a Growth Factor Signaling Loop in Cancer Stem Cells. *Cancer Discov.* **2019**, *9*, 1574–1589. [CrossRef]
195. Ji, W.; Yu, Y.; Li, Z.; Wang, G.; Li, F.; Xia, W.; Lu, S. FGFR1 promotes the stem cell-like phenotype of FGFR1-amplified non-small cell lung cancer cells through the Hedgehog pathway. *Oncotarget* **2016**, *7*, 15118–15134. [CrossRef]
196. Servetto, A.; Kollipara, R.; Formisano, L.; Lin, C.C.; Lee, K.M.; Sudhan, D.R.; Gonzalez-Ericsson, P.I.; Chatterjee, S.; Guerrero-Zotano, A.; Mendiratta, S.; et al. Nuclear FGFR1 Regulates Gene Transcription and Promotes Antiestrogen Resistance in ER+ Breast Cancer. *Clin. Cancer Res.* **2021**, *27*, 4379–4396. [CrossRef] [PubMed]
197. Karajannis, M.A.; Vincent, L.; Direnzo, R.; Shmelkov, S.V.; Zhang, F.; Feldman, E.J.; Bohlen, P.; Zhu, Z.; Sun, H.; Kussie, P.; et al. Activation of FGFR1beta signaling pathway promotes survival, migration and resistance to chemotherapy in acute myeloid leukemia cells. *Leukemia* **2006**, *20*, 979–986. [CrossRef] [PubMed]
198. Lee, H.; Kim, M.; Lim, J.; Kim, Y.; Han, K.; Cho, B.S.; Kim, H.J. Acute myeloid leukemia associated with FGFR1 abnormalities. *Int. J. Hematol.* **2013**, *97*, 808–812. [CrossRef] [PubMed]
199. Marinelli, L.M.; Romain, J.T.; Ehman, W., Jr.; Ortega, V.; Velagaleti, G.; Gibbons, T.F.; Nazario-Toole, A.; Holmes, A.R. Myeloid/Lymphoid Neoplasm with FGFR1 Rearrangement Presenting with Polycythemia Vera and T-cell Acute Lymphoblastic Leukemia. *Cancer Genet.* **2023**, *276-277*, 43–47. [CrossRef]
200. Chew, N.J.; Lim Kam Sian, T.C.C.; Nguyen, E.V.; Shin, S.-Y.; Yang, J.; Hui, M.N.; Deng, N.; McLean, C.A.; Welm, A.L.; Lim, E.; et al. Evaluation of FGFR targeting in breast cancer through interrogation of patient-derived models. *Breast Cancer Res.* **2021**, *23*, 82. [CrossRef]
201. Ding, S.; Chen, X.; Shen, K. Single-cell RNA sequencing in breast cancer: Understanding tumor heterogeneity and paving roads to individualized therapy. *Cancer Commun.* **2020**, *40*, 329–344. [CrossRef] [PubMed]
202. Guo, Q.; Zhou, Y.; Xie, T.; Yuan, Y.; Li, H.; Shi, W.; Zheng, L.; Li, X.; Zhang, W. Tumor microenvironment of cancer stem cells: Perspectives on cancer stem cell targeting. *Genes Dis.* **2024**, *11*, 101043. [CrossRef] [PubMed]
203. Knuchel, S.; Anderle, P.; Werfelli, P.; Diamantis, E.; Ruegg, C. Fibroblast surface-associated FGF-2 promotes contact-dependent colorectal cancer cell migration and invasion through FGFR-SRC signaling and integrin alphavbeta5-mediated adhesion. *Oncotarget* **2015**, *6*, 14300–14317. [CrossRef]
204. Stefanova, D.; Olszewski, D.; Glitscher, M.; Bauer, M.; Ferrarese, L.; Wust, D.; Hildt, E.; Greber, U.F.; Werner, S. FGF receptor kinase inhibitors exhibit broad antiviral activity by targeting Src family kinases. *Cell. Mol. Life Sci.* **2024**, *81*, 471. [CrossRef] [PubMed]
205. Meyiah, A.; Khan, F.I.; Alfaki, D.A.; Murshed, K.; Raza, A.; Elkord, E. The colorectal cancer microenvironment: Preclinical progress in identifying targets for cancer therapy. *Transl. Oncol.* **2025**, *53*, 102307. [CrossRef]
206. Calabrese, C.; Poppleton, H.; Kocak, M.; Hogg, T.L.; Fuller, C.; Hamner, B.; Oh, E.Y.; Gaber, M.W.; Finklestein, D.; Allen, M.; et al. A perivascular niche for brain tumor stem cells. *Cancer Cell* **2007**, *11*, 69–82. [CrossRef]
207. Fumarola, C.; Cretella, D.; La Monica, S.; Bonelli, M.A.; Alfieri, R.; Caffarra, C.; Quaini, F.; Madeddu, D.; Falco, A.; Cavazzoni, A.; et al. Enhancement of the anti-tumor activity of FGFR1 inhibition in squamous cell lung cancer by targeting downstream signaling involved in glucose metabolism. *Oncotarget* **2017**, *8*, 91841–91859. [CrossRef]
208. Zhou, C.; Chen, T.; Xie, Z.; Qin, Y.; Ou, Y.; Zhang, J.; Li, S.; Chen, R.; Zhong, N. RACK1 forms a complex with FGFR1 and PKM2 and stimulates the growth and migration of squamous lung cancer cells. *Mol. Carcinog.* **2017**, *56*, 2391–2399. [CrossRef]
209. Liu, J.; Chen, G.; Liu, Z.; Liu, S.; Cai, Z.; You, P.; Ke, Y.; Lai, L.; Huang, Y.; Gao, H.; et al. Aberrant FGFR Tyrosine Kinase Signaling Enhances the Warburg Effect by Reprogramming LDH Isoform Expression and Activity in Prostate Cancer. *Cancer Res.* **2018**, *78*, 4459–4470. [CrossRef]
210. Ye, Y.; Yang, F.; Gu, Z.; Li, W.; Yuan, Y.; Liu, S.; Zhou, L.; Han, B.; Zheng, R.; Cao, Z. Fibroblast growth factor pathway promotes glycolysis by activating LDHA and suppressing LDHB in a STAT1-dependent manner in prostate cancer. *J. Transl. Med.* **2024**, *22*, 474. [CrossRef]
211. Du, X.; Wang, Q.R.; Chan, E.; Merchant, M.; Liu, J.; French, D.; Ashkenazi, A.; Qing, J. FGFR3 stimulates stearoyl CoA desaturase 1 activity to promote bladder tumor growth. *Cancer Res.* **2012**, *72*, 5843–5855. [CrossRef]
212. Zhao, Q.; Lin, X.; Wang, G. Targeting SREBP-1-Mediated Lipogenesis as Potential Strategies for Cancer. *Front. Oncol.* **2022**, *12*, 952371. [CrossRef]
213. Jung, J.H.; Yang, D.Q.; Song, H.; Wang, X.; Wu, X.; Kim, K.P.; Pandey, A.; Byeon, S.K. Characterization of Lipid Alterations by Oncogenic PIK3CA Mutations Using Untargeted Lipidomics in Breast Cancer. *OMICS* **2023**, *27*, 327–335. [CrossRef]
214. Oh, K.; Yoo, Y.J.; Torre-Healy, L.A.; Rao, M.; Fassler, D.; Wang, P.; Caponegro, M.; Gao, M.; Kim, J.; Sasson, A.; et al. Coordinated single-cell tumor microenvironment dynamics reinforce pancreatic cancer subtype. *Nat. Commun.* **2023**, *14*, 5226. [CrossRef]

 https://doi.org/10.3390/cancers18010089

15. Hwang, W.L.; Jagadeesh, K.A.; Guo, J.A.; Hoffman, H.I.; Yadollahpour, P.; Reeves, J.W.; Mohan, R.; Drokhlyansky, E.; Van Wittenberghe, N.; Ashenberg, O.; et al. Single-nucleus and spatial transcriptome profiling of pancreatic cancer identifies multicellular dynamics associated with neoadjuvant treatment. *Nat. Genet.* **2022**, *54*, 1178–1191. [CrossRef]
16. Zhang, J.; Liu, Y.; Xia, L.; Zhen, J.; Gao, J.; Atsushi, T. Constructing heterogeneous single-cell landscape and identifying microenvironment molecular characteristics of primary and lymphatic metastatic head and neck squamous cell carcinoma. *Comput. Biol. Med.* **2023**, *165*, 107459. [CrossRef] [PubMed]
17. Krook, M.A.; Bonneville, R.; Chen, H.Z.; Reeser, J.W.; Wing, M.R.; Martin, D.M.; Smith, A.M.; Dao, T.; Samorodnitsky, E.; Paruchuri, A.; et al. Tumor heterogeneity and acquired drug resistance in FGFR2-fusion-positive cholangiocarcinoma through rapid research autopsy. *Cold Spring Harb. Mol. Case Stud.* **2019**, *5*, a004002. [CrossRef]
18. Jin, S.; Guerrero-Juarez, C.F.; Zhang, L.; Chang, I.; Ramos, R.; Kuan, C.H.; Myung, P.; Plikus, M.V.; Nie, Q. Inference and analysis of cell-cell communication using CellChat. *Nat. Commun.* **2021**, *12*, 1088. [CrossRef] [PubMed]
19. Reck, M.; Kaiser, R.; Mellemgaard, A.; Douillard, J.Y.; Orlov, S.; Krzakowski, M.; von Pawel, J.; Gottfried, M.; Bondarenko, I.; Liao, M.; et al. Docetaxel plus nintedanib versus docetaxel plus placebo in patients with previously treated non-small-cell lung cancer (LUME-Lung 1): A phase 3, double-blind, randomised controlled trial. *Lancet Oncol.* **2014**, *15*, 143–155. [CrossRef]
20. Angevin, E.; Lopez-Martin, J.A.; Lin, C.C.; Gschwend, J.E.; Harzstark, A.; Castellano, D.; Soria, J.C.; Sen, P.; Chang, J.; Shi, M.; et al. Phase I study of dovitinib (TKI258), an oral FGFR, VEGFR, and PDGFR inhibitor, in advanced or metastatic renal cell carcinoma. *Clin. Cancer Res.* **2013**, *19*, 1257–1268. [CrossRef] [PubMed]
21. Andre, F.; Bachelot, T.; Campone, M.; Dalenc, F.; Perez-Garcia, J.M.; Hurvitz, S.A.; Turner, N.; Rugo, H.; Smith, J.W.; Deudon, S.; et al. Targeting FGFR with dovitinib (TKI258): Preclinical and clinical data in breast cancer. *Clin. Cancer Res.* **2013**, *19*, 3693–3702. [CrossRef]
22. Babina, I.S.; Turner, N.C. Advances and challenges in targeting FGFR signalling in cancer. *Nat. Rev. Cancer* **2017**, *17*, 318–332. [CrossRef]
23. Perera, T.P.S.; Jovcheva, E.; Mevellec, L.; Vialard, J.; De Lange, D.; Verhulst, T.; Paulussen, C.; Van De Ven, K.; King, P.; Freyne, E.; et al. Discovery and Pharmacological Characterization of JNJ-42756493 (Erdafitinib), a Functionally Selective Small-Molecule FGFR Family Inhibitor. *Mol. Cancer Ther.* **2017**, *16*, 1010–1020. [CrossRef] [PubMed]
24. Rodon, J.; Damian, S.; Furqan, M.; Garcia-Donas, J.; Imai, H.; Italiano, A.; Spanggaard, I.; Ueno, M.; Yokota, T.; Veronese, M.L.; et al. Pemigatinib in previously treated solid tumors with activating FGFR1-FGFR3 alterations: Phase 2 FIGHT-207 basket trial. *Nat. Med.* **2024**, *30*, 1645–1654. [CrossRef] [PubMed]
25. Sootome, H.; Fujita, H.; Ito, K.; Ochiiwa, H.; Fujioka, Y.; Ito, K.; Miura, A.; Sagara, T.; Ito, S.; Ohsawa, H.; et al. Futibatinib Is a Novel Irreversible FGFR 1-4 Inhibitor That Shows Selective Antitumor Activity against FGFR-Deregulated Tumors. *Cancer Res.* **2020**, *80*, 4986–4997. [CrossRef]
26. Wainberg, Z.A.; Enzinger, P.C.; Kang, Y.K.; Qin, S.; Yamaguchi, K.; Kim, I.H.; Saeed, A.; Oh, S.C.; Li, J.; Turk, H.M.; et al. Bemarituzumab in patients with FGFR2b-selected gastric or gastro-oesophageal junction adenocarcinoma (FIGHT): A randomised, double-blind, placebo-controlled, phase 2 study. *Lancet Oncol.* **2022**, *23*, 1430–1440. [CrossRef]
27. Siefker-Radtke, A.O.; Currie, G.; Abella, E.; Vaena, D.A.; Kalebasty, A.R.; Curigliano, G.; Tupikowski, K.; Andric, Z.G.; Lugowska, I.; Kelly, W.K. FIERCE-22: Clinical activity of vofatamab (V) a FGFR3 selective inhibitor in combination with pembrolizumab (P) in WT metastatic urothelial carcinoma, preliminary analysis. *J. Clin. Oncol.* **2019**, *37*, 4511. [CrossRef]
28. Beck, A.; Goetsch, L.; Dumontet, C.; Corvaia, N. Strategies and challenges for the next generation of antibody-drug conjugates. *Nat. Rev. Drug Discov.* **2017**, *16*, 315–337. [CrossRef] [PubMed]
29. Kim, S.B.; Meric-Bernstam, F.; Kalyan, A.; Babich, A.; Liu, R.; Tanigawa, T.; Sommer, A.; Osada, M.; Reetz, F.; Laurent, D.; et al. First-in-Human Phase I Study of Aprutumab Ixadotin, a Fibroblast Growth Factor Receptor 2 Antibody-Drug Conjugate (BAY 1187982) in Patients with Advanced Cancer. *Target. Oncol.* **2019**, *14*, 591–601. [CrossRef]
30. Lau, D.K.; Collin, J.P.; Mariadason, J.M. Clinical Developments and Challenges in Treating FGFR2-Driven Gastric Cancer. *Biomedicines* **2024**, *12*, 1117. [CrossRef]
31. Maecker, H.; Jonnalagadda, V.; Bhakta, S.; Jammalamadaka, V.; Junutula, J.R. Exploration of the antibody-drug conjugate clinical landscape. *MAbs* **2023**, *15*, 2229101. [CrossRef]
32. Pozniak, M.; Porebska, N.; Krzyscik, M.A.; Sokolowska-Wedzina, A.; Jastrzebski, K.; Sochacka, M.; Szymczyk, J.; Zakrzewska, M.; Otlewski, J.; Opalinski, L. The cytotoxic conjugate of highly internalizing tetravalent antibody for targeting FGFR1-overproducing cancer cells. *Mol. Med.* **2021**, *27*, 46. [CrossRef]
33. Wei, Q.; Li, P.; Yang, T.; Zhu, J.; Sun, L.; Zhang, Z.; Wang, L.; Tian, X.; Chen, J.; Hu, C.; et al. The promise and challenges of combination therapies with antibody-drug conjugates in solid tumors. *J. Hematol. Oncol.* **2024**, *17*, 1. [CrossRef]
34. Lu, J.; Jiang, F.; Lu, A.; Zhang, G. Linkers Having a Crucial Role in Antibody-Drug Conjugates. *Int. J. Mol. Sci.* **2016**, *17*, 561. [CrossRef]

235. Sommer, A.; Kopitz, C.; Schatz, C.A.; Nising, C.F.; Mahlert, C.; Lerchen, H.G.; Stelte-Ludwig, B.; Hammer, S.; Greven, Schuhmacher, J.; et al. Preclinical Efficacy of the Auristatin-Based Antibody-Drug Conjugate BAY 1187982 for the Treatment FGFR2-Positive Solid Tumors. *Cancer Res.* **2016**, *76*, 6331–6339. [CrossRef]
236. Su, Z.; Xiao, D.; Xie, F.; Liu, L.; Wang, Y.; Fan, S.; Zhou, X.; Li, S. Antibody-drug conjugates: Recent advances in linker chemist *Acta Pharm. Sin. B* **2021**, *11*, 3889–3907. [CrossRef]
237. Zhou, L.; Lu, Y.; Liu, W.; Wang, S.; Wang, L.; Zheng, P.; Zi, G.; Liu, H.; Liu, W.; Wei, S. Drug conjugates for the treatment of lu cancer: From drug discovery to clinical practice. *Exp. Hematol. Oncol.* **2024**, *13*, 26. [CrossRef]
238. Porebska, N.; Knapik, A.; Pozniak, M.; Krzyscik, M.A.; Zakrzewska, M.; Otlewski, J.; Opalinski, L. Intrinsically Fluoresce Oligomeric Cytotoxic Conjugates Toxic for FGFR1-Overproducing Cancers. *Biomacromolecules* **2021**, *22*, 5349–5362. [CrossRe [PubMed]
239. Dasgupta, A.; Sofias, A.M.; Kiessling, F.; Lammers, T. Nanoparticle Delivery to Tumours: From EPR and ATR Mechanisms Clinical Impact. *Nat. Rev. Bioeng.* **2024**, *2*, 714–716. [CrossRef] [PubMed]
240. Li, J.; Lu, J.; Guo, H.; Zhou, J.; Wang, S.; Jiang, K.; Chai, Z.; Yao, S.; Wang, X.; Lu, L.; et al. A pentapeptide enabled AL38 liposome-based glioma-targeted therapy with immune opsonic effect attenuated. *Acta Pharm. Sin. B* **2021**, *11*, 283–299. [CrossRe [PubMed]
241. Sindhwani, S.; Syed, A.M.; Ngai, J.; Kingston, B.R.; Maiorino, L.; Rothschild, J.; MacMillan, P.; Zhang, Y.; Rajesh, N.U.; Hoang, et al. The entry of nanoparticles into solid tumours. *Nat. Mater.* **2020**, *19*, 566–575. [CrossRef]
242. Wilhelm, S.; Tavares, A.J.; Dai, Q.; Ohta, S.; Audet, J.; Dvorak, H.F.; Chan, W.C.W. Analysis of nanoparticle delivery to tumou *Nat. Rev. Mater.* **2016**, *1*, 16014. [CrossRef]
243. Ramanathan, R.K.; Korn, R.L.; Raghunand, N.; Sachdev, J.C.; Newbold, R.G.; Jameson, G.; Fetterly, G.J.; Prey, J.; Klinz, S.C Kim, J.; et al. Correlation between Ferumoxytol Uptake in Tumor Lesions by MRI and Response to Nanoliposomal Irinotecan Patients with Advanced Solid Tumors: A Pilot Study. *Clin. Cancer Res.* **2017**, *23*, 3638–3648. [CrossRef]
244. Danhier, F. To exploit the tumor microenvironment: Since the EPR effect fails in the clinic, what is the future of nanomedicine? *Control Release* **2016**, *244*, 108–121. [CrossRef]
245. Hare, J.I.; Lammers, T.; Ashford, M.B.; Puri, S.; Storm, G.; Barry, S.T. Challenges and strategies in anti-cancer nanomedicir development: An industry perspective. *Adv. Drug Deliv. Rev.* **2017**, *108*, 25–38. [CrossRef]
246. Perez-Herrero, E.; Fernandez-Medarde, A. Advanced targeted therapies in cancer: Drug nanocarriers, the future of chemotherap *Eur. J. Pharm. Biopharm.* **2015**, *93*, 52–79. [CrossRef]
247. Stylianopoulos, T.; Jain, R.K. Combining two strategies to improve perfusion and drug delivery in solid tumors. *Proc. Natl. Aca Sci. USA* **2013**, *110*, 18632–18637. [CrossRef] [PubMed]
248. Chauhan, V.P.; Stylianopoulos, T.; Martin, J.D.; Popovic, Z.; Chen, O.; Kamoun, W.S.; Bawendi, M.G.; Fukumura, D.; Jain, R. Normalization of tumour blood vessels improves the delivery of nanomedicines in a size-dependent manner. *Nat. Nanotechn* **2012**, *7*, 383–388. [CrossRef]
249. Mi, Y.; Hagan, C.T.t.; Vincent, B.G.; Wang, A.Z. Emerging Nano-/Microapproaches for Cancer Immunotherapy. *Adv. Sci.* **201** *6*, 1801847. [CrossRef] [PubMed]
250. Alijaj, N.; Moutel, S.; Gouveia, Z.L.; Gray, M.; Roveri, M.; Dzhumashev, D.; Weber, F.; Meier, G.; Luciani, P.; Rossler, J.K.; et Novel FGFR4-Targeting Single-Domain Antibodies for Multiple Targeted Therapies against Rhabdomyosarcoma. *Cancers* **202** *12*, 3313. [CrossRef] [PubMed]
251. Dong, Z.; Yin, Y.; Luo, J.; Li, B.; Lou, F.; Wang, Q.; Zhou, Q.; Ye, B.; Wang, Y. An FGFR1-Binding Peptide Modified Liposome f siRNA Delivery in Lung Cancer. *Int. J. Mol. Sci.* **2022**, *23*, 8380. [CrossRef] [PubMed]
252. Wang, Y.; Liu, D.; Zhang, T.; Xia, L. FGF/FGFR Signaling in Hepatocellular Carcinoma: From Carcinogenesis to Recer Therapeutic Intervention. *Cancers* **2021**, *13*, 1360. [CrossRef] [PubMed]
253. Junyaprasert, V.B.; Thummarati, P. Innovative Design of Targeted Nanoparticles: Polymer-Drug Conjugates for Enhanced Canc Therapy. *Pharmaceutics* **2023**, *15*, 2216. [CrossRef] [PubMed]
254. Parashar, A.K.; Saraogi, G.K.; Jain, P.K.; Kurmi, B.; Shrivastava, V.; Arora, V. Polymer-drug conjugates: Revolutionizin nanotheranostic agents for diagnosis and therapy. *Discov. Oncol.* **2024**, *15*, 641. [CrossRef]
255. Jin, Z.; Al Amili, M.; Guo, S. Tumor Microenvironment-Responsive Drug Delivery Based on Polymeric Micelles for Precisio Cancer Therapy: Strategies and Prospects. *Biomedicines* **2024**, *12*, 417. [CrossRef]
256. Cong, X.; Chen, J.; Xu, R. Recent Progress in Bio-Responsive Drug Delivery Systems for Tumor Therapy. *Front. Bioeng. Biotechn* **2022**, *10*, 916952. [CrossRef]
257. Liu, J.; Huang, Y.; Kumar, A.; Tan, A.; Jin, S.; Mozhi, A.; Liang, X.J. pH-sensitive nano-systems for drug delivery in cancer therap *Biotechnol. Adv.* **2014**, *32*, 693–710. [CrossRef]
258. Vogel, A.; Sahai, V.; Hollebecque, A.; Vaccaro, G.M.; Melisi, D.; Al Rajabi, R.M.; Paulson, A.S.; Borad, M.J.; Gallinson, D.; Murph A.G.; et al. An open-label study of pemigatinib in cholangiocarcinoma: Final results from FIGHT-202. *ESMO Open* **2024**, *9*, 10348 [CrossRef]

259. Bekaii-Saab, T.S.; Valle, J.W.; Van Cutsem, E.; Rimassa, L.; Furuse, J.; Ioka, T.; Melisi, D.; Macarulla, T.; Bridgewater, J.; Wasan, H.; et al. FIGHT-302: First-line pemigatinib vs gemcitabine plus cisplatin for advanced cholangiocarcinoma with FGFR2 rearrangements. *Future Oncol.* **2020**, *16*, 2385–2399. [CrossRef]

260. Siefker-Radtke, A.O.; Necchi, A.; Park, S.H.; Garcia-Donas, J.; Huddart, R.A.; Burgess, E.F.; Fleming, M.T.; Rezazadeh Kalebasty, A.; Mellado, B.; Varlamov, S.; et al. Efficacy and safety of erdafitinib in patients with locally advanced or metastatic urothelial carcinoma: Long-term follow-up of a phase 2 study. *Lancet Oncol.* **2022**, *23*, 248–258. [CrossRef]

261. Loriot, Y.; Matsubara, N.; Park, S.H.; Huddart, R.A.; Burgess, E.F.; Houede, N.; Banek, S.; Guadalupi, V.; Ku, J.H.; Valderrama, B.P.; et al. Erdafitinib or Chemotherapy in Advanced or Metastatic Urothelial Carcinoma. *N. Engl. J. Med.* **2023**, *389*, 1961–1971. [CrossRef]

262. Pant, S.; Schuler, M.; Iyer, G.; Witt, O.; Doi, T.; Qin, S.; Tabernero, J.; Reardon, D.A.; Massard, C.; Minchom, A.; et al. Erdafitinib in patients with advanced solid tumours with FGFR alterations (RAGNAR): An international, single-arm, phase 2 study. *Lancet Oncol.* **2023**, *24*, 925–935. [CrossRef] [PubMed]

263. Borad, M.J.; Schram, A.M.; Kim, R.D.; Kamath, S.D.; Sahai, V.; Dotan, E.; Kelley, R.K.; Ponz-Sarvise, M.; Oh, D.-Y.; Yachnin, J.; et al. Updated dose escalation results for ReFocus, a first-in-human study of highly selective FGFR2 inhibitor RLY-4008 in cholangiocarcinoma and other solid tumors. *J. Clin. Oncol.* **2023**, *41*, 4009. [CrossRef]

264. Subbiah, V.; Sahai, V.; Maglic, D.; Bruderek, K.; Toure, B.B.; Zhao, S.; Valverde, R.; O'Hearn, P.J.; Moustakas, D.T.; Schonherr, H.; et al. RLY-4008, the First Highly Selective FGFR2 Inhibitor with Activity across FGFR2 Alterations and Resistance Mutations. *Cancer Discov.* **2023**, *13*, 2012–2031. [CrossRef]

265. Zheng, X.; Wang, H.; Deng, J.; Yao, M.; Zou, X.; Zhang, F.; Ma, X. Safety and efficacy of the pan-FGFR inhibitor erdafitinib in advanced urothelial carcinoma and other solid tumors: A systematic review and meta-analysis. *Front. Oncol.* **2022**, *12*, 907377. [CrossRef] [PubMed]

266. Schauwvlieghe, P.P.; Peeters, F.; Strijbos, M.; Vulsteke, C. The unjustified fear of erdafitinib-induced central serous retinopathy. *Ann. Oncol.* **2025**, *36*, 1001–1002. [CrossRef]

267. Meric-Bernstam, F.; Bahleda, R.; Hierro, C.; Sanson, M.; Bridgewater, J.; Arkenau, H.T.; Tran, B.; Kelley, R.K.; Park, J.O.; Javle, M.; et al. Futibatinib, an Irreversible FGFR1-4 Inhibitor, in Patients with Advanced Solid Tumors Harboring FGF/FGFR Aberrations: A Phase I Dose-Expansion Study. *Cancer Discov.* **2022**, *12*, 402–415. [CrossRef] [PubMed]

268. Li, P.; Wu, L.; Song, Z. Data mining and safety analysis of FGFR tyrosine kinase inhibitors based on the FAERS database. *Sci. Rep.* **2025**, *15*, 26150. [CrossRef]

269. Chandana, S.R.; Frisch, A.; Mendoza, S.; Sinniah, R.S.; Crysler, O.; Banga, R.; Perkins, D.E. Management and mechanism of calciphylaxis in a patient treated with the FGFR inhibitor pemigatinib-a case report. *J. Gastrointest. Oncol.* **2024**, *15*, 478–484. [CrossRef]

270. Saverno, K.; Zimmerman Savill, K.M.; Brown-Bickerstaff, C.; Kotomale, A.; Rodriguez, M.; Feinberg, B.; Ren, H.; Blecker, M.; Kim, R. Real-world use of pemigatinib for the treatment of cholangiocarcinoma in the US. *Oncologist* **2025**, *30*, oyae204. [CrossRef]

271. Gile, J.J.; Ou, F.S.; Mahipal, A.; Larson, J.J.; Mody, K.; Jin, Z.; Hubbard, J.; Halfdanarson, T.; Alberts, S.R.; Jatoi, A.; et al. FGFR Inhibitor Toxicity and Efficacy in Cholangiocarcinoma: Multicenter Single-Institution Cohort Experience. *JCO Precis. Oncol.* **2021**, *5*, 1228–1240. [CrossRef]

272. Benjamin, D.J.; Mita, A.C. FGFR-Altered Urothelial Carcinoma: Resistance Mechanisms and Therapeutic Strategies. *Target. Oncol.* **2025**, *20*, 1–11. [CrossRef]

273. Liu, J.; Jin, Q.; Xia, J.; Liu, L.; Liu, Y.; Jiang, L. Computational Insights into the Resistance of the Gatekeeper Mutation V564F of FGFR2 Against Pemigatinib. *J. Comput. Biophys. Chem.* **2024**, *23*, 891–900. [CrossRef]

274. Chell, V.; Balmanno, K.; Little, A.S.; Wilson, M.; Andrews, S.; Blockley, L.; Hampson, M.; Gavine, P.R.; Cook, S.J. Tumour cell responses to new fibroblast growth factor receptor tyrosine kinase inhibitors and identification of a gatekeeper mutation in FGFR3 as a mechanism of acquired resistance. *Oncogene* **2013**, *32*, 3059–3070. [CrossRef] [PubMed]

275. Mahapatra, S.; Kar, P. Computational biophysical characterization of the effect of gatekeeper mutations on the binding of ponatinib to the FGFR kinase. *Arch. Biochem. Biophys.* **2024**, *758*, 110070. [CrossRef] [PubMed]

276. Krook, M.A.; Lenyo, A.; Wilberding, M.; Barker, H.; Dantuono, M.; Bailey, K.M.; Chen, H.Z.; Reeser, J.W.; Wing, M.R.; Miya, J.; et al. Efficacy of FGFR Inhibitors and Combination Therapies for Acquired Resistance in FGFR2-Fusion Cholangiocarcinoma. *Mol. Cancer Ther.* **2020**, *19*, 847–857. [CrossRef]

277. Zhou, Y.; Xiang, S.; Yang, F.; Lu, X. Targeting Gatekeeper Mutations for Kinase Drug Discovery. *J. Med. Chem.* **2022**, *65*, 15540–15558. [CrossRef]

278. Lin, Q.; Chen, X.; Qu, L.; Guo, M.; Wei, H.; Dai, S.; Jiang, L.; Chen, Y. Characterization of the cholangiocarcinoma drug pemigatinib against FGFR gatekeeper mutants. *Commun. Chem.* **2022**, *5*, 100. [CrossRef]

279. Yue, S.; Li, Y.; Chen, X.; Wang, J.; Li, M.; Chen, Y.; Wu, D. FGFR-TKI resistance in cancer: Current status and perspectives. *J. Hematol. Oncol.* **2021**, *14*, 23. [CrossRef]

280. Ko, B.; He, T.; Gadgeel, S.; Halmos, B. MET/HGF pathway activation as a paradigm of resistance to targeted therapies. *Ann Transl. Med.* **2017**, *5*, 4. [CrossRef]
281. Herrera-Abreu, M.T.; Pearson, A.; Campbell, J.; Shnyder, S.D.; Knowles, M.A.; Ashworth, A.; Turner, N.C. Parallel RNA interference screens identify EGFR activation as an escape mechanism in FGFR3-mutant cancer. *Cancer Discov.* **2013**, *3*, 1058–1071. [CrossRef] [PubMed]
282. Georgescu, M.M. PTEN Tumor Suppressor Network in PI3K-Akt Pathway Control. *Genes Cancer* **2010**, *1*, 1170–1177. [CrossRef]
283. Lau, D.K.; Jenkins, L.; Weickhardt, A. Mechanisms of acquired resistance to fibroblast growth factor receptor targeted therapy. *Cancer Drug Resist.* **2019**, *2*, 568–579. [CrossRef]
284. Pearson, A.; Smyth, E.; Babina, I.S.; Herrera-Abreu, M.T.; Tarazona, N.; Peckitt, C.; Kilgour, E.; Smith, N.R.; Geh, C.; Rooney, C. et al. High-Level Clonal FGFR Amplification and Response to FGFR Inhibition in a Translational Clinical Trial. *Cancer Discov.* **2016**, *6*, 838–851. [CrossRef]
285. Sakthianandeswaren, A.; Christie, M.; D'Andreti, C.; Tsui, C.; Jorissen, R.N.; Li, S.; Fleming, N.I.; Gibbs, P.; Lipton, L.; Malaterre, J.; et al. PHLDA1 expression marks the putative epithelial stem cells and contributes to intestinal tumorigenesis. *Cancer Res.* **2011**, *71*, 3709–3719. [CrossRef]
286. Fearon, A.E.; Carter, E.P.; Clayton, N.S.; Wilkes, E.H.; Baker, A.M.; Kapitonova, E.; Bakhouche, B.A.; Tanner, Y.; Wang, J.; Gadaleta, E.; et al. PHLDA1 Mediates Drug Resistance in Receptor Tyrosine Kinase-Driven Cancer. *Cell Rep.* **2018**, *22*, 2469–2481. [CrossRef] [PubMed]
287. DiPeri, T.P.; Zhao, M.; Evans, K.W.; Varadarajan, K.; Moss, T.; Scott, S.; Kahle, M.P.; Byrnes, C.C.; Chen, H.; Lee, S.S.; et al. Convergent MAPK pathway alterations mediate acquired resistance to FGFR inhibitors in FGFR2 fusion-positive cholangiocarcinoma. *J. Hepatol.* **2024**, *80*, 322–334. [CrossRef] [PubMed]
288. Bockorny, B.; Rusan, M.; Chen, W.; Liao, R.G.; Li, Y.; Piccioni, F.; Wang, J.; Tan, L.; Thorner, A.R.; Li, T.; et al. RAS-MAPK Reactivation Facilitates Acquired Resistance in FGFR1-Amplified Lung Cancer and Underlies a Rationale for Upfront FGFR-MEK Blockade. *Mol. Cancer Ther.* **2018**, *17*, 1526–1539. [CrossRef]
289. Peng, R.; Chen, Y.; Wei, L.; Li, G.; Feng, D.; Liu, S.; Jiang, R.; Zheng, S.; Chen, Y. Resistance to FGFR1-targeted therapy leads to autophagy via TAK1/AMPK activation in gastric cancer. *Gastric Cancer* **2020**, *23*, 988–1002. [CrossRef]
290. Settembre, C.; Fraldi, A.; Medina, D.L.; Ballabio, A. Signals from the lysosome: A control centre for cellular clearance and energy metabolism. *Nat. Rev. Mol. Cell Biol.* **2013**, *14*, 283–296. [CrossRef]
291. Ortega-Batista, A.; Jaen-Alvarado, Y.; Moreno-Labrador, D.; Gomez, N.; Garcia, G.; Guerrero, E.N. Single-Cell Sequencing: Genomic and Transcriptomic Approaches in Cancer Cell Biology. *Int. J. Mol. Sci.* **2025**, *26*, 2074. [CrossRef]
292. Varghese, A.M.; Patel, J.; Janjigian, Y.Y.; Meng, F.; Selcuklu, S.D.; Iyer, G.; Houck-Loomis, B.; Harding, J.J.; O'Reilly, E.M.; Abou-Alfa, G.K.; et al. Noninvasive Detection of Polyclonal Acquired Resistance to FGFR Inhibition in Patients With Cholangiocarcinoma Harboring FGFR2 Alterations. *JCO Precis. Oncol.* **2021**, *5*, 44–50. [CrossRef]
293. Li, Y.; Qiu, X.; Wang, X.; Liu, H.; Geck, R.C.; Tewari, A.K.; Xiao, T.; Font-Tello, A.; Lim, K.; Jones, K.L.; et al. FGFR-inhibitor-mediated dismissal of SWI/SNF complexes from YAP-dependent enhancers induces adaptive therapeutic resistance. *Nat. Cell Biol.* **2021**, *23*, 1187–1198. [CrossRef] [PubMed]
294. Gupta, M.; Concepcion, C.P.; Fahey, C.G.; Keshishian, H.; Bhutkar, A.; Brainson, C.F.; Sanchez-Rivera, F.J.; Pessina, P.; Kim, J.Y.; Simoneau, A.; et al. BRG1 Loss Predisposes Lung Cancers to Replicative Stress and ATR Dependency. *Cancer Res.* **2020**, *80*, 3841–3854. [CrossRef]
295. Kurashima, K.; Kashiwagi, H.; Shimomura, I.; Suzuki, A.; Takeshita, F.; Mazevet, M.; Harata, M.; Yamashita, T.; Yamamoto, Y.; Kohno, T.; et al. SMARCA4 deficiency-associated heterochromatin induces intrinsic DNA replication stress and susceptibility to ATR inhibition in lung adenocarcinoma. *NAR Cancer* **2020**, *2*, zcaa005. [CrossRef]
296. Xue, Y.; Morris, J.L.; Yang, K.; Fu, Z.; Zhu, X.; Johnson, F.; Meehan, B.; Witkowski, L.; Yasmeen, A.; Golenar, T.; et al. SMARCA4/2 loss inhibits chemotherapy-induced apoptosis by restricting IP3R3-mediated Ca^{2+} flux to mitochondria. *Nat. Commun.* **2021**, *12*, 5404. [CrossRef]
297. Yu, L.; Wu, D. SMARCA2 and SMARCA4 Participate in DNA Damage Repair. *Front. Biosci.* **2024**, *29*, 262. [CrossRef]
298. Kotagiri, S.; Blazanin, N.; Xi, Y.; Han, Y.; Qudratullah, M.; Liang, X.; Wang, Y.; Pandey, P.; Mazhar, H.; Lam, T.N.; et al. Enhancer reprogramming underlies therapeutic utility of a SMARCA2 degrader in SMARCA4 mutant cancer. *Cell Chem. Biol.* **2024**, *31*, 2069–2084.e2069. [CrossRef]
299. Jo, H.; Shim, K.; Kim, H.U.; Jung, H.S.; Jeoung, D. HDAC2 as a target for developing anti-cancer drugs. *Comput. Struct. Biotechnol. J.* **2023**, *21*, 2048–2057. [CrossRef] [PubMed]
300. Minisini, M.; Mascaro, M.; Brancolini, C. HDAC-driven mechanisms in anticancer resistance: Epigenetics and beyond. *Cancer Drug Resist.* **2024**, *7*, 46. [CrossRef] [PubMed]
301. Bogatyrova, O.; Mattsson, J.S.M.; Ross, E.M.; Sanderson, M.P.; Backman, M.; Botling, J.; Brunnstrom, H.; Kurppa, P.; La Fleur, L.; Strell, C.; et al. FGFR1 overexpression in non-small cell lung cancer is mediated by genetic and epigenetic mechanisms and is a determinant of FGFR1 inhibitor response. *Eur. J. Cancer* **2021**, *151*, 136–149. [CrossRef]

https://doi.org/10.3390/cancers18010089

2. Etcheverry, A.; Aubry, M.; de Tayrac, M.; Vauleon, E.; Boniface, R.; Guenot, F.; Saikali, S.; Hamlat, A.; Riffaud, L.; Menei, P.; et al. DNA methylation in glioblastoma: Impact on gene expression and clinical outcome. *BMC Genom.* **2010**, *11*, 701. [CrossRef]
3. Butler, M.; Pongor, L.; Su, Y.T.; Xi, L.; Raffeld, M.; Quezado, M.; Trepel, J.; Aldape, K.; Pommier, Y.; Wu, J. MGMT Status as a Clinical Biomarker in Glioblastoma. *Trends Cancer* **2020**, *6*, 380–391. [CrossRef]
4. Das, S.; Hayden, J.; Sullivan, T.; Rieger-Christ, K. The Roles of miRNAs in Predicting Bladder Cancer Recurrence and Resistance to Treatment. *Int. J. Mol. Sci.* **2023**, *24*, 964. [CrossRef]
5. Bogale, D.E. The roles of FGFR3 and c-MYC in urothelial bladder cancer. *Discov. Oncol.* **2024**, *15*, 295. [CrossRef] [PubMed]
6. Hrabeta, J.; Belhajova, M.; Subrtova, H.; Merlos Rodrigo, M.A.; Heger, Z.; Eckschlager, T. Drug Sequestration in Lysosomes as One of the Mechanisms of Chemoresistance of Cancer Cells and the Possibilities of Its Inhibition. *Int. J. Mol. Sci.* **2020**, *21*, 4392. [CrossRef] [PubMed]
7. Trybus, W.; Trybus, E.; Krol, T. Lysosomes as a Target of Anticancer Therapy. *Int. J. Mol. Sci.* **2023**, *24*, 2176. [CrossRef] [PubMed]
8. Chen, T.; Lin, X.; Lu, S.; Li, B. V-ATPase in cancer: Mechanistic insights and therapeutic potentials. *Cell Commun. Signal* **2024**, *22*, 613. [CrossRef]
9. Pal, S.K.; Rosenberg, J.E.; Hoffman-Censits, J.H.; Berger, R.; Quinn, D.I.; Galsky, M.D.; Wolf, J.; Dittrich, C.; Keam, B.; Delord, J.P.; et al. Efficacy of BGJ398, a Fibroblast Growth Factor Receptor 1-3 Inhibitor, in Patients with Previously Treated Advanced Urothelial Carcinoma with FGFR3 Alterations. *Cancer Discov.* **2018**, *8*, 812–821. [CrossRef]
0. Xu, N.; Yao, Z.; Shang, G.; Ye, D.; Wang, H.; Zhang, H.; Qu, Y.; Xu, F.; Wang, Y.; Qin, Z.; et al. Integrated proteogenomic characterization of urothelial carcinoma of the bladder. *J. Hematol. Oncol.* **2022**, *15*, 76. [CrossRef]
1. Khorsandi, S.E.; Dokal, A.D.; Rajeeve, V.; Britton, D.J.; Illingworth, M.S.; Heaton, N.; Cutillas, P.R. Computational Analysis of Cholangiocarcinoma Phosphoproteomes Identifies Patient-Specific Drug Targets. *Cancer Res.* **2021**, *81*, 5765–5776. [CrossRef] [PubMed]
2. Gonzalez-Medina, A.; Vila-Casadesus, M.; Gomez-Rey, M.; Fabregat-Franco, C.; Sierra, A.; Tian, T.V.; Castet, F.; Castillo, G.; Matito, J.; Martinez, P.; et al. Clinical Value of Liquid Biopsy in Patients with FGFR2 Fusion-Positive Cholangiocarcinoma During Targeted Therapy. *Clin. Cancer Res.* **2024**, *30*, 4491–4504. [CrossRef]
3. Nakamura, H.; Arai, Y.; Totoki, Y.; Shirota, T.; Elzawahry, A.; Kato, M.; Hama, N.; Hosoda, F.; Urushidate, T.; Ohashi, S.; et al. Genomic spectra of biliary tract cancer. *Nat. Genet.* **2015**, *47*, 1003–1010. [CrossRef]
4. Hwang, S.; Woo, S.; Kang, B.; Kang, H.; Kim, J.S.; Lee, S.H.; Kwon, C.I.; Kyung, D.S.; Kim, H.P.; Kim, G.; et al. Concordance of ctDNA and tissue genomic profiling in advanced biliary tract cancer. *J. Hepatol.* **2025**, *82*, 649–657. [CrossRef] [PubMed]
5. Lu, Y.; Li, X.; Zhao, K.; Shi, Y.; Deng, Z.; Yao, W.; Wang, J. Proteomic and Phosphoproteomic Profiling Reveals the Oncogenic Role of Protein Kinase D Family Kinases in Cholangiocarcinoma. *Cells* **2022**, *11*, 3088. [CrossRef] [PubMed]
6. Iqbal, S.; Karim, M.R.; Mohammad, S.; Mathiyalagan, R.; Morshed, M.N.; Yang, D.C.; Bae, H.; Rupa, E.J.; Yang, D.U. Multiomics Analysis of the PHLDA Gene Family in Different Cancers and Their Clinical Prognostic Value. *Curr. Issues Mol. Biol.* **2024**, *46*, 5488–5510. [CrossRef]
7. Diehl, F.; Schmidt, K.; Choti, M.A.; Romans, K.; Goodman, S.; Li, M.; Thornton, K.; Agrawal, N.; Sokoll, L.; Szabo, S.A.; et al. Circulating mutant DNA to assess tumor dynamics. *Nat. Med.* **2008**, *14*, 985–990. [CrossRef]
8. Razavi, P.; Li, B.T.; Brown, D.N.; Jung, B.; Hubbell, E.; Shen, R.; Abida, W.; Juluru, K.; De Bruijn, I.; Hou, C.; et al. High-intensity sequencing reveals the sources of plasma circulating cell-free DNA variants. *Nat. Med.* **2019**, *25*, 1928–1937. [CrossRef]
9. Bettegowda, C.; Sausen, M.; Leary, R.J.; Kinde, I.; Wang, Y.; Agrawal, N.; Bartlett, B.R.; Wang, H.; Luber, B.; Alani, R.M.; et al. Detection of circulating tumor DNA in early- and late-stage human malignancies. *Sci. Transl. Med.* **2014**, *6*, 224ra224. [CrossRef]
20. Wang, C.; Zhang, M.; Zhao, J.; Li, B.; Xiao, X.; Zhang, Y. The prediction of drug sensitivity by multi-omics fusion reveals the heterogeneity of drug response in pan-cancer. *Comput. Biol. Med.* **2023**, *163*, 107220. [CrossRef]
21. Eckhart, L.; Lenhof, K.; Rolli, L.M.; Lenhof, H.P. A comprehensive benchmarking of machine learning algorithms and dimensionality reduction methods for drug sensitivity prediction. *Brief. Bioinform.* **2024**, *25*, bbae242. [CrossRef]
22. Cai, Z.; Poulos, R.C.; Liu, J.; Zhong, Q. Machine learning for multi-omics data integration in cancer. *iScience* **2022**, *25*, 103798. [CrossRef]
23. Unger, M.; Kather, J.N. Deep learning in cancer genomics and histopathology. *Genome Med.* **2024**, *16*, 44. [CrossRef] [PubMed]
24. Elemento, O.; Leslie, C.; Lundin, J.; Tourassi, G. Artificial intelligence in cancer research, diagnosis and therapy. *Nat. Rev. Cancer* **2021**, *21*, 747–752. [CrossRef]
25. Kline, A.; Wang, H.; Li, Y.; Dennis, S.; Hutch, M.; Xu, Z.; Wang, F.; Cheng, F.; Luo, Y. Multimodal machine learning in precision health: A scoping review. *npj Digit. Med.* **2022**, *5*, 171. [CrossRef]
26. Vanguri, R.S.; Luo, J.; Aukerman, A.T.; Egger, J.V.; Fong, C.J.; Horvat, N.; Pagano, A.; Araujo-Filho, J.A.B.; Geneslaw, L.; Rizvi, H.; et al. Multimodal integration of radiology, pathology and genomics for prediction of response to PD-(L)1 blockade in patients with non-small cell lung cancer. *Nat. Cancer* **2022**, *3*, 1151–1164. [CrossRef]
27. Huang, S.C.; Pareek, A.; Seyyedi, S.; Banerjee, I.; Lungren, M.P. Fusion of medical imaging and electronic health records using deep learning: A systematic review and implementation guidelines. *npj Digit. Med.* **2020**, *3*, 136. [CrossRef]

 https://doi.org/10.3390/cancers18010089

328. Waqas, A.; Tripathi, A.; Ramachandran, R.P.; Stewart, P.A.; Rasool, G. Multimodal data integration for oncology in the era of de neural networks: A review. *Front. Artif. Intell.* **2024**, *7*, 1408843. [CrossRef] [PubMed]
329. Qu, L.; Chen, X.; Wei, H.; Guo, M.; Dai, S.; Jiang, L.; Li, J.; Yue, S.; Chen, Z.; Chen, Y. Structural insights into the potency ar selectivity of covalent pan-FGFR inhibitors. *Commun. Chem.* **2022**, *5*, 5. [CrossRef]
330. Liu, D.; Liu, H.; Gan, J.; Zeng, S.; Zhong, F.; Zhang, B.; Zhang, Z.; Zhang, S.; Jiang, L.; Wang, G.; et al. LY2874455 ar Abemaciclib Reverse FGF3/4/19/CCND1 Amplification Mediated Gefitinib Resistance in NSCLC. *Front. Pharmacol.* **202** *13*, 918317. [CrossRef] [PubMed]
331. Mouron, S.; Manso, L.; Caleiras, E.; Rodriguez-Peralto, J.L.; Rueda, O.M.; Caldas, C.; Colomer, R.; Quintela-Fandino, M.; Buer M.J. FGFR1 amplification or overexpression and hormonal resistance in luminal breast cancer: Rationale for a triple blockade ER, CDK4/6, and FGFR1. *Breast Cancer Res.* **2021**, *23*, 21. [CrossRef]
332. Kollmannsberger, C.; Britten, C.D.; Olszanski, A.J.; Walker, J.A.; Zang, W.; Willard, M.D.; Radtke, D.B.; Farrington, D.L.; Be McGuinn, K.M.; Patnaik, A. A phase 1 study of LY3076226, a fibroblast growth factor receptor 3 (FGFR3) antibody-drug conjuga in patients with advanced or metastatic cancer. *Investig. New Drugs* **2021**, *39*, 1613–1623. [CrossRef] [PubMed]
333. Kallus, S.; Englinger, B.; Senkiv, J.; Laemmerer, A.; Heffeter, P.; Berger, W.; Kowol, C.R.; Keppler, B.K. Nanoformulations anticancer FGFR inhibitors with improved therapeutic index. *Nanomedicine* **2018**, *14*, 2632–2643. [CrossRef]
334. Waheed, I.; Ali, A.; Tabassum, H.; Khatoon, N.; Lai, W.F.; Zhou, X. Lipid-based nanoparticles as drug delivery carriers for canc therapy. *Front. Oncol.* **2024**, *14*, 1296091. [CrossRef]
335. Im, J.H.; Buzzelli, J.N.; Jones, K.; Franchini, F.; Gordon-Weeks, A.; Markelc, B.; Chen, J.; Kim, J.; Cao, Y.; Muschel, R.J. FGF2 alte macrophage polarization, tumour immunity and growth and can be targeted during radiotherapy. *Nat. Commun.* **2020**, *11*, 406 [CrossRef] [PubMed]
336. Akhand, S.S.; Liu, Z.; Purdy, S.C.; Abdullah, A.; Lin, H.; Cresswell, G.M.; Ratliff, T.L.; Wendt, M. Pharmacologic Inhibition FGFR Modulates the Metastatic Immune Microenvironment and Promotes Response to Immune Checkpoint Blockade. *Canc Immunol. Res.* **2020**, *8*, 1542–1553. [CrossRef]
337. Palakurthi, S.; Kuraguchi, M.; Zacharek, S.J.; Zudaire, E.; Huang, W.; Bonal, D.M.; Liu, J.; Dhaneshwar, A.; DePeaux, K.; Gowas M.R.; et al. The Combined Effect of FGFR Inhibition and PD-1 Blockade Promotes Tumor-Intrinsic Induction of Antitum Immunity. *Cancer Immunol. Res.* **2019**, *7*, 1457–1471. [CrossRef] [PubMed]
338. Tian, M.; Wei, J.S.; Shivaprasad, N.; Highfill, S.L.; Gryder, B.E.; Milewski, D.; Brown, G.T.; Moses, L.; Song, H.; Wu, J.T.; et Preclinical development of a chimeric antigen receptor T cell therapy targeting FGFR4 in rhabdomyosarcoma. *Cell Rep. Me* **2023**, *4*, 101212. [CrossRef]
339. Sullivan, P.M.; Kumar, R.; Li, W.; Hoglund, V.; Wang, L.; Zhang, Y.; Shi, M.; Beak, D.; Cheuk, A.; Jensen, M.C.; et al. FGFR4-Targete Chimeric Antigen Receptors Combined with Anti-Myeloid Polypharmacy Effectively Treat Orthotopic Rhabdomyosarcoma. *M Cancer Ther.* **2022**, *21*, 1608–1621. [CrossRef]
340. Nguyen, A.; Johanning, G.; Shi, Y. Emerging Novel Combined CAR-T Cell Therapies. *Cancers* **2022**, *14*, 1403. [CrossRef]
341. Chakrabarti, S.; Finnes, H.D.; Mahipal, A. Fibroblast growth factor receptor (FGFR) inhibitors in cholangiocarcinoma: Curre status, insight on resistance mechanisms and toxicity management. *Expert. Opin. Drug Metab. Toxicol.* **2022**, *18*, 85–98. [CrossRe [PubMed]
342. Peng, M.; Li, H.; Cao, H.; Huang, Y.; Yu, W.; Shen, C.; Gu, J. Dual FGFR and VEGFR inhibition synergistically restrain hexokina 2-dependent lymphangiogenesis and immune escape in intrahepatic cholangiocarcinoma. *J. Gastroenterol.* **2023**, *58*, 908–92 [CrossRef]
343. Chan, S.L.; Schuler, M.; Kang, Y.-K.; Yen, C.-J.; Edeline, J.; Choo, S.P.; Lin, C.-C.; Okusaka, T.; Weiss, K.-H.; Macarulla, T.; et A first-in-human phase 1/2 study of FGF401 and combination of FGF401 with spartalizumab in patients with hepatocellul carcinoma or biomarker-selected solid tumors. *J. Exp. Clin. Cancer Res.* **2022**, *41*, 189. [CrossRef]
344. Kassab, J.G.; Meeks, R.H.; de Riese, W.T.W. Current Clinical Aspects of Androgen Deprivation Therapy for Locally Advanced an Metastatic Prostate Cancer: A Scoping Review for Urologists and Medical Providers. *Res. Rep. Urol.* **2024**, *16*, 187–193. [CrossRe
345. West, J.; Adler, F.; Gallaher, J.; Strobl, M.; Brady-Nicholls, R.; Brown, J.; Roberson-Tessi, M.; Kim, E.; Noble, R.; Viossat, Y.; et al. survey of open questions in adaptive therapy: Bridging mathematics and clinical translation. *Elife* **2023**, *12*, e84263. [CrossRe [PubMed]
346. Grisay, G.; Turco, F.; Litiere, S.; Fournier, B.; Patrikidou, A.; Gallardo, E.; McDermott, R.; Alanya, A.; Gillessen, S.; Tombal, EORTC 2238 "De-Escalate": A pragmatic trial to revisit intermittent androgen deprivation therapy in the era of new androge receptor pathway inhibitors. *Front. Oncol.* **2024**, *14*, 1391825. [CrossRef]
347. Ito, Y.; Yamada, D.; Kobayashi, S.; Sasaki, K.; Iwagami, Y.; Tomimaru, Y.; Asaoka, T.; Noda, T.; Takahashi, H.; Shimizu, et al. The combination of gemcitabine plus an anti-FGFR inhibitor can have a synergistic antitumor effect on FGF-activati cholangiocarcinoma. *Cancer Lett.* **2024**, *595*, 216997. [CrossRef]

https://doi.org/10.3390/cancers1801008

348. Morelli, M.P.; Cascone, T.; Troiani, T.; De Vita, F.; Orditura, M.; Laus, G.; Eckhardt, S.G.; Pepe, S.; Tortora, G.; Ciardiello, F. Sequence-dependent antiproliferative effects of cytotoxic drugs and epidermal growth factor receptor inhibitors. *Ann. Oncol.* **2005**, *16*, iv61–iv68. [CrossRef]
349. Subbiah, V.; Iannotti, N.O.; Gutierrez, M.; Smith, D.C.; Feliz, L.; Lihou, C.F.; Tian, C.; Silverman, I.M.; Ji, T.; Saleh, M. FIGHT-101, a first-in-human study of potent and selective FGFR 1-3 inhibitor pemigatinib in pan-cancer patients with FGF/FGFR alterations and advanced malignancies. *Ann. Oncol.* **2022**, *33*, 522–533. [CrossRef]
350. Chmiel, P.; Geca, K.; Rawicz-Pruszynski, K.; Polkowski, W.P.; Skorzewska, M. FGFR Inhibitors in Cholangiocarcinoma-A Novel Yet Primary Approach: Where Do We Stand Now and Where to Head Next in Targeting This Axis? *Cells* **2022**, *11*, 3929. [CrossRef] [PubMed]
351. Valle, J.; Wasan, H.; Palmer, D.H.; Cunningham, D.; Anthoney, A.; Maraveyas, A.; Madhusudan, S.; Iveson, T.; Hughes, S.; Pereira, S.P.; et al. Cisplatin plus gemcitabine versus gemcitabine for biliary tract cancer. *N. Engl. J. Med.* **2010**, *362*, 1273–1281. [CrossRef] [PubMed]
352. Vogel, A.; Ducreux, M.; ESMO Guidelines Committee. ESMO Clinical Practice Guideline interim update on the management of biliary tract cancer. *ESMO Open* **2025**, *10*, 104003. [CrossRef] [PubMed]
353. Ramjeesingh, R.; Chaudhury, P.; Tam, V.C.; Roberge, D.; Lim, H.J.; Knox, J.J.; Asselah, J.; Doucette, S.; Chhiber, N.; Goodwin, R. A Practical Guide for the Systemic Treatment of Biliary Tract Cancer in Canada. *Curr. Oncol.* **2023**, *30*, 7132–7150. [CrossRef] [PubMed]
354. Xiao, J.F.; Caliri, A.W.; Duex, J.E.; Theodorescu, D. Targetable Pathways in Advanced Bladder Cancer: FGFR Signaling. *Cancers* **2021**, *13*, 4891. [CrossRef]
355. Ascione, C.M.; Napolitano, F.; Esposito, D.; Servetto, A.; Belli, S.; Santaniello, A.; Scagliarini, S.; Crocetto, F.; Bianco, R.; Formisano, L. Role of FGFR3 in bladder cancer: Treatment landscape and future challenges. *Cancer Treat. Rev.* **2023**, *115*, 102530. [CrossRef]
356. Sergei, B.; Pavel, D.; Aigul, G.; Firyuza, B.; Ilmira, N.; Ilshat, M.; Aida, A.; Refat, K.; Natalia, A.; Elena, S.; et al. Inhibition of FGFR2-Signaling Attenuates a Homology-Mediated DNA Repair in GIST and Sensitizes Them to DNA-Topoisomerase II Inhibitors. *Int. J. Mol. Sci.* **2020**, *21*, 352. [CrossRef]
357. SenthilKumar, G.; Fisher, M.M.; Skiba, J.H.; Miller, M.C.; Brennan, S.R.; Kaushik, S.; Bradley, S.T.; Longhurst, C.A.; Buehler, D.; Nickel, K.P.; et al. FGFR Inhibition Enhances Sensitivity to Radiation in Non-Small Cell Lung Cancer. *Mol. Cancer Ther.* **2020**, *19*, 1255–1265. [CrossRef]
358. Siefker-Radtke, A.O.; Matsubara, N.; Park, S.H.; Huddart, R.A.; Burgess, E.F.; Ozguroglu, M.; Valderrama, B.P.; Laguerre, B.; Basso, U.; Triantos, S.; et al. Erdafitinib versus pembrolizumab in pretreated patients with advanced or metastatic urothelial cancer with select FGFR alterations: Cohort 2 of the randomized phase III THOR trial. *Ann. Oncol.* **2024**, *35*, 107–117. [CrossRef]
359. Buckarma, E.; De La Cruz, G.; Truty, M.; Nagorney, D.; Cleary, S.; Kendrick, M.; Borad, M.; Graham, R.P.; Gores, G.; Smoot, R. Impact of FGFR2 gene fusions on survival of patients with intrahepatic cholangiocarcinoma following curative intent resection. *HPB* **2022**, *24*, 1748–1756. [CrossRef]
360. Tsujie, M.; Iwai, T.; Kubo, S.; Ura, T.; Hatano, E.; Sakai, D.; Takeda, Y.; Kaibori, M.; Kobayashi, T.; Katanuma, A.; et al. Fibroblast growth factor receptor 2 (FGFR2) fusions in Japanese patients with intrahepatic cholangiocarcinoma. *Jpn. J. Clin. Oncol.* **2021**, *51*, 911–917. [CrossRef]
361. Goyal, L.; Shi, L.; Liu, L.Y.; Fece de la Cruz, F.; Lennerz, J.K.; Raghavan, S.; Leschiner, I.; Elagina, L.; Siravegna, G.; Ng, R.W.S.; et al. TAS-120 Overcomes Resistance to ATP-Competitive FGFR Inhibitors in Patients with FGFR2 Fusion-Positive Intrahepatic Cholangiocarcinoma. *Cancer Discov.* **2019**, *9*, 1064–1079. [CrossRef]
362. Neumann, O.; Burn, T.C.; Allgauer, M.; Ball, M.; Kirchner, M.; Albrecht, T.; Volckmar, A.L.; Beck, S.; Endris, V.; Goldschmid, H.; et al. Genomic architecture of FGFR2 fusions in cholangiocarcinoma and its implication for molecular testing. *Br. J. Cancer* **2022**, *127*, 1540–1549. [CrossRef]
363. Pacini, L.; Jenks, A.D.; Lima, N.C.; Huang, P.H. Targeting the Fibroblast Growth Factor Receptor (FGFR) Family in Lung Cancer. *Cells* **2021**, *10*, 1154. [CrossRef] [PubMed]
364. Chang, M.M.; Wu, S.Z.; Yang, S.H.; Wu, C.C.; Wang, C.Y.; Huang, B.M. FGF9/FGFR1 promotes cell proliferation, epithelial-mesenchymal transition, M2 macrophage infiltration and liver metastasis of lung cancer. *Transl. Oncol.* **2021**, *14*, 101208. [CrossRef]
365. Yang, Z.; Liang, S.Q.; Saliakoura, M.; Yang, H.; Vassella, E.; Konstantinidou, G.; Tschan, M.; Hegedus, B.; Zhao, L.; Gao, Y.; et al. Synergistic effects of FGFR1 and PLK1 inhibitors target a metabolic liability in KRAS-mutant cancer. *EMBO Mol. Med.* **2021**, *13*, e13193. [CrossRef] [PubMed]
366. Manchado, E.; Weissmueller, S.; Morris, J.P.t.; Chen, C.C.; Wullenkord, R.; Lujambio, A.; de Stanchina, E.; Poirier, J.T.; Gainor, J.F.; Corcoran, R.B.; et al. A combinatorial strategy for treating KRAS-mutant lung cancer. *Nature* **2016**, *534*, 647–651. [CrossRef] [PubMed]

 https://doi.org/10.3390/cancers18010089

367. Giacomini, A.; Taranto, S.; Rezzola, S.; Matarazzo, S.; Grillo, E.; Bugatti, M.; Scotuzzi, A.; Guerra, J.; Di Trani, M.; Presta, M.; et al. Inhibition of the FGF/FGFR System Induces Apoptosis in Lung Cancer Cells via c-Myc Downregulation and Oxidative Stress. *Int. J. Mol. Sci.* **2020**, *21*, 9376. [CrossRef]
368. Cevatemre, B.; Bulut, I.; Dedeoglu, B.; Isiklar, A.; Syed, H.; Bayram, O.Y.; Bagci-Onder, T.; Acilan, C. Exploiting epigenetic targets to overcome taxane resistance in prostate cancer. *Cell Death Dis.* **2024**, *15*, 132. [CrossRef]
369. Kommalapati, A.; Tella, S.H.; Borad, M.; Javle, M.; Mahipal, A. FGFR Inhibitors in Oncology: Insight on the Management of Toxicities in Clinical Practice. *Cancers* **2021**, *13*, 2968. [CrossRef]
370. Schwartz, G.; Darling, J.O. Practical Management of Adverse Events Associated With FGFR Inhibitors for Cholangiocarcinoma for the Advanced Practice Provider. *J. Adv. Pract. Oncol.* **2024**, 1–27. [CrossRef]
371. Meric-Bernstam, F.; Hollebecque, A.; Furuse, J.; Oh, D.-Y.; Bridgewater, J.A.; Shimura, M.; Anderson, B.; Hangai, N.; Wacheck, V.; Goyal, L. Management (mgmt) of futibatinib-associated adverse events (AEs) in patients (pts) with advanced cancers: Results of a pooled analysis. *J. Clin. Oncol.* **2023**, *41*, 586. [CrossRef]
372. Robertson, A.G.; Kim, J.; Al-Ahmadie, H.; Bellmunt, J.; Guo, G.; Cherniack, A.D.; Hinoue, T.; Laird, P.W.; Hoadley, K.A.; Akbani, R.; et al. Comprehensive Molecular Characterization of Muscle-Invasive Bladder Cancer. *Cell* **2017**, *171*, 540–556.e525. [CrossRef] [PubMed]
373. Ouyang, Y.; Ou, Z.; Zhong, W.; Yang, J.; Fu, S.; Ouyang, N.; Chen, J.; Xu, L.; Wu, D.; Qian, J.; et al. FGFR3 Alterations in Bladder Cancer Stimulate Serine Synthesis to Induce Immune-Inert Macrophages That Suppress T-cell Recruitment and Activation. *Cancer Res.* **2023**, *83*, 4030–4046. [CrossRef]
374. Okato, A.; Utsumi, T.; Ranieri, M.; Zheng, X.; Zhou, M.; Pereira, L.D.; Chen, T.; Kita, Y.; Wu, D.; Hyun, H.; et al. FGFR inhibition augments anti-PD-1 efficacy in murine FGFR3-mutant bladder cancer by abrogating immunosuppression. *J. Clin. Investig.* **2024**, *134*, e169241. [CrossRef] [PubMed]
375. White, B.S.; Sindiri, S.; Hill, V.; Gasmi, B.; Nah, S.; Gartner, J.J.; Prickett, T.D.; Li, Y.; Gurusamy, D.; Robbins, P.; et al. Specific recognition of an FGFR2 fusion by tumor infiltrating lymphocytes from a patient with metastatic cholangiocarcinoma. *J. Immunother. Cancer* **2023**, *11*, e006303. [CrossRef] [PubMed]
376. Dadgar, N.; Arunachalam, A.K.; Hong, H.; Phoon, Y.P.; Arpi-Palacios, J.E.; Uysal, M.; Wehrle, C.J.; Aucejo, F.; Ma, W.W.; Melenhorst, J.J. Advancing Cholangiocarcinoma Care: Insights and Innovations in T Cell Therapy. *Cancers* **2024**, *16*, 3232. [CrossRef]
377. Tian, M.; Wei, J.S.; Cheuk, A.T.; Milewski, D.; Zhang, Z.; Kim, Y.Y.; Chou, H.C.; Liu, C.; Badr, S.; Pope, E.G.; et al. CAR T-cells targeting FGFR4 and CD276 simultaneously show potent antitumor effect against childhood rhabdomyosarcoma. *Nat. Commun.* **2024**, *15*, 6222. [CrossRef]
378. Zanuso, V.; Tesini, G.; Valenzi, E.; Rimassa, L. New systemic treatment options for advanced cholangiocarcinoma. *J. Liver Cancer* **2024**, *24*, 155–170. [CrossRef]
379. Pang, Z.; Lu, M.M.; Zhang, Y.; Gao, Y.; Bai, J.J.; Gu, J.Y.; Xie, L.; Wu, W.Z. Neoantigen-targeted TCR-engineered T cell immunotherapy: Current advances and challenges. *Biomark. Res.* **2023**, *11*, 104. [CrossRef]
380. He, Q.M.; Wei, Y.Q.; Tian, L.; Zhao, X.; Su, J.M.; Yang, L.; Lu, Y.; Kan, B.; Lou, Y.Y.; Huang, M.J.; et al. Inhibition of tumor growth with a vaccine based on xenogeneic homologous fibroblast growth factor receptor-1 in mice. *J. Biol. Chem.* **2003**, *278*, 21831–21836. [CrossRef]
381. Kono, M.; Komatsuda, H.; Yamaki, H.; Kumai, T.; Hayashi, R.; Wakisaka, R.; Nagato, T.; Ohkuri, T.; Kosaka, A.; Ohara, K.; et al. Immunomodulation via FGFR inhibition augments FGFR1 targeting T-cell based antitumor immunotherapy for head and neck squamous cell carcinoma. *Oncoimmunology* **2022**, *11*, 2021619. [CrossRef]
382. Mittendorf, E.A.; Lu, B.; Melisko, M.; Price Hiller, J.; Bondarenko, I.; Brunt, A.M.; Sergii, G.; Petrakova, K.; Peoples, G.E. Efficacy and Safety Analysis of Nelipepimut-S Vaccine to Prevent Breast Cancer Recurrence: A Randomized, Multicenter, Phase III Clinical Trial. *Clin. Cancer Res.* **2019**, *25*, 4248–4254. [CrossRef] [PubMed]
383. Cebon, J.S.; Gore, M.; Thompson, J.F.; Davis, I.D.; McArthur, G.A.; Walpole, E.; Smithers, M.; Cerundolo, V.; Dunbar, P.R.; MacGregor, D.; et al. Results of a randomized, double-blind phase II clinical trial of NY-ESO-1 vaccine with ISCOMATRIX adjuvant versus ISCOMATRIX alone in participants with high-risk resected melanoma. *J. Immunother. Cancer* **2020**, *8*, e000410. [CrossRef] [PubMed]
384. Li, F.; Deng, L.; Jackson, K.R.; Talukder, A.H.; Katailiha, A.S.; Bradley, S.D.; Zou, Q.; Chen, C.; Huo, C.; Chiu, Y.; et al. Neoantigen vaccination induces clinical and immunologic responses in non-small cell lung cancer patients harboring EGFR mutations. *J. Immunother. Cancer* **2021**, *9*, e002531. [CrossRef] [PubMed]

https://doi.org/10.3390/cancers18010089

MDPI AG
Grosspeteranlage 5
4052 Basel
Switzerland
Tel.: +41 61 683 77 34

Cancers Editorial Office
E-mail: cancers@mdpi.com
www.mdpi.com/journal/cancers

www.ingramcontent.com/pod-product-compliance
Lightning Source LLC
Chambersburg PA
CBHW061605220926
51850CB00005B/426
9783725870684